Dictionary of Paper

5th Edition

Edited by
Michael Kouris

©1996
Atlanta, Georgia

The Association assumes no liability or responsibility in connection with the use of this information or data, including, but not limited to, any liability or responsibility under patent, copyright, or trade secret laws. The user is responsible for determining that this document is the most recent edition published.

Within the context of this work, the author(s) may use as examples specific manufacturers of equipment. This does not imply that these manufacturers are the only or best sources of the equipment or that TAPPI endorses them in any way. The presentation of such material by TAPPI should not be construed as an endorsement of or suggestion for any agreed upon course of conduct or concerted action.

Table of Contents

iii

Dedication

Michael Kouris, after serving TAPPI as the Editor-in-Chief of *Tappi Journal* since 1961, "retired" in 1989 only to continue his work with the journal as the Editor Emeritus. When asked in 1992 to assist in the revision of the dictionary, he readily agreed, having been a part of the committee which completed the 1980 revision. Michael was soon asked to form a full revision committee which became the core of this document. Michael devoted many hours of reading, writing, and personal contact with individuals throughout the world and without his guidance and personal effort, this dictionary would not be what it is today.

On behalf of TAPPI I would like to dedicate this fifth revision of the *Dictionary of Paper* to Michael Kouris, *Tappi Journal* Editor Emeritus and extend the thanks of the industry for his dedication and effort.

Matthew J. Coleman, CAE
Director of Publications
TAPPI

Introduction

The previous four editions of this dictionary (1940, 1951, 1965, 1980) all carried with them one fundamental, unshakable precept, namely that the dictionary itself would be published as a bound book, printed on paper, and distributed and used as dictionaries had been for over two centuries. This edition, however, was begun as a revision of the 1980 version with the intent to publish it in an electronic format with the print-on-paper version being secondary to that intent. As it progressed, it became obvious that both versions would be necessary to meet the needs of this industry. The American Forest and Paper Association (AF&PA), who held the rights to the Dictionary as the successor society to the API, graciously agreed to let TAPPI have the copyright to the dictionary for the purpose of revision and continuation of this long-standing and useful publication.

The past decade has proved even to the most ostrich-like among us that change is not a buzzword, but a reality. The concept of an electronic dictionary residing on the distributed base of Personal Computers used throughout the industry which could be updated easily was the underlying motivation for this revision. What became obvious as the revision committee met was that the terminology base needed to be more closely aligned with existing sources such as the *Thesaurus of Pulp and Paper Terms*, as well as the many glossaries from specific segments of the paper and related industries. Glossaries from many associations and industries have been incorporated into this dictionary, expanding the coverage and hence, the usefulness of this product.

The initial revision project was begun with the intent of having an updated electronic version available in one year. As the revisions began to flow in, it became apparent that the terminology had grown and usage had changed enough to warrant holding up the initial publication until a more thorough revision could be completed. This was done, and what you have now, whether on your screen or on the printed page, is the product of the work of many dedicated individuals who devoted many hours to this.

This dictionary will not have to wait for the better part of two decades for further revision though. Since it is to be available in electronic format, TAPPI will be releasing updated electronic versions when sufficient numbers of changes have been submitted to the committee. Suggestions, corrections, and additions should be sent to:

Dictionary Revision Committee
TAPPI
P.O. Box 105113
Atlanta, GA 30348-5113

With your help, the sixth edition of the *Dictionary of Paper* can be done faster, easier, and even better than this one.

Matthew J. Coleman, CAE
Director of Publications, TAPPI

Acknowledgements

Without the invaluable assistance of the following people, the fifth edition could not have been done. TAPPI is pleased to recognize and applaud their voluntary efforts to update, revise, and expand the *Dictionary of Paper.*

Terry N. Adams
Rajai H. Attala
Robert A. Bareiss
David J. Bentley, Jr.
Terry L. Bliss
Donald B. Brewster
Leroy H. Busker
Philip C. Clark
Richard L. Davis
G. Kenneth Fekete
C. Douglas Foran
William J. Frederick, Jr.
William S. Fuller
Frank R. Hamilton
Alfred H. Jaehn
Otto J. Kallmes
Edward G. Kelleher
T. Kent Kirk
Charles P. Klass
William O. Kroeschell
Joseph A. Kurdin
Edward P. Laumer
Norman Liebergott
Robert G. Lucas
Ellen R. McCrady
James P. McNamee
Jennifer Marron
John W. Mercer
James L. Minor
Geoffrey Long
Frank W. Lorey
Gordon W. Mitchell

Diane L. Murdock
Keith D. Otto
John E. Pinkerton
Jeffery H. Pulowski
Richard A. Reece
R. Heath Reeves
John H. Schulz
William E. Scott
W.B.A. Sharp
Richard J. Spangenberg
Allan M. Springer
Edward Strazdins
Benjamin A. Thorp
J. Robert Wagner
John W. Walkinshaw

Permission to use material was obtained from the following:

American Society for Testing and Materials (ASTM)

American Society for Quality Controls (ASQC)

Fibre Box Association (FBA)

Paper Shipping Sack Manufacturers' Association

Institute of Scrap Recycling Industries, Inc.

The following association assisted by revising a section of this book:

American Forest and Paper Association

A

ABACA—A naturally occurring fiber found in the stem of the abaca plant, a member of the banana family, *Musa textilis*. The fiber is also called manila hemp, and is used extensively in the manufacture of marine cordage, abrasive backing papers, tea bags, and other products requiring high tensile strength.

ABIETIC ACID—A resin acid formed by the isomerization of pinewood rosin (q.v.) at high temperature. See also ROSIN ACIDS.

ABRASION—(1) The scratching or wearing away of a sheet of paper or paperboard, either through contact with another sheet of paper or paperboard or with some other object. See also SCUFFING. (2) The wear on manufacturing equipment, particularly paper machine fabrics due to filler pigments and other components of the paper furnish as the sheet is being formed; also, the wear caused by the finished paper in the subsequent operations as, for example, on printing plates.

ABRASION RESISTANCE—The durability of paper or paperboard when subjected to the abrading action of an eraser, emery cloth, or the like. It is commonly evaluated in terms of the rate of loss of weight when subjected to a specified abrading material operating under a definite load, rate, and pattern of movement.

ABRASIVE FIBER—A grade of vulcanized fiber (q.v.) developed as a supporting base for abrasive grit for both disc and drum sanders. It is tough, resilient, with high tearing resistance and a smooth surface to allow uniform adhesive distribution. It is made in thicknesses of 0.010 to 0.040 inch (0.25 to 1 mm).

ABRASIVENESS—The property of a substance expressing the degree to which it abrades or wears away another surface by friction. This property is of importance in both paper manufacture and paper uses. Filler pigments should have low abrasiveness to minimize wear on paper machine fabrics, etc. Both filler and coating pigments can contribute to abrasiveness in paper causing undue wear in printing or converting operations. Emery paper affords an example of extreme abrasiveness, whereas lens tissue is made to have minimum abrasiveness.

ABRASIVE PAPERS—Relatively light to heavy chemical woodpulp or rope fiberpapers, coated on one side with an abrasive material. In the coated abrasive industry, basis weight is expressed in terms of the 24 x 36 – 480 ream. Fourdrinier sheets are generally used for hand and small mechanical sanders: kraft sheets in 50–60 pound weights for flint papers and in 67–70 lb weights for emery papers; alpha-cellulose sheets (40, 70, and 90 lb) for garnet, aluminum oxide, and silicon carbide lines of handsanding or light-duty machine sanding material. Where greater strength is desired, lamination to lightweight cloth is resorted to, using fourdrinier kraft and cylinder rope sheets in 110-lb weights for drum sander covers and in 130-lb weights for belt, drum, disc, or other mechanical sanding operations. Heavyweight materials are generally available with flint, garnet, aluminum oxide, and silicon carbide coatings. For wet sanding operations, fourdrinier papers in the 48 to 80 lb range are used. Such papers usually contain waterproofing additives and abrasive coatings consisting of aluminum oxide, silicon carbide, orgarnet particles. See EMERY PAPER; FINISHING PAPER; FLINT PAPER; GARNET PAPER; GLASS PAPER; POUNCING PAPER; TANNING PAPER.

ABSOLUTE HUMIDITY—The ratio of the actual moles of water vapor per mole of vapor-free air to the moles of water vapor per mole of vapor-free air at saturation. See HUMIDITY, RELATIVE HUMIDITY.

ABSORBENCY—The property of a material that causes it to take up liquid with which it is in contact. Several measures of absorbency are: (a) the time required for the material to take up a specified volume of liquid, (b) the rate of rise of liquid along a vertical strip dipping into the liquid, (c) the area of a specimen wetted in a specified time, (d) the quantity of liquid taken up by a completely saturated specimen. The

1

method of measurement selected depends on the specific use of the paper.

ABSORBENT PAPERS—Soft, loosely felted papers that readily absorb water solutions or liquid chemicals. They are not sized with water-repellent agents, but may be treated with materials that enhance their wet strength. They include blotting, filter, matrix, and toweling papers, and base papers for the manufacture of vegetable parchment, artificial leather, vulcanized fiber, and many other processed papers.

ABSORPTIVE CAPACITY—See ABSORBENCY.

ACCEPT CHIP—See CHIP SIZE.

ACCEPTS—Material accepted by a separator. Usually consists of mostly good fibers, but may contain small quantities of debris.

ACCORDION FOLD—Paper with two or more parallel folds, adjacent folds in opposite directions. Also called fan fold.

ACCORDION PLEATS—See CORRUGATIONS (DEFECT).

ACCOUNT BOOK PAPER—See LEDGER PAPER.

ACCURACY—The extent to which the average of a series of repeat measurements made on a single unit of material differs from the true value of material. See also PRECISION.

ACETOCELL PROCESS—See SOLVENT PULPING.

ACID ALUM—See ALUM.

ACID DEPOSITION—The wet or dry deposition of low pH material that usually contains SO_x or NO_x as part of the deposit.

ACID DYES—A class of aniline dyes, so named because they are in the form of the sodium salt of dye acids. As a class they have much greater solubility and less tinctorial value than basic dyes but are much more resistant to light fading; on mixed furnishes they give more even dyeings than basic or direct dyestuffs. They have no direct affinity for cellulose, being mordanted to the fiber in stock dyeing by the presence of size and alum. Acid dyes are also extensively used in surface coloring.

ACID-FREE PAPER—(1) A wrapping or protective paper for applications where paper acidity would be harmful to the material in contact with the paper. See ANTI-TARNISH. (2) A permanent record paper where the lack of acidity will prevent its premature deterioration.

ACID-INSOLUBLE ASH—See ASH, ACID-INSOLUBLE.

ACIDITY—(1) In aqueous solution, the condition wherein the concentration of hydrogen ions exceeds that of hydroxyl ions. (2) In paper, the condition that results in an acid solution when the paper is treated or extracted with water. In testing paper for acidity, the specimen is extracted with water at a definite temperature, and the extract is tested to determine its pH value or is titrated to determine the total amount of acid present.

ACID-PROOF PAPER—An industrial or packing type paper colored with dyes which do not discolor when exposed to acid compounds.

ACID-RESISTANT PAPER—(1) Paper that has been treated to resist the action of acids or acid fumes. (2) Paper specially dyed with colors which are resistant to acids and their fumes. Manila, kraft, and rope wrapping papers may be used.

ACID-RESISTING FELT—An asphalt-saturated and coated rag felt used over concrete or wood subbases where liquids, particularly acids, are present.

ACID SIZE—A rosin size which contains a considerable portion of unsaponified but emulsified free rosin. If dilution of such size produces a milky emulsion, it is known as white size. See ROSIN SIZE.

ACID-SOLUBLE IRON—See IRON, ACID-SOLUBLE.

ACID STABLE SIZE—A wax or wax-rosin emulsion which is not coagulated by acidic conditions. See WAX EMULSION.

ACID SULFITE PROCESS—The process of pulping fibrous raw materials, such as wood, in an acidic liquor containing a high percentage of free sulfur dioxide. The sulfite base may be calcium, sodium, magnesium, or ammonia. The fibers produced by this process are lower in lignin content than other chemical pulping processes. This process has declined in usage in the last decade.

ACID TREATMENT—A treatment, most often sulfur dioxide, but in some cases sulfuric or hydrochloric acid, applied to pulp (a) after a final chlorine dioxide, hypochlorite, or peroxide stage to destroy residual bleaching agent and remove metal ions, which prevents brightness loss and avoids fiber degradation, and to create a pH favorable for brightness and stability (alkaline pulps tend to yellow on exposure to light and air); (b) as a pretreatment before an oxygen stage, which tends to improve both brightness and physical strength, probably by removing metal ions from pulp; (c) between bleaching stages, such as peroxide-hydrosulfite, to eliminate the washing steps by creating acid conditions necessary for the hydrosulfite stage, and to destroy residual peroxide by reduction, which would otherwise consume costly hydrosulfite; and (d) in place of a chelation stage in a total chlorine compound free sequence with peroxide. In this application the pH is very important.

ACM—Active chlorine multiple (q.v.).

ACOUSTICAL ABSORPTION COEFFICIENT—See SOUND-ABSORPTION COEFFICIENT.

ACOUSTICAL BOARD—A building product used on ceilings to absorb sound or as a thermal insulator. It is commonly made in the same manner and from the same material as low-density fibrous insulating board, i.e., from wood, straw, bagasse, cornstalks, and similar materials refined to a coarse pulp, formed into sheets one-half to one inch in thickness on a wet machine or a modified fourdrinier machine, and dried in a tunnel drier. Acoustical boards are commonly made in 1/2–3/4 and 1-inch thicknesses, usually by gluing together two or more thinner boards, and are commonly cut into tiles in relatively small sizes, such as one foot square, with beveled edges. Holes and textured surfaces are frequently cut into the surface to increase sound absorption. The boards usually weigh 500–650 pounds per 1000 square feet (1/2-inch basis) and are of soft, porous structure.

ACOUSTICAL TRANSMISSIVITY—See SOUND TRANSMISSIVITY.

ACTIVATED CARBON—Either granular or powdered carbon that is used to take up undesirable materials from air or water.

ACTIVATED SLUDGE—Sludge floc produced in raw or settled wastewater by the growth of bacteria and other organisms in the presence of dissolved oxygen and accumulated in sufficient concentration by returning floc previously formed.

ACTIVATED SLUDGE LOADING—The pounds of biochemical oxygen demand (BOD) in the applied liquid per unit volume of aeration capacity or per pound of activated sludge per day.

ACTIVATED SOLIDS—The combination of organisms in wastewater solids produced in the presence of dissolved oxygen in activated sludge treatment. See ACTIVATED SLUDGE.

ACTIVE ALKALI—The sum of the sodium hydroxide and sodium sulfide concentrations in white liquor (q.v.) expressed as equivalent concentrations of sodium oxide, Na_2O.

ACTIVE CHLORINE MULTIPLE—See KAPPA FACTOR.

ACTIVE OXYGEN—A term used to express oxidizing bleaching chemicals in terms of equivalent amounts of oxygen, especially for peracids. In this case, only one oxygen in the –O–O– linkage of a peracid is considered active or available for bleaching.

ACTIVE SULFUR—See REDUCIBLE SULFUR.

ACTUAL WEIGHT—The actual weight as contrasted with the nominal weight (q.v.) of a ream or a number of reams of paper or a bundle or a number of bundles of board.

ACTUATORS—See CD ACTUATORS.

ACUTE TOXICITY—Immediate or short-term toxicity. Sometimes used to mean toxicity which causes the death of an organism. Used in conjunction with test procedures for water quality. Contrast with CHRONIC TOXICITY; see also TOXICITY.

ADAMS COORDINATES—See OPPONENT COLOR SCALES (L, a, b).

ADAPTIVE CONTROL—A control system which can adjust its parameters to compensate for variations in the controlled process.

ADDING-MACHINE PAPER—A writing or tablet paper furnished in small rolls for use on adding machines, calculators and the like, and characterized by a lint-free surface, uniform caliper, smooth finish, and good printability. It is normally made in a basis weight of 16 pounds (17 x 22 inch – 500).

ADDITIVE PRIMARIES—Red, green, and blue light which, added together, generate white.

ADDITIVES—See PAPER ADDITIVES.

ADDRESS-LABEL PAPER—A writing or book paper commonly used in the preparation of address labels for periodicals. It is usually made from chemical wood and/or mechanical pulps in white and colors in a basis weight of 16 pounds (17 x 22 inch – 500) or 40 pounds (25 x 38 inch – 500). This grade is also referred to as addressing-machine paper.

ADDRESSOGRAPH ROLLS—Narrow rolls of writing papers used for the preparation of mailing lists, mailing labels, etc. Also called mailing, listing, addressing, and proofing rolls.

ADHESIVE—Any substance which can bond two or more bodies or surfaces. Adhesives can be 100% (hot melts) or carried in an aqueous or non-aqueous solvent. They can be activated by removal of the solvent, by heating or cooling or by chemical reaction. Typical adhesives used in papermaking, coating, and converting are animal glues, gelatins, starches, dextrins, resins, latexes, caseins, silicates, asphaltic materials, waxes, and various thermoplastic and thermosetting materials.

ADHESIVE FELT—A heavy bogus paper, usually gray, which is often used as a backing or stiffener in leather or vinyl products, such as pocketbooks and cap visors.

ADHESIVE GLASSINE TAPE—A gummed glassine paper, normally with a basis weight of 25 pounds (24 x 36 inches – 500). It is used in mending books, music sheets, currency, etc., and is supplied in rolls of narrow width and small diameter.

ADHESIVENESS—(1) The strength of the bond between gummed paper tape and kraft paper. (2) The quality of the bond between two sheets of heavy paper or paperboard joined by a specified adhesive.

ADHESIVE PAPER—(1) Paper coated with any of several types of adhesive such as water-activated (see GUMMED PAPER), solvent-activated (see GUMMED WATER-RESISTANT TAPE), heat-activated (see HEAT-SEALING PAPER), or pressure-sensitive (see PRESSURE- SENSITIVE PAPER). (2) Paper to which an adhesive is to be applied. See GUMMING PAPER.

ADSORBABLE ORGANICALLY BOUND HALOGEN (AOX)—A measure of the total amount of organochlorine, (reported as Cl), in effluent. In this test, effluent is treated with activated carbon and all organic material adsorbed onto the carbon. The carbon is burned and all organochlorine converted to hydrochloric acid which is adsorbed in an aqueous solution and measured by microcoulometry.

ADVANCED WASTE TREATMENT—Any treatment method or process employed following biological treatment (1) to increase the removal of pollution load, (2) to remove substances that may be deleterious to receiving waters or the environment, (3) to produce a high-quality effluent suitable for reuse in any specific manner or for discharge under critical conditions. The term tertiary treatment is commonly used to denote advanced waste treatment methods.

AERATED POND—A natural or artificial wastewater treatment pond in which mechanical or diffused-air aeration is used to supplement the oxygen supply.

AERATORS—Mechanical devices used to add dissolved oxygen to wastewater. They can be either of two types, surface or submerged.

AEROBIC—Requiring, or not destroyed by, the presence of free elemental oxygen.

AEROBIC BIOLOGICAL OXIDATION—The process by which a substance is metabolized by aerobic bacteria.

AEROBIC DIGESTER—Digestion of suspended organic matter by means of aeration.

AEROBIC PROCESS—A process which is carried out in the presence of oxygen.

AF&PA—The American Forest & Paper Association (AF&PA) represents approximately 500 member companies and related trade associations that grow, harvest, and process wood and wood fiber, manufacture pulp, paper, and paperboard products from both virgin and recovered fiber, and produce solid wood products. As a single national association, AF&PA represents a vital national industry which accounts for over 7% of the total U.S. manufacturing output.

AFI—American Forest Institute.

A-FLUTE—See FLUTE.

AFTER DRYER SECTION—A term commonly used to describe the dryer section following the size press and before the calender. It is designed to provide capacity adequate to remove the water added to the sheet by surface sizing. The after dryer section is sometimes referred to simply as the "after section."

AGALITE—A gray, natural filler, similar to talc, although less soapy. It is a hydrated magnesium silicate. Talc and asbestine are sometimes called agalite.

AGATE MARBLE PAPER—A paper made from chemical woodpulp and used for end leaves in books. It features a distinctive surface decorated with colored striations resembling the banded appearance of agate glass. See MARBLE PAPER.

AGING—See PERMANENCE.

AGITATOR—A mechanical device used to create shear in the slurry for the purpose of keeping the suspended solid components in suspension, evening out consistency fluctuations, mixing multiple streams of stock, or blending fillers, chemicals, or dyes with the stock. Most agitators are either propeller agitators or circulating pump agitators. See also MID-FEATHER AGITATOR; SIDE ENTRY AGITATOR; TOP ENTRY AGITATOR.

AIR BELLS—Surface defects in paper sometimes called blisters or foam marks. See AIR BUBBLES; BLISTER; FOAM MARKS.

AIR BRUSH—(1) Small pencil-shaped spray gun that uses watercolor pigment for tonal effects. (2) A retouching technique.

AIR-BRUSH COATING—Coating sprayed onto a paper web by means of air pressure, which atomizes the liquid mixture of coating pigment and adhesive. Air-brush coating has been replaced by a variety of coating methods. See also AIR-KNIFE COATING.

AIR BUBBLES—(1) Entrained air between layers of paper which accumulates at the entrance of an ingoing nip. (2) The imprint or marks left in the finished sheet by the bubbles upon bursting.

AIR CAPS—Devices that dry paper by blowing high velocity (up to approximately 101 m/sec) hot air (up to approximately 425°C) on the paper web. Most commonly used in conjunction with a Yankee (q.v.) or through-air-dryer (q.v.).

AIR COOLED—Descriptive of a paper web in the manufacturing process, cooled by contact with air after leaving the heated dryers and before being wound into a roll.

AIR DECKLE—See JET DECKLE.

AIR-DRIED BOARD—A board generally made on a wet machine and air dried, as contrasted to drying by steam on a paper machine. The term is also applied to clay or similarly coated board, dried on festoon dryers. Obsolete as a commercial practice.

AIR-DRIED PAPER—Paper dried by contact with air, either at normal or elevated temperatures, as distinguished from machine-dried paper, where the drying is accomplished by contact with heated rolls. Obsolete as a commercial practice. See BARBER DRYING; DRYER CYLINDERS; FESTOON DRYING; LOFT DRYING.

AIR-DRIED WRITINGS—Typical writing papers with or without a cockle finish produced by drying in heated air (such as festoon drying) instead of in contact with heated cylinders. The range of basis weights is from 13 to 36 pounds (17 x 22 inches – 500) (49–135 grams per square meter). Obsolete as a commercial practice.

AIRDRY—Paper having its moisture content (usually 3–9%) in equilibrium with the atmospheric conditions to which it is exposed, or pulp containing 10% moisture. Percentages are calculated on the ovendry weight.

AIR ENTRAINMENT—The forcing of air between a web and roller or between layers of a winding roll on nonporous webs run at higher speeds. Air entrainment on rollers can cause loss of traction. Air entrainment in rolls can cause a variety of defects.

AIR FLOAT DRYER—Banks of hot air impingement nozzles positioned on each side of the sheet, allowing the sheet to float between the top and bottom banks as the coating or web printing ink is dried.

AIR FLOTATION—In paper finishing, the reduction of contact between a web and roller due to air entrainment.

AIR IMPACTER—A device used in early aqueous coating to promote coating smoothness. Generally obsolete because blade coating produces satisfactory smoothness and levelness.

AIR-KNIFE COATING—A method of coating using the so-called air knife which acts on the principle of a doctor blade and uses a thick, flat jet of air for removing the excess coating from a wet, freshly coated web of paper or paperboard.

AIR LAID—The process of forming a web using air as a fluid to convey and lay down fibers on a forming drum or screen.

AIR-MAIL PAPER—A lightweight opaque writing or typewriting paper made of cotton and/or chemical woodpulps in a basis weight range of 5 to 9 pounds (17 x 22 inches – 500) and designed to minimize mailing costs. See ONIONSKIN; MANIFOLD PAPER.

AIR PAD—The process of using air pressure to move a chemical out of a railcar or truck. When air is used to unload a chemical like chlorine, the air must be dried because moisture will cause chlorine to be extremely corrosive.

AIR PERMEABILITY—The property of a sheet that allows the passage of air when a pressure difference exists across the boundaries of the specimen. It is evaluated by obtaining the rate of flow of air through a specimen under specified experimental conditions. The ability of a package to "breathe" depends on the air permeability of the packaging material, and the permeability of the material to various gases (odors, volatile flavorings, etc.) probably correlates with air permeability; this would not be true for gases, such as water vapor, for which the packaging material has a strong affinity. See POROSITY; VAPOR PERMEABILITY.

AIR POLLUTION—The presence of contaminants in the air in concentrations that prevent the normal dispersive ability of the air and that interfere directly or indirectly with one's health, safety, or comfort or with the full use and enjoyment of one's property.

AIR QUALITY STANDARDS—The prescribed level of pollutants in the outside air that cannot be exceeded legally during a specified time in a specified geographical area.

AIR SHEAR BURST—A roll defect consisting of a machine direction shear failure generated inside a roll winding on a surface winder.

AISLE SIDE—The side of a machine which is most accessible to operators because it is not encumbered by the drive mechanism which is located on the opposite side of the machine. The aisle side normally provides a space for the controls and easier access to the process, and so is also referred to as the operating side. Adjacent machines are usually oriented so operating sides are adjacent thus providing an aisle; i.e., aisle side.

AKD—See ALKYL KETENE DIMER.

ALABASTER—See ANHYDRITE.

ALBUM BOARD—A thick pasted or unpasted album paper. See ALBUM PAPER.

ALBUM PAPER—A cover paper principally used for making photographic albums. It is made in smudge-proof solid colors such as black and gray, in basis weights of 50, 65, and 80 pounds (20 x 26 inches – 500). It is characterized by a soft surface which will take paste

without cockling and freedom from any impurities which might discolor photographic prints. See BLACK ALBUM PAPER.

ALCELL PROCESS—See SOLVENT PULPING.

ALFA—See ESPARTO.

ALGORITHM—A mathematical statement for turning process operational needs into a communication vehicle to implement computer control. Algorithms provide a precise mathematical statement for use in programming computer controls to perform tasks previously done by process operators.

ALIASING—Misrepresentation of information due to the generation of false images or signals by use of inadequate resolution or too low "sampling" frequency (undersampling) of original image or signal. In printing, low dot resolution produces ragged edges or "staircase" at the edge of characters, lines, or shapes. In process control, undersampling of measurements, by sampled data (digital) control systems, generates spurious "beat" frequency signals in time or in space (e.g., paper machine cross direction) which, if acted upon by the controller, will increase variation of the original signal.

ALIGNING PAPER—(1) A typical lightweight map or chart paper made from cotton and/or bleached chemical woodpulps, and characterized by a smooth surface, good printing quality, excellent erasability, uniform formation, low hygroexpansivity, and in some cases, high transparency for reproduction purposes. It is normally made in a basis weight range of 9 to 16 pounds (17 x 22 inches – 500). (2) A typewriting paper, creped to provide extensibility, and coated on one side with a special adhesive so that strips thereof can be typewritten and stretched to provide uniform right-hand margins on manuscripts which are to be reproduced photographically.

ALIGNMENT—A term referring to the status of the ideal geometry of the components of a process machine. Normally, all functional components are specified to be level and square to the machine centerline within some specified dimensional or angular tolerance. See MIS-ALIGNMENT.

ALKALI CELLULOSE—A compound resulting from the treatment of cellulose with sodium hydroxide, in which many of the hydroxyl hydrogens are replaced by sodium ions. Its preparation is a stage in the manufacture of viscose (q.v.).

ALKALINE EXTRACTION STAGE (E)—An intermediate stage in a multistage bleaching process, which removes colored components rendered alkali-soluble in the oxidative bleaching stages. An alkaline extraction is normally placed after a chlorination stage, whereby the lignin and other color-contributing components react with the chlorine; and, while partially soluble in the acidic chlorination solution, the chlorinated compounds easily dissolve in a warm alkaline solution and are removed. An extraction step is also placed before the final chlorine dioxide stage to complete the lignin removal. For an effective extraction, the temperature and pH of the reaction must be high. Typical conditions are 70°C, 10–12% consistency, 90-minute reaction time, end pH around 11.

ALKALINE FILLERS—Fillers which give rise to an alkaline reaction in the presence of water or which react with acids. Calcium carbonate (q.v.) is the most common alkaline filler.

ALKALINE PAPERMAKING—Term referring to paper manufacture at pH values of 7.0 or higher. In practice, the term refers most often to papermaking at pH values of 8.0–8.4 due to the use of calcium carbonate filler, which buffers at the pH. Papermaking at pH 7.0 is also called "neutral papermaking." Conversion from acid papermaking conditions to alkaline papermaking most often requires the use of different chemical additives than were employed under acid papermaking conditions. Therefore, such conversions are significant undertakings.

ALKALINE PEROXIDE MECHANICAL PULP (APMP)—A chemimechanical pulping process, where the chemical impregnation of the wood chips is carried out by alkaline peroxide, prior to refining at atmospheric conditions. The process is best suited for low-density hardwoods, like aspen.

ALKALINE PULPING—See ALKALINE PULPING PROCESS; KRAFT PULPING.

ALKALINE PULPING PROCESS—The process of pulping fibrous raw material, such as wood, in an alkaline liquor. When sodium hydroxide is the only chemical used, this specific alkaline process is called soda pulping. When a combination of sodium sulfide and sodium hydroxide are the active pulping chemicals, the process is termed kraft pulping. A wide range of unbleached and bleached products can be made with the kraft process, the most common alkaline process.

ALKALINE SIZING—A process or procedure for introducing water resistance into paper and board at pH values in excess of 7.0 in the stock at the point of sheet formation.

ALKALINITY—(1) In aqueous solution, the condition wherein the concentration of hydroxyl ions exceeds that of hydrogen ions. (2) In paper, the condition that results in an alkaline solution when the paper is extracted with water. In testing paper for alkalinity, the specimen is extracted with water at a definite temperature, and the extract is tested to determine its pH value or is titrated to determine the total amount of alkali present.

ALKALIPROOF—Resistant to change in color on contact with alkaline materials. This term is applied to paper and to papermaking additives.

ALKALIPROOF PAPER—Paper resistant to alkali, used in packaging alkaline materials such as soaps and adhesives. The alkali resistance is sometimes tested by wetting the paper with a one percent solution of sodium hydroxide or 40 degrees Baume sodium silicate. It is made with a variety of furnishes, primarily semi-bleached and fully bleached chemical wood.

ALKALI-STAINING—The tendency of a paper to stain in contact with alkaline solutions. Freedom from alkali-staining is important in soap wrappers, for example.

ALKENYL SUCCINIC ACID ANHYDRIDE (ASA)—Alkenyl succinic acid anhydride is used as a sizing (q.v.) agent. It is made by reacting alpha-olefins with maleic anhydride. The material is usually liquid, chemically reactive, and must be emulsified at the mill just before adding it to the paper machine. Cationic polymers (most commonly starches) are often used as stabilizers and retention aids for emulsified ASA particles. ASA can react with the hydroxyl (–OH) groups of cellulose, bond with other components by forming a salt, or develop purely physical bonds with fibers through H-bonds, dipole, and Lewis acid-base type reactions. See CELLULOSE REACTIVE SIZES.

ALKYL KETENE DIMER (AKD)—Alkyl ketene dimer (AKD) is a sizing (q.v.) agent. It is a complex molecule made from a mixture of palmitic and stearic acids. It can partially react with cellulose (or starch) hydroxyl (–OH) groups to form an ester bond, while leaving hydrophobic groups on the fiber surface. The hydrophobic groups resist the penetration of polar liquids, such as water or dilute lactic acid, into the interior of paper. AKD, a low melting product, is relatively stable and is supplied to mills as an aqueous cationic dispersion. See CELLULOSE REACTIVE SIZES.

ALPHA-CELLULOSE—The portion of a cellulosic material that is insoluble in 9.45% sodium hydroxide solution, after the material has been previously swollen with 17.5% sodium hydroxide. This portion is considered to be the highest molecular weight and purest cellulose. The determination of alpha-cellulose by this experimental method is applicable primarily to pulps and to papers made from cotton or chemical wood fibers. For papers containing lignin, coatings, fillers, etc., certain corrections must be made.

ALPHA PRINTING PAPER—A printing or writing paper made from esparto pulp. The term is a misnomer being derived from the German word "alfa," meaning esparto. It should not be confused with paper made from alpha pulp or alpha cellulose. See ESPARTO PAPER. As a grade of paper, the term alpha printing paper would not be recognized by most manufacturers or buyers today.

ALPHA-PROTEIN—See SOYBEAN PROTEIN.

ALPHA PULP—Chemically treated woodpulp having greater than 90% alpha-cellulose (q.v.).

ALPHA WRITING PAPER—See ALPHA PRINTING PAPER. As a grade of paper, the term alpha writing paper would not be recognized by most manufacturers or buyers today.

ALUM—Papermakers' alum is hydrated aluminum sulfate, $Al_2(SO_4)^3$ (14-18)H_2O.

ALUMINUM CHLORIDE—An inorganic salt ($AlCl_3$) used to precipitate the size in photographic and other special papers where alum (aluminum sulfate) is undesirable because of the sulfate radical.

ALUMINUM-COATED PAPER—See ALUMINUM PAPER.

ALUMINUM-DUSTED PAPER—See ALUMINUM PAPER.

ALUMINUM PAPER—A base paper of ordinary wrapping weight coated with aluminum powder. It is sometimes made by incorporating the powder in the paper at the beater or in a size press. In the manufacture of aluminum-coated paper, flake aluminum powder (not too coarse mesh) is incorporated with casein or other aqueous sizing vehicle or with lacquers in organic solution to form a coatable composition. In aqueous vehicles, it may be applied by a brush or other coating means; in organic vehicles, it is usually applied by a roll or doctor knife. After coating, the product may be calendered or embossed. The appearance is typical of aluminum bronze powder coatings. Now an obsolete method of manufacturing aluminum paper, this grade was used for a variety of wrapping purposes, particularly for wrapping food products

9

and tobacco. See FOIL LAMINATE; METALLIC COATING; SILVER-LABEL PAPER.

ALUM SPOTS—Imperfections in papers caused by undissolved crystals of alum, which are crushed and fall out in the drying or finishing operation.

AMBIENT AIR QUALITY—That quality which air has at any given time and place. Can be a synonym for "normal" or usual air quality.

AMERICAN FOREST & PAPER ASSOCIATION—See AF&PA.

AMMONIA BASE LIQUOR—A sulfite pulping liquor with ammonium bisulfite as the primary ingredient.

AMMONIUM BISULFITE—NH_4HSO_3. A crystalline compound, freely soluble in water, used in the ammonia base sulfite pulping process.

AMPERE—A unit of electrical current equal to the current produced by a potential difference of one volt across a resistance of one ohm.

AMPHOTERIC POLYELECTROLYTES—See POLYELECTROLYTES.

ANAEROBIC BIOLOGICAL TREATMENT—Any treatment method or process utilizing an aerobic or facultative organisms (q.v.) in the absence of air for the purpose of reducing the organic matter in wastes or organic solids settled out of wastes. Commonly referred to as an aerobic digestion or sludge digestion when applied to the treatment of sludge solids.

ANALYSIS OF VARIANCE (ANOVA)—A basic statistical technique for analyzing experimental data. It subdivides the total variation of a data set into meaningful component parts associated with specific sources of variation to test a hypothesis on the parameters of the model or to estimate variance components.

ANALYTICAL FILTER PAPER—A specially prepared filter paper used for chemical analytical work. There are two principal types: qualitative and quantitative. High retentiveness of fine precipitates and high filtering rate are important characteristics. In addition, quantitative filter paper must have a very low ash content, which is achieved by washing with hydrochloric and hydrofluoric acids. Some grades of both types are used in analysis by paper chromatography. See FILTER PAPER.

ANGIOSPERMS—Plants having their seeds in an enclosed ovary, such as hardwoods, esparto, bagasse, flax, jute, and cotton. See GYMNOSPERMS.

ANGLE, ANGLE-CUT, OR ANGULAR—Cut at an angle to the machine direction. Envelope blanks are usually cut in this manner.

ANGLE OF CONTACT—The angle between the surface of paper and a line tangent to the surface of a drop of water placed thereon, at the intersection of the two surfaces. It indicates the resistance of the paper surface to wetting, and is useful in studies of the ruling and writing qualities of paper.

ANHYDRITE—A naturally occurring mineral which is calcium sulfate ($CaSO_4$). Also known as alabaster.

ANILINE DYES—A term originally applied to dyes derived from aniline including the first synthetic dyestuffs known. Today the term is used broadly to designate synthetic organic dyes and pigments (whether or not they are derived from aniline), as distinguished from animal and vegetable coloring materials (usually organic), natural earth pigments (usually inorganic), and artificial inorganic pigments. See ACID DYES; BASIC DYES; DIRECT DYES.

ANILINE PRINTING—See FLEXOGRAPHIC PRINTING.

ANILOX ROLLERS—Steel or ceramic rollers with an etched surface, used to meter inks or adhesive application.

ANIMAL PARCHMENT—See PARCHMENT (1).

ANIMAL SIZE—Gelatinous material of the type of glue or gelatin (q.v.), used as a size in paper-making.

ANIMAL SIZED—A term applied to any paper sized with gelatin or glue, either in the beater or by tub sizing. In general, only high-grade writing, bonds, ledgers, etc., are animal sized. See SURFACE SIZED; TUB-SIZED

ANION—A negatively charged ion.

ANIONIC CONTAMINANTS—Soluble anionic polymers that react with cationic polymers, such as wet-strength resins or dry-strength resins, to neutralize them and preclude their retention on fibers. Excessive anionic contaminants also impair the efficiency of very high molecular weight cationic retention aids. The amount of anionic contaminants can be monitored by measuring the cationic demand (q.v.) and adding cationic promoters, or alum, to offset the excess anionic character of the contaminants. The terms "anionic trash," "interfering substances," and "dissolved contaminants" are also used to refer to anionic contaminants. See ANIONIC TRASH; CATIONIC DEMAND; ELECTROKINETIC CHARGE TITRATION; WET END CHEMISTRY.

ANIONIC DEMAND—See COLLOID TITRATION.

ANIONIC POLYMERS—Polymer molecules having numerous anionic sites. Examples are polyacrylic acid and carboxymethylcellulose (CMC).

ANNOUNCEMENTS—Plain or paneled papers or cards which are cut to size or folded so as to fit envelopes made from the same paper, and sold in sets. The term includes greeting cards, business and social stationery, weddings, etc.

ANNUAL RING—In wood and bark, a growth layer of one year as seen in cross section.

ANODIC PROTECTION—A technique to reduce the corrosion rate of a metal by polarizing it into a passive (q.v.) region where dissolution rates are low. The highly polarized condition, produced with very little current flow, is maintained by a constant applied potential.

ANOVA—Analysis of variance.

ANPA—The American Newspaper Publishers Association.

ANSI—The American National Standards Institute. This organization is concerned with the development of national technical standards, and representation of the United States in the development of international technical standards.

ANTIBLOCKING AGENTS—Materials used in a coating formulation or as an overcoating to prevent sticking of one paper sheet to another or any other object within a specified temperature–humidity range.

ANTI-BLOW-BOX—Device used to prevent the paper from separating, or blowing from the dryer fabric. Normally placed behind the dryer fabric, it induces a low level vacuum to keep the sheet on the fabric.

ANTICHLOR—A chemical used to remove traces of free chlorine, hypochlorite, or chlorine from materials bleached with these substances. Sodium bisulfite ($NaHSO_3$) and sulfur dioxide are common antichlors.

ANTIFALSIFICATION PAPER—See SAFETY PAPER.

ANTIFOAMING AGENTS—See DEFOAMER.

ANTIFRICTION SLIDE—A device that utilizes antifriction bearings to minimize the normal friction developed between sliding parts

ANTIQUE BOOK PAPER—A paper with an antique finish (q.v.) usually cream-white or natural in color, but also made in other shades. It is frequently used for novels where a specific bulk for a given weight is required. It is usually made of bleached chemical woodpulp with a preponderance of short-fibered pulp, although mechanical pulp may also be used.

ANTIQUE BRISTOL—A typical bristol with a very low finish.

ANTIQUE COVER PAPER—Cover paper with an antique finish.

ANTIQUE EGGSHELL PAPER—See ANTIQUE FINISH (2).

ANTIQUE FINISH—(1) A rough finish obtained by operating with reduced pressure at the wet presses and at the calender stacks. It is generally considered to be rougher than eggshell and perceptibly rough to sight and touch. (2) The term "antique" is also used to denote a rougher finish when used as a prefix to another term for finish such as Antique-Eggshell, Antique-Vellum. Antique finish is distinguished from eggshell finish by the design of the feltmarks on the surface, which is made up of relatively large hills and valleys that are long and narrow, running with the grain direction, whereas the design of the surface of an eggshell paper is made up of smaller, rounder hills and valleys that are not as definitely aligned with the grain direction, and the paper presents a smoother surface appearance.

ANTIQUE GLAZED PAPER—A paper which has a smooth, glossy finish on one side and an antique finish on the other.

ANTIQUE PAPER—See ANTIQUE FINISH.

ANTIRUST PAPERS—Papers which of themselves will not facilitate the oxidation of the metal being packaged. Such papers must not only be free from acid and reducible sulfur compounds but other materials as well, which could cause metal discoloration by means other than oxidation. These papers are commonly referred to as anticorrosive, acid-free, noncorrosive, or antitarnish paper (q.v.).

ANTISTATIC AGENTS—Materials used in or on a paper or plastic to minimize accumulation of static electricity and thus attraction of airborne foreign matter or of one surface for another.

ANTITARNISH BOARD—See ANTITARNISH PAPER.

ANTITARNISH PAPER—A paper made in various weights from various furnishes and used for wrapping coins, silverware, aluminum goods, leaded glass, hardware, razor blades, needles, and other tarnishable articles. It must be relatively free from acids, alkalies, and from reducible sulfur compounds which yield sulfides. The term was originated for a lightweight sheet or tissue made from rags and carefully bleached and washed, which was used for wrapping silverware, steel tools, cutlery, and other polished steel hardware. Later the term was applied to wrapping paper made from sulfite or sulfate pulp free from acid or sulfur compounds. Copper salts or other inhibitors, including vapor phase (V.P.I.) are sometimes used in treating the paper when used for wrapping silverware or leaded glass.

ANTITARNISH TISSUES—See ANTI-TARNISH PAPER.

AOX—An acronym for adsorbable organic halide, a test to determine the total quantity of halide(s) present. It consists of adsorption of all organic halide containing material onto activated carbon followed by destruction of the carbon to determine the total quantity of halide(s) present. See also ADSORBABLE ORGANICALLY BOUND HALOGEN.

APA—The American Pulpwood Association or the American Plywood Association.

API—The American Paper Institute, once the trade association of pulp paper and paperboard manufacturers in the United States, now incorporated in the American Forest and Paper Association. Also an acronym for the American Petroleum Institute.

APMP—Alkaline peroxide mechanical pulp.

APPARENT DENSITY—The apparent weight per unit volume. It is often calculated by dividing the basis weight by the thickness, though it must be recognized that the numerical value thus obtained depends on the definition of the ream (q.v.). Consistent numerical values can be obtained by using in every case the basis weight in metric units (grams per square meter) and the thickness in millimeters. See APPARENT SPECIFIC VOLUME; SPECIFIC GRAVITY.

APPARENT SPECIFIC VOLUME—The volume per unit mass of paper, i.e., the reciprocal of apparent density (q.v.).

APPEARANCE—The effect upon the sense of sight resulting from observation of the color, brightness, finish, cleanliness, formation, and other visible characteristics of papers or boards.

APPLE AND PEAR WRAPS—An MF or MG tissue paper used for wrapping or packing fruit. It is usually manufactured from sulfate or sulfite-mechanical pulp furnish. During the process of manufacture, the sheet is impregnated with a colorless, odorless, and tasteless mineral oil of U.S.P. standard, a satisfactory technical grade of oil, or other approved oil base treatment that can be relied upon not to affect the flavor of the fruit on which the wraps are used. The oil is incorporated to prevent decay when the fruit is subjected to long storage periods. See FRUIT WRAPS.

APPROACH FLOW SYSTEM—The components located immediately prior to the paper machine headbox, usually from the fan pump or the primary paper machine cleaner feed pump, to the headbox. Also called the thin stock circuit, because the stock has been diluted to headbox consistency at the fan pump, or almost to headbox consistency at the cleaner feed pump.

APRICOT PAPER—(1) See FRUIT WRAPS. (2) A paper with a light moderately dull yellowish orange color.

A-PRINTING—One of several trade classifications for printing papers made from mechanical pulps, e.g., A-1, A-2, B, C. See GROUNDWOOD PRINTING PAPER.

APRON—Originally an oil cloth attached to the headbox and extending over the fourdrinier fabric from the breast roll to the first slice and up on the sides to the deckle strap. This arrangement allowed the stock to flow from the headbox onto the fabric. The slice, or top lip of the headbox, would meter the flow against the apron to the fourdrinier fabric. Modern headboxes employ a metal apron or bottom lip that usually has its downstream edge at the breast roll centerline, and with a clearance of approximately 1/16 to 1/4 inch above the fabric. The metal apron, which is usually stainless steel extends between the side plates of the downstream part of the headbox. Metal aprons are made adjustable in the machine direction so as to control drainage at the point that the paper stock contacts the moving fabric. With metal aprons, this may also be accomplished by proper positioning of the top slice lip.

AQUATINT—An intaglio process of etching on copper or steel plates with a grained resist, producing an effect resembling a drawing in water color, sepia, or India ink. The plate is printed on a handpress in a limited edition. This process is not used commercially.

AQUATONE PRINTING—A type of planographic printing in which the copy is transferred photographically to a zinc plate coated with sensitized gelatin. The exposed plate is developed in alcohol and water and baked, leaving the light-fixed bichromate image on the plate. Its chief difference from the collotype process (q.v.) is that a halftone screen of very great fineness is used, i.e., from 200- to 400-line screen, which provides up to 160,000 dots to the square inch.

ARABINAN—A hemicellulose consisting of arabinose monomeric units. In woody plants, a 1 5 linked ALPHA-L-ARABINAN is associated with other pectic substances in the middle lamella.

ARABINOGALACTAN—A hemicellulose composed of arabinose and galactose monomeric units. A $1 \rightarrow 3m$ $1 \rightarrow 6$ linked arabinogalactan is a minor constituent of most softwoods, but appears in significant quantities in larch. There is also a $1 \rightarrow 4$ linked galactan with arabinose substitution at C-6 associated with pectic substances.

ARACHNE—A machine used to produce stitch-through nonwovens.

ARAMID—A manufactured fiber in which the fiber-forming substance is a long-chain synthetic polyamide in which at least 85% of the amide linkages are attached directly to two aromatic rings.

ARCHIVAL PAPER—Paper with exceptional durability and high resistance to aging, intended for long-term storage.

AREA BONDED—A nonwoven that is bonded throughout its entire area and which characteristically is stiffer than a print or pattern bonded nonwoven.

ARMATURE PAPER—See IMITATION JAPANESE PAPER.

ART COVER—A special decorative cover paper, used for announcements, greetings, Valentine cards, and the like.

ARTICULATING PAPER—A paper impregnated with a vegetable wax mixed with a blue orred nontoxic coloring matter which is used by dentists in adjusting the "bite" of teeth. It gives a clear mark on porcelain, natural teeth, and metals (wet or dry) on the slightest pressure. It is from 0.004 to 0.005 inch in thickness and is supplied in sheets or rolls.

ARTIFICIAL LEATHER PAPER—The base paper used in the manufacture of artificial leather and made from a variety of woodpulp, rag, or rope furnishes in weights ranging from 60 to 200 pounds or more (24 x 36 inches – 500). It may contain a rubber latex applied either prior to or after sheet formation or may be a saturating paper intended for impregnation with a latex composition.

ARTIFICIAL PARCHMENT—A simulated parchment usually made from well hydrated long-fibered woodpulp or cotton fiber (and of-

ten with chemical additives) and characterized by a mottled or wild formation, fairly high density, crisp "feel," and moderate grease-proofness. It is used primarily for greeting cards, announcements, diplomas, and the like. It should not be confused with vegetable parchment. See also IMITATION PARCHMENT; VEGETABLE PARCHMENT.

ARTISTS ILLUSTRATION BOARD—A bristol, specially finished and suitable for pencil drawings, pen sketching, or watercolor work. Important properties are finish, color, rigidity, and freedom from warping.

ARTISTS PAPERS AND BOARDS—See ART PAPER; ARTISTS ILLUSTRATION BOARD; CHARCOAL DRAWING PAPER; DRAWING BOARD; DRAWING PAPER; ILLUSTRATION BOARD; JAPANESE DECORATING PAPER; SKETCHING PAPER; WATERCOLOR PAPER.

ART PAPER—(1) A high-grade drawing or artists' paper with a highly finished and smooth surface, which may be obtained by supercalendering or coating. Close formation and smooth surface are its most important characteristics. See also DRAWING PAPER. (2) A coated paper having a high finish to take halftones. (3) A United Kingdom term for a body paper or board which has been coated on one or both sides with mineral substances, such as china clay, which produces a smooth surface capable of taking, if desired, a very high finish; it is used primarily for printing from fine screen halftone blocks. (4) A fancy figured paper used by bookbinders for the end sheets and fly leaves in books.

ART PARCHMENT—A hard-sized, heavy sheet of cotton fiber and/or chemical woodpulp, with color, appearance, and surface reminiscent of the old vellums. It is particularly adapted for diplomas and documents where a heavy paper is desired. See also DOCUMENT PARCHMENT.

ART VEGETABLE PARCHMENT—Similar to art parchment (q.v.) but manufactured from vegetable parchment.

ASA—See ALKENYL SUCCINIC ACID ANHYDRIDE.

ASAM PROCESS—See SOLVENT PULPING.

ASBESTINE—A mineral (nearly pure magnesium silicate in fibrous form), occurring in the eastern part of the United States and southern Germany, which is intermediate in physical properties between talc and asbestos. It is used as a paper filler especially in blotting papers and in boards. It is sometimes called agalite.

ASBESTOS—White asbestos (chrysolite) is a naturally occurring fibrous crystalline magnesium silicate. It occurs in various combinations as white, grayish, or green veins of smooth nontubular fibers which may be separated readily by mechanical treatment. It has fair acid and good heat resistance and may be spun or woven. In addition to the white variety, blue asbestos (crocidolite) and brown asbestos (amosite) are used by the industry. The former is an iron sodium silicate and the latter is an iron magnesium silicate. The blue and brown forms are quite harsh and are difficult to spin or weave. Its use is rapidly decreasing due to risks of a lung disease (asbestosis) that occurs when fibers are inhaled.

ASBESTOS CEMENT BOARD—Obsolete term. See ASBESTOS LUMBER.

ASBESTOS DIAPHRAGM PAPER—Obsolete term. A paper made from long, high-purity grades of asbestos fibers used principally as a membrane in electrolytic cells.

ASBESTOS ELECTRICAL INSULATION PAPER—Obsolete term. (1) An inorganic base paper made of specially processed white asbestos fibers in thicknesses of 0.003 to 0.009 inch; the paper was treated with plastics and/or laminated to improve its mechanical properties. This paper was used for heat-resistant electrical insulations for the higher temperature ranges. (2) A paper made of specially processed white asbestos fibers combined with certain organic fillers to improve its mechanical strength in thicknesses of 0.005 to 0.015 inch. This paper was used primarily as a heat-resistant electrical insulation in combination with other dielectrics.

ASBESTOS LUMBER—Obsolete term. A homogeneous sheet of asbestos and cement formed under pressure. It weighed approximately 124 pounds per cubic foot. It was manufactured in flat sheets in various sizes up to 4 by 8 feet and in corrugated sheets in sizes up to 3.5 by 11 feet. It was used as a fire- and corrosion-resistant building material. This material has not been used for a number of years and the term has only historical significance.

ASBESTOS MILLBOARD—Obsolete term. An asbestos board, made in various densities and with various fire-resisting properties, in thicknesses of 1/32 to 3/4 of an inch. The furnish contained up to 20% of filler and sizing. It was used where relatively thin sheets or boards were required for protection against fire, heat, and acid fumes, such as fireproof linings for floors, partitions, ceilings; it was used also in ranges, stoves, grates, etc.

ASBESTOS PAPER—Obsolete term. A sheet of asbestos fibers, from 0.0015 to 0.0625 of an inch thick. For special purposes, it could be laminated to greater thicknesses. The furnish consisted mainly of asbestos fiber of varying qualities, depending upon the use to which the paper was to be put. A small amount of sizing was used, but sometimes the amount of sizing and filler was as great as approximately 15% in papers for special uses. Asbestos paper was used as an insulating material where minimum thickness was required, principally as a protection against heat and as a fire retardant between floors, walls, and ceilings.

ASBESTOS ROLL BOARD—Obsolete term. A heavy asbestos board in roll form, 3/32 to 1/8 of an inch in thickness, composed of asbestos fiber with a small amount of filler and sizing. It was used as a protection against heat and as a fire retardant between floors, walls, and ceilings.

ASBESTOS ROOFING FELT—Obsolete term. Asphalt-impregnated asbestos felt with or without a coating which was on one or both sides. Felts which were impregnated only were generally of either 15 or 32 pounds per square weight. Impregnated and coated felts most commonly encountered weighed 20, 35, and 55 pounds per square but could vary with the manufacturer. Asbestos felt impregnated with coal tar was also marketed in a weight of 15 pounds per square. In all cases, a square refers to sufficient material (108 sq. ft.) to cover one square of roof area. Packaging was in rolls varying from one to four squares depending upon weight.

ASBESTOS SLATERS FELT—Obsolete term. A high-grade asbestos felt, impregnated with asphalt, and used as a sheathing and as a liner under tile or slate shingles. A roll of 324 square feet weighed approximately 45 pounds.

ASBESTOS WALLBOARD—Obsolete term. A sheet building material, composed of asbestos fiber and cement, which was formed on a wet machine as a solid, homogeneous, flexible sheet. It was usually repressed on a hydraulic press after it left the wet machine. It was manufactured in thicknesses of 1/8 and 3/16 inch, but it could be laminated to form greater thicknesses. A cubic foot weighed approximately 120 pounds. Its principal use was for interior and exterior finishes for homes, farm buildings, and industrial structures.

ASBESTOS WATERPROOFING FELT—Obsolete term. A strong, asphalt-impregnated asbestos felt once used in waterproofing work where a thick and durable felt was required.

AS-FIRED BLACK LIQUOR—Black liquor after final evaporation and after the addition of (1) makeup chemicals, (2) recycled precipitator catch, and (3) boiler bank ash hopper catch, and economizer ash hopper catch. Black liquor flowing in the ring header (the manifold surrounding the recovery boiler delivering black liquor to the individual liquor guns) and through the black liquor guns.

ASH—The inorganic residue after igniting a specimen of wood, pulp, or paper so as to remove combustible and volatile compounds.

ASH, ACID-INSOLUBLE—The part of the ash in pulp that is insoluble in hydrochloric acid. Excessive amounts of acid-insoluble ash in a pulp may cause undesirable abrasiveness in fine papers made therefrom.

ASHLESS—Practically free from ash. See ANALYTICAL FILTER PAPER.

A-SIZES OF PAPER—ISO R216 standard metric sizes of paper, based on a rectangular sheet A0, which is 841 x 1149 mm, and is one square meter. The sides have a ratio of 1:1.4142. Succeeding sizes are made by cutting the sheet in half, width-wise. The A4 size is 297 x 210 mm, close to 8.5- x 11-inch letterhead size. B-sizes cover poster size paper and C-sizes cover envelopes.

ASPHALT—A naturally occurring mixture of hydrocarbons or a similar product prepared from coal tar, which is widely used in the paper industry in the manufacture of roofing and waterproof papers.

ASPHALT DISPERSION—A process for dispersing asphalt contaminants in paper and board by submitting a 30% consistency slurry of fiber in water to elevated temperatures and pressures in a continuous digester for 3 to 5 minutes, refining, and ejecting through a blow valve. The process does not remove the asphalt but disperses it so particles are not readily visible.

ASPHALTED (ASPHALTING) BOARD—A paperboard used by the building trade as roofing or siding, for wrapping, and for automobile panels. It is made of woodpulp, wastepapers, or a combination of the two, without sizing to permit impregnation with asphalt or to allow for combining with other paperboard or paper using asphalt as an adhesive. Generally 0.009 of an inch (9 points) in thickness, or heavier, it is readily absorbent on one or both sides. Stiffness and strength are important characteristics.

16

ASPHALT EMULSION—A stable dispersion of asphalt, water, clay, and other materials which is used for internal sizing of unbleached kraft papers and certain paperboards. Treatment with such emulsions increases the resistance of paper to water and to water vapor.

ASPHALT FELT—A roofing felt (q.v.) saturated with asphalt. The original dry felt weighs from 26 to 52 pounds per 480 square feet, and the saturated felt weighs from 13 to 30 pounds per 100 square feet. The percentage of asphalt saturant ranges from approximately 140 to 160% on the weight of the original dry felt. This felt is usually marketed in 36-inch rolls, containing 432 square feet for the lighter weights and 216 square feet for the heavier weights. The principal uses of asphalt-saturated felts are as sheathing papers under roof shingles and sidings and particularly in the construction of built-up roofs, wherein several thicknesses are cemented to each other by mopping with hot asphalt. Such a roof is sometimes topped with a capsheet, which consists essentially of an asphalt-saturated felt coated with talc on the weather side only. See TARRED FELT.

ASPHALTING PAPER—A kraft paper used as the base in the manufacture of asphalt papers. The basis weight is usually 30 pounds (24 x 36 inches – 500).

ASPHALT-LAMINATED PAPER—Two or more sheets of paper bonded together with one or more layers of asphalt. The paper may be plain or creped in one or two directions. It also may be reinforced.

ASPHALT PAPERS—A general term which includes papers saturated, coated, or laminated with asphalt or other bituminous material. See ASPHALT SHEATHING PAPER; DUPLEX ASPHALT PAPER; HOUSE SHEATHING PAPER; ROOFING PAPER; SHEATHING PAPER; WRAPPING PAPER.

ASPHALT SATURATION—Saturation of the paper with molten asphalt to provide resistance to water or water vapor or both. The most commonly used devices for asphalt saturation are size presses or two roll saturators. The saturator and the circulation system must be heated to a temperature that will prevent solidifying of the asphalt. The sheet to be saturated must be absorbent; it may be made of either cellulose or glass fibers. Asphalt saturation is a common process in the manufacture of roofing paper and shingles. In most cases, asphalt saturation is an off machine converting process.

ASPHALT SHEATHING PAPER—A paper saturated with asphalt, the furnish of which is sulfate pulp and news (recovered papers). The finished saturated sheet is quite smooth and contains asphalt up to the extent of approximately 80–100% of the original paper. It is usually sold in rolls of 36-inch width and containing 500 square feet, weighing 25–30 pounds. The caliper varies from 10 to 14 points. Its important characteristic is waterproofness.

ASPHALT SLATERS FELT—See ASPHALT FELT.

ASSIGNABLE CAUSES—Causes of variation that have a definite cause or set of causes. (Also often called "special" causes.)

ASSIMILATIVE CAPACITY—The capacity of a natural body of water to receive: (a) wastewaters without deleterious effects; (b) toxic materials, without damage to aquatic life or humans who consume the water; (c) BOD within prescribed dissolved oxygen limits.

ASTM—The American Society for Testing and Materials, a scientific and technical organization formed for the development of standards of characteristics and performance of materials, products, systems and services, and the promotion of related knowledge.

ATLAS PAPER—A kind of map paper designed for atlas printing, and characterized by excellent lithographic printing quality. See also MAP PAPER.

ATOMIC ABSORPTION SPECTROPHOTOMETRY—An analytical process in which a substance in solution is thermally atomized, and its absorption of radiation at different wave lengths is measured. The technique is used as a sensitive and rapid means for quantitative analyses, especially of metallic elements.

ATTAPULGITE CLAY—A naturally occurring mineral which is a complex form of hydrated aluminum silicate. It has absorptive properties and is used in specialty paper coatings and other applications.

ATTRIBUTES DATA—The mere counting, or conversion of counts to proportions or percentages, of the presence or absence of some characteristic or attribute in the units, items, or areas being examined.

AUSTENITIC STAINLESS STEEL—Corrosion-resistant alloys containing chromium and >8% nickel that take up the austenitic crystal structure.

AUTOCHROME PRINTING PAPER—A coated paper suitable for multi-color printing.

AUTOGRAPHIC REGISTER PAPER—See REGISTER BOND.

AUTOMATIC CONTROL—A combination of instruments control a unit process efficiently and safely.

AUTOMATIC PROCESS CONTROL—A combination of sensors, control algorithms, and final control elements which automatically adjusts process set points for feedback and/or feedforward control. The control algorithms may be implemented in a distributed control system (DCS) or in a computer.

AUTOMOBILE BOARD—A paperboard used in automobile body construction, largely as panels, either plain, embossed, coated, decorated, or to be covered with fabrics. It is generally made of woodpulp, reclaimed paper stock, or a combination of both on a cylinder machine or one of the recently developed former-type machines. It ranges in caliper from 0.040 to 0.500 of an inch. This board is rigid, resistant to blows, to abuse, and to penetration by water and other liquids. It does not soften at high temperatures. Asphalt and resins are commonly used as the waterproofing agents.

AUTOMOBILE PANEL BOARD—See AUTOMOBILE BOARD.

AUTOMOBILE TIRE ROLL—See AUTOMOBILE TIRE WRAP.

AUTOMOBILE TIRE WRAP—MF, MG, or WF fourdrinier or cylinder kraft or kraft-rope paper, usually of basis weights of 60 and 90 pounds (24 x 36 inches – 500), which is dry or water finished and may sometimes be decorated with various trademarks or designs. It may also be duplex, one side of which is colored. It is cut in strips approximately two-and-one-half inches wide and in small rolls to be used with automatic tire-wrapping machines for wrapping tires to protect the tires from mechanical injury during shipment or storage. Strength and resistance to abrasion are important properties, and the colors should be fast to light.

AUTO TIRE WRAP—See also AUTOMOBILE TIRE WRAP.

AUTOTYPE PAPER—A bond or writing paper similar to register bond. It is usually made from chemical woodpulps in basis weights of 9 to 16 pounds (17 x 22 inches – 500) and is characterized by a smooth printing surface and good manifolding qualities. It is used for multi-copy automatic register applications. See also REGISTER BOND.

AVAILABLE (ACTIVE) CHLORINE—See BLEACHING POWDER.

AVERAGE—See MEAN.

AVERAGE CHART (X CHART)—A control chart in which the subgroup average, X-bar, is used to evaluate the stability of the process level.

AXIAL—In papermaking, in the direction of a roll's core.

AZURE LAID WRITING PAPER—A writing paper with laid mark and light blue color. Azure wove differs only in having a wove finish.

B

B—(1) A mark indicating a degree of roughness in the finish of superfine drawing papers. "B" is rough and "B.B." is double rough. (2) A designation used frequently in connection with groundwood papers to denote grades in which the woodpulp content is wholly unbleached.

BABY DRYER—A single dryer cylinder of small diameter.

BACK—(1) In boards composed of plies of different stocks, the side of better quality or finish is usually referred to as the top; the other side of the board is known as the back. (2) The drive side of the paper machine.

BACK DRUM—The winding drum that is farthermost from the winder operator when approaching the winder from the discharge side of a two-drum winder.

BACKING BOARD—A thick paperboard placed in the backs of mirrors, pictures, etc., or used for partitions in furniture. It is generally made on a cylinder machine of paper stock, is pasted or unpasted, and is rigid to prevent warping. It may be lined or colored on one side to give the desired finish.

BACKING PAPER—(1) A paper that strengthens the flong or mold of alternated sheets. It is used as a mold for stereotype work; pastes down easily; and is usually brown. See STEREOTYPE BACKING. (2) An unprinted hanging paper sometimes used on a wall before the printed hanging paper is applied. For other uses of this term, see BOOK BACK LINER; STENCIL-BACKING SHEET; TYMPAN PAPER.

BACKING ROLL—A large-diameter roll or cylinder which is an integral part of the inverted blade, puddle blade, and roll coaters. The purpose of the backing roll is to firmly support the web as the coating color is applied on the other side of the web. In the case of the inverted blade and puddle blade coaters, excess coating is scraped off the surface of the paper by a thin steel blade which is positioned tangentially to the web while in contact with the backing roll. In the case of roll coaters, a pre-metered amount of coating is applied to a web by another roll pressing against the backing roll.

BACKING ROLL MARK—A mark on a coated sheet caused by a buildup or mark on a backing roll (q.v.). This area of the sheet receives a different amount of coating from the rest of the sheet.

BACKING UP—Printing the other side of a previously printed sheet.

BACK LINER—The liner on the back side of a multi-ply board. It is usually of a different grade of stock from the top liner (q.v.), and often a different grade of stock from the middle liner(s) (filler ply).

BACK-LINING PAPER—See BOGUS PAPER; BOOK BACK LINER.

BACK MARK—The ridge or mark in sheets of paper which have been loft-dried on poles or lines. It is also called pole mark or stick mark. The terms are not in current usage.

BACKSIDE—(1) In reference to paper, that edge of the web of paper which is produced away from the operating side of the paper machine. (2) In reference to a paper machine, that side of a paper machine on which the drive mechanism is located.

BACKTENDERS FRIEND—See HARDNESS TESTER (PAPER).

BACKTRAP—Ink transfer from a previously printed surface back to the blanket of a subsequent print station.

BACK TRAP MOTTLE—Irregular, blotchy print caused by backtrapping.

BACK WATER—See WHITE WATER.

BACON PAPER OR WRAPPER—A glassine, greaseproof, or vegetable parchment paper, or a laminated product made from these papers and other materials, used for wrapping bacon. When made from greaseproof papers, the basis weight is usually 30 to 40 pounds (24 x 36 inches – 500); when made from vegetable parchment, the basis weight is usually 27 to 40 pounds (24 x 36 inches – 500).

BACT—An acronym for Best Available Control Technology.

BACTERIAL COUNT—Enumeration of bacteria in a definite weight of paper or board; all organisms, both pathogenic and nonpathogenic, which will grow under the conditions set up by the technique used, should be included.

BAD SPLICE—A defect in a mill splice, such as an exposed part of the glue line, a tail not removed, or webs not parallel.

BAFFLE AERATOR—An aerator wherein baffles are provided to cause turbulence and minimize short-circuiting.

BAG—A flexible container made of paper, foil, plastic film, or combinations thereof or similar materials usually applied to consumer size (smaller) packages. See BAG PAPER; SHIPPING SACK.

BAGASSE—The crushed stalks of the sugar cane after the sugar has been extracted. It is pulped for pulp and paper use.

BAG (DEFECT)—A web defect existing in a web process where part of the web has been previously stretched and unable to return to its original length. The result is that the tensioned web has areas that are locally slack with no capacity to transmit tension. If the web is strong enough, an increase of the average web tension may allow enough of the bag to be under tension to permit the web to pass through the process without incident.

BAGGY—Having a puffed-out sag or bulge.

BAGGY ROLL—A web that developed a "stretched" or "slack" area during a roll winding process because the web was wound over an area of higher caliper or greater roll diameter than adjacent areas in the same roll. The

baggy area which has been stretched beyond its elastic limit is noted when the web is unwound at low web tension. See DEFECTS.

BAG LINERS—(1) The inner liners of duplex or multiwall bags used to package products requiring special protection. Various types of papers, films, foils, or combinations of these materials are used, depending on the protection desired, i.e., against the transmission of moisture, grease, odor, or flavor. Typical liners are asphalt, waxed and greaseproof papers, foils, free films, vegetable parchments, and film coated papers. The typical basis weight range is from 25 to 100 pounds (24 x 36 inches – 500). (2) Liners of paper in bags of burlap or cotton to prevent leakage and exclude dampness and dust. These are usually of creped kraft and are sometimes waterproofed with asphalt. See LINER.

BAG PAPER—Paper used in the manufacture of bags or sacks with the choice of paper dependent upon the quality and quantity of the product intended to be packaged and subsequent bag filling operation, handling, transporting, and storage requirements. Examples of bag papers and their intended end uses are: (a) grocery bag and grocery sack paper—unbleached or bleached kraft paper varying in basis weight in lighter weights used for grocery bags and heavier weights used for grocery sacks; (b) multiwall kraft paper—unbleached, semibleached, colored, or bleached kraft paper having high strength characteristics defined by specifications which relate to the ability of the paper and finished bag to perform its heavy-duty packaging function.

BAKERS WRAP—A lightweight wrapping paper used for wrapping baked goods. It is usually made from bleached chemical woodpulps in basis weights of 20 to 25 pounds (24 x 36 inches – 500). It may be either machine finished or machine glazed, the larger percentage being the latter. Significant properties include a reasonable strength, a high finish, a good white color, and, in most cases, a machine-mark stripe.

BAKING CUP PAPER—A grease-resistant or dry-waxed sheet of paper used in the manufacture of baking cups.

BALANCED-DRAFT BOILER—A balanced-draft boiler is one in which the actions of a fan pressurizing the inlet combustion air and a fan-inducing draft on the outlet combustion gases of the boiler are combined to keep near-atmospheric pressure in the boiler.

BALANCED THRUST—Where the mechanical load on a refiner is equal on both sides of the rotating element.

BALANCE SHEET—A sheet of resin-impregnated paper used to maintain symmetrical stresses in a laminate in order to eliminate excessive warpage.

BALE—A wire- or strap-bound bundle of fibrous raw material usually 90% air dry. Bale sizes and weights vary with the nature of the bundled material.

BALER BAG—A large sack designed to carry a quantity of filled smaller bags.

BALING—(1) Forming a bale of pulp, rags, wastepaper, or other materials by compression in a baling press and banding with straps of metal or other material. (2) A method of packing paper, in which the paper is covered with moisture resistant, heavy weight kraft paper, baled under pressure, and protected at the top, bottom and corners by boards; metal bands are then fastened around the bale.

BALING PAPER—A general term applied to any heavy paper used for covering bales to protect the contents. It may be reinforced with cloth fabric (cheesecloth) which is pasted to one side after it has been coated with asphalt, or it may be crinkled, corrugated, or treated with asphalt.

BALSTON'S PAPER—A drawing paper.

BAMBOO—A giant woody grass, often reaching a height of forty feet or more, found in the tropical and subtropical regions of the eastern hemisphere. It also has been grown successfully in certain parts of the southern United States. The fibers closely resemble those from straws in many of their characteristics. Its fibers have an average length of 2.4 mm., thus standing between softwood and hardwood fibers.

BAND STOCK—A paper used for banding bolts of cloth, hosiery, etc. It may be a coated or uncoated sheet made of chemical or mechanical woodpulp, suitable for printing. Strength is an important characteristic.

BANK NOTE PAPER—High-grade bond paper designed for bank notes, stock certificates, currency, and the like. It is usually made from high-quality cotton pulps, to which highly purified chemical woodpulps may sometimes be added. It is characterized by high physical strength, durability, and good engraving quality, and is generally made in basis weights of 20 to 24 pounds (17 x 22 inches – 500). See CURRENCY PAPER.

BANK PAPER—High-grade writing paper made of cotton and/or bleached chemical woodpulp with bond characteristics and durability.

BANK STOCK—See BANK NOTE PAPER.

BANQUET-TABLE COVER PAPER—Any type of paper which is used in place of cloth as a table covering. It may be plain, embossed, colored, or printed in any number of designs.

BARBED NEEDLES—Needles used in needle punching fibers to mechanically interlock the fibers when producing needle punched nonwoven.

BARBER DRYING—A method of air drying, where the moist paper, after tub sizing, is passed over a series of driven carrier rolls arranged in two tiers, one placed four to six feet above the other, so that the sheet travels successively in approximately vertical passes alternately over an upper tier roll and then down around a bottom tier roll and back up to the next top roll, etc. The dryer mechanism for these rolls is arranged to allow control of tension on the sheet as it is dried by heated air blown over the mov-

21

ing sheet. This treatment imparts a cockle finish to bond papers. It is so called from the name of the inventor. Archaic term for this process.

BARBERS' HEADREST PAPER—Usually a machine-glazed book or absorbent paper resembling toweling, which may be striped or decorated and which is put up in rolls and used on barbers' chairs. The basis weight ranges from 20 to 25 pounds (24 x 36 inches – 500).

BAR CODE—Vertical line pattern on merchandise, scanned by optical sensors, and used for automated product identification and merchandise control.

BARIUM CARBONATE—A chemical compound ($BaCO_3$) obtained either from the naturally occurring mineral or by chemical reaction, and used as a coating pigment, usually in combination with other pigments.

BARIUM SULFATE—A chemical compound ($BaSO_4$) obtained either from the natural mineral barytes or by chemical reaction and used as a filler and as a coating pigment, especially for photographic papers either alone or in combination with other pigments. The artificial product is called blanc fixe, fast white, pearl white, or permanent white.

BARK—The outer covering of woody stems or roots of plants and trees. Bark is composed of inner living bark and outer dead bark. While bark is composed of some very short fibers, it is not used as a source of papermaking fibers primarily because of the low yield and high resin content. It is usually removed from round wood and burned or processed into mulch.

BARK CONTENT—The amount of bark remaining on chips after debarking and chipping. Most pulp mills require less than 1% bark content (by weight) in chips to avoid pulp dirt and appearance problems. Bark has a low pulp yield and high resin content. The resin does not dissolve in the pulping process and creates a brown to yellow residual in pulp which sticks to surfaces of pulp machinery.

BARKING—Use discouraged. See DEBARKING.

BARK SPECKS—Dark specks in paper caused by bark fragments in the pulp.

BARREL LINER—Paper used in lining barrels to prevent leaking, to retain freshness, to prevent the contents from sticking to the barrel, or to exclude odors and dust. It is usually made from chemical woodpulps in basis weights of 25 pounds (24 x 36 inches – 500) or heavier, which may or may not be waxed, creped, or otherwise treated, depending upon the nature of the product to be packed. Vegetable parchment is also used for this purpose. The lining is shaped to the side wall of the barrel and includes a sewn, pasted, or inserted bottom as well as a top (cap paper).

BARRELLING—A defective area in a paper-filled roll, also referred to as a hollow roll—typically for a supercalender. This phenomenon occurs when the paper fill has too low a coefficient of friction for the fill pressure used in the roll manufacturing process and the fill is squeezed away from the shaft instead of being compressed in the same direction as the roll axis.

BARRIER MATERIAL—(1) The coating applied to a substrate to make the substrate resistant to the passage of moisture vapor, gases, water, or other liquids including oils. (2) The term also applies to the product itself, such as paper, paperboard, or plastic film that is treated, coated, or laminated so as to provide resistance to the passage of substances given in (1).

BARRING—(1) Longitudinal bars usually observed in supercalender filled rolls which can cause severe stack vibrations. The excitation source can be machine direction barring in the web itself or it can be caused by the filled rolls if they were subjected to high nip loads without roll rotation. Barring in paper filled embossing rolls is associated with speed differentials during spur gear tooth engagement, as the number of bars match the number of teeth in the spur gear drive. (2) A machine direction basis weight variation in the sheet, usually caused by pressure pulsations in the paper machine approach flow system. The sheet has noticeable, regularly spaced, light and dark streaks that extend across the machine direction.

BARYTES—Naturally occurring barium sulfate (q.v.). It is also called heavy spar.

BASE COATING—A coat applied by a variety of methods, dried, and then overcoated with a second-down top coat. Base coating is also known as precoating and prime coating.

BASE PAPER—See BODY STOCK.

BASE STOCK—See BODY STOCK; COATING RAW STOCK; HANGING RAW STOCK.

BASEWAD PAPER—A dense paper used in the base of a shotgun shell to help hold the paper tube to the brass wall of the shell. It is made in thicknesses of about 0.010 to 0.012 of an inch. See WADSTOCK.

BASIC DYES—A class of insoluble coloring materials derived from aniline and commonly used in the form of dispersed pigments for tinting white papers. These materials are also converted into water-soluble salts in which form they are used for coloring various grades of paper where light fastness is not an important consideration. Basic dyes are also used in oil-soluble or spirit-soluble forms for such products as hectograph inks, typewriter ribbons, "carbon paper" coatings, stamp pad inks, etc.

BASIC SIZE—A certain sheet size recognized by buyers and sellers as the one from which its basis weight is determined. Initially, it was that size which printed, folded, and trimmed most advantageously. Some of the specifications for basic sizes now in use are as follows (the first two numbers are the dimensions of a sheet in inches; the last number is the number of sheets per ream):

Bible:	25 x 38 – 500
Blanks:	22 x 28 – 500
Blotting:	19 x 24 – 500
Bond:	17 x 22 – 500
Book:	25 x 38 – 500
Box cover:	20 x 26 – 500
	24 x 36 – 500
	25 x 38 – 500
Cover:	20 x 26 – 500
Glassine:	24 x 36 – 500
Gummed:	25 x 38 – 500
Hanging:	24 x 36 – 480
Index bristol:	15.5 x 30.5 –500
Ledger:	17 x 22 –500
Manifold:	17 x 22 –500
Manuscript cover:	18 x 31 – 500
Mill bristol:	22.5 x 28.5 – 500
	22.5 x 35 – 500
Mimeograph:	17 x 22 – 500
News:	24 x 36 – 500
Offset:	25 x 38 – 500
Postcard:	22.5 x 28.5 – 500
Poster:	24 x 36 – 500
Railroad manila:	17 x 22 – 500
Tag:	22.5 x 28.5 – 500
	24 x 36 – 500
Tissues:	24 x 36 – 500
Tough check:	22 x 28 – 500
Waxing:	24 x 36 – 500
Waxing tissue:	24 x 36 – 500
Wedding bristol:	22.5 x 28.5 – 500
Wrapping:	24 x 36 – 500
Wrapping tissues:	24 x 36 – 480
Writing:	17 x 22 – 500

The weight of tissue is sometimes given for a ream of 480 sheets (20 x 30).

23

BASIS WEIGHT—In the United States, the weight in pounds of a ream cut to a specified basic size (q.v.). The number of sheets in a ream is usually 500 (480 sheets for hanging paper and wrapping tissues); the Federal Government generally specifies 1000 sheets for all papers. In most countries, the basis weight is expressed in grams per square meter (grammage). For paperboards, basis weight is most often expressed in pounds per 1000 square feet. The basis weight of wallpaper is expressed in ounces, and this may be converted into pounds per ream (24 in. x 36 in. – 500) by multiplying the ounce weight by four and adding two. A basis weight of 30 pounds on this ream size means 30 pounds per 3000 square feet. The ream size for writing paper is 17 in. x 22 in. – 500, which is approximately 1300 square feet. This ream size is sometimes referred to as "sub." For example, "sub 30" means 30 pounds per 1300 square feet. See EQUIVALENT WEIGHT; SUBSTANCE NUMBER.

BASIS WEIGHT PROFILE—A graph indicating the variation of basis weight per ream across the width of the web.

BASIS WEIGHT VARIATION—The local peak-to-peak deviation of basis weight from the average basis weight per ream of paper among a population of samples or across the width of web.

BASKET LINER—(1) Paper used for lining baskets for shipping fruits, vegetables, mushrooms, etc. It is usually a kraft, sulfite, or vegetable parchment sheet in a wide variation of basis weights. The principal requirements are strength and waterproofness. (2) A paperboard liner made from reclaimed paper stock, which may be plain or colored. This is generally cut to size and the edges are joined for insertion into the basket. Suitable circles for the top and bottom are also supplied.

BAST FIBERS—Fibers obtained from the inner bark or phloem of fibrous plants. Examples of bast fibers harvested annually and used in papermaking are flax, hemp, jute, kenaf, kozo, and mitsumata.

BAT—An acronym for Best Available Technology.

BATCH BLEACHING PROCESS—A process whereby a reaction vessel is filled with pulp and bleaching chemicals. The chemicals are distributed throughout the pulp by mixing within the vessel. The vessel may be heated and pressurized. At the end of the reaction period, the entire contents are discharged for washing and for the next bleaching operation.

BATCH PROCESS—Refers to a discontinuous pulping or bleaching process. Large vessels are filled with chips or pulp, treated with chemicals and/or heat, and held for a time. Each vessel load is considered a batch. Most mills are converting to more cost-effective continuous processes.

BATCH PULPER—A pulper that operates on a discontinuous or cyclic basis. Typically, the batch pulping cycle consists of charging the pulper with dry furnish components and water, pulping, and extracting the stock.

BATEA—Best Available Technology Economically Achievable.

BATHROOM TISSUES—A term applied by some manufacturers to resale rolls of toilet paper or facial tissues.

BATTERY BOARD—A paperboard used by the manufacturers of dry storage batteries. It is usually made of woodpulp, reclaimed paper stock, or a combination of both, 0.030 of an inch or more in thickness. Stiffness and freedom from metallic or electrically conducting substances are important properties.

BATTERY PAPER—A paper used between the plates of storage batteries to absorb excess moisture from the paste which is used in making such plates. It may be made of reclaimed paper stock, chemical or mechanical woodpulps, or even rags. The paper is made on a fourdrinier machine, with a very rough finish requiring practically no calendering. The basis weights range from 35 to 90 pounds (24 x 36 inches – 500). The sheet has an alkaline reaction; it has fair strength, but high absorbency is the chief characteristic.

BATTERY PASTING PAPER—See BATTERY PAPER.

B.B. NOTE PAPERS—(1) Paper used in the manufacture of blank books. Writings, bonds, ledgers, book, or lower grades may be used, provided they have good writing qualities. (2) An abbreviation for black-bordered note papers.

BCTMP—Bleached chemithermomechanical pulp.

BEADING—The operation of forming the edges of paper or board into a tight roll as for reinforcing the lips of some drinking cups.

BEAMING PAPER OR BOARD—High finish and high strength kraft paper or board used for heavy-duty wrapping of textiles and fibers. The term originated in the textile industry to describe paper or board used for wrapping strands of silk on the "beam" prior to weaving.

BEATER—A machine consisting of a tank or "tub" usually with a partition or "midfeather" and containing a heavy roll revolving against a bedplate. Both roll and bedplate may contain horizontal metal bars set on edge. The beater may be "furnished" by either (a) pumping stock slurry from a pulper or (b) adding pulp or wastepaper slowly with sufficient water so that the mass may circulate and pass between the roll and the bedplate. The primary function of the beater is to initiate the development of the fibers to achieve specific sheet properties by cutting, bruising, fibrillating, and hydrating the fibers. Fillers, dyestuffs, and sizing materials may be added to the beater and thus incorporated with the paper stock. Beaters are generally used on a batch basis for fiber treatment, but can be used on a continuous basis for slush wet or dry lap pulp.

BEATER ADDITIVE—Any nonfibrous material such as a coloring or sizing agent, and adhesive, etc., which is added to the furnish to im-

prove the processing and final properties of the paper. See BEATER ADHESIVE; WET-END ADDITIVE.

BEATER ADHESIVE—A beater additive (q.v.) which is added to the papermaking furnish prior to sheet formation for the primary purpose of improving strength properties.

BEATER COLORED—See BEATER DYEING.

BEATER DYEING—A paper coloring method by which dyes are added to the fibers in the beater or at any stage in the preparation of the furnish up to the headbox (q.v.), as opposed to surface coloring (q.v.) after the sheet has been formed.

BEATER ENGINEER—See BEATER OPERATOR.

BEATER LOADING—(1) The process of adding a filler to the stock in the beater or to the pulp furnish prior to sheet formation. (2) The operation of adding pulp, pigment, etc., to the beater.

BEATER MAN—Usually a helper in the beater room; the person who actually loads and empties the beaters.

BEATER OPERATOR—The person in charge of the beater room or the operation of the beaters in the stock preparation area. Also called a beater engineer.

BEATER ROOM—Archaic term for the room or area of the mill where the beaters are located, and also where the stock is refined and blended. The term is used today occasionally to refer to the location where the equivalent functions are performed; this area would fall under the more generally preferred modern term of the stock preparation area.

BEATER SIZING—See INTERNAL SIZING.

BEATING—(1) That portion of the pulp refining operation carried out in the beater. See BEATER; REFINING. (2) A generic term for any mechanical fiber development by mechanical treatment.

BEDSTEAD-WRAPPING PAPER—A kraft paper of average strength, in basis weights of 30 pounds (24 x 36 inches – 500) and heavier, used for wrapping beds to protect the paint and finish from damage in transit.

BEER FILTER PAPER—A long-fibered filter paper used as a medium in a centrifuge for filtering beer. The medium is capable of being washed and reused. It is also called filter masse.

BEER MAT BOARD—See COASTER BOARD.

BENDER—A term applied to boxboard regarding its ability to be bent or folded on a score without rupture of the top liner. A full bending board, when properly scored, may be folded 180° without showing pronounced failure of the top liner. A semibending sheet may be folded 90°.

BENDING CHIP—A paperboard, which is almost gray or gray-brown in color, that is used for the manufacture of folding cartons. It is made of reclaimed paper stock and, by definition, must endure a single fold to 180° without breaking or separating the plies.

BENDING NUMBER—A measure of the bending quality of paperboard to determine its suitability for conversion into folded cartons without a scoreline rupture.

BENDING STRENGTH—The strength properties of paperboard that affect its bending qualities. This property is affected by the tensile strength of the top liner, the ability of the sheet to delaminate at the score, and the moisture content of the sheet. See BENDER.

BENT BLADE COATING—Coating is metered with an unbeveled blade that is at a tangent to the sheet and backing roll.

BENTONITE—A naturally occurring clay mineral (also called montmorillonite) that has high absorptive and colloidal properties. Bentonites are used as paper fillers, pitch dispersants, and in specialty coatings.

BETA-CELLULOSE—The portion of cellulosic material that dissolves in the 9.45% sodium hydroxide solution under the conditions of the alpha-cellulose determination, but which is reprecipitated on acidification of the alkaline solution. See ALPHA-CELLULOSE.

BETA GAUGE—An instrument for the measurement and control of basis weight through the application of radioactive isotopes.

BEVELED BLADE COATING (STIFF BLADE COATING)—Coating is metered by a blade that has an angled bevel at the tip.

B-FLUTE—See FLUTE.

BIAXIALLY ORIENTED FILM—Plastic film subjected to lateral and longitudinal stretching followed by heat setting to provide molecular alignment in these directions for improved stabilization.

BIBLE PAPER—A lightweight opaque paper designed for the printing of bibles, missals, encyclopedias, rate books, and the like. Bible paper is normally made from bleached chemical woodpulps, and/or cotton, flax, and linen fibers. Low bulk, opacity, permanence, and durability are important qualities. Opacity is provided through heavy loading with titanium dioxide and/or other pigments, and basis weights usually run in the range of 14 to 30 pounds (25 x 38 inches – 500). The paper is also called India Bible, India Oxford Bible, or Cambridge Bible especially in England where it was first developed.

BIBS—Paper bibs used for lobster bibs, dental bibs, and baby bibs. The back of the bib is often polyethylene-coated; the paper may be embossed or creped.

BIBULOUS PAPER—See ABSORBENT PAPERS.

BICHROMATE—Colloidal plate coating made light-sensitive by ammonium bichromate.

BICONSTITUENT FIBER—The preferred term is matrix fiber (q.v.).

BILLBOARD PAPER—See POSTER PAPER; NON-FADING POSTER PAPER.

BILLHEAD PAPER—A bond or writing paper used for the printing of bills, statements and the like where handwritten or typewritten entries are made. It is generally made from chemical woodpulps in basis weights of 16 to 24 pounds (17 x 22 inches – 500).

BILLING MACHINE PAPER—A ledger-type paper made of chemical woodpulps in basis weights of 24 to 36 pounds (17 x 22 inches – 500) and designed for accounting machine use. It is characterized by a smooth printing surface and good physical strength.

BILL PAPER—See BILLHEAD PAPER.

BILL POSTER BLANKING PAPER—See BLANKING PAPER.

BILL STRAPS—A general term applied to strips of kraft, manila, bond, or ledger paper [basis weight, 35 to 50 pounds (24 x 36 inches – 500)] used to fasten currency notes into packages. The strips may be colored and printed to denote different denominations of bills; they are usually gummed on one end.

BILLY CLUB—A stick which is used to strike and sound out the hardness of a roll.

BIMETAL PLATE—See MULTIMETAL PLATE.

BINDER—A material used to cause substances to bond or adhere. (1) In the paper industry, binders are used widely to cause fibers to bond together, coatings to adhere to paper, or as laminants. (2) In nonwovens, binders are used to bond fibers together at fiber cross-over points.

BINDER ADD-ON—The percent binder applied to a nonwoven as calculated by:

$$\% \text{ Binder Add-On} = \text{Binder Weight} \times 100\%/ \text{Fiber Weight}$$

BINDER CONTENT—The percent binder applied to a nonwoven as calculated by:

$$\% \text{ Binder Content} = \text{Binder Weight} \times 100\%/ (\text{Fiber Weight} + \text{Binder Weight})$$

BINDER MIGRATION—(1) The movement of binder to the outer surface of a nonwoven during drying which can cause the nonwoven to be binder lean in the center and binder rich on the surface. See also BINDER. (2) In paper coating, the movement of natural or synthetic binders in the coating as it dries. Migration may contribute to mottle.

BINDERS BOARD—A single-ply, solid board used principally for the binding of books. It is made on a wet machine from a base stock of mixed papers and is tunnel or platen dried. It ranges in thickness from 30 to 300 points (0.030 to 0.300 of an inch). Important properties are smoothness, uniformity, high density, stiffness, and strength. It should be free of objectionable odors.

BINDERS WASTE—The trimmings of books obtained from the cutting machine of the bindery. This is, generally, clean white paper of good quality and commands a premium in the recovered paper market. This grade is included in either the categories of coated soft white shavings or in hard white shavings, depending on whether the paper is coated.

BINDING—Method of attaching pages into books or leaflets. Loose-leaf binding methods include ring, spine, clamp, or comb binding. Permanent binding methods include saddle stitching (usually stapling), side stitching, sewn stitching, with individual sections, and perfect binding, using only glue to bind the pages.

BINDING TAPE—See GUMMED SEALING TAPE.

BIOBLEACHING—The treatment of woodpulp to reduce color or to reduce chemical demand in chemical bleaching; the latter has been referred to as "bleach boosting."

BIOCHEMICAL OXYGEN DEMAND (BOD)— (1) The quantity of oxygen used in the biochemical oxidation of organic matter in a specified time, at a specified temperature, and under specified conditions. (2) A standard test used in assessing wastewater characteristics. See also FIRST STAGE BIOCHEMICAL OXYGEN DEMAND.

BIOCIDES—Chemical agents with the capacity to kill biological life forms. Bactericides, insecticides, pesticides, etc., are examples.

BIOCONTROL—Controlling unwanted organisms using biotechnology. For example, fungi that stain wood chips can be controlled by prior introduction of colorless fungi that consume the food base in the wood.

BIODEGRADABLE—Any organic material which is capable of being converted by certain bacteria into basic elements or compounds, such as carbon dioxide and water. Most paper products are considered to be biodegradable.

BIODEGRADATION—The conversion of a substance by microorganisms into basic elements or compounds, such as carbon dioxide and water.

BIOLOGICAL CONVERSION OR CONTROL—The transformation of industrial and municipal organic wastes including cellulose by microorganisms. Such wastes are subsequently treated chemically to prevent pollution of the receiving waters.

BIOLOGICAL WASTEWATER TREATMENT—Forms of wastewater treatment in which bacterial or biochemical action is intensified to stabilize, oxidize, and nitrify the unstable organic matter present. Intermittent sand filters, contact beds, trickling filters, and activated sludge processes are examples.

BIOMASS—The total amount of living material in a particular habitat or area; in fiber raw material sourcing, the total weight or volume of commercial and non-commercial vegetation.

BIOPULPING—Treatment of wood chips with a lignin-degrading fungus prior to pulping. With mechanical pulping, the fungal pretreatment can result in significant refiner energy savings, depending on the fungus used. With chemical pulping, the fungal pretreatment reportedly can give chemicals and energy savings, depending on the fungus used.

BIOTECHNOLOGY—Any technique that uses living organisms (or parts of organisms) to make or modify products, to improve plants or animals, or to develop microorganisms for specific use. The development of materials that mimic molecular structures or functions of living organisms are also included.

BIT—Basic unit of digital information processing. See also BYTE.

BITE—See TOOTH.

BIT MAP—The digital representation of a page.

BLACK ALBUM PAPER—A fairly heavy cover-type, antique finish paper used for the manufacture of photographic albums. It is heavily dyed in a deadblack shade and is usually made of chemical and/or mechanical woodpulps in basis weights ranging from 60 to 110 pounds (24 x 36 inches – 500). It is characterized by good punching and folding qualities, fairly hard sizing to resist the cockling effects of paste, and freedom from any impurities which might affect photographic prints.

BLACKENING—A darkening (or glassining) of an area in a web which has been subjected to above normal pressure, heat and/or moisture in the nip of a calendering process, which can be caused by a localized expansion of a filled paper roll during the supercalendering operation. Blackening is accompanied by a decrease in the opacity of a sheet. Compare CRUSHING.

BLACK-LINE PAPER—A chemically treated paper, similar to blueprint paper, in which the developed design appears as black lines on a white background.

BLACK LIQUOR—The spent pulping liquid after the alkaline pulping process is complete. It is separated from the pulp in brownstock washers. Black liquor contains both dissolved organic compounds from the wood and inorganic compounds from the wood and original cooking liquor. After the washers, dilute black liquor is concentrated and burned to recover the inorganic chemicals for reuse and to provide energy for mill processes. Obtained from the pulp-washing system. See also AS-FIRED BLACK LIQUOR; CONCENTRATED BLACK LIQUOR; INTERMEDIATE BLACK LIQUOR; WEAK BLACK LIQUOR.

BLACK LIQUOR GUN—The component of a recovery boiler through which black liquor enters the furnace. The end of the gun is fitted with special nozzles that break the black liquor flow into a spray and distribute the spray throughout the lower furnace volume.

BLACK LIQUOR OXIDATION—A system for introducing air or oxygen into a black liquor stream for the purpose of oxidizing sodium sulfide in the black liquor to less volatile sodium/sulfur compounds. The purpose of black liquor oxidation is to avoid generation of total reduced sulfur (q.v.) during combustion.

BLACK NEEDLE PAPER—See NEEDLE PAPER.

BLACK PHOTO PAPER—A paper which is used to protect or wrap sensitized photographic materials. It must be free from pinholes and from chemicals or other materials harmful to a photographic emulsion.

BLACK POSITIVE PAPER—See BLACK PHOTO PAPER.

BLACKPRINT PAPER—See BLACK-LINE PAPER.

BLACK-PRINT PAPER—See BLACK-LINE PAPER.

BLACK WATERPROOF PAPER—A high-grade sheathing and insulating paper manufactured from strong jute or kraft paper stock, saturated and coated with special asphalts. It is manufactured in weights from 35 to 60 pounds per 500 square feet. It is used as an insulator under roofings, sheathings, and under or between floors.

BLACK WRAPPING PAPER—A wrapping paper, made of sulfite, jute, or kraft pulp or a combination thereof. It is used in a wide range of basis weights for wrapping and decorative purposes. For photographic purposes, it is light-proof.

BLADE COATER—See BLADE COATING.

BLADE COATING—A method of coating using a flexible blade set at an adjustable angle against a web of paper or board supported by a soft, usually rubber-covered backing roll. Also called trailing blade coating and flexible blade coating.

BLADE CUT—A sharp cut (or near cut) running in a straight line, parallel to the direction of web travel.

BLADED BOX—See DRAINAGE SHOE.

BLADE MARKS—See BLADE STREAKS.

BLADE METERING SIZE PRESS—A metering size press in which a blade is used as the metering device to control the amount of wet film applied to the sheet. In most blade metering size presses, the applicator used is a short dwell coater.

BLADE SCRATCH—A very fine hair-like indentation in the coating surface (less than 1/8 inch wide). The length varies from a few feet to several hundred feet in the machine direction. It usually appears less opaque than the general coated area when viewed by transmitted light.

BLADE STREAKS—Broad indentations in the coating surface (1/8 inches or wider). Their length varies from a few feet to several hundred feet in the machine direction. The center of indentation usually appears less opaque than the general coating area when viewed by transmitted light. The edges of indentation may or may not appear more opaque than the general coating area.

BLANC FIXE—See BARIUM SULFATE.

BLANK BOOK PAPER—A general term for any grade of paper which is to be used in the manufacture of blank books.

BLANKET—A flexible material that can readily be conformed to curved or irregular surfaces. In printing, the rubberized or plastic-coated fabric covering a roller from which ink is transferred to paper or other substrate in any offset printing process.

BLANKET MARK—See SUCTION-BLANKET MARK.

BLANKET SMASH—A defect in the blanket resulting from irreversible compression of the blanket caused by foreign matter or by excessive thickness of paper or other substrate passing through the press.

BLANKET TO BLANKET PERFECTING—A press configuration where paper is run between two offset blankets, each acting as the impression cylinder for the other. This yields one-pass, two-side printed work.

BLANKING PAPER—A plain, uncoated poster paper (q.v.) used for covering or as a border for billboards. It has the same characteristics as poster paper, except that printing or lithographic qualities are not essential.

BLANK NEWS—A grade of reclaimed paper stock consisting of cuttings or sheets of unprinted white newsprint, used as a furnish component especially in the manufacture of newsprint or paperboard.

BLANKS—A term applied to a class of paperboard ranging in thickness from 0.012 to 0.078 of an inch, with corresponding basis weights of 120 to 775 pounds (22 x 28 inches – 500). They may be either single-ply fourdrinier board, multi-ply cylinder board, or laminations of these. The liner may be made of deinked stock, clean shavings, bleached or unbleached ground-wood, or chemical pulps. The surface may be either coated or uncoated, white or colored. Blanks are generally made to produce maximum stiffness and surface smoothness. They are used for various purposes where stiffness and good printing qualities are required as in bus and subway signs, window displays, etc. A blank is also a flat sheet of corrugated or solid fiberboard that has been cut, slotted, and scored so that when folded and joined it will form a box.

BLASTING PAPER—(1) A paper used by miners for lining a drilled hole or for enclosing black powder or other explosives. It is a high machine- or water-finished sheet, usually specified in heavyweights, from 60 to 90 pounds or more (24 x 36 inches – 500), in rolls 9 to 15 inches wide and containing approximately 5 pounds of paper. This paper is made from chemical woodpulp or rope. It is impregnated with wax or oil to produce a high water resistance. (2) See DYNAMITE-SHELL PAPER.

BLEACH—In the paper industry, an oxidizing or reducing agent used to remove color from pulp so that it has a high brightness.

BLEACHABILITY—A qualitative term used to describe the relative ease with which a pulp can be bleached.

BLEACHABLE GRADE PULPS—Kraft pulps cooked to a 3% to 5% lignin content (kappa no. 15 to 35, k no. 12 to 24) are referred to as bleachable grade pulps.

BLEACH DEMAND—The amount of bleaching chemicals required to remove the lignin from the pulp. There are several tests to indicate the amount of chemicals required. Most of them are based on the reaction of residual lignin with chlorine or potassium permanganate. A pulp with a high bleaching chemical demand indicates a "hard" or high lignin pulp. See CHLORINE NUMBER; HYPO NUMBER; KAPPA NUMBER; PERMANGANATE NUMBER.

BLEACHED BOARD—A general term covering any board (q.v.) composed of 100% bleached fiber.

BLEACHED CHEMITHERMOMECHANICAL PULP (BCTMP)—Chemithermomechanical pulp (CTMP) bleached to a higher brightness, (80% + GE%).

BLEACHED CORRUGATING MATERIAL—Usually a thin corrugating material used for making specialty wraps and advertising displays. It is made in white and colors from bleached chemical woodpulp.

BLEACHED MANILA-LINED CHIPBOARD—A boxboard used principally for folding cartons. It is made with a reclaimed paper stock (chip base) back and a white manila vat liner. Good bending qualities and a surface well adapted to printing in fancy color designs are essential properties.

BLEACHED PACKAGING PAPERBOARD—A paperboard (q.v.) made from a furnish of approximately 85% virgin bleached chemical pulp.

BLEACHING—A chemical process for whitening fibers, yarns, fabrics, and other materials. See also PULP BLEACHING.

BLEACHING CHEMICAL DEMAND—See BLEACH DEMAND.

BLEACHING LIQUOR—A solution of a bleaching agent applied to the pulp.

BLEACHING POWDER—Calcium hypochlorite. The quality of bleaching powder is calculated on the basis of a standard bleaching powder containing 35% by weight of available chlorine.

BLEACHING SEQUENCE—A series of stages with washing between stages, each stage optimized to contribute to the whitening process by solubilization, removal and whitening of remaining colored materials (chemical pulps), or selective destruction of colored groups (mechanical pulps). Each stage has a brightness ceiling which cannot be exceeded without damaging the fibers.

BLEACHING STAGE—See BLEACHING SEQUENCE.

BLEACHING STAGE (C)—See CHLORINATION STAGE (C).

BLEACHING STAGE (D)—See CHLORINE DIOXIDE BLEACHING STAGE (D).

BLEACHING STAGE (E)—See ALKALINE EXTRACTION STAGE (E).

BLEACHING STAGE (Eo)—See OXIDATIVE EXTRACTION STAGE (Eo).

BLEACHING STAGE (H)—See HYPOCHLORITE BLEACHING STAGE (H).

BLEACHING STAGE (O)—See OXYGEN DELIGNIFICATION (O).

BLEACHING STAGE (P)—See HYDROGEN PEROXIDE STAGE (P).

BLEACHING STAGE (Pa)—See PERACETIC ACID.

BLEACHING STAGE (Q)—See CHELATION STAGE (Q).

BLEACHING STAGE (Z)—See OZONE STAGE (Z).

BLEACH LIQUOR STRENGTH—Expresses, usually in units of weight per unit volume, the concentration of bleaching reagents—usually grams/liter (gpl).

BLEACH REQUIREMENT—See BLEACH DEMAND.

BLEACH-RESISTANT PAPERS—Colored papers containing dyestuffs or coloring materials which are resistant to decolorization by common bleaching agents, ink eradicators, etc.

BLEACH SCALE—A pearly, light-brown, brittle spot in paper, caused by insoluble bleach residues.

BLEED—(1) To arrange for printed matter (usually an illustration) to extend somewhat beyond the trim position of the page on one or more edges, to ensure that the edge of the illustration will be flush with the trimmed edge. (2) To cut into printed matter, either by plan or accident, when trimming printed sheets. (3) The illustration or page planned to bleed. (4) The printed trim from a bleed page. See also BLEED TRIM.

BLEEDING—(1) The dissolving out of color from paper or pulp by water, oil, or other liquid. (2) Discoloration of the surface of paper or board, or contiguous materials, by the migration of components of asphalt. (3) Staining of contiguous objects by colored papers, generally in the presence of moisture.

BLEED TRIM—Recovered paper consisting of clean, dry, printed, all-white book-paper shavings. This is no longer considered to be a separate recovered paper grade; if it is coated paper, it would be coated book stock; if uncoated, it would be sorted white ledger.

BLEND CHEST—A chest that receives stock from multiple sources, and blends the stock into a more homogeneous slurry before passing it downstream. The incoming stock may vary in consistency, flow rate, fiber composition, degree of refining, etc. Blend chests are also used to mix-in papermaking additives.

BLENDED STOCK—Stock produced by blending several types of fibers, typically in very specific desired proportions, in the blend chest. Blended stock, generally, is sent to the machine chest, but may also be used for sweetener stock for the saveall (q.v.).

BLENDING—In the processing of fibers, blending refers to dispersing of one or more fiber types into an intimate homogeneous mixture.

BLIND DRILLED PRESSURE ROLL—A blind drilled roll used as a pressure roll (q.v.) in a tissue machine. See BLIND DRILLED ROLL; CRESCENT FORMER; PRESSURE ROLL.

BLIND DRILLED ROLL—The elastomeric composition (see COVERED ROLL) on these rolls is drilled with a pattern of small-diameter holes that do not go through the cover, to provide a void volume capacity for removing liquid pressed out of the paper web and the felt in the pressure nip. They are used primarily in transverse press designs. The holes are typically 2.0 mm to 3.5 mm (0.078 inches to 0.138 inches) in diameter with a total open area of 20% to 24%. The holes are drilled to two depths, the deepest being about one-half the cover thickness. Depending on the particular application, the hardness of covers on press rolls range from 4 P&J to 25 P&J (q.v.), and on pressure rolls from 30 P&J to 45 P&J. Since the metal body of blind drilled press rolls is not drilled, as is done with suction rolls (q.v.), they may be run under higher operating pressures.

BLIND EMBOSSING—An uninked raised image in paper caused by die stampling (q.v.) or intaglio printing (q.v.).

BLISS BOX—An important construction for corrugated shipping containers (q.v.) in which three pieces, the body and two ends, are formed and sealed together, usually at the plant where the container is filled. The cover can be an integral part of the body or a separate piece may be formed as a cover.

BLISTER—(1) An elevation of part of the surface of the sheet or of the surface ply or of the coating, enclosing air or moisture vapor between the surface and the rest of the sheet. (2) A rapid test for papers such as glassine or greaseproof made from highly hydrated pulps; such papers develop blisters when exposed to the flame of a match or cigarette lighter just enough to char the sheet very slightly.

BLISTER CUT—A cut occurring when an excess of paper, caused by a full area in the web that has accumulated as a blister at the entrance of a nip in a calender stack, is carried through the nip in a folded condition.

BLOCKING—Undesirable adhesion of adjacent layers of materials in rolls or sheets. It may be caused by temperature, pressure, humidity, coating materials, internal sizing agents, or a combination of these.

BLOCKING RESISTANCE—See BLOCKING.

BLOODPROOF PAPER—A strong wrapping paper, usually in a basis weight of 35 to 50 pounds (24 x 36 inches – 500), which is hard sized with wax or a wax emulsion, or otherwise treated to make it resistant to blood and meat juices.

BLOOD RESISTANCE—Ability of paper, especially butcher's wrap, to withstand penetration by blood when in contact with freshly cut meat.

BLOOD SOAKING PAPER—Generally, a multi-ply tissue that lines meat and poultry trays to absorb the food juices.

BLOTTING BOARD—A heavyweight blotting paper.

BLOTTING PAPER—An unsized paper used wherever absorptivity is the required characteristic or where soft spongy paper is needed, even though the absorptivity is of secondary importance. It is made from rag, cotton linters, chemical or mechanical woodpulp, or mixtures of these. The paper is porous, bulky, of low finish, and possesses little strength. The normal basis weights range from 60 to 140 pounds (19 x 24 inches – 500). Some grades are made with a smooth machine finish, which makes them suitable for printing with coarse-screen halftones. See ENAMELED BLOTTING.

BLOW-BOX—See ANTI-BLOW-BOX.

BLOWDOWN—Boiler feedwater contains small amounts of compounds that do not volatilize when the water is converted to steam. In order to prevent buildup of these compounds in the boiler, a small percentage of the boiler water, in the section of the boiler where volatilization occurs, is removed as blowdown on a periodic or continuous basis. This blowdown contains the non-volatiles in a much higher concentration than in the feedwater.

BLOW-OUT—A defect in a supercalender cotton filled roll or other elastomer covered nip roll that has locally overheated and thermally degraded to the point that a section of that roll cover has melted or burned a hole in the cover. This hole most often extends from the internal hot spot to the outer surface of the roll cover.

BLOW-THROUGH—The portion of steam used to heat a dryer cylinder that passes, or "blows" through the cylinder without condensing.

BLUE CARPET PAPER OR BOARD—See INDENTED BOARD.

BLUE PASTERBOARD—A heavyweight dry sheathing paper of a light blue color. It is used for lining walls and as a sheathing. A roll containing 250 square feet weighs approximately 30 pounds.

BLUEPRINT PAPER—A paper produced from the base stock (see above) by treatment with chemicals, such as potassium ferricyanide, and iron salts, such as the oxalates and tartrates. When developed, the light-exposed areas turn blue. On an ordinary print there are white lines on a blue background. In some cases, a negative is made from the original line drawing by making a brownprint in which the lines are white on a black or brown background. This brownprint or negative is then used as a subject in blueprinting, and the final blueprint has blue lines on a white background. This paper is used to make copies or reproductions of drawings and documents.

BLUEPRINT PAPER (BASE STOCK)—A paper used for the manufacture of blueprint paper. It is usually made from cotton fiber pulp, but bleached chemical woodpulps and mixtures of cotton and chemical woodpulps are sometimes used. The basis weight ranges from 12 to 30 pounds (17 x 22 inches – 500). It is a well-formed sheet with a fairly smooth surface, good wet tensile strength, and, though well-sized, of uniform absorbency. This paper must be free from chemicals that would affect the sensitizing materials. Minimum change in dimension when the paper is wet is an essential property.

BLUE ROSIN SHEATHING PAPER—Same as red rosin sheathing paper (q.v.), except for the color.

BLUE SPOTS—See COLOR SPOTS.

BLUE TRACING PAPER—A tracing paper having a blue color. See TRACING PAPER.

BOARD—See PAPERBOARD.

BOARD AND BOX LINING—(1) A grade of groundwood paper generally containing 65% to 75% of mechanical woodpulp with the balance usually unbleached chemical woodpulp, though occasionally bleached chemical woodpulp is used. Its principal use is in covering chipboard before manufacture into inexpensive setup boxes. Cleanliness and sufficient sizing for good pasting qualities are its important properties. The principal basis weight is 32 pounds (24 x 36 inches – 500). (2) A general term for papers used for the same purpose including a free sheet, 40 to 60 pounds (25 x 38 inches – 500), and plain or coated book papers, particularly when printing is employed.

BOARD LINER—(1) See LINERS. (2) A machine for pasting a paper liner to a boxboard.

BOD—See BIOCHEMICAL OXYGEN DEMAND.

BODE PLOT—A graphical method of presenting the frequency response characteristics of a control system.

BOD LOAD—The BOD content, usually expressed in pounds per unit of time, of wastewater passing into a waste treatment system or to a body of water.

BODY PAPERS—Papers used by paper converters for gumming, coating, finishing, etc.

BODY STOCK—The base stock or coating raw stock for plain or decorated coated papers and boards. It may be uncoated or precoated on the paper machine. It is also used in connection with industrial papers before they are treated. Because of the wide range that body stock may cover and also because it is usually made to order under special specifications, it cannot be described as containing certain amounts of any particular kind of pulp, nor is there any way of referring to weights and colors. It is also termed base paper.

BOGUS—Manufactured principally from old papers or inferior or low-grade stock in imitation of papers or paperboards using a higher quality of raw material.

BOGUS BACK LINING—See BOGUS PAPER.

BOGUS BRISTOL—A bristol board usually made on a cylinder machine. It may be solid, or different stocks may be used for the filler and liners. The furnish consists of overissue news, blank news, bleached sulfite, soft white shavings, and hard white shavings in varying amounts, according to the quality being made. One or both liners or the entire board may be white or colored. If colored, the liners are usually either calender stained or tinted in the beaters. The regular weights of bogus bristols are: 2 ply—90 pounds; 3 ply—120 pounds; 4 ply—140 pounds (22 1/2 x 28 1/2 inches – 500). When used for various ticket purposes, the term "ticket bristol" or "colored ticket bristol" is applied to colored bogus bristols. Colored bristols usually have an overissue news or chip filler. It is sometimes called B bristol.

BOGUS CORRUGATING MEDIUM—A corrugating medium (q.v.) made entirely or predominantly from recovered paper. This term has been largely replaced by recycled corrugating medium (q.v.).

BOGUS DRAWING PAPER—A paper of good texture and with sufficient tooth to take charcoal, soft pencil, crayon, or water colors quickly. It is sized and finished slightly but without gloss, and it is made from 100% reclaimed paper stock on a cylinder or fourdrinier machine. It is usually 0.010 of an inch in thickness and is sold in sizes 9 x 12, 12 x 18, and 24 x 36 inches, all on a basis weight of 90 pounds (24 x 36 inches – 500).

BOGUS DUPLEX—A paper calender-dyed on one side.

BOGUS KRAFT—See IMITATION KRAFT PAPER.

BOGUS LINING PAPER—Paper made from recovered paper in thicknesses of 9 points (0.009 inches) and up, and used for lining shipping cases.

BOGUS MANILA—A paper used to replace sulfite or kraft manila paper when strength and quality are not essential. It is commonly made from recovered paper and colored to represent manila.

BOGUS MEDIUM—See RECYCLED CORRUGATING MEDIUM.

BOGUS MILL WRAPS—See MILL WRAPPER.

BOGUS PAPER—A paper that is used for the same purposes as book back liner (q.v.) but which is made of paper of inferior strength. This paper does not meet the minimum specifications for book back liner as set forth in the "Official Minimum Manufacturing Standards and Specifications for Textbooks."

BOGUS PASTING PAPERS—Bogus papers suitable for adhesion to another paper, board, or other material.

BOGUS SATURATING PAPER—A term applied to those bogus papers which are prepared as a vehicle to carry and hold various tars or asphalt compounds. See SATURATING PAPER; ROOFING PAPER.

BOGUS SCREENINGS—Papers manufactured largely from old screenings and other recovered papers, producing a sheet of similar characteristics to the original papers, but generally having less strength and varying in appearance.

BOGUS TAG—A tag board made of overissue news or mixed papers, usually having a white or colored liner on each side to furnish a printing or writing surface. It is not a standardized product, generally being made to meet some special requirement and specifications. See TAG BOARD.

BOGUS WRAPPING PAPER—An absorbent and bulky wrapping paper made from recovered paper. Screenings or chemical woodpulp are sometimes added in small percentages. The paper may be finished and dyed to resemble unbleached wrapping paper. It is used where strength requirements are not high and, in some papermaking countries, where virgin fiber furnish is too costly. Also referred to as bogus wrapping.

BOILER EFFICIENCY—In its simplest definition, boiler efficiency is the percentage of the fuel heat content that is transferred to the steam product. It takes into account heat losses such as that required to heat combustion air, dry wet fuels, radiation, convection, etc.

BOILER FEEDWATER—Mill water or returned condensate that has been treated to reduce its tendency to corrode the boiler tube materials and reduce its potential to leave deposits on the boiler tube walls.

BOILER LOAD—A measure of boiler throughput. For most steam generators, it is the steam flow rate compared to the maximum continuous rating (MCR) of the unit. For recovery boilers, it can also be the throughput of black liquor dry solids compared to the throughput of solids at the maximum rated capacity.

BOILING POINT RISE—The increase in the boiling point of a solution over that of the solvent due to dissolved constituents.

BOND CIRCULAR PAPER—A writing or printing paper used for advertising bond issues and the like. It is made from chemical woodpulps in basis weights ranging from 16 to 32 pounds (17 x 22 inches – 500), and is characterized by a vellum finish, high white color, good folding, and fast ink-drying qualities. See OPAQUE CIRCULAR.

BONDING—In nonwovens, the process of stabilizing fibers to produce nonwovens by chemical, mechanical, or thermal means.

BONDING STRENGTH—The force with which fibers adhere to each other within a sheet, or with which a coating or film adheres to the surface of a sheet, or with which plies in a paperboard or laminated sheet adhere to each other. See PICKING; PLY ADHESION.

BOND PAPER—Originally a cotton-content writing or printing paper designed for the printing of bonds, legal documents, etc., and distinguished by superior strength, performance, and durability. The term is now also applied to papers for less demanding applications such as letterheads, business forms, and social correspondence papers. Because of the broad-spectrum applications involved, bond paper is now made from cotton and/or chemical woodpulps in basis weights ranging from 13 to 24 pounds (17 x 22 inches – 500) and includes cut-size papers, parent or folio sizes, and rolls. Typical properties include printability, erasability, whiteness, cleanliness, freedom from fuzz, uniform finish, and good formation.

BONEDRY—Moisture free. See OVENDRY.

BOOK—See PLATE FINISH; PLATER BOOK.

BOOK-BACKING PAPER—See BOOK BACK LINER.

BOOK BACK LINER—Usually a kraft paper either with creped or flat finish, which is glued to the backbone of sewed books to bind the signature and to keep the backbone of the book free from the backbone of the cover. If sewn books require the strongest possible construction, they are made "tight backed" whereby the book backliner is glued to the case back liner. The "Official Minimum Manufacturing Standards and Specifications for Textbooks" stipulates that materials for lining "shall be of suitable weight and quality and shall have a bursting strength (Mullen test) of not less than 45 pts. per square inch, TAPPI Test Method T 403 om-85."

BOOK BASIS—25 x 38 inches – 500. See BASIC SIZE.

BOOK-BINDERS PAPER—See END-LEAF PAPER.

BOOKBOARD—See BINDERS BOARD.

BOOK BULK—The overall thickness of a given number of sheets of printing paper under a pressure of 35 lb/in.^{2} (2.46 kg/cm^2).

BOOK-COVER PAPER—(1) A white or colored cover paper used for case-bound books. In this form it may be plain, pigment or plastic coated, embossed, decorated, or otherwise embellished. It is generally made from chemical woodpulps in basis weights ranging from 40 to 80 pounds (20 x 26 inches – 500). (2) Any paper used as a jacket for a case-bound book. (3) Any heavy kraft paper used as a protective covering for school and library books. See BOOK JACKET; BOOK WRAPPER; COVER PAPER.

BOOK END PAPER—See END-LEAF PAPER.

BOOK FASHION—A method of sorting paper; about one-half of the area of one side of the sheets is inspected for a distance of several inches down the pile; these are then rolled back to their original position on the pile, and the other half of the sheets is inspected. A slip of paper is inserted whenever an imperfection is found, and—as the perfect sheets are transferred to a new pile—the defective sheets are removed.

BOOK JACKET—Paper used by printers to fold over bound books for protective and advertising purposes. See BOOK-COVER PAPER; BOOK WRAPPER.

BOOKKEEPING-MACHINE PAPER—Bond or ledger paper cut into form of single sheets suitable for use in bookkeeping machines. See STATEMENT LEDGER.

BOOKLET—A small book, commonly bound in paper covers.

BOOKLET COVER—See BOOK-COVER PA-PER; COVER PAPER.

BOOK LINING—(1) See END-LEAF PAPER. (2) A hard-sized book paper which is used to line chipboard or strawboard.

BOOK-MATCH BOARD—See MATCH-STEM STOCK.

BOOK PAPER—(1) A general term for a group of coated and uncoated papers (exclusive of newsprint) suitable for the graphic arts. Bookpapers are made from all types of virgin and reclaimed pulps and mixtures thereof, in basis weights usually ranging from 30 to 100 pounds (25 x 38 inches – 500). They are characterized by a wide variety of surface finishes (e.g., antique, eggshell, machine, English, supercalendered, dull-coated, matte-coated, glossy-coated), with good formation, printability, and cleanliness; (2) a generic term encompassing the above and related grades (e.g., tablet, envelope, converting base) which are made by so-called "book paper mills."

BOOK PAPER (COATED)—A coated printing paper designed for advertising matter, magazines, books, pamphlets, brochures, and general printing applications. The base paper for coated book paper is generally made from virgin or reclaimed chemical woodpulps, or both, and is then coated on or off the paper machine with various pigment formulations to provide brightness, opacity, printability, etc. Coated book paper is commonly supercalendered and its finish ranges from dull matte to high gloss. It is normally made in a basis weight range of 30 to 150 pounds (25 x 38 inches – 500). See COATING; MACHINE COATED.

BOOK PAPER (UNCOATED)—An uncoated printing paper designed for advertising matter, magazines, books, pamphlets, brochures, and general printing applications. The term is also applied to tablet, envelope, converting papers, and other grades made by so-called "book papermills." Book paper is made from virgin and/or reclaimed chemical woodpulp, or mechanical woodpulp, generally in basis weights ranging from 30 to 100 pounds (25 x 38 inches – 500). It is characterized by a variety of finishes such as high bulk, antique, eggshell, machine, English, supercalendered, and embossed.

BOOK SHAVINGS—See SHAVINGS.

BOOK STOCK—The term used in the 1940s when books and magazines were the predominant furnish for deinking. See DEINKED-PAPER STOCK.

BOOK WRAPPER—Usually a long-fibered paper either coated or uncoated, with good strength and printing surface, used as a jacket for a bound book. The term may also be applied to printed paper used as loose covers on hard bound books. It is also called book-cover paper.

BOOK-WRAPPING PAPER—See BOOK WRAPPER.

BOOT BOARD—See SHOE BOARD.

BOROHYDRIDE—See SODIUM BOROHYDRIDE.

BOTTLE-CAP BOARD—A sanitary food paperboard used in the form of circular discs for bottle stoppers. It is made of chemical or mechanical woodpulp or mixtures of these, hard sized and suitable for waterproofing by wax impregnation or barrier coating, and has a surface adapted to color printing. The normal thicknesses range from 0.040 to 0.048 of an inch. Important properties are rigidity, uniformity in thickness, the quality of giving a clean edge when die-cut, and cleanliness. See COVER-CAP BOARD; HOOD-CAP.

BOTTLE LABELLING PAPER—A special paper designed for bottle labels. It is usually made from chemical woodpulps, in basis weights of 40 to 60 pounds (25 x 38 inches – 500) and is characterized by good printability, non-blocking qualities, and (in some cases) oil and water resistance. See also LABEL PAPER.

BOTTLE WRAPPING PAPER—Any one of several grades (including glassine, kraft wrapping, etc.) used for protective wrapping of bottles.

BOTTOM BOARD FELT—See PRESS FELT.

BOTTOM FABRIC—See TWIN WIRE FORMING FABRICS.

BOTTOM FELT—See PRESS FELT.

BOTTOM FELTED PRESS—See PRESS SECTION.

BOWED ROLL—A curved roll used to spread out the paper web and to eliminate wrinkles as the web travels through the paper machine. Bowed rolls (also termed spreader rolls) are found primarily on size press (q.v.) units and at the dry end of the machine on slitters and winders. These rolls consist of a heavy curved shaft containing a series of small freely rotating metal sleeves or segments mounted on bearings. On most bowed rolls, a flexible tube made from an elastomeric composition covers the sleeves forming a continuous surface to support the paper web.

BOWL—See FILLED ROLL.

BOWL GLAZING—Obsolete terms. Replaced by CALENDERING; GLOSS CALENDER; HOT CALENDER.

BOW-WAVE FINISH—A fancy two-toned ripple finish in stratum formation, produced by a plater. The paper varies in caliper, the thinner portions having the darker tone.

BOXBOARD—A general term designating the paperboard used for fabricating boxes. It may be made of woodpulp or paper stocks or any combinations of these and may be plain, lined, or clay coated. Terminology used to classify boxboard grades is normally based upon the composition of the top liner, filler, and back liner. Thus, patent coated news, kraft back, has a patent coated (q.v.) top, a news filler, and a kraft back liner. Double manila-lined news has a manila (q.v.) furnish on the top and back sides with a news filler. Typical examples are as follows: clay-coated news, patent coated news,

manila back, patent coated back, clay coated kraft, manila-lined news, manila-lined chip, manila-lined kraft, and news lined chip. See FOLDING BOXBOARD; SETUP BOXBOARD.

BOX CLIPPINGS—A grade of reclaimed paper stock also called boxboard cuttings consisting of new cuttings of paperboard grades used in the manufacture of folding and setup boxes and similar boxboard products.

BOX COMPRESSION TEST—The resistance which an empty, but sealed, paperboard box offers to a compressive force applied perpendicular to parallel faces or panels of the box. The compression strength of a box may be expressed as the maximum resistance prior to failure, although the usual practice is to express the compression strength as the maximum force sustained within specified deformation ranges, the latter differing, depending on the direction of loading relative to box opening.

BOX-COVER PAPER—A paper used to cover paper boxes. It may be plain, antique, embossed, ink embossed, glazed, flint glazed, coated, or printed to add to its usefulness or attractiveness. It is generally made of chemical woodpulps, although ground wood and rag pulps may be included in the furnish. Strength requirements are nominal for setup box use but must be sufficient for scoring and 180° folds in folding cartons and boxes.

BOXED WRITINGS—A trade term applied to "cut-size" (i.e., 8 1/2 x 11 inches, 8 1/2 x 13 inches, and 8 1/2 x 14 inches) bond, writing, onionskin, and related papers packed in convenient 500-sheet boxes for office use. Also referred to as boxed papers.

BOX ENAMEL PAPER—Papers used for the inside of boxes containing food or meat or crates containing celery, lettuce, or other vegetables, to keep the products fresh through the retention of moisture and to protect the contents from dirt or other contamination. Kraft paper, vegetable parchment, waxed paper, and waterproof papers are the grades most frequently used, although other grades may be used for the purpose. See CANDY-BOX; CRATE LINERS (2).

BOX STAY TAPE—A tape used in reinforcing the edges and corners of setup boxes and corrugated cartons. The paper is made from kraft, rope, etc. usually in basis weights of 35 to 100 pounds (24 x 36 inches – 500). It is furnished in rolls of narrow width and often gummed, and it may be made in colors. The tape may be cloth-lined or laminated to glass fiber reinforcing strands.

BOX WRAP PAPER—See BOX-COVER PAPER.

BPCT—Best Practical Control Technology.

BPCTCA—Best Practical Control Technology Currently Available.

BRAILLE PRINTING PAPER—Paper used in the braille process in which the paper is embossed to form a well-organized pattern of raised dots which form characters or letters so that the blind may read by touch. Usually the paper is embossed or "printed" wet. Normally a good grade of chemical woodpulp is used in the manufacture of braille paper in basis weights of 32 to 36 pounds (17 x 22 inches – 500). Significant properties include smooth surface, good elongation, and high tensile strength.

BRDA—Formerly, the Boxboard Research and Development Association, its name has been changed to the Recycled Paperboard Technical Association.

BREAD-BAG PAPER—A bleached or unbleached chemical or chemical-mechanical pulp paper used for bread bags. It is machine finished or machine glazed and is usually specified in basis weights of 25 to 30 pounds (24 x 36 inches – 500). The paper is generally made with a metal roll mark (stripe) or a felt mark (stripe) of varying designs.

BREAD LABEL—A paper for the manufacture of labels used by bakers to trademark bread. A lightweight paper or any book paper coated on one side may be used.

BREAD WRAPPERS—A paper used for wrapping bread, often made opaque by the use of materials, such as titanium dioxide, giving good opacity despite waxing. Two means are used for supplying opacity to the sheet: *viz.*, beater filling and surface coating. Basis weights of the base stock range normally from 21 to 25 pounds (24 x 36 inches – 500); the wrapper, after printing and waxing, has a weight of 30 to 35 pounds. It is sold in sheets to fit the bread loaf for handwrapping, or in rolls for automatic wrapping machines. The wax forms a self-sealing wrapper. See WAXED GLASSINE; WAXED PAPER.

BREAK—A term used to denote a complete rupture of a web of paper or paperboard during manufacture or some subsequent operation which uses rolls of paper. Such breaks are generally spliced and marked on a finished roll by a protruding flag.

BREAKER—In electrical systems, a breaker is a mechanical protective device that can isolate a portion of an electrical system. It is usually equipped with sensors that will isolate and protect equipment on a downstream portion of an electrical system if conditions warrant.

BREAKER STACK—A mid-paper machine calender (q.v.) usually located before the on-machine coater.

BREAKING—(1) The operation of passing a gummed paper over the edges of a square bar which cracks the layer of adhesive and reduces or eliminates the tendency to curl. (2) The operation of bending paper to facilitate feeding it to a printing press.

BREAKING LENGTH—The length of a strip of paper, usually expressed in meters, which would break of its own weight when suspended vertically. It is a value calculated from the tensile strength, the width of the tensile specimen, and the basis weight of the sheet in expressing paper and especially pulp testing results. See TENSILE STRENGTH.

BREAKPOINT CHLORINATION—Addition of chlorine to water or wastewater until the chlorine demand has been satisfied and further additions result in a residual that is directly proportional to the amount added beyond the breakpoint.

BREAST ROLL—A large-diameter roll around which the forming fabric passes at the headbox just at or behind the point where the stock is delivered to the fabric by the headbox nozzle. Breast rolls are made either of corrosion-resistant metal or are covered with an elastomeric composition or fiberglass composite.

BRIGHT ENAMELS—Papers coated on one side only and highly polished by calendering or brush polishing and sometimes both. They are chiefly used for labels.

BRIGHTENERS—Bleaching agents, such as peroxides, hydrosulfites, and borohydrides, which alter the colored elements in high-yield pulps to render them colorless without removing them, thus retaining the yield advantage of these pulps.

BRIGHTENING (MECHANICAL PULPS)— The object in brightening mechanical pulps is to decolorize or brighten the lignin or other coloring components without dissolving them so that their yield advantage is not lost. Mechanical pulps are brightened by treatment with either an oxidizing (hydrogen peroxide) or a reducing (sodium hydrosulfite) agent or, for higher brightness, a peroxide-hydrosulfite sequence which alters the chromophoric groups in fibers so that their light absorption is diminished.

BRIGHTNESS MEASUREMENT—See BRIGHTNESS (OF PAPER); GE BRIGHTNESS.

BRIGHTNESS (OF PAPER)—(1) The reflectivity of pulp, paper, or paperboard for specified blue light measured under standardized conditions on a particular instrument designed and calibrated for this purpose. (2) C.I.E. Relative. See LUMINOUS REFLECTIVITY.

BRIGHTNESS REVERSION—All cellulosic material yellow (lose brightness) with age. The rate is influenced by the chemical composition of the pulp and by environmental conditions such as temperature, alkalinity or acidity, presence or absence of oxygen, humidity, quality, and intensity of illumination. Pulps of high lignin content that have been whitened only by reductive treatment are particularly sensitive to brightness reversion (loss of brightness). See POST COLOR NUMBER; REDUCTANTS.

BRIGHTNESS STABILITY—The resistance of a pulp to brightness reversion (q.v.).

BRISTLE MARKS—Indentations in the surface of brush-coated paper which are pressed there by bristles that have come off coating brushes and adhered to the calender roll. See BRUSH MARKS.

BRISTOL BOARD—See BRISTOLS.

BRISTOLS—A general term for a solid or laminated heavyweight printing paper made to a thickness of 0.006 in. or higher. The name is derived from the original pasted rag content board made in Bristol, England. See BOGUS BRISTOL; CARDBOARD; INDEX BRISTOL; MILL BRISTOL; WEDDING BRISTOL.

BRITISH THERMAL UNIT (Btu)—A British thermal unit is the amount of heat required to raise one pound of water one Fahrenheit degree from 59.5 to 60.5°F. It is used to express the heating value of fuels or the enthalpy of steam. 1 Btu = 252 calories.

BRITTLENESS—That property of a material which causes it to break or fail when deformed by bending. It is of practical interest only when the deformation producing failure is small. For example, a sheet of paper which has undergone severe degradation as a result of aging exhibits brittleness. It cracks and breaks when bent only slightly.

BROAD FOLD—(1) A term denoting that, after folding, the grain runs with the shorter dimension of the paper. See LONG FOLD. (2) A sheet of printing paper folded so as to make the pages wider than the usual shape, or of greater width than depth; an oblong fold or page, as distinguished from an upright page.

BROADLEAF—An adjective used to describe hardwood, or deciduous trees which have relatively broad, flat leaves such as oak, maple, poplar, beach, birch, and eucalypt.

BROADSIDE—A large printed sheet, intended as a circular, folded into a size convenient for mailing. It is distinguished from a folder by the fact that its printed matter runs across the sheet, regardless of the fold.

BROCADE PAPER—(1) Paper with heavy embossing, such as a cover or box paper. (2) Marbled paper with a brocade-like pattern.

BROKE—Paper after forming which is not suitable for its end use. When produced, broke can be wet or dry, coated or uncoated, on-machine or off-machine, calender broke, winder broke, coater broke, or trim. Machine trim, while generally identical to the main sheet at that point in the process, is usually referred to as trim if it is continuously produced, to differentiate it from other broke, which is only sporadically produced.

BROKEN—A mill term for broke (q.v.) or waste generated during operation. The term is not in current usage.

BROKEN CARTON—A quantity of paper or paperboard less than a full carton.

BROKEN CASE—A quantity of paper or paperboard less than a full case.

BROKEN EDGES—Edges of paper or paperboard in sheets or rolls that have been broken or ruptured through faulty operations or rough handling before, during, or after shipment.

BROKEN REAM—A quantity of paper less than a full ream.

BROKE PULPER—A pulper intended to operate on broke (q.v.) only. Broke pulpers may be located under the paper machine (See UTM BROKE PULPER), or at a remote location. Remotely located broke pulpers may handle broke from a variety of locations on the paper machine, or broke that has been stored or accumulated for some time, and they may operate continuously or in a batch mode. Many mills pulp broke in their furnish pulpers (q.v.). See OFF-MACHINE BROKE PULPER; UTM BROKE PULPER.

BROKE SCREEN—A screen used to remove casual debris from broke streams before the broke rejoins the main stock stream. A broke screening system may contain primary, secondary, tertiary, etc., broke screens. See also SCREENING SYSTEM.

BROKE SYSTEM—A system of tanks, pits, consistency regulators, agitators, screens, and cleaners that store and convert wet broke (q.v.) or dry broke (q.v.) into usable stock. See BROKE.

BROMIDE PHOTOGRAPHIC PAPER—A photographic paper base coated with an emulsion in which the photosensitive material is primarily silver bromide. It is the most popular type of printing paper for enlarging because its high sensitivity to light allows short exposure times. See PHOTOGRAPHIC PAPER.

BRONZE CREPE—Crepe paper which has been coated with gold, silver, or copper metallics. See CREPE PAPER.

BRONZE FLECK—See BRONZE SPECKS.

BRONZE PAPER—A paper or board coated on one or both sides with a composition consisting of a finely divided metallic powder and a binder of pyroxylin, casein, glue, etc. See ALUMINUM PAPER; METALLIC COATING.

BRONZE SPECKS—Particles of metal picked up during processing or from the press rolls. Bronze specks may also occur as a dendritic growth, which is characteristic and is known as a bronze fleck. These terms are now considered to be obsolete.

BRONZING—The process of producing gold or silver effects in printing by means of dry metallic powders. The design is printed using a slightly colored sticky ink called gold size. While still wet, the printed sheets are passed through the bronzing machine where they are dusted with the metallic powder. The powder sticks to the printed areas and is brushed off the unprinted areas. The bronzing machine is usually run in tandem with an offset press.

BRONZING OF INK—The phenomenon in which an ink develops a metallic surface reflection on drying. It is caused by "leafing" or orientation of dichroic pigment crystals to the surface of the ink film. Bronzing mainly occurs with certain reds, violets, and blues, and with blacks toned with violet or blue pigments. It is sometimes called iridescence.

BROWNPRINT PAPER—A lightweight, translucent paper designed for brownprint coatings (e.g., ferric ammonium oxalate or citrate and silver nitrate). This paper is normally made of cotton pulp in basis weights of 12 to 24 pounds (17 x 22 inches – 500), and is used for the negatives from which positive blueprints are made. It is characterized by high wet-tensile strength, good wet-rub resistance, hard internal and surface sizing, and in most cases considerable durability and permanence. See also BLUEPRINT PAPER.

BROWN PULP—A ground woodpulp made from wood which is steamed before grinding. Obsolete practice.

BROWNSTOCK—(1) The slurry of pulp and spent liquor after discharging from the digester and after washing. It is brown because of the remaining lignin in the fibers. Higher kappa number (higher lignin) pulps are, in general, darker, regardless of the chemical pulping process. (2) Unbleached chemical pulps.

BROWNSTOCK WASHING—After discharge from the digester, the pulp and spent cooking liquor are separated using either a drum or diffusion washer. The washing is usually done in several countercurrent stages. For example, in a three-stage washing system, clean water is used only in the third (last) stage of washing. The second stage is washed with the third-stage effluent; similarly, the first stage, with second-stage effluent. The effluent from the first stage of washing is termed weak black liquor. It is used for dilution in the digester or sent to the chemical recovery system for evaporation and burning in the recovery boiler. Washing is a function of both the displacement of free liquor from the fiber slurry and diffusion of the dissolved lignin from inside the fiber wall.

BROWN WRAPPING PAPER—A paper usually made from unbleached kraft pulp. It is sold in various roll widths and diameters and in sheets. The minimum basis weight is 20 pounds (24 x 36 inches – 500). Since it is used for wrapping purposes of all kinds, strength is the most important property.

BRUSH COATER—See BRUSH COATING.

BRUSH COATING—The process of applying a semifluid mixture of the pigment and binder to a sheet by means of a revolving cylindrical brush and smoothing the coating so applied by means of oscillating flat brushes that contact the coated sheet as it is being drawn forward while held tightly on a moving rubber apron or a revolving drum.

BRUSH ENAMEL PAPER—A paper coated on one or both sides and brush polished previous to calendering to produce a smooth, even, and brilliant surface. It is used largely for cigar labels, illustrations, and box coverings.

BRUSH-FINISH COATING—A paper coating which is given an especially high polish by running the dried or partially dried coated paper over a revolving drum provided with six or more rapidly revolving cylinder brushes which contact the coated surface of the sheet.

BRUSHING KRAFT—An MF fourdrinier unbleached kraft sheet, about 50 pounds in basis weight (24 x 36 inches – 500), used as windbreak protection for young plants, such as melons. It derives its name from the brush used to hold it in place.

BRUSH MARKS—Those marks left on the surface of coated paper by the brushes used in spreading the coating material. They may be due to defects in the brushes, in the coating material, or in the adjustment or operation of the brushing machinery.

BRUSH POLISHING—The operation of polishing a coated paper, the coating of which has a high wax content, by means of cylindrical brushes revolving at a higher peripheral speed than the speed of the paper.

Btu—British thermal unit (q.v.).

BUBBLE—See AIR BUBBLES; BLISTER.

BUBBLE COATING—A white, opaque coating containing little or no pigment, consisting of a film of casein, starch, other binders, or mixtures thereof, in which minute air pockets are dispersed.

BUFF COPYING PAPER—A copying paper colored buff, a moderate shade of yellow. See COPYING PAPER.

BUFFER PAD—Dunnage (q.v.) used to secure rolls of paper in railcars.

BUFFING PAPER—An abrasive paper (q.v.) which is coated with flint grains and is used in buffing operations in the leather industry where a smooth, velvety finish is desired. Buffing paper is usually made on a cylinder machine in a basis weight of 130 pounds (24 x 36 inches – 500).

BUILDING BOARD—A general term for large boards used in the building trade and manufactured from organic or inorganic raw materials, or both. The term is broader than wallboard and includes structural, acoustical, and core materials as well as coverings for walls and partitions.

BUILDING PAPER—A general term applied to a class of papers used in general construction work. They are used in building construction for sheathing and under flooring and may be converted to such products as roofing, sheath-

ing, and tarred or asphalt-coated vapor barrier. They are also used in the manufacture of rock wool, mineral wool, and fiberglass insulation batts.

BUILT-IN STRAIN—See DRIED-IN STRAIN.

BUILT-IN STRESS—See DRIED-IN STRAIN.

BULK—(1) The thickness of a pile of a specified number of sheets under a specified pressure. (2) The apparent specific volume of a sheet of paper when in a pile under a definite pressure. (3) Goods or cargo not in packages or containers or goods unpacked (loose) within a container. (4) A large box used to contain a volume of product.

BULK DENSITY—The mass of chips or other material that occupies a specified unit of volume, such as pounds per cubic foot or kilograms per cubic meter. The bulk density of particles depends on the degree of compaction, particle size distribution, moisture content, and the specific gravity of the material. This is an important parameter for designing the volumetric size of storage silos and digesters.

BULK INDEX—See APPARENT SPECIFIC VOLUME; BULK.

BULKING BOARD—A general term indicating that the board is subjected to little or no calender pressure and is lighter in weight per point of thickness or count than board that has been calendered.

BULKING BOOK PAPER—A paper having unusually high bulk per unit of ream weight. It is made from a variety of furnishes, selected and blended to give this property. Some grades are made of cotton-linters pulp, rag pulp, esparto, or chemical woodpulps; other grades contain a large percentage of mechanical woodpulp mixed with chemical woodpulps. There is little or no loading (fillers). This paper is usually made to specification as to finish or caliper, or both. See NOVEL PAPER.

BULKING DUMMY—A dummy (q.v.) of blank sheets of paper made up to determine the ac-

43

tual bulk of a given number of sheets to be used in the manufacture of a book or other printed matter.

BULKING NUMBER—The number of sheets of printing paper required to bulk 1 inch (2.54 cm) in overall thickness.

BULKING PAPER—See BULKING BOOK PAPER.

BULKING PRESSURE—The pressure under which the bulk of paper is measured.

BULKING THICKNESS—See BULK.

BULK MODULUS—The rate of pressure applied to the fractional change in volume. See COMPRESSIBILITY.

BULK OF PULP—The apparent specific volume, calculated by dividing the single-sheet thickness in thousandths of a millimeter by the basis weight in grams per square meter.

BULKY—Of a sheet, lacking in compactness, having a light weight for a given thickness.

BUNCHED PLATING—A method of plating a pile of paper between plating boards which consists in running it through a plater in such a way that both sides have the same finish.

BUNCH PLATER FINISH—A kind of paper finish produced when a pile of paper, usually from one to one-and-one-half inches in thickness, is placed between zinc plates or pressboards (without any fabric) and passed through the plater press. The process gives a uniform surface on both sides of the paper, slightly smoother than the original finish.

BUNDLE—(1) A unit of board measure, weighing 50 pounds. The number of sheets varies with the size and the caliper. See COUNT; REGULAR NUMBER. (2) A shipping unit of paper, as well as a packing method.

BURKEITE—A double salt of sodium carbonate and sodium sulfate which precipitates from black liquor when it is concentrated above about 50% total dry solids. Often found as a soluble deposit on black liquor evaporator surfaces.

BURLAP FINISH—(1) A finish resembling the texture of burlap cloth. It is produced by using sheets of burlap between the sheets of paper in a plater book. The same effect may be produced by attaching burlap to calender rolls and running the paper through these rolls or by using rolls etched to simulate burlap on an embossing machine. (2) A finish of hanging paper produced by embossing.

BURLAP-LINED PAPER—A wrapping paper, usually made of kraft, pasted to a loosely woven burlap cloth. It is used where an exceptionally strong, waterproof, and tear-resisting wrapper is required. It is usually asphalted.

BURN—To expose a printing plate with a high-intensity light source.

BURNISHED FINISH—Another term for glazed finish. It is sometimes restricted to flint and friction glazing.

BURNOUT—An inspection method for coating uniformity in which a film of alcohol is applied to the coated sheet surface. The alcohol is allowed to evaporate, the sheet is heated and the fibers absorbing the alcohol are charred while the coating remains white.

BURNT—(1) Of papers, overdried and brittle. (2) Of pulps, overheated and darkened in the cooking process.

BURR—A tool used to recondition pulpstone surfaces by burring, trueing, or sharpening. There are three groups of tools in this class: hand-operated burr sticks, mechanical-feed burr lathes, and hydraulic-feed burr lathes.

BURST—(1) A rupture in the web that does not extend to the edge. (2) A web rupture within a wound roll caused by excessive web tension, usually due to nonuniform web caliper profiles which result in nonuniform nip pressures during a surface winding operation. See BURSTING STRENGTH.

BURST FACTOR—The numerical value obtained by dividing the bursting strength in grams per square centimeter by the basis weight in grams per square meter. See BURST RATIO; PERCENT POINTS.

BURST INDEX—The numerical value obtained by dividing the burst strength in kilopascals by the basis weight in grams per square meter.

BURSTING STRENGTH—A measure of the ability of a sheet to resist rupture when pressure is applied to one of its sides by a specified instrument, under specified conditions. It is largely determined by the tensile strength and extensibility of the paper or paperboard. Testing for bursting strength is very common although its value, except for limited, specific purposes is questionable. See BURST FACTOR; POINTS PER POUND.

BURST RATIO—The bursting strength in points per pound (q.v.).

BUSH ROLL—See BURR.

BUSINESS PAPERS—Papers used in the home and office for business purposes. Examples of business papers include reprographic paper, computer paper, stationery, and file folders, etc. Business papers are made from chemical woodpulps, mechanical pulps, recycled fiber, cotton, or a combination thereof.

BUTCHERS D.F—See DRY-FINISH BUTCHERS WRAP.

BUTCHERS MANILA—A paper similar in nature to dry-finish butchers, except that it has a steam or water finish and is usually made in manila color.

BUTCHERS PAPER—See DRY-FINISH BUTCHERS WRAP.

BUTCHERS WRAP—See DRY-FINISH BUTCHERS WRAP.

BUTTED SPLICE—A butted joint which is formed by trimming the ends of two webs of paper, placing them end to end and pasting a strip over and under to make a continuous web without overlapping. See SPLICE. (Also called Butt splice.)

BUTTER-BAG PAPER—A vegetable parchment or a greaseproof paper used for butter-bags.

BUTTER-BOX LINER—A paper, similar in nature and weight to butter paper (q.v.), used as a liner for butter tubs. The basis weight is usually about 27 pounds (24 x 36 inches – 500).

BUTTER CHIP BOARD—Paperboard normally manufactured from bleached fibers and used to hold an individual pat of butter as served in restaurants. See SPECIAL FOOD BOARD.

BUTTER PAPER—A paper used for wrapping butter. Vegetable parchment or uncalendered grease-resistant paper (plain or waxed) is usually used. Important properties are grease-proofness and high wet strength. The usual basis weights are 27 pounds for vegetable parchment and 30 pounds for dry-waxed greaseproof paper (24 x 36 inches – 500).

BUTTER PARCHMENT PAPER—See BUTTER PAPER.

BUTTER WRAPPERS—See BUTTER PAPER.

BUTTON SPECKS—Specks in rag-content papers caused by small pieces of buttons that appear in the finished sheet as light-colored, powdered spots.

BUTT ROLL—(1) A narrow roll on the ends of a winding set that will be discarded. Butt rolls occur when the total width of an order results in a trim that is too wide to be handled by the trim conveying system. Also called a cookie (q.v.). (2) The paper left on a core after the end of a print job.

BYTE—A unit of digital information; a character. May contain between 8 and 32 bits (q.v.) of information.

C

C 1 S—Coated on one side of the sheet.

C 1 S LABEL—See LABEL PAPER.

C 1 S LITHO—See LABEL PAPER.

C 2 S—Coated on two sides of the sheet.

CABLE-INSULATING PAPER—See CABLE PAPER.

CABLE MARKING PAPER—Usually a lightweight twisting paper cut into narrow strips for incorporation in wire cables or ropes as a mark of identification. It is usually colored and is frequently printed to aid in identification. For furnish and other properties, see TWISTING PAPER.

CABLE PAPER (TURN INSULATION)—A paper suitable for use as insulation on wire or cable. It is usually made from manila rope or sulfate pulp. Because of the narrow widths which may be used (1/32 inch minimum), it must have sufficient strength to withstand the high-speed bending and winding operations both during manufacture and use of the cable or wire. The thicknesses range from 0.0006 to 0.0085 inch. The paper should be free from foreign materials and must have high dielectric strength and low power factor. These are especially important at higher voltages. Good stability under dry heat and subsequent treatment with insulating liquids, high machine-direction tensile strength and cross-machine direction tearing strength are necessary.

CAD—See COMPUTER AIDED DESIGN.

CAKE BOARD—A paperboard upon which cakes are placed after baking or a layer board between rows of cakes. It may be a manila-lined or solid-bleached board and is stiff, clean, sanitary, and free from fuzz. Waxed corrugated board is also used.

CAKE-PAN LINER PAPER—See PAN LINER.

CAKE-WRAPPER PAPER—See BREAD WRAPPERS. The only difference is that cake-wrappers are almost always transparent.

CALCIUM CARBONATE—A chemical compound ($CaCO_3$), occurring in nature, usually from sea deposition, or obtained commercially by grinding limestone or by chemical precipitation. Calcite and aragonite are the two principal crystalline types with calcite being the thermodynamically stable form. Chalk (or ground limestone) is a naturally occurring form, containing some impurities, used in papermaking and coating. Precipitated calcium carbonate is used because it has higher purity than the natural product. The precipitated product may be manufactured by precipitation of milk of lime with carbon dioxide gas or sodium carbonate, or by precipitation from calcium chloride-sodium carbonate reactors. Calcium carbonate is used both as a filler and as a coating pigment.

CALCIUM HYPOCHLORITE—See HYPOCHLORITES.

CALCIUM SULFATE—A chemical compound of general formula $CaSO_4 \cdot H_2O$. The compound is very slightly soluble in water and is used primarily as a filler pigment. Where the material exists in nature, it may be in the form of anhydrite ($CaSO_4$) or as gypsum ($CaSO_4 \cdot 2H_2O$). Precipitated calcium sulfate is known as crown filler ($CaSO_4 \cdot 2H_2O$). Now rarely used in papermaking or coating.

CALCIUM SULFITE—A chemical ($CaSO_3$), prepared by interaction of sulfurous acid and calcium hydroxide. Now rarely used as a filler or coating pigment.

CALENDER—A set or "stack" of vertically oriented rolls designed to impart uniform nip pressure to level the caliper profile and/or smoothness and/or surface texture of a web. The rolls can be chilled-iron, hardened steel, or variations of soft-nips which can be composites of polymeric compounds or, in the case of supercalenders, filled rolls (q.v.) of compressed cellulose fibers. See also BREAKER STACK;

EMBOSSING CALENDER; FINISHING STACK; GLOSS CALENDER; HOT CALENDER; MACHINE CALENDER; SOFT-NIP CALENDER; SUPERCALENDER.

CALENDER BLACKENED—See CALENDER CRUSHED.

CALENDER BOX—A device used for application of calender sizing (q.v.). It consists of an inlet pipe, a flow control valve, and a pond formed by a reinforced rubber lip contacting the machine calender roll and a dam to control pond level; the overflow over the dam is drained back to a recirculation tank. The sizing material is carried on the roll surface into the calender nip where it is applied to the sheet surface. One or more boxes may be lifted to one side of the calender stack, and the same number to the other side if equal pick up is desired.

CALENDER BROKE—See BROKE.

CALENDER COLORED—See CALENDER DYED.

CALENDER CRUSHED—Having the fibers pushed out of position and the formation disturbed by excessive pressure in calendering, usually with a decrease in opacity, surface mottling, and duller color. See BLACKENING; CRUSHED.

CALENDER-CRUSH FINISH—A crushed paper which has been calendered with a water finish. It is used for such papers as express mill wrappers and for such boards as imitation pressboard.

CALENDER CUTS—A cut in a web which is usually caused by a wrinkle going through the nip of a calender. A specific type of calender cut can be caused by a human hair on a web passing through a metal to metal calender nip.

CALENDER DYED—Dyed or stained at the calender rolls. The dye solution is supplied from calender boxes to the calender rolls, which

transfer it to either one or both sides of the paper or the board.

CALENDERED PAPER—See CALENDER FINISHED.

CALENDER–EMBOSSING—See EMBOSSING CALENDER.

CALENDER FINISHED—A term applied to any paper with a surface glazed by means of calenders; it does not include plate finish but refers to machine finish, English finish, supercalendered, and calendering.

CALENDER–GLOSS—See GLOSS CALENDER; HOT CALENDER.

CALENDER–HOT—See HOT CALENDER.

CALENDERING—The process of passing a web through a pressurized nip (or multiplicity of nips) for the purpose of leveling the normal profile variations in the web while compacting and smoothing the substrate. See CALENDER; GLOSS CALENDER; HOT CALENDER; MACHINE CALENDER; SOFT-NIP CALENDER; SUPERCALENDER.

CALENDER–MACHINE—See MACHINE CALENDER.

CALENDER MARKED—Having marks caused by defects on the calender rolls or foreign material that has adhered to the rolls.

CALENDAR-PAD BOND PAPER—A bond type sheet having the characteristics of ordinary bond paper but, in addition, having bulk specifications required by calendar-pad manufacturers.

CALENDAR PAPER—See CALENDAR STOCK.

CALENDER-ROLL PAPER—A soft, unsized, non-acid cotton paper used in the manufacture of rolls for supercalenders and embossing machines. The paper is cut into discs or octagons and pressed on a shaft under very high pressures. The rolls are finally turned down and polished on a lathe. In England, this is termed bowl paper. See FILLED ROLL.

CALENDER ROLLS—See BREAKER STACK; CALENDER; EMBOSSING CALENDER; FILLED ROLL; FINISHING STACK; GLOSS CALENDER; HOT CALENDER; MACHINE CALENDER; SOFT-NIP CALENDER; SUPERCALENDER.

CALENDER SCABS—A mark on a sheet of paper or board caused by a particle of pigment or of fiber that adheres to a roll in the calender stack and embosses its shape into the surface of the sheet.

CALENDER SCALES—Small particles of pigment or other foreign materials that gather on the calender roll and are pressed onto the paper.

CALENDER SECTION—See BREAKER STACK; CALENDER; FINISHING STACK; FOURDRINIER MACHINE; GLOSS CALENDER; HOT CALENDER; MACHINE CALENDER; SOFT-NIP CALENDER.

CALENDER SIZING—The application of an emulsified wax size or solution of a starch, alginate, polyvinyl alcohol, or other film-forming adhesive to paper or paperboard at the calender, usually for the purpose of improving surface properties, such as sizing, smoothness, porosity, and printability. The paper or board may be treated on one or both sides.

CALENDER–SOFT-NIP—See SOFT-NIP CALENDER.

CALENDER SPOTS—Glazed or indented spots, often translucent, resulting from small flakes or pieces of paper adhering to the calender rolls or carried through the nips on the sheet.

CALENDER STACK—See BREAKER STACK; CALENDER; FINISHING STACK; MACHINE CALENDER; SUPERCALENDER.

CALENDER-STACK CRUMBS—Small flakes of paper which collect at the calender of the paper machine and which are probably formed by the crushing action of the calender upon localized thick areas of the paper sheet.

CALENDER STAINING—See CALENDER DYED.

CALENDAR STOCK—(1) A printing paper used in the manufacture of calendars. It is typically a coated or uncoated offset grade made in basis weights of 50 to 70 pounds (25 x 38 inches – 500) with good printability and non-curling qualities. (2) A paperboard used for certain types of wall or desk calendars.

CALENDER STREAKS—Continuous streaks of darkened paper occurring parallel to the grain, caused by uneven pressing and drying preliminary to calendering.

CALENDER VELLUM FINISH—A finish with the same surface characteristics as vellum except that the surface is smoother. It is produced by calender rolls. See PLATER VELLUM FINISH; VELLUM FINISH.

CALENDER–WINDER—See COMBINING CALENDER.

CALF PAPER—Colored embossed paper which imitates leather. It is usually used for bookbinding.

CALIBRATION—The comparison of a measurement instrument or system of unverified accuracy to a measurement instrument or system of known accuracy to detect any variation from the required performance specification.

CALIPER—Thickness.

CALIPER GAUGE (ROLL)—See ROLL CALIPER GAUGE.

CALIPER SHEAR BURST—This machine direction (MD) shear failure occurs inside a winding roll, and results from excessive cross-machine differential in MD layer-to-layer displacement resulting from non-uniform winding nip mechanics. This failure is often associated with "U-shaped" crepe wrinkles found at a ridge or high caliper area of the winding web.

CALORIE—The amount of heat required to raise one gram of water one Celsius degree from 14.5 to 15.5°C.

CAMBIUM—The cambium layer, which separates the bark from the wood, is that portion of the tree from which growth in diameter originates.

CAMBRIC FINISH—Obsolete term. See EMBOSSING.

CAMBRIDGE BIBLE PAPER—See BIBLE PAPER.

CAMBRIDGE INDIA PAPER—A name for a grade of Bible paper (q.v.).

CAMERON GAP—A roll structure measurement test which determines the winding strain by severing the outer layer and measuring the resulting gap.

CAMPAIGN BRISTOL—(1) A coated post card stock which originally derived its name from its use in political campaign advertising. It is normally made in a 0.010-in. thickness with a basis weight equivalent to 120 pounds (22.5 x 28.5 inches – 500). (2) A specialty bristol paper. See COATED POST CARD STOCK.

CAN—See FIBER CAN.

CANADIAN FREENESS TESTER—See FREENESS. (140)

CANARY WRITING—See RAILROAD MANILA.

CAN BOARD—Paperboard used to make spiral wound and convolute cans with general qualities of high internal bonding strength and stiffness. (1) For liquid-tight containers used in food packaging the paperboard is highly sized and is usually made of bleached chemical pulps. See SPECIAL FOOD BOARD. (2) For dry containers a wide variety of paperboard grades are used. (3) For other specific purposes, various paperboards may be used in conjunction with foil or plastics.

CANDY-BAG PAPER—Paper usually made from bleached chemical woodpulp. It may have an MF or MG finish and may be embossed or decorated with varying designs. It is sold in various width rolls of jumbo diameter to manufacturers of candy bags. The basis weights vary from 25 to about 40 pounds (24 x 36 inches – 500). Characteristics are strength, cleanliness, and appearance.

CANDY-BAR WRAPPERS—Glassine, glazed, waved, or plain sheets which are usually printed and which are used as sanitary protection for candy bars. Printed wrappers are frequently coated with a lacquer or varnish for high gloss.

CANDY-BOX DIVIDERS—See CHOCOLATE DIVIDERS AND LAYER BOARD; DIVIDERS.

CANDY-BOX LINER—A paper used for lining candy boxes, made of bleached chemical woodpulp furnishes, which is dry waxed. The principal requirement is a smooth, dull finish.

CANDY-CUP PAPER—Glassine paper used for forming fluted cups for packaging individual pieces of candy.

CANDY PAPER—See CANDY-BAR WRAPPERS; CANDY-BOX LINER; CANDY-CUP PAPER; CANDY TWISTING TISSUE.

CANDY-SLAB PAPER—A special type of release-coated vegetable parchment paper used as a candy tray or worktable liner. It provides high strength and resistance to penetration by sugars, fats, oils, or moisture which create adhesion problems in the manufacture of candy.

CANDY TWISTING TISSUE—A tissue paper which is also called kiss paper, used for wrapping candy kisses, saltwater taffy, candy bars, photographic film, gum, etc. It is made from unbleached or bleached chemical woodpulp and is usually waxed. Generally it is made on a single-cylinder paper machine but it is sometimes made on fourdriniers. It is usually made

49

in basis weights of 12, 16, 17, and 18 pounds (24 x 36 inches – 480). It has a high tensile strength in the machine direction and a high tearing strength in the cross direction. The sheet is soft and raggy, takes a good finish when waxed, and retains its twist.

CANDY WRAPPERS—See CANDY-BAR WRAPPERS.

CANISTER—A receptacle made from paperboard or various combinations of paperboard, paper, films or metal foil which may have metal ends of circular or rectangular shape. It is used for dry products weighing not over five pounds net. See CONTAINER; FIBER CAN; FIBER DRUM.

CANS—Heated metal cylinders used to dry and cure nonwovens by surface contact.

CANVAS NOTE PAPER—A note paper embossed to resemble canvas.

CAPACITOR—An electromagnetic device that stores electrical energy by charging plates separated with a dielectric material. The energy exchange is reversible and the direction of flow depends upon the voltage applied. Capacitors are used to improve the power factor (q.v.) of an electrical system.

CAPACITOR PAPER—There are two classes of capacitor paper, i.e., dry and wet. DRY—A paper used as a dielectric between foils of insulating liquid-filled condensers. This paper is made on a fourdrinier machine, usually from sulfate pulp or hemp. It is used in thicknesses of 0.002 to 0.001 inch. The important properties are chemical purity, uniformity of thickness, good formation, and freedom from foreign matter, particularly conducting particles; low power factor characteristics are important for paper intended for alternating current capacitors and high insulation resistance for direct current applications. WET—A paper used as a spacer between foils of electrolytic capacitors. This paper is made from sulfate pulp, cotton, or hemp. Thicknesses range from about 0.0006 to 0.004 inch. The important properties are po-

rosity, absorption of electrolytic liquids, and chemical purity, especially freedom from soluble chlorides.

CAPACITY—The capacity of a machine is the output expressed in pounds or in tons per day when under full operation. The actual output for a period of a week, month, or longer, divided by the capacity for that period, gives the percent of capacity. See PRACTICAL MAXIMUM CAPACITY.

CAP BOARD—A paperboard used for making caps for closure of the ends of cans or tubes. The term formerly described a cylinder board with a straw pulp liner on one side. Currently cap board ranges, depending on use requirements, from a cylinder board made from reclaimed paper stock and/or woodpulp to a solid bleached kraftboard made on a fourdrinier machine. It ranges in thicknesses from 0.016 to 0.40 of an inch. See BOTTLE-CAP BOARD.

CAP PAPER—(1) See BOTTLE-CAP BOARD. (2) See PLANT-CAP PAPER. (3) A bogus paper used in the manufacture of paper caps for use in toy pistols. (4) See BARREL LINER.

CAPPING PAPER—See PLANT-CAP PAPER.

CAPSHEET—See ASPHALT FELT.

CARBIDE COATED DRUM—A drum coated with a rough metalized surface, normally with tungsten carbide because of its hardness. The roughness of this surface may be specified to be between 200 and 550 micro-inches. This coating is often applied to other web support rolls in a winding process to improve traction.

CARBONATE PAPER—A printing paper formerly used largely in the magazine publishing field. Replaced to a large extent by the advent of machine-coated papers, it is currently of lesser importance and is generally sold to the commercial printing trade. It is made of chemical woodpulps heavily loaded with calcium carbonate, generally unsized, with an English or supercalendered finish and of good color and

50

opacity. It is usually made in basis weights of from 40 to 70 pounds (25 x 38 inches – 500). See also CIGARETTE PAPER.

CARBONATES—A term often applied to alkaline fillers (q.v.). See also CALCIUM CARBONATE.

CARBON BLACK—A term covering all of the carbon products in the colloidal range of particle size produced by the thermal decomposition of hydrocarbon products. Each of the five principal processes develops its own characteristic carbon: channel or impingement carbon black, furnace carbon black, lamp black, thermal carbon black, and acetylene carbon black.

CARBON-BLACK BAG PAPER—A specialty type bag paper made from MF fourdrinier unbleached kraft, in various basis weights and in many types, black or dark gray in color. It is used for packaging and shipping carbon black.

CARBON COATING—See CARBON PAPER.

CARBONIZED ROLLS—Carbon paper in roll form.

CARBONIZING PAPER—An uncoated grade of paper made from bleached or unbleached chemical pulps or mixtures of unbleached chemical and mechanical pulps. The paper is the raw stock to be surface coated on one or two sides with a carbon dope (solvent or wax). This sheet is usually made in basis weights of 4 to 28 pounds (20 x 30 inches – 500). Significant properties include uniformity of surface and caliper, freedom from pinholes, close formation, high density, strength, nonporosity, and ability to take carbon inks without penetration and to release these carbon inks subsequently under pressure or impact.

CARBONLESS PAPER—(1) A reproduction paper coated on one side with a waxy carbon-like impact-sensitive, mechanical transfer coating and used primarily in producing typewritten multiple copies. (2) A chemical transfer reproduction paper as above where one side of the paper is coated with a receptor coating. (3) A

chemical transfer reproduction paper where the donor and receptor materials are coated together on one side of the sheet and impact ruptures the donor capsules to produce an image. Also called self-contained carbonless paper and NCR (no carbon required) paper.

CARBON MONOXIDE—(CO) Pollutant species found in flue gases due to incomplete combustion of carbon-containing fuels. Often used as an indicator of incomplete combustion.

CARBON PAPER—(1) A coated paper used for making duplicate copies with pencil, pen, typewriter, or business machine commonly known as carbon or duplicate copies. It may range in weight from 4 to 28 pounds (20 x 30 inches – 500) or, rarely, even higher. It may be coated on one side (semicoated) or both sides (full-coated) with a mixture of carbon black or some other coloring matter in a vehicle, which may be wax or some oil-soluble substance according to the intended use. The application is generally made by means of a coating machine with heated rolls revolving in melted inks, or a rotary or rotogravure printing press only printing on certain sections of the paper. It may be supplied in different finishes such as intense writing, medium writing, or hard writing. Varieties include pencil, billing machine, adding machine, typewriter, hectograph, and lithographic transfer. (2) In photography, a paper coated with gelatin and a pigment.

CARBON SPOTS—Spots caused by fragments of cinders or coal dust. Cinder specks is a synonym.

CARBON STEEL—An alloy of iron, carbon, and manganese: most steels have a carbon content of 0.1 to 1% and a manganese content of 0.5 to 2%. The physical properties of carbon steel can be controlled by heat treatment processes.

CARBONYL COMPOUNDS—A bleached pulp tends to be less susceptible to brightness reversion if the carbonyl groups ($>C=0$) on the cellulose molecules are reduced to hydroxyl groups. Oxidation of cellulose, resulting in the formation of aldehydes and ketones, tends to enhance brightness reversion.

CARBONYL GROUPS—See CARBONYL COMPOUNDS.

CARBORUNDUM PAPER—See ABRASIVE PAPERS.

CAR CARD—See CAR-SIGN BOARD.

CARDBOARD—A general term applied to board 0.006 of an inch or more in thickness, where stiffness is the paramount characteristic. The word cardboard as used by the public is too vague to be technical. In the paper industry the term board is generally used in combination with words indicating its character or use. See BLANKS; BRISTOLS; POSTCARD BRISTOL; RAILROAD BOARD; TOUGHCHECK; TRANSLUCENTS; etc.

CARDBOARD FINISH—Obsolete term. See EMBOSSING.

CARDBOARD LININGS—Papers used by some board mills as a lining for chipboard or newsboard. This grade may range from groundwood and manila or bleached manila to high-grade book or special fancy printed paper.

CARDBOARD MIDDLES—(1) Rough coarse boards made from mixed wastepapers and mechanical pulp and used as fillers for cardboards. (2) Chipboards or newsboards, cylinder made, nonfolding, and with fairly smooth surfaces, to which are laminated cardboard lining paper for the production of special surfaced boards. Basis weight schedules used in folding boxboards generally apply.

CARD CLOTHING—Metallic band-saw-like wire or fillet staple-like wire punched through a backing that is used to cover the rolls of cards. In the processing of man-made fibers, emery is sometimes used.

CARD INDEX BRISTOL—See INDEX BRISTOLS.

CARDING—The process of opening, disentangling, blending and cleaning fibers using rolls clothed or covered with card clothing. The two

actions that can take place are working and stripping.

CARDS—A term usually applied to the sizes cut from various kinds of boards (usually bristols). Their use is indicated by prefixing another word, such as business, postal, visiting, wedding, etc. The word card as used by the public is too vague to be technical. In the paper industry, the term board is generally used in combination with words indicating its character or use. See BLANKS; BRISTOLS.

CAR LINER—A heavy paper or paperboard made from kraft pulp, chemical woodpulp screenings, or wastepaper stock, on a fourdrinier or cylinder machine, for lining freight cars and protecting the contents from dirt and abrasion against the sides of the car. It is specified in varying weights depending upon the nature of the goods loaded in the car. It may also be a laminated sheet with asphalt as a binder.

CARLOAD LOT—(1) The minimum amount of paper required for individual freight-car shipment at the carload rate of freight. It ranges from 36,000 to 100,000 pounds, depending upon the freight classification zone. The minimum for bulkier papers is less. (2) The amount customarily shipped by mills in one freight car. Most shipments exceed the minimum carload lot.

CAR MAT PAPER—A soft, bulky paper used by car mechanics to protect the inside of the car from oil and grease.

CARPET BROWN—See CARPET FELT.

CARPET FELT—A kind of soft, thick, and spongy building paper, plain or indented, used for inserting between floor boards and carpets. It may be made on a cylinder or fourdrinier machine from paper stock, roofing rags, etc. It is also called carpet brown, carpet lining, or deadening felt.

CARPET LINING—See CARPET FELT.

CARPET-LINING BOARD—A paperboard used as a pad placed under a carpet. It is made of chipboard with a soft finish or of indented board to provide a cushion. See INDENTED BOARD.

CARPET YARN—See TWISTING PAPER.

CARRYOVER—Non-process matter carried from one process or one part of a process to another by the main process stream. For recovery boilers, it is the large char and smelt particulate carried by the flue gas into the convective section of the boiler.

CAR-SIGN BOARD—A board used primarily for advertising posters in railroad trains, buses, etc. It is a special grade of blanks, manufactured on a cylinder board machine, using chiefly wastepaper as a middle, with liners of chemical woodpulp and/or shavings. It is regularly clay-coated and supercalendered on one side. The basis weight is about 200 pounds (22 x 28 inches – 500). It is usually 0.018 and 0.021 of an inch in thickness and is cut in the following sizes: 22.5 x 42, 22.5 x 42.5, and 34 x 43. Significant properties are a high degree of stiffness and an especially smooth finish on the coated side for the purpose of high-grade printing or lithography.

CARTON—(1) A general term loosely used to indicate (a) a folding paperbox, (b) a rigid setup box, or (c) a fiberboard shipping container. (2) A general term usually applied to a folding paperboard box as distinguished from a setup or rigid box or a shipping container. See FOLDING PAPER BOX. (3) A shipping unit of paper which usually weighs 125 to 150 pounds, equivalent to one-fourth of a case. (4) A corrugated container designed to accommodate sheets of fine paper up to 38 x 50 inches in size with overall weight not to exceed 150 pounds.

CARTON LABELS—Labels used on paper cartons, as contrasted to case or skid labels.

CARTON-LINER PAPER—A paper placed inside a carton to give added protection to the contents. It may be greaseproof, glassine, or waxed paper. The liner may be sealed for additional protection. See CASE-LINING PAPER; CRATE LINERS.

CARTON SEALER—A device used to glue the top cover of a carton to the main body of the carton.

CARTON-SEALING PAPER—Any paper used as a wrapper for cartons containing cereals, crackers, salt, or other foodstuffs and designed to protect the packaged material from contamination. A heat-sealing waxed paper or a lacquered paper may be used. The basis weight will depend upon the size of the carton and the desire of the user. It is usually printed or decorated with a design.

CARTON STOCK—Any paperboard normally made and used for the manufacture of folding paper boxes (q.v.).

CARTRIDGE PAPER—(1) A paper used to form the tube section of a shotgun shell. The furnish may consist of rag, flax, or chemical woodpulps, or high-grade reclaimed paper stock in various combinations. It is usually made on a cylinder machine, from 0.008 to 0.012 of an inch in thickness. It is usually unsized and lightly calendered for pasting, and is absorbent to facilitate waxing for waterproofness. Uniform weight, caliper, and density are important. It may be colored or plain, and the color may be solid or duplex in nature. Other important properties include tensile strength, stretch, and stiffness. See also BASE WAD PAPER. (2) A groundwood, manila, or chemical woodpulp sheet used as a wrapper for stick dynamite. See BLASTING PAPER; DYNAMITE-SHELL PAPER.

CASCADE CONTROL—The output of one controlled process cascade to one or more additional processes, making the downstream processes dependent upon (and controlled by) the upstream variables.

CASCADE EVAPORATOR—A device containing a rotating paddle wheel used to contact black liquor with hot recovery boiler flue gases in order to evaporate water from the black liquor.

CASE—See SHIPPING CONTAINER; see also CASE LOT.

CASE HARDENED LIGNIN—When thermoplastic lignin is heated to temperatures above the glass transition point, it will become liquid. If cooled to atmospheric conditions, it will coat the individual fibers and form a hard casing around them.

CASEIN—The acid-coagulable protein of skim milk obtained as a by-product of the dairy industry used in the sizing of paper. Now rarely used as a coating binder.

CASE LABELS—Large labels used on the outside of cases of paper. Labels also use bar coding.

CASE-LINING PAPER—Wrapping paper used for lining the inside of packing cases. Normally it is a heavyweight kraft wrapping paper; duplex asphalt and dry-waxed papers are also used. See also CARTON-LINER PAPER; CRATE LINERS.

CASE LOT—A quantity of flat paper or paperboard usually wrapped in packages (if writing paper) and enclosed in a fiberboard or wooden box. The term is indeterminate, but usually means a quantity of from 500 to 600 pounds or four cartons.

CASH REGISTER PAPER—A bond or writing paper made from chemical wood and/or mechanical pulps in a basis weight range of 12 to 16 pounds (17 x 22 inches – 500). It is converted into narrow small diameter "receipt" and "detail" rolls for use in cash registers. See also ADDING MACHINE PAPER.

CASING PAPER—See CASE-LINING PAPER.

CASKET PAPER—A paper used underneath the cloth lining of a casket. It is made of chemical and mechanical woodpulp, in basis weights of 40 and 50 pounds (24 x 36 inches – 500). The color is usually black, although it may be made in blue and green.

CAST-COATED PAPER—A paper or board, the coating of which is allowed to harden or set while in contact with a finished casting surface. In general, cast-coated papers have, in general, a high gloss. For printing, they may be made (a) with greater ink receptivity than supercalendered coated papers, (b) suitable for gloss inks, or (c) with highly impervious surface coatings.

CAST COATING—A term applied to the process in which the coated paper is pressed against a solid surface while the coating is in a highly plastic condition. For most coatings a steam-heated drum is used. When dried, the finish is similar to the contacted surface.

CASTOR OIL TEST—A measure of the receptivity to oil-based inks of easily permeable printing papers; it determines the time required for a drop of castor oil to be absorbed and produce a translucent spot.

CATALOG PAPER—A lightweight coated or uncoated printing paper designed for use in the printing of various types of catalogs, directories, etc. It is usually made from chemical wood, mechanical, or reclaimed pulps in basis weights of 19 to 45 pounds (25 x 38 inches – 500) and is characterized by good printability and opacity.

CATHODIC PROTECTION—A technique to reduce the corrosion rate of a metal surface by making it the cathode in an electrochemical cell.

CATION—A positively charged ion.

CATION EXCHANGE—The reversible exchange of positive ions between functional groups of the ion exchange medium and the solution in which the solid is immersed. Used as a wastewater treatment process for removal of cations, e.g., calcium.

CATIONIC DEMAND—Cationic demand (CD) is a quantitative measure of the available anionic charge of furnish components (e.g., fiber carboxyl groups, recycled starches, polyphosphates, anionic sizing materials, residual lignins, and miscellaneous anionic contaminants) that can be neutralized by reaction with a cationic polymer. Cationic demand is defined as the amount of cationic polymer required to reach zero charge per given amount of stock. It can be expressed in several different ways, such as percent cationic polymer on fiber weight, milliequivalent charge per liter of stock, or pounds cationic polymer per ton of ovendry stock. There is no standard procedure for determining the cationic demand of a furnish. Variations in the procedure occur in the cationic polymer titrants, pH conditions, polymer: fiber contact times, and endpoint detection method. These all can influence the amount of cationic polymer necessary to adjust the furnish to zero charge. In any form, it is a useful parameter for assessing the cationic requirements of a furnish, and the measurement is widely utilized in troubleshooting paper machine problems and optimizing paper machine performance. See WET END CHEMISTRY, ELECTROKINETIC CHARGE MEASUREMENT, and ANIONIC CONTAMINANTS.

CATIONIC POLYMERS—Polymer molecules that possess numerous cationic sites. Examples are cationic starch and cationic polyacrylamide. See PROMOTERS.

CAUSTIC—Sodium hydroxide (NaOH) often referred to as caustic soda (q.v.).

CAUSTIC CRACKING—Anodic environmental (i.e., stress corrosion) cracking of steel due to exposure to an alkaline solution. Also called caustic embrittlement.

CAUSTIC EMBRITTLEMENT—See CAUSTIC CRACKING.

CAUSTICITY—Ratio of the sodium hydroxide to the sum of sodium hydroxide and sodium carbonate (active alkali) in white liquor. Often used as the ratio of sodium hydroxide to total titratable alkali (q.v.) in white liquor (q.v.).

CAUSTICIZER—An agitated tank where the causticizing reaction of lime with green liquor (q.v.) is carried out. A causticizing line often consists of three or more causticizers.

CAUSTICIZING EFFICIENCY—Ratio of the sodium carbonate converted to caustic to the initial sodium carbonate in green liquor (q.v.). Very often used interchangeably with causticity (q.v.).

CAUSTIC SODA—A general term for sodium hydroxide (NaOH). It is a crystalline substance used in alkaline pulping, in the alkaline (caustic) extraction stage of bleaching, and to create alkaline conditions in mill processes. It dissolves readily in water and is highly corrosive.

CAVITATION—Degradation of a material due to the formation and collapse within a liquid of cavities or bubbles on the metal surface (e.g., of pump impellers).

C BLEACHING STAGE—See CHLORINATION STAGE (C).

CCD—Initials of charge coupled device, an image capture device which digitizes the input for further digital processing.

C CHART—See COUNT CHART.

CD—An abbreviation meaning cross direction or cross machine direction of the machine, web or nonwoven.

CD ACTUATORS—A series of mechanical devices staged across the front of a headbox to individually respond to control signals from the CD control system to adjust the headbox slice (q.v.) opening to compensate for cross machine (CD) basis weight variations.

CD CONTROLS—Cross machine direction measurement and control of paper machine variables such as basis weight and moisture to provide better product uniformity and quality. Occurs perpendicular to machine direction (MD) controls.

CD CURL—See CURL.

CEDARIZED PAPER—Paper treated with cedar oil and used for wrapping where protection from vermin or moths is desired.

CEILING BOARDS—See ACOUSTICAL BOARD; WALLBOARD.

CELERY-BLEACHING PAPER—A heavyweight kraft wrapping paper or a laminated asphalt paper of basis weight from 80 to 100 pounds (24 x 36 inches – 500) which is used for covering celery during growth to bleach the stalk. The principal requirements for this paper are stiffness, high water resistance, and good wet-tensile strength.

CELERY WRAPPER—A vegetable parchment paper used for wrapping celery stalks for merchandising.

CELL—Term applied to the simple structural units of plant materials which includes wood fibers, vessel elements, and other diverse elements having various forms and functions.

CELL LIQUOR—A term used to denote "chlorate cell liquor," a solution containing sodium chlorate and sodium chloride. In addition to chlorine and sodium hydroxide, it is one of the principal products of electrolysis of sodium chloride in some types of cells. The cell liquor is used in some of the processes that manufacture chlorine dioxide in pulp mills.

CELLULASE—An enzyme that catalyzes the hydrolysis of cellulose to glucose and glucooligomers. "Cellulase" actually is a complex of enzymes consisting of endo-Beta-1, 4-glucanase, cellobiohydrolase, and Beta-glucosidase which, in concert, hydrolyze cellulose to glucose.

CELLULOSE—The main structural constituent of woody plants, occurring widely elsewhere in the vegetable kingdom. Chemically it is a linear polysaccharide of beta (1→4) linked D-glucose units. The molecular weight varies with the source, but is frequently over 1,000,000 g/mole. Wood cellulose is the material remaining after a large portion of the lignin and certain carbohydrates have been removed from wood cell walls by pulping and bleaching.

CELLULOSE I—The crystalline form of cellulose that is the native cellulose produced throughout the plant kingdom.

CELLULOSE II—The crystalline form of cellulose that is predominantly produced after chemical regeneration, for example, after mercerization or xanthation.

CELLULOSE III—The crystalline form of cellulose that is obtained after treatment of cellulose with anhydrous liquid ammonia.

CELLULOSE IV—The crystalline form of cellulose that is obtained by heating cellulose II or III, or by regenerating cellulose at high temperature.

CELLULOSE FILMS—Nonfibrous cellulosic sheets, usually transparent, manufactured from cotton or wood pulp by formation of soluble derivatives, followed by regeneration from solutions of the derivatives which are passed through a slit into a coagulating bath. The viscose process is the most common example: cellulose is treated with sodium hydroxide and carbon disulfide to form a solution of sodium cellulose xanthate. Cellulose film is regenerated from the xanthate in an acid bath. Cellulose films may be plasticized, coated, or dyed and are used extensively in wrapping. Films may also be made by evaporation of solvent from solutions of cellulose esters and ethers. These films are not cellulose but the unchanged derivatives, for example, cellulose acetate or ethyl cellulose.

CELLULOSE LACQUER—See LACQUER.

CELLULOSE NAPKINS—Multi-ply napkins made from facial tissue stock. See PAPER NAPKINS.

CELLULOSE REACTIVE SIZES—Sizing agents of the alkyl ketene dimer (AKD) or the alkenyl succinic acid anhydride (ASA) type develop ester bonds with cellulose fibers or the starch that is adsorbed on the fiber surface. Although such reaction sites may be few, these materials do develop moderate to hard sizing and are especially useful for developing sizing in $CaCO_3$ filled (alkaline, neutral) paper grades. See NEUTRAL SIZING; PERMANENT PAPERS.

CELLULOSE WADDING—(1) Water-formed creped paper wadding. A material consisting of fibers of chemical woodpulp loosely matted into a sheet formed on a Yankee cylinder or fourdrinier machine and creped off a Yankee drier. It is used in bleached grades as an absorbent material in hospitals and for sanitary purposes; in unbleached and treated grades for packaging, thermal, acoustical, and other applications. It is available in single- or multiple-ply sheet form either plain or embossed, and is frequently backed with various papers which provide certain required characteristics not available in the cellulose wadding grades alone. The crepe ratio, when expressed as 100 times the difference between drier and reel speed divided by reel speed, varies from approximately 70 to approximately 200 which is equivalent to a crepe ratio range of 1.7 to 3.2 when expressed as drier speed divided by reel speed. Basis weights after creping vary from approximately 7 pounds to over 20 pounds (24 x 36 inches – 500). Many desirable properties are available depending on the end use and these are: softness, cleanliness, high rate of absorption, high liquid retention, bulk, fluffiness, water resistance, flame resistance, good acoustical sound-deadening properties, non-abrasiveness, and low lint count or nondusting properties where needed. (2) Dry-formed wadding. A material made from shredded woodpulps or paper or from various textile fibers or combinations of these. The web may be air formed (see NONWOVEN FABRIC) or produced with carding or garnetting equipment such as found in the textile industry. It is used for cushioning, packaging, acoustical, and thermal insulation, for sanitary purposes and other applications. It may be treated with an adhesive or binder either homogeneously or on the surface. Thickness may range from 1/8 inch to several inches and density may vary from 1

pound to several pounds per cubic foot. As is true of water-formed creped paper wadding, many properties may be imparted to dry-formed wadding depending on the end use.

CELL WALL—A more or less rigid membrane enclosing the protoplast of a cell. In higher plants composed of polysaccharides, chiefly cellulose, and other organic and inorganic substances. Term has triple usage: (1) cell wall of an individual cell, (2) partition between two cells composed of intercellular substance and two walls belonging to the two adjacent cells, (3) primary or secondary wall layer.

CEMENT-SACK PAPER—A strong, flexible kraft paper, either flat or extensible, used in the manufacture of multiwall sacks, for packaging cement, lime and similar products. This paper is normally manufactured against specifications which define minimum physical properties that permit the finished multiwall sack to perform its heavy-duty packaging function. Special sheets may be used as an innerlining, which serves as a water- or moisture-proofing medium. The lining may consist of waxed, asphalt-laminated, or plastic-coated paper, or plastic film. See SHIPPING SACK KRAFT PAPER.

CENTER STOCK—See FILLER (3).

CENTER-SURFACE WINDER—A winding process in which the prime motivating forces to wind a roll consist of the sum of the efforts of one or more surface winding drums and a centerwind at the core.

CENTER WIND—A winding system whereby the rotational torque is provided at the core or axis of a winding roll. See SURFACE WIND; WINDER.

CENTERWIND ASSIST—A surface winder such as a supercalender wind-up, reel, or re-reeler that has a centerwind drive added to enhance the control of wound-in-tension (q.v.) in the winding process.

CENTRIFUGAL CLEANER—A conical or cylindrical-conical pressure vessel that develops an elevated internal centrifugal force field when liquid, gas, or slurry is fed tangentially into it. This force field is the basis for a specific gravity and hydraulic drag-based separation, which is commonly used to remove cubical or spherical debris whose specific gravities differ sufficiently from that of the carrier fluid. The term centrifugal cleaner is generic, in the sense that it applies to forward cleaners, reverse cleaners, through flow cleaners, and multifunctional centrifugal cleaners. Prior to about 1970, only forward cleaners were used in the paper industry, so historically the term centrifugal cleaner was applied only to what is now known as a forward cleaner, for stock processing applications. As variations on this design and application were developed, and eventually moved into everyday mill use, the term centrifugal cleaner came to mean all types of centrifugal cleaners. If the carrier fluid is water or stock, it is also called a cyclone, hydraulic cyclone, or a hydrocyclone; if the carrier fluid is a gas, it is also called a gas cyclone.

CENTRIFUGAL CLEANER SYSTEM—A system which uses multiple stages of centrifugal cleaners to remove debris with a minimal fiber loss. The term can apply to forward cleaners, reverse cleaners, through-flow cleaners, or combination cleaners.

CENTROID WAVELENGTH—Wavelength of center of gravity of area under the curve of spectral function.

CERAMIC TRANSFER PAPER—See DECALCMANIA PAPER.

CEREAL-BOX LINERS—See CARTON-LINER PAPER.

CEREAL-BOX WRAPPERS—See CARTON-SEALING PAPER.

CEREAL STRAW—Stalks of various grasses such as barley, oat, rice, rye, or wheat used as pulping raw materials. See NONWOOD FIBERS.

C-FLUTE—See FLUTE.

CHADLESS PAPER TAPE—A type of perforator tape (q.v.) wherein the punching is incomplete, i.e, the tiny paper circles (chads) still adhere to the body of the tape. This type of punching is generally used in applications where information pre-printed on the tape needs to be read after punching.

CHADS—Tiny circular pieces of paper which result from punching operations, as for example in "punched" tape. See also CHADLESS PAPER TAPE.

CHAIN GRINDER—A machine for producing mechanical pulp, ground wood. It consists of a rotating pulpstone against which debarked logs are pressed and reduced to pulp. The debarked logs are fed from the top and are pressed against the rotating stone by a continuous chain. The speed of the chain determines the pressure between the wood and the stone. See also GRINDING.

CHAIN LINES—(1) The more widely spaced watermark lines which run with the grain in laid paper, caused by the "chain wires" (also called twists or tying wires), which are twisted around the laid wires to tie them together. They are usually about one inch apart. (2) Markings resembling impressions of rope or a chain which are formed in paper as the result of unequal stresses. They appear in the calendered sheet and are generally due to irregular formation, i.e., heavy and light streaks in the machine direction.

CHAIN MARKS—See CHAIN LINES (1).

CHALK—See CALCIUM CARBONATE.

CHALKING—(1) A printing defect, in which ink pigment may be easily rubbed from the surface of paper. (2) A condition encountered in some papers where fine particles of pigment leave the sheet during the finishing, converting, printing operation, or subsequent use.

CHALK OVERLAY PAPER—A lightweight paper with a heavy coating of chalk. It is not exactly a printing paper, although it is used to assist in the printing of halftones. The pressman first takes an impression of the halftone on overlay paper and then, with the aid of a bath of calcium hypochlorite, etches away all the chalk not touched by ink. The etched overlay paper is then used in the make-ready to produce the desired effect. The total thicknesses of paper and coating are usually between 0.0085 and 0.014 of an inch. Some is coated only on one side and some on two sides, according to the preference of the printer. The coating is special in its formulation and requires special supervision during manufacture. The finished paper is uniform in thickness and smoothness and sufficiently strong to withstand the etching operation.

CHALK PAPER—A paper used to wrap chalk or crayons to protect the hands from soiling. In most cases, a colored paper is used which is light enough to conform to the chalk; there are no definite specifications.

CHALK TRANSFER PAPER—See CHALK OVERLAY PAPER.

CHALKY APPEARANCE—Not glossy, either as a result of the nature of the coating materials or of the coating process.

CHANCE CAUSES—See COMMON CAUSES.

CHARCOAL BOOK PAPER—A term applied to charcoal drawing paper (q.v.) when used as a book paper.

CHARCOAL DRAWING PAPER—A drawing paper especially suited for use with charcoal sticks or pencils. It is usually a cotton fiber content sheet in basis weight of 60 to 70 pounds (25 x 38 inches – 500). A surface suitable for "taking" charcoal and good erasability are necessary characteristics.

CHARCOAL KRAFT PAPER—A strong kraft paper, usually in the heavier weights, used in the manufacture of bags or sacks for holding coal, charcoal, or briquettes. It is also called fuel-sack paper.

CHARCOAL PAPER—See CHARCOAL DRAWING PAPER.

CHARGE EQUIVALENCY—A situation where an equal number of anionic and cationic groups react to produce a neutral system. See ELECTROKINETIC CHARGE TITRATION; STOICHIOMETRY.

CHARGE MEASUREMENT—See ELECTROKINETIC CHARGE MEASUREMENT.

CHARGE TITRATION—See ELECTROKINETIC CHARGE TITRATION.

CHART PAPER—(1) See MAP PAPER. (2) A paper having bond or ledger paper characteristics. It is made from cotton fiber and/or chemical woodpulps and is usually tub sized, the better grades with starch or glue. The basis weights range from 9 to 28 pounds (17 x 22 inches – 500). Because most of it is printed, good printing qualities with low expansion and contraction are important. Sizing, a smooth surface, good erasability, and, in some cases, transparency and formation are other significant properties. Examples are profile, cross-section, logarithmic, plane profile, coordinate, and isometric. (3) A paper sometimes called recording instrument paper and used for that purpose in circular flat form or as rolls for strip chart instruments. See STRIP CHART PAPER; METER PAPER.

CHECK BOARD—See CHECK BOOK COVER.

CHECK BOOK COVER—A dense rigid board lined with a plain white surface on one side and either a plain or usually embossed colored paper on the other. Usually referred to as an embossed 14-ply news-filled board, 50 points thick.

CHECK PAPER—A bond or ledger paper, made of cotton or chemical woodpulps, or mixtures thereof, and used for checks. It is usually made in basis weights of 20 to 24 pounds (17 x 22 inches – 500). This paper may or may not be treated, either in the beater or on the surface, with chemicals or dyes or both to make chemical or mechanical alteration difficult or impossible. Strength, writing qualities, and, in some cases, sensitivity to chemical and me-

59

chanical erasure are significant properties. See also COUNTER CHECK PAPER; SAFETY PAPER.

CHEESE MANILA—A paper used as an outer wrapper for brick cheese, the cheese first being wrapped in foil or other material. It is manufactured from chemical and mechanical woodpulps in a basis weight of about 90 pounds (24 x 36 inches – 500), is medium sized and water finished to give a well-closed sheet with a smooth, hard finish. The color closely resembles that of cheese.

CHEESE WRAPPERS—Vegetable parchment, glassine, greaseproof, foil-laminated and waxed papers used for wrapping cheese, the types of paper depending upon the nature of the product.

CHELATING AGENT—A chemical compound, e.g., EDTA (ethylenediaminetetraacetic acid), DTPA (diethylenetriaminepentaacetic acid), commonly used to deactivate or remove transition metal ions prior to oxidative or reductive bleaching. The transition metal ions of concern in bleaching are manganese, iron, cobalt and copper, which have significant catalytic activity that reduces bleaching efficiency.

CHELATION STAGE (Q)—A stage that usually combines an acid wash with a chelating agent for removal or deactivation of transition metal ions from a pulp prior to bleaching.

CHEMICAL BAG PAPER—See SHIPPING SACK KRAFT PAPER.

CHEMICAL COAGULATION—The destabilization and initial aggregation of colloidal and finely divided suspended matter by the addition of a floc-forming chemical.

CHEMICAL CONSUMPTION—In bleaching, amount of bleaching chemical consumed by a given weight of pulp during a bleaching stage.

CHEMICAL DEBARKING—Removal of bark from woody stems after the tree has been treated with chemicals that loosen the bark, usually by

destroying the cambium (growing) layer. See PEELING; DEBARKING.

CHEMICAL FINISHING—A method of finishing in which chemicals are incorporated.

CHEMICALLY BONDING—A method of bonding using either adhesive-like materials known as binders or solvents.

CHEMICAL MANILA WRITING—A writing paper made from chemical woodpulp and having a "manila" color.

CHEMICAL OXYGEN DEMAND (COD)—A measure of the oxygen-consuming capacity of inorganic and organic matter present in water or wastewater. It is expressed as the amount of oxygen consumed from a chemical oxidant in a specific test. It does not differentiate between stable and unstable organic matter and thus does not necessarily correlate with biochemical oxygen demand. Also known as OC and DOC, oxygen consumed and dichromate oxygen consumed, respectively.

CHEMICAL PAPER—Paper made entirely from chemically prepared woodpulp, usually bleached.

CHEMICAL PRECIPITATION—(1) Precipitation induced by addition of chemicals. (2) The process of softening water by the addition of lime or lime and soda ash as the precipitants.

CHEMICAL PULP—See BLEACHABLE GRADE PULPS; CHEMICAL WOODPULP; PULPING.

CHEMICAL WOODPULP—Pulp obtained by using both chemical and heat energy to dissolve the lignin bond between fibers in wood. The primary chemical processes are the sulfate (kraft), sulfite, and soda.

CHEMIGROUNDWOOD—A pulping process in which debarked logs segments are treated with sodium sulfite and sodium carbonate at high temperatures prior to grinding them against an abrasive drum. Chemigroundwood pulps are obtained in higher yield than chemical pulps, but are generally lower in strength and brightness.

CHEMIMECHANICAL PULP (CMP)—An ultra-high-yield mechanical pulping process, where the wood chips are impregnated with sodium sulfite, heated to elevated temperatures in a vapor phase digester for a sufficiently long time to accomplish sulfonation of the lignin. Such pulp has improved physical strength, but is low in opacity. It is used as a reinforcement pulp in the furnish.

CHEMITHERMOMECHANICAL PULP (CTMP)—Pulp produced from chemically impregnated wood chips, by means of pressurized refining at high consistencies.

CHESTNUT BOARD—A board formerly made from chestnut pulp but more recently from pulp of other hardwood species. Today the term is applied to a range of brown boards made essentially from southern hardwood semichemical pulp. Ninepoint fourdrinier grades for corrugating medium are sometimes referred to as chestnut medium. Cylinder grades of solid fiber with dense formation, high stiffness, and rigidity are made in thicknesses of 18 points and up (with lamination of 250 points) for various industrial uses.

CHEVIOT—(1) A term used to describe the appearance of paper made with a small percentage of deeply colored fibers added to the basic lighter colored furnish to give a granite effect. (2) A term sometimes used to designate lightweight colored antique paper used for the facing of boards. See GRANITE PAPER.

CHEVIOT BOX COVERING—A box-cover paper mainly used by the paper box trade to cover cardboard or chipboard setup boxes. It is made of mechanical and chemical pulps and contains a small percentage of colored fibers. It is used chiefly in basis weights of 35 to 40 pounds (24 x 36 inches – 500). Gray is the predominant color used, but other colors including light green, light blue, brown, tan, pink, and yellow are also manufactured.

CHEVRON–GROOVING—A type of pattern machined onto the surface of winder drums or web supporting rolls to provide traction or spreading.

CHICKEN TRACKING—See FILM SPLIT PATTERN.

CHILI WRAPPER—A greaseproof or parchment paper used to wrap chili-like food products, tamales, etc. Greaseproofness and wet tensile strength are important properties.

CHILLED-IRON ROLL—A type of cast iron roll with large amounts of carbon (1.7% to 4%) and possibly other alloying constituents. It derives its name from the casting process, i.e., when the melted cast-iron is poured into a mold the exterior surfaces are cooled faster (or chilled) so the separation of graphite is retarded. This skin of dissolved carbon and iron forms a very hard carbide, Fe_3C (cementite), or white iron. The core remains as a more ductile gray iron with precipitated graphite.

CHINA CLAY—A term originally applied to the beneficiated kaolin mined in Europe but now applied to all beneficiated kaolin. See CLAY; KAOLIN.

CHINA GRASS—As a raw material for papermaking, the term China grass will be found in quite old references. See RAMIE.

CHINA PAPER—A soft, waterleaf (unsized) paper made in China from bamboo fiber. It has a pale yellow color and a very fine texture. The usual size is 56 by 27 inches. It is used by (engraved) plate printers to pull proofs. It is also called Chinese paper or India paper. As a grade of paper, the term China paper would not be recognized by most manufacturers or buyers today.

CHIP—A paperboard made from paper stock—usually mixed papers. See also CHIPS.

CHIPBOARD—A paperboard used for many purposes that may or may not have specifications of strength, color, or other characteristics. It is normally made from paper stock with a relatively low density in thicknesses of 0.006 of an inch and up. Lightweight grades are made on both the fourdrinier and cylinder machines, the heavier weights on cylinder machines only. It may be a filled sheet or a solid sheet. (a) Combination chipboard has paper stock as a base or center and is vat lined on one or both sides with a different grade or stock—usually of a higher grade and possessing a smoother and better appearing surface. News grade of mechanical pulp, blank news, etc., are used for the vat liner. See also FILLED BOARD. (b) Solid chipboard is unlined and is made of paper stock throughout. See SOLID BOARD.

CHIP CLASSIFIER—A testing device to measure the chip size distribution of a sample of chips. Several trays fitted with trays with round holes or slots are shaken for about 10 minutes and the weight percentage of chips retained on each tray is determined. A common series of trays used in many countries are: 45 mm round hole, retains overlength chips; 7 to 10 mm slot, retains overthick chips; 5 to 7 mm round hole, retains accept chips; 3 mm round hole, retains pin chips; and pan; retains fines.

CHIP DECAY—A gradual decomposition of wood chips caused by a variety of fungi that thrive at the expense of the fiber portion of wood.

CHIPPER—A machine consisting essentially of a revolving disc equipped with sharp knives, set approximately in a radial direction, which cuts pulp wood and sawmill waste into chips, diagonal to the grain. "Whole tree chippers" process an entire tree, including bark, branches, and leaves.

CHIPPING—In intaglio printing, a void or large line break in a solid area caused by lack of ink transfer, or ink loss after printing.

CHIPS—Wood particles produced from debarked stems, logs, and wood products residuals by multi-knife chippers. Chippers are designed to produce a particle size distribution that matches the pulp mill needs. Most mills target a 1/2 to 3/4 inches (13 to 20 mm) chip length and a 6-

to 8-mm maximum chip thickness. The distribution of chips sizes depends on the stability of the wood infeed, chipper maintenance, and the type of wood. When a high amount of overlength, overthick, pin chips, and fines are produced, the distribution must be narrowed with subsequent chip screening to prevent pulp mill operating and product quality problems. Bark content of chips should be less than 1% for bleachable pulp products.

CHIP SIZE—A chipper produces a distribution of chip sizes. The majority of the chips are near the target size and are called accept chips. Chips that are not sized correctly are either too small (fines) or too large (oversize). Oversize can be either too long (overlength chips) or too thick (overthick chips). Overthick chips are generally greater than 7 to 10 mm thick. Small chips are narrow and thin, but still about the right length (pin chips) or too short (fines).

CHIP STORAGE—The storage of chips in piles, bins constructed of wood or concrete—or concrete lined with wood—or steel. Pile storage systems have a capacity of several days or weeks of inventory. Bins and silos hold only a few hours of inventory and are used to meter chips just ahead of the digesters.

CHLORATE—See SODIUM CHLORATE.

CHLORATE CELL LIQUOR—See CELL LIQUOR.

CHLORIDE PHOTOGRAPHIC PAPER—A photographic paper base coated with an emulsion in which the photosensitive material is silver chloride. Since the sensitivity of these papers is quite low, they are used only for contact printing. Hence, they are sometimes referred to as contact papers. See also PHOTOGRAPHIC PAPER.

CHLORINATION—(1) The process by which chlorine is introduced into an organic compound in replacement of a hydrogen atom. The process of chlorination may require the presence of a catalyst or ultraviolet light. Chlorination is frequently carried out in an inert solvent. (2) In pulping, the treatment of wet pulp, with a compound containing available chlorine, as a step in removing unwanted non-cellulosic matter and bleaching the pulp. See PULP BLEACHING.

CHLORINATION (MIXTURE)—A chlorination stage where chlorine and chlorine dioxide are added to the pulp slurry prior to the bleaching tower. The amount of chlorine dioxide in the mixture may range from a very small amount up to 100% of the total available chlorine charge.

The advantages of chlorine dioxide additions in the C bleaching stage (q.v.) are reported to be viscosity maintenance, lower caustic requirement in the extraction stage, less brightness reversion, increased temperature tolerance of the C stage, which permits shorter retention times and effluent recycling from the high-temperature D stages (q.v.), and reduced formation of organic chlorine compounds.

CHLORINATION (SEQUENTIAL)—A chlorination stage whereby chlorine dioxide (or chlorine) is mixed into a heated unbleached pulp stock, followed by chlorine (or chlorine dioxide) addition without washing between these treatments.

CHLORINATION STAGE (C)—The first stage of some bleaching sequences in which the chlorine and lignin react to form orange-colored chlorolignins that are partly water-soluble in the washing step and readily alkali soluble in the next extraction stage. The chlorination stage may be considered an extension of pulping in that is continues the delignification process which makes it possible to brighten the pulp by oxidation in the third and subsequent bleaching stages. Chlorination is most commonly done in upflow towers at about 3% consistency, and at temperatures from 20°C to 60°C for a 10- to 60-minute reaction time. Medium consistency (10 to 12% consistency) chlorination stages are also used. Chlorine bleaching is being phased out because of environmental concerns about the potential aquatic toxicity and bioaccumulation of chlorinated organic compounds in pulp and pulp mill effluents.

CHLORINE—Cl_2. A major bleaching chemical used in the chlorination stage and in the manufacture of hypochlorite for use in the hypochlorite bleaching stage. Chlorine usually arrives at the pulp mill as a dry liquid and is converted to a gas through vaporization by heat. Chlorine gas is usually dispersed in water or chlorine dioxide solution immediately before being injected into the pulp slurry. Elemental chlorine is also produced to varying degrees in chlorine dioxide generators. It can be isolated for use in the chlorination or hypochlorite bleaching stages.

CHLORINE DEMAND—(1) The difference between the amount of chlorine added to water or wastewater and the amount of residual chlorine remaining at the end of a specified contact period. The demand for any given water varies with the amount of chlorine applied, time of contact, and temperature. (2) Bleach demand (q.v.) in terms of elemental chlorine.

CHLORINE DIOXIDE—ClO_2. A chemical that selectively oxidizes lignin and wood resin without adverse reactions with carbohydrate components under optimized conditions, thus making it possible to bleach pulp to 90+ brightness with minimum loss in pulp strength. In bleach plants, chlorine dioxide is produced by reduction of sodium chlorate to chlorine dioxide by either hydrochloric acid, salt, methanol, or sulfurous acid in a strongly acid medium. It is used as an oxidizing agent in one or several stages of a multistage bleaching sequence. See CHLORINE DIOXIDE MANUFACTURE.

CHLORINE DIOXIDE BLEACHING STAGE (D)—An oxidative bleaching stage in a multistage bleaching sequence. To produce high and stable brightness, it is usually done twice with an alkaline extraction in between. It can also be used in the first stage of a bleaching sequence. A D (ClO_2) stage is conducted under controlled conditions optimized to promote its reactivity with lignin and other colored constituents of the pulp with minimum carbohydrate degradation, thus preserving pulp strength. Optimum D conditions are as follows:

D0—1.0 to 2.3% ClO_2 on pulp (depending on pulp type), at 40°C to 60°C, 3 to 10% consistency, 30 to 60 minutes retention, pH 2.5 to 3.

D1—0.5 to 1.0% ClO_2 on pulp (depending on pulp type), at 70°C, 10 to 12% consistency, 3 hour treatment time, end pH 3.5 to 4.0, preferably 3.8.

D2—0.2 to 0.4% ClO_2 on pulp, at 70°C, 10 to 12% consistency, 3-hour treatment time, end pH 4 to 6.

CHLORINE DIOXIDE MANUFACTURE— Chlorine dioxide is an explosive gas that cannot be prepared safely in undiluted form nor can it be liquefied for transport and storage; therefore it is manufactured on site. The processes that have been developed generate chlorine dioxide in dilute aqueous solution by reducing sodium chlorate in a strong acid solution with a reducing agent, usually sulfur dioxide, methanol, sodium/hydrogen chloride, and hydrogen peroxide.

CHLORINE NUMBER—A measure which indicates how much chlorine is absorbed in 15 minutes at 20°C by 100 g of pulp containing 55 g of water in a specially designed apparatus. It is linearly related to the lignin content in pulps below 75% yield, since most of the chlorine is consumed by the lignin. For kraft pulp: % lignin content = 0.8 x chlorine number. It reflects the degree of delignification of a pulping process, and provides a good estimate of the bleach requirement in the first stage.

CHLORINE PROCESS—A soda-chlorine process for pulping straw which incorporates four stages: (a) alkaline pretreatment; (b) gas chlorination; (c) alkaline wash; and (d) hypochlorite bleaching.

CHLORINE REQUIREMENT—Bleach demand (q.v.) in terms of elemental chlorine.

CHLORITE—One of the oxygen-containing chlorine compounds formed when chlorine dioxide reacts with pulp in proportions that are highly dependent on the pH of the solution and on the lignin concentration of the pulp. The chlorite ion (ClO_2) is unreactive toward lignin, but in acid media the chlorous acid formed readily oxidizes lignin. See SODIUM CHLORITE.

CHLORO-BROMIDE PAPER—A photographic paper base coated with an emulsion in which the photosensitive material is primarily silver chloride with a small proportion of silver bromide. It is faster than chloride paper by a factor of ten and, although most often used for contact printing, is sometimes used for enlarging. See PHOTOGRAPHIC PAPER.

CHLOROLIGNIN—A highly colored, alkali-soluble and partially water-soluble chlorinated phenolic material formed from the reaction of chlorine and the ligneous component of chemical pulp.

CHOCOLATE BOARD—A paperboard used to make boxes or cartons to hold chocolate. There are two types of this board—a setup boxboard colored brown on one side about 0.044 of an inch in thickness, and a folding boxboard colored brown, of various calipers. The boards conform to the usual requirements of setup and folding boxboard, the chief characteristics being the color in imitation of chocolate.

CHOCOLATE-DIPPING PAPER—A highly glazed or vegetable parchment paper on which chocolate candies are placed after they have been dipped. The basis weight generally ranges from 75 to 85 pounds (24 x 36 inches – 500). See CANDY-SLAB PAPER.

CHOCOLATE DIVIDERS AND LAYER BOARD—A paperboard used to separate individual pieces or groups of chocolate candies or one layer from another. It is made of manila-lined board or of chipboard colored to imitate chocolate, in various thicknesses, 0.008 and 0.015 of an inch being generally used. It has a smooth surface free from fuzz and is nonbending.

CHOCOLATE WRAPPING PAPER—A vegetable parchment or greaseproof paper used for wrapping cakes of chocolate.

CHOP—A bundle of fibers that have never been separated from each other, and later refined or otherwise cut into a roughly cubical shape. Chop can be thought of as small lengths of shives or splinters.

CHROMATICITY—That part of a color specification that is given by dominant wavelength and purity or, alternatively, the tri-chromatic coefficients. See CHROMATICITY COORDINATES; COLOR.

CHROMATICITY COORDINATES—Ratio of each tristimulus value of a color to the sum of tristimulus values. Chromaticity coordinates in the CIE system of color specification are designated x, y, and z.

CHROMATIC PAPER—A mottled paper having several colors on the surface or different colored fibers in the same sheet. It is used chiefly for box papers.

CHROMATOGRAPHIC PAPER—A filter paper (q.v.) which has the properties of uniformity, texture, and porosity required for use in paper chromatography.

CHROMATOGRAPHY—Chemical analysis of mixtures of solutions through selective absorption of the substances in solution by suitable absorbing materials.

CHROMO BOARD—See CHROMO PAPER.

CHROMOLITHO PAPER—See CHROMO PAPER.

CHROMO PAPER—A term applied to any paper or board which is particularly suited to accept colored printing. It is usually defined as a coated paper of high quality with surface characteristics such that it will take many colors. Surface characteristics believed to enhance colored printing include smoothness, uniformity of ink receptivity, high total reflectance, and neutrality of shade (true "white" rather than tinted). This term is not normally used in the United States.

CHROMOPHORES—Colored compounds containing a basic chromogenic structure, usually a conjugated system of double bonds that absorb radiation in the ultraviolet and visible region. The light-absorbing material in pulp is mainly in the lignin or lignin products resulting from reactions occurring during pulping. Brightening of mechanical pulps is accomplished by decolorizing the lignin without removing it; i.e., the chromophoric groups are altered to eliminate their absorption in the visible range of the wavelength, thus increasing the whiteness of the pulp.

CHRONIC TOXICITY—Toxicity resulting from exposure to a toxin over an extended period of time. Sometimes used to mean toxicity which affects the health and well-being of an organism but does not result in its death. A common test now often carried out on wastewater samples. Contrast with ACUTE TOXICITY; see also TOXICITY.

CHUCK—The roll centering and gripping device on a paper roll stand.

CIE—International Commission on Illumination. In French, Commission Internationale de l'Eclairage.

CIE COLOR SYSTEM—The application of the CIE system to describe any particular color producing three-color coordinates called tristimulus values. The tristimulus values are related to the way a human observer sees a color, and thus provide a means for representing instrumentally the human response to a color signal.

CIE COORDINATES—See CHROMATICITY COORDINATES.

CIE L*, a*, b* (CIELAB)—See OPPONENT COLOR SCALES.

CIE OBSERVER—See STANDARD OBSERVER; SUPPLEMENTARY OBSERVER.

CIGAR-BAND PAPER—A coated one-side paper of good strength and finish to take gold and colored printing, embossing, and die-cutting.

The basis weight is from 45 to 60 pounds (25 x 38 inches – 500). It is used in the manufacture of cigar bands.

CIGAR-BOX BOARD—A paperboard from which a box is made to hold cigars. It is a chipboard, usually pasted to form a sheet about 140 points in thickness and has a surface to which a cover paper can be readily adhered. Its chief characteristic is stiffness.

CIGARETTE MOUTHPIECE PAPER—A soft, flexible paper, made of flax and/or bleached chemical woodpulps, of a basis weight of 80 to 90 pounds (25 x 38 inches – 500), used for the tips of cigarettes.

CIGARETTE PAPER—A strong tissue paper of close, uniform texture, free from pinholes, used as a wrapper for tobacco in the manufacture of cigarettes. It is generally made from flax pulp, and contains no size. This paper may be either combustible or noncombustible. Combustible paper (q.v.), also called free burning paper, contains from about 15% to about 30% of calcium carbonate filler. All cigarette paper is made from highly beaten stock which contributes to its high strength and uniform formation but which, at the same time, necessitates the addition of large percentages of filler to provide the desired porosity. The rate of burning of the cigarette is controlled largely by the porosity of the paper, which depends, among other factors, upon the percentage of filler in the sheet, and by the size and shape of the filler particles. The filler also contributes importantly to the whiteness and opacity of the paper. Noncombustible paper (q.v.) has little or no filler added. The normal weight of cigarette paper for consumption in the American industry is about 20 to 22 grams per square meter, and the calcium carbonate content is about 25%. Cigarette paper for roll-your-own cigarettes may weigh as little as 15 to 18 grams per square meter. Significant properties include strength, stretch, filler content, texture, color, opacity, and porosity.

CIGAR LABELS—See CIGAR-BAND PAPER.

CINDER SPECKS—See CARBON SPOTS.

CIRCULAR CHART PAPER—See METER PAPER.

CIRCULAR CUTTER—Obsolete term. See CUT-SIZE CUTTER; CUTTER; DOUBLE FLYKNIFE CUTTER; DUPLEX CUTTER; FOLIO CUTTER; PRECISION CUTTER.

CIRCULATING LOAD—The energy consumption rate with stock circulating through a refiner, with the plates backed off, such that the stock is being circulated, but not refined; typical units are horsepower-hour or kilowatt-hour. Also called the no-load power or the idling load.

CLAMP MARKS—Marks in paper produced by the clamps which hold it in position for guillotine trimming.

CLARIFICATION—Removal of suspended solids from water, usually by flotation or sedimentation, although the term has also been applied to the use of strainers for the same purpose. See also DISSOLVED AIR FLOTATION CLARIFIER.

CLARIFIED WATER—Water which has been treated in a clarifier (q.v.) to remove most of the suspended solids. Clarified water is generally not potable, but is among the cleanest process water available, after fresh water, and non-contact cooling water (q.v.).

CLARIFIERS—(1) Devices used to remove suspended solids from wastewater by either settling or flotation. (2) In chemical recovery, tanks used to clarify green or white liquor by allowing the entrained particles to settle out of the main upward flow and be removed as a slurry from the bottom of the tank.

CLAY—A natural, earthy, fine-grained material, primarily aluminum silicate, which is plastic when wet and becomes hard when dried under heat. In papermaking, clays are used for paper coatings and fillers. See BENTONITE; CHINA CLAY; KAOLIN; PAPER CLAY.

CLAY-COATED BLANKS—A paperboard with a clay coating on either one or both sides. The clay coating may have been applied onto the board at the time of manufacture or a white or colored clay-coated paper may be pasted onto an uncoated board surface. It is made in standard thicknesses and used for all types of printing such as calendars, cutouts, advertising cards. Significant properties are excellence of printing surface, color permanence, and a high degree of whiteness and brightness. For basis weight and thickness schedule, see BLANKS.

CLAY-COATED BOXBOARD—A grade of paperboard that has been clay coated on one or both sides to obtain whiteness and smoothness. It is characterized by brightness, resistance to fading, and excellence of printing surface. Colored coatings may also be used and the body stock for coating may be any variety of paperboard.

CLAY-COATED SOLID BLEACHED SULFATE—See CLAY-COATED BOXBOARD.

CLAY-FILLED PAPER—Paper containing an appreciable amount of clay as a filler, especially as distinguished from paper filled with other inorganic white pigments.

CLAY LUMP—A lump of pigment or dried coating color embedded in the surface of a coated sheet and sufficiently large to cause difficulties in subsequent use of the paper.

CLAY-SACK PAPER—See SHIPPING SACK KRAFT PAPER.

CLEANING SHOWERS—See SHOWERS.

CLEANSING TISSUE—See FACIAL TISSUE.

CLEAR WATER—See CLEAR WHITE WATER.

CLEAR WHITE WATER—White water that is relatively low in suspended solids, to the point that it is relatively clear; usually produced late in the filtration cycle by a disc or vacuum drum filter. Also called clear water.

CLOSED-CYCLE BLEACH PLANT—A term used to signify no water pollution emanating from a bleach plant.

CLOSE FORMATION—The formation of a sheet that is uniform and free from a wild (q.v.) or porous appearance when viewed by transmitted light.

CLOTH-CENTERED PAPER—Duplex paper with a core or center of linen or canvas.

CLOTH FINISH—A finish produced by impressing the weave of a cloth into the paper, such as crash, linen, burlap, etc.

CLOTHING—In papermaking, the term applies to paper machine forming fabrics, press felts and dryer fabrics. See DRYER FABRICS; FABRICS AND FELTS; FORMING FABRICS; PRESS FELTS.

CLOTH-LINED BOARD—A general term descriptive of any paperboard upon which cloth is pasted. The board should have a surface to which the cloth will adhere, and it should be rigid and nonwarping.

CLOTH-LINED PAPER—Cloth used for tags by manufacturers of work clothing and for printed matter and envelopes where resistance to repeated handling and exposure to water, grease, and dirt are important, such as the covers of shop manuals and maps for outdoor use.

CLOTH PAPER—See CLOTH-LINED PAPER.

CLOTH-WINDING BOARD—A paperboard used for winding cloth in bolts. It is made of pasted chip, 0.200 of an inch or more in thickness, and is stiff with a smooth surface but without any special sizing or coating. Before the board is ready for winding, the edges and ends are usually covered with paper tape. Stiffness and compactness are important characteristics.

CLOUD BOX COVERING—A mottled effect box-cover paper usually used by the paper box trade to cover cardboard or chipboard setup boxes. It contains a substantial portion of mechanical pulp mixed with unbleached and bleached or colored chemical woodpulps. A cloud effect is developed by the use of a secondary headbox on the paper machine, which produces an uneven dispersion of the bleached or colored pulps in the upper surface of the sheet. This paper is used chiefly in basis weights of 35 and 40 pounds (24 x 36 inches – 500). Gray is the predominant color, but it is also supplied in blue, green, brown, and black with white or multicolored cloud effects.

CLOUD EFFECT (CLOUDY)—Unevenness in look-through. See WILD.

CLOUD FINISH—A cloudlike effect produced by dropping water suspended white pulp onto colored paper in the process of formation on the fourdrinier fabric.

CLOUDY WATER—See CLOUDY WHITE WATER.

CLOUDY WHITE WATER—White water that is relatively high in suspended solids, compared to clear white water; usually produced early in the filtration cycle, by a disc filter or vacuum drum filter. Cloudy white water from a drum or disc filter is generally much leaner than paper machine white water. Also called cloudy water. See also WHITE WATER.

CLUSTER RULE—A regulation proposed in December 1993 by the U.S. Environmental Protection Agency which would limit discharges of BOD5, TSS, COD, color, AOX, dioxins/furans, chloroform, acetone, methyl ethyl ketone, methylene chloride and certain chlorinated phenolic compounds and emissions of "hazardous air pollutants" such as methanol, chloroform, and chlorine. The proposed regulation would replace existing EPA effluent guidelines for all mills and establish new "Maximum Achievable Control Technology" emission standards for chemical wood pulp mills.

CMC—A commonly used abbreviation for carboxymethyl cellulose, an additive used in special papers for the improvement of sizing quality.

CMP—Chemimechanical pulp (q.v.).

CMT—Corrugating medium test (concora test—q.v.).

CO—Carbon monoxide (q.v.).

COAGULATION—In water and wastewater treatment, the destabilization and initial aggregation of colloidal and finely divided suspended matter by the addition of a floc-forming chemical or by biological processes. See CHEMICAL COAGULATION.

COAGULATION BASIN—A basin used for the coagulation of suspended or colloidal matter, with or without the addition of a coagulant, in which the liquid is mixed gently to induce agglomeration with a consequent increase in settling velocity of particulates.

COAL SPECKS—See CINDER SPECKS.

COAL WRAP—Strong wrapping paper with moisture-resisting characteristics which is used for wrapping pressed coal.

COARSE CLEANERS—Also called high-density cleaners, coarse cleaners are used to separate heavy debris, trash, and abrasive materials from the main fiber flow through the processing plant. These cleaners are a type of hydrocyclone, large in diameter, and having a trap/chamber at the bottom to collect the separated, high specific gravity debris, which is discharged discontinuously. The main purpose is to protect downstream equipment from high abrasive wear. These cleaners typically operate at 2–4% consistency. See also CENTRIFUGAL CLEANERS; FORWARD CLEANERS.

COARSE PAPERS—Papers used for industrial purposes, as distinguished from those used for cultural or sanitary purposes. See INDUSTRIAL PAPERS; SANITARY TISSUE.

COASTER BOARD—Heavy paperboard, made of woodpulp or reclaimed paper stock, that is reasonably absorbent with a minimum of warping (or distortion) when wet or after drying following moisture absorption. It has a reasonably good finish for printing. The thicknesses usually range from 0.060 to 0.135 of an inch. Since excessive grain or fiber direction in an absorbent die-cut piece of board usually contributes to warping, the most successful coaster stock has been made on wet machines with the board air or steam dried. Coaster boards made on cylinder machines are pasted to obtain the required thickness. It is also called beer plaque and beer mat board.

COAT WEIGHT—Weight of coating material present per unit area, usually expressed as pounds per ream.

COAT WEIGHT PROFILE (CROSS DIRECTION)—The coat weight distribution across the width of the sheet.

COATED—A term applied to paper and paperboard, the surface of which has been treated with clay or some other pigment and adhesive mixture, or other suitable material, to improve the finish with respect to printing quality, color, smoothness, opacity, or other surface properties. The term is also applied to lacquered and varnished papers.

COATED ART PAPER—A paper used for high-grade printing work, especially in halftone printing, where definition and detail in the handling of shading and highlights are important. It is usually a high-grade coated paper having a high brightness and a glossy, highly uniform printing surface.

COATED BLANKS—See CLAY-COATED BLANKS.

COATED BLOTTING PAPER—See ENAMELED BLOTTING PAPER.

COATED BOARD—A general term indicating a paperboard which is clay coated. This board is used for box making and miscellaneous purposes. See CLAY-COATED BLANKS; CLAY-COATED BOXBOARD; SOLID BLEACHED SULFATE.

COATED BOND PAPER—A bond paper coated on one side or on both sides and used when the strength of a bond paper and a smooth printing surface are required. Sometimes high-grade art papers are made in this manner.

COATED BOOK PAPER—See BOOK PAPER (COATED).

COATED BOOK STOCK—Consists of coated bleached sulfite or sulfate papers, printed or unprinted in sheets, shavings, guillotined books, or quire waste. A percentage of papers containing fine groundwood may be included (Institute of Scrap Recycling Industries, Inc. scrap specifications for 1994).

COATED BOX-COVERING PAPER—A paper similar to box-cover paper except that it has been coated on one side with a clay coating. A high gloss lacquer maybe applied over the clay coating to enhance the appearance. Strength not always essential but must have good folding and bending qualities.

COATED BRISTOL—See CAMPAIGN BRISTOL; COATED POSTCARD STOCK.

COATED BROKE—Broke after a coater, whether on the paper machine or off. It is sometimes treated separately from uncoated broke, particularly if the coating is difficult to break down during repulping, or if the coating is colored or functional.

COATED CHROMOLITHO PAPER—A paper used for high-grade multi-color lithographic work. It is usually a glossy, high-quality paper, coated on one side and sized for the lithographic

COATED CHROMOTYPE PAPER—A paper similar to coated chromolitho, used for printing from type, but not necessarily sized for the lithographic process.

COATED COVER PAPER—Paper in weights used for cover paper but with a coated surface. It is usually coated on both sides and has either a dull or high finish. See COVER PAPER.

COATED GLASSINE—(1) Glassine coated with a film applied from an organic solvent; it is transparent, water resistant, and water-vapor resistant. The last property depends on the type of lacquer used. Another property which is often important in solvent-coated papers is heat sealing. See HEAT-SEALING PAPER. (2) Glassine coated with a hot-melt coating which is applied from a molten bath without solvents. See DRY COATING.

COATED INDEX BRISTOL—A regular chemical woodpulp index bristol which is coated on two sides, usually with a finished weight of 110 and 140 pounds (25 x 30 inches – 500), with a special coating suitable for use on gelatin-type duplicating machines; its surface is also suitable for letterpress and offset printing, and for pencil and ink writing. It is made in white and a variety of colors.

COATED LITHOGRAPH PAPER—A paper used in lithographic printing, especially where there are a large number of colors used in the picture or design reproduction. It is sized for the lithographic process and is frequently coated only on one side. The basis weights usually range from 40 to 60 pounds (25 x 38 inches – 500).

COATED MAGAZINE PAPER—A coated printing paper used for magazines, periodicals, and the like. The term is applied to a broad range of "on-machine" and "off-machine" coated stocks produced from mechanical and/or chemical woodpulps in basis weights ranging from 26 to 60 pounds (25 x 38 inches – 500). This grade is also referred to as periodical publishing paper.

COATED MOTTLE—A small scale variation of gloss of a coated sheet which can be detected by viewing the surface in specular reflection. The pattern somewhat resembles that found on a piece of galvanized metal. See MOTTLE.

COATED OFFSET PAPER—A grade of coated paper, with a high resistance to picking, coated on one or two sides, and sized the same as coated lithograph paper, suitable for use in offset printing. The basis weights usually range from 26 to 100 pounds (25 x 38 inches – 500).

COATED ONE SIDE—See SINGLE COATED.

COATED PAPER—Any paper which has been coated. This term covers a wide range of qualities, basis weights, and uses.

COATED PLAYING-CARD STOCK—See PLAYING-CARD STOCK.

COATED POSTCARD STOCK—A heavyweight card stock, coated on one or both sides, and used in the manufacture of "picture post cards." It is usually made from chemical wood and/or reclaimed pulps in thicknesses of 0.008 in. to 0.011 in. and is well sized for pen and ink writing.

COATED PRINTING PAPER—See BOOK PAPER (COATED).

COATED SECONDS—Imperfect coated sheets which have been removed from perfect sheets by sorting, but which are still usable. See SECONDS.

COATED TAG—A tag board to which has been applied a coating of clay or some other coating material.

COATED TOUGH CHECK—A tag board coated, usually on two sides, with clay or other coating. It is regularly made in size 22 x 28; 3-, 4-, and 6-ply; corresponding to 0.012, 0.018, and 0.024 of an inch in thickness. The basis weight runs about 11 pounds to the point (22 x 28 inches – 500). Tags, hat checks, tickets, identification cards, and other printed forms which are subject to considerable handling are frequently made of coated tough check, which combines the qualities of unusual strength, fine printing surface, and brilliant colors.

COATING—(1) A term applied to the layer of pigment and adhesive substances which has been applied to the surface of paper or paperboard to create a new surface. (2) A term applied to the film of substances, usually clear, used as a barrier or other functional covering on the surface of paper or paperboard. (3) See SURFACE SIZED; FILM-COATED. (4) The operation of applying a coating formulation of any of the above types to the surface of a sheet. See AIR KNIFE COATING; BLADE COATING; BRUSH-FINISH COATING; CAST COATING; FLOW-ON COATING; KNIFE COATING; MACHINE COATING; POLISHED DRUM COATING; PRINT-ON COATING; ROLL COATING; SPRAY COATING; WIRE-WOUND ROD COATER for various types of coating operations.

COATING CLAY—Any clay suitable for coating paper, generally characterized by smaller particle size and higher brightness than filler clay. See CLAY; COATING GRADE CLAY; KAOLIN.

COATING COLOR—The coating mixture in suspension or slurry form which is applied to the surface of the paper or paperboard in the coating process. It includes the pigments, adhesives, dyestuffs, modifiers, and the liquid medium (usually water) required to carry and apply the components to the paper.

COATING GRADE CLAY—A refined clay, usually kaolin, that meets specifications for use in paper coating. Freedom from grit, correct particle size, good color and brightness, low viscosity, and purity of mineral type are included in the requirements. See PAPER CLAY.

COATING LUMP—A discolored, shiny, hard, and brittle spot in coated paper.

COATING PILING—In offset printing (q.v.), an accumulation on the blanket of material from coated papers or boards. It can be caused by loose material, particles partially bonded, actual pick outs of coating, or coating loosened or dissolved by the dampening solution. The milking or piling can occur either in the image, or the non-image areas or both.

COATING RAW STOCK—Any paper used as a base paper for coating. The type of paper varies with its ultimate use. See BODY STOCK.

COATING ROLL—A covered roll used in a coating (q.v.) operation to apply the coating chemicals to the surface of the sheet. These rolls are covered with a medium hardness elastomeric composition designed to resist the coating chemicals.

COBB TEST—A method for measuring the water absorptiveness of sized paper and paperboard, by determining the weight of water absorbed through one surface from a given volume applied.

COCKLE—(1) A puckered condition of the sheet resulting from non-uniform drying and shrinking. It usually appears on paper that has had very little restraint during drying. See COCKLE FINISH. (2) Macro-scale roughness of paper caused by stresses induced in the web during the drying process.

COCKLE CUT—A cut resulting from the passage of highly cockled paper through the nip of a press or calender.

COCKLE FINISH—A ripple-like finish caused by shrinkage during drying under little or no tension. It may be caused deliberately or inadvertently and is frequently desired, in varying degrees, in some grades of writing papers.

CODE PRESSURE—The maximum pressure that a dryer cylinder, or other pressure vessel is approved to operate. This is determined according to the "code" of the governing authority where the cylinder is installed.

COEFFICIENT OF FRICTION—The ratio of the frictional force to the force, usually gravitational, acting perpendicular to the two surfaces in contact. This coefficient is a measure of the relative difficulty with which the surface of one material will slide over an adjoining surface of itself or of another material. The static or starting coefficient of friction (μs) is related to the force measured to begin movement of the surfaces relative to each other. The kinetic or sliding coefficient (μk) is related to the force measured in sustaining this movement.

COEXTRUSION—The simultaneous extrusion of two or more molten polymers to form a multilayer product.

COFFEE-BAG PAPER—A strong printable paper used in the manufacture of coffee bags. It is usually made from bleached or unbleached chemical pulps. It may be calendered, coated, ribbed or fluted and/or embossed. The usual basis weights are from 30 to 50 pounds (24 x 36 inches – 500).

COFFEE FILTER PAPER—A lightweight, high wet-strength, porous paper used in certain types of coffee percolators and automatic coffee makers to remove grounds from the coffee extract.

COGENERATION—The simultaneous production of both electric power and steam, such as can be accomplished using a boiler together with a steam turbine generator, a gas turbine together with a heat recovery steam generator with or without a steam turbine, or similar combinations.

COILS—Paper slit to a desired width from a roll of paper and rewound on cores for use in all kinds of machines using a continuous roll of paper. Coils generally refer to small rolls such as adding machine and cash register rolls.

COIN WRAP—A strong paper, usually kraft, of uniform caliper used in coin-wrapping machines or for manual wrapping of coins. It is manufactured in various colors and must be suitable for simple printing.

COLD CAUSTIC PULP—Pulp produced by treatment of wood with caustic soda solution at room temperature and atmospheric pressure prior to mechanical defiberizing. (Note: the term may also be applied to dissolving grade pulps that have had a cold caustic extraction stage in the bleach sequence. These pulps have a high purity, or alpha-cellulose (q.v.) content, as the result of the extraction.)

COLD FLOW—See CREEP.

COLD GRINDING—A method of preparing mechanical pulp in which the temperature of the pulp in the grinder pits is controlled by the use of large volumes of cold water. This process is no longer in commercial use. See also HOT GRINDING.

COLD PRESSED—Loft-dried paper which has been pressed in a hydraulic press.

COLD PRESSED FINISH—An extreme antique finish, usually applied to a heavyweight ledger paper. Such paper is used for diplomas, drawing purposes, maps, etc.

COLD PRESSING—A bonding operation in which no heat is applied.

COLD SODA PULP—See COLD CAUSTIC PULP.

COLD WAXED BOARD—See WAXED BOARD.

COLLAR CIRCLES—A double manila-lined chipboard which will bend around the shape of the collar. Appearance is an important property. It is used for protection in shipment of shirts by the factory or laundry.

COLLECTING—Accumulation of loose material or particles on offset blankets.

COLLOID—A partially solubilized or dispersed material having at least one dimension of size 10^{-5} to 10^{-7} centimeters.

COLLOID MILL—A machine for dispersing or mixing a solid or liquid in a liquid. Its essential feature is the relative motion, usually at high speed, of two very closely spaced surfaces, which produces intense shearing stresses in the liquid and the solid particles flowing between these surfaces. The shearing stresses are due to viscous forces in the liquid, rather than to a grinding action between the moving surfaces.

COLLOID TITRATION—Colloid titration procedures can be used to monitor anionic contaminants in fiber furnishes, to assess the cationic requirements of a furnish, or to determine the net charge of a papermaking furnish of furnish fraction. The quantities determined are designated as the cationic demand (q.v.), the anionic demand, or the colloid titration ratio. The general procedure requires adding excess cationic or anionic polymers of known charge equivalency and then back titrating the excess electric charge with a "standard" polymer of opposite charge. For example, the cationic demand can be found by adding excess known cationic polymer and then back titrating the excess polymer with a "standard" anionic polymer. Anionic demand is found by adding excess anionic polymer and back titrating with a standard cationic polymer. The sign of the electric charge of the furnish can be derived from the ratio of the two quantities. Colloid titration gives results that are appreciably different from the cationic demand determined with a microelectrophoresis (q.v.) endpoint, although both procedures have the same objective. See CATIONIC DEMAND and ELECTROKINETIC CHARGE TITRATION.

COLLOTYPE PAPER—A paper used for printing from collodion or glass. Any paper or cardboard can be used that has a high bonding strength, so that it will not peel.

COLLOTYPE PRINTING—A type of planographic printing in which a gelatin-coated glass or metal plate is used as the printing surface. Glass plates maybe used on flat-bed presses and grained zinc or aluminum plates on rotary or offset presses. The process is based on the fact that, when a soluble bichromate salt is added to a gelatin coating, the ability of water to swell the gelatin decreases with increasing exposure to light. The coating is exposed through a negative and is then soaked in water. The uneven swelling in the surface of the exposed gelatin coating causes it to form into a pattern of very fine wrinkles. Glycerin and salts are added to the last rinse to keep the coating damp. When a greasy ink is applied to the damp coating, the wrinkled surface accepts ink in proportion to the amount of light exposure it has received. In lightly exposed areas, the ink adheres to the coating as small, widely spaced, specks or grains of ink. With increasing exposure, the grains become larger and more closely spaced until they fill in solid. Collotype is a halftone

process but the fine random grain pattern of the ink causes it to resemble continuous-tone photography. Papers used are offset, bonds, ledgers, vellums, cover, bristols, and practically all papers or boards that are tub sized. Coated and enamelled papers are also occasionally used. It may be used for postcards, art reproductions, displays, broadsides, posters, etc. This process is also known as albertype, artotype, heliotype, lichtdruck, phototype, and hydrotype.

COLOR—(1) That property of a substance which determines the nongeometrical part of the visual sensation experienced by an observer who views the substance. The color of a specimen depends upon the spectral character of the illuminant, on the geometrical and other conditions of illuminating and viewing the specimen, on the spectral reflectivity of the specimen, and on the characteristics of the observer's eyes. Hence, the only characteristic of a specimen which is the same under all conditions of observation and for all observers is its spectral reflectivity. Knowledge of the spectral reflectivity of a specimen permits calculation of its CIE color specification, i.e., specification of dominant wavelength, purity, and luminous reflectivity. (2) The suspension or slurry of the materials for use in the pigment coating of paper. See CIE COLOR; COATING COLOR.

COLOR-BRIGHTNESS TESTER—A photoelectric instrument designed to measure the reflectance of light at various points distributed across the visible spectrum in terms of a reflectance standard.

COLORED LABEL—A label paper (q.v.). The fiber furnish for this paper ranges from mechanical pulp to chemical woodpulp free from mechanical pulp. The finish may vary from a dull to a high, glossy finish, and the basis weights range from 50 to 70 pounds (25 x 38 inches – 500). It is usually coated on one side and is made in a wide variety of colors.

COLORED LEDGER SORTED—Consists of printed or unprinted sheets, shavings, and cuttings of colored or white sulfite or sulfate ledger, bond, writing, and other papers that have a similar fiber and filler content. This grade must be free of treated, coated, padded, or heavily printed stock (Institute of Scrap Recycling Industries, Inc. scrap specifications for 1994).

COLORED SCHOOL PAPER—See CONSTRUCTION PAPER.

COLOR FASTNESS—The property of a paper, dye, or dyed paper to retain its color in normal storage or use or to resist changes in color when exposed to light, heat, or other deleterious influences. See PERMANENCE.

COLORIMETRIC PURITY—The ratio of the luminosity of the spectrum color to the luminosity of the mixture of illuminant and spectrum color which matches the color of the specimen viewed under the illuminant alone. Luminosity is the brightness sensation produced by unit intensity of light. Purity is also referred to as saturation or depth of color, e.g., a deep red has a higher purity than a pastel red.

COLORING PIGMENT—See PIGMENT.

COLOR LAKE—An artificial pigment prepared by precipitating a dye upon some base such as alumina, barium sulfate, clay, etc. The term pulp color is also applied to these lakes but is sometimes extended to cover pigments sold in the paste form, such as the chrome yellows. Dry lakes are made by drying and grinding color lakes.

COLOR OK SHEETS—Paper bearing the printed colors which have been approved as standard colors for any ink color-matching operation.

COLOR REVERSION—See BRIGHTNESS REVERSION.

COLOR SEPARATION—The process of separating color originals into the primary printing components for color reproduction.

COLOR SHEETS—Paper carrying standard ink colors for each color used in a multicolor printing job.

COLOR SPECIFICATION—The quantitative description of a color. The color specification of the International Commission on Illumination consists of a statement of dominant wavelength, purity, and luminous reflectivity under standardized conditions. The color of papers is often specified in terms of the trichromatic coefficients x, y, and z, and sometimes in terms of matching certain standardized color chips, or colored papers previously designated as standards. Less exact approximations, such as requirements for reflectance at one or two wavelengths, are sometimes used in purchase descriptions.

COLOR SPOTS—Variously colored spots or specks (blue, red, black, etc.) in the sheet. They may be undispersed particles of color pigment or undissolved dyes or they may also be caused by color reactions with rosin, pitch particles, or other elements of the furnish which may give rise to a colored froth or foam that accumulates back of the slices. When this dries and drops over onto the wire, many different kinds of specks are produced which mar the appearance of the finished sheet.

COLOR STRIPPING—Color stripping differs from bleaching in that it is only for removing colors in the colored and tinted papers, and not for removing residual lignin, i.e., in bleaching. Although many bleaching chemicals will remove colors, a few colors are resistant to these chemicals.

COLOR VARIATION—(1) In printing, a term used to describe changes in color of printing; or changes in the density of color which may be caused by variations in the amount of ink accepted by paper or by the amount of ink fed to the paper. (2) In papermaking, the coating formulation is called the coating "color." In this context, "color variation" means variation in the shade of coating formulation. It results in a deviation in paper shade or color from standard.

COLUMN STRENGTH—The maximum compressive force which a plane, unsupported, rectangular test piece of corrugated board, standing on its edge, can withstand without failure. A test, often called short column test, for this property of edgewise compression strength is known as the edge crush test (ECT).

COMBINATION BOARD—A general term designating a board made on a cylinder machine wherein one or both of the outer plies are of a different raw material and/or color than the middle ply, whereas a plain or solid board has the same material throughout. Where three stocks are used, or where two outside liners of the same stock are vat lined onto a filler of a different stock, this is customarily indicated in the name of the board. Buff patent-coated news, white patent-coated news, bleached manila-lined chip, etc., indicate two different kinds of stock in each board and also indicate but one liner. This is also indicated by the words "single-lined," either together at the end of the board name or separated; bleached manila-lined chip, single-lined or single-bleached manila-lined chip both mean chip center and back, vat lined with one liner of bleached manila furnish. However, bleached manila-lined chip means exactly the same thing. When the top and bottom liners of a board are both of the same kind of stock, this is indicated by the words "double-lined" or double." Colored liners are more or less definitely indicated in the name of the grade. Most boxboards and many paper board grades are combination boards. See also BOXBOARD; LINED BOARD.

COMBINATION CLEANER—A type of centrifugal cleaner that removes both low and high specific gravity debris via separate outlets in the same unit. Typically, the heavy debris is removed as the apex end of the conical section, and the light debris is removed through a small diameter outlet tube located inside the accepted stock outlet, or the high-specific-gravity debris outlet, or both. Combination cleaners first became available to the paper industry in the late 1960s. Also known as multifunction cleaners. See also CENTRIFUGAL CLEANER; FORWARD CLEANER; REVERSE CLEANER; THROUGH FLOW CLEANER.

COMBINATION PAPERBOARD—Paperboard manufactured from a combination of fibers from various grades of recovered paper with the predominant portion of the total furnish being recycled fibers. Now also commonly referred to as recycled paperboard (q.v.).

COMBINATION STAGE—A bleaching stage in which either at the beginning of the stage or after an interval, a second bleaching agent is added to obtain special effects or to intensify the bleaching reaction.

COMBINED BOARD—A term used to designate that two or more boards have been joined together by an adhesive. For example, a corrugated board or a solid fiberboard is a combined board after having been passed through the corrugating or pasting machine.

COMBINED-CYCLE COGENERATION—Two or more thermodynamic power cycles combined to convert the energy in fuel to electric power and steam, usually more efficiently. An example would be firing natural gas in a gas turbine to produce electric power, using the exhaust gases in a heat recovery steam generator to produce high-pressure steam, and then passing the steam through a steam turbine generator to produce additional electric power.

COMBINING CALENDER—The calender-winder (combining calender) is used in the tissue industry to combine two or more tissue webs for subsequent facial or toilet tissue converting operations. It is composed of one or more single nip calenders between the multiple roll unwinds and the conventional two-drum winder.

COMBUSTIBLE PAPER—A cigarette paper that has been impregnated with a nitrate to control its burning properties.

COMBUSTION PRODUCTS—Products that result from the burning of a material.

COMMERCIAL BLOTTINGS—See BLOTTING PAPER.

COMMERCIAL MATCH—The duplication of a paper in a mill run which does not exactly match the sample but which is close enough to be considered acceptable.

COMMERCIAL TISSUES—See WRAPPING TISSUE.

COMMERCIAL WOVE ENVELOPE—A term originally applied to many uncoated papers made from chemical woodpulps for conversion into envelopes. It is now largely applied to a commodity-type envelope base stock as distinguished from bond, cotton content, kraft, manila, papeterie, and other grades. It is generally made in white and colors in 20 and 24 pound weights (17 x 22 inches – 500), and is often referred to simply as wove envelope.

COMMERCIAL WRITINGS—Writing papers used in the conduct of business generally, such as tablet, bond, writing, and ledger grades.

COMMON CAUSES—Causes of variation that are inherent in a process over time. (Also often called "chance" or "random" causes.)

COMMON IMPRESSION PRESS—A flexo or offset press which has only one, large impression cylinder, but several blanket or plate cylinders.

COMMUNICATION PAPERS—See PRINTING AND WRITING PAPERS.

COMPACTION—In nonwovens, the process of introducing crimp into nonwovens to enhance flexibility, draping, and softness properties.

COMPARTMENT ROLLS—Very small rolls of writing paper which are used on bank proof machines in the processing of bank checks. They are usually converted from tablet or register bond stock and are rewound into single-ply or carbonized two-ply forms.

COMPOSITE CAN—See FIBER CAN.

COMPOSITE COVERS—These are rock hard (90–92 SHORE D) fiber-reinforced plastic roll covers and are being used in increasing numbers to replace granite rolls (q.v.) in press positions, and filled rolls in calenders. See SHORE DUROMETER.

COMPOSITE FELT—See LAMINATED FELT.

COMPOSITION BOOK PAPER—See TABLET PAPER.

COMPOSITION ROPE AND KRAFT SHIPPING SACK PAPER—A paper containing less than 75% manila rope fibers and used in the construction of single wall (and sometimes double wall) paper shipping sacks. It is usually made on a cylinder machine.

COMPRESSIBILITY—The percentage decrease in caliper of the sheet produced by an arbitrarily specified increase in load. The conditions under which the determinations are made must be completely specified. This definition must be distinguished from the definition of compressibility in engineering practice, which is the ratio of the fractional change in volume to the pressure producing that change in volume. This property is of considerable importance in several uses of paper, notably in printing and bookbinding.

COMPRESSION (STRENGTH) RESISTANCE—The resistance of paperboard or containers to compressive stresses. See BOX COMPRESSION TEST; COLUMN STRENGTH; EDGEWISE CRUSH RESISTANCE; FLAT CRUSH RESISTANCE; RING CRUSH TEST; SHORT SPAN COMPRESSION TEST.

COMPRESSION WOOD—An abnormal type of wood, occurring as a rule on the lower, or compression, side of branches and of leaning tree trunks of all coniferous species. Typical compression wood can be identified in logs by the presence of markedly eccentric annual growth rings. Compression wood has a higher lignin content and a lower cellulose content than normal wood. See also TENSION WOOD.

COMPRESSIVE STRENGTH—A property of a solid material that indicates its ability to withstand a compressive load; resistance to crushing.

COMPUTER AIDED DESIGN—Use of computers and computer programs in the design of buildings or any manufactured object. The computer drawings may range in complexity from simple two-dimensional line drawings to complex, three-dimensional depictions showing motion or flow.

COMPUTER CONTROL—A process utilizing a special purpose computer to make appropriate adjustments to a continuous digester or other papermaking operation in order to accomplish a smooth transition from one operating condition to another.

COMPUTER PRINTOUT PAPER—A lightweight bond paper made from chemical wood and/or mechanical or reclaimed pulp, which is used on line printers associated with digital computers. It is usually made in basis weights ranging from 12 to 16 pounds (17 x 22 inches – 500) and is put up in the form of small rolls or fan folded sets—with or without carbon paper interleaving. It is also referred to as computer readout or computer output paper. See also REGISTER BOND.

COMPUTING MACHINE PAPER—A general term for any paper used on computing machines, but it usually refers to bonds and ledgers.

CONBUR TEST—See IMPACT TEST FOR SHIPPING CONTAINERS.

CONCENTRATED BLACK LIQUOR—Black liquor after final evaporation or concentration process and before addition of makeup chemicals or recycled precipitator catch. Often referred to as "strong" black liquor or "virgin" black liquor.

CONCENTRATOR—An evaporator (q.v.) supplied with fresh steam used in the final stage of concentration of an aqueous solution.

CONCORA TEST—Flat crush resistance (q.v.) of corrugating medium (CMT) test. A method for measuring the crush resistance of a laboratory-fluted strip of corrugating medium for estimating the potential flat crush resistance of corrugated board fabricated from that medium.

CONCRETE CURING PAPER—A reinforced, waterproof paper designed to retard the drying of concrete (curing) by covering the surface immediately after the initial set. The paper is constructed of two kraft plies, usually wet strength, laminated with asphalt, in which reinforcing material is imbedded, or polyethylene-coated wet-strength kraft laminated with reinforcing glass scrim.

CONDENSATE—Steam which has condensed, or turned into, liquid.

CONDENSATE COEFFICIENT—The heat transfer coefficient from steam inside a steam-heated cylinder, through the condensate layer, to the inner diameter of the cylinder. Normal units are Btu\(h-ft^2-°F) or Watts/(m^2-°K)

CONDENSER—An indirect contact heat exchanger used to condense the steam from the lowest pressure effect or body of a multiple effect evaporator (q.v.) using water as a cooling medium.

CONDENSER PAPER—A paper used as a spacer between foils of electrolytic capacitors. This paper is made from sulfate pulp, cotton, or hemp. Thicknesses range from about 0.0006 to 0.004 inch. The important properties are porosity, absorption of electrolytic liquids, and chemical purity, especially freedom from soluble chlorides. See CAPACITOR PAPER.

CONDENSER TISSUE—See CAPACITOR PAPER.

CONDITIONING—Exposure of paper to accurately controlled and specified atmospheric conditions, so that its moisture content reaches equilibrium with the surrounding atmosphere.

CONE PAPER—See TUBE PAPER.

CONFECTIONERS' PAPER SPECIALTIES—Various forms of paper, cut in different sizes from sulfite bond, and bearing die-cut or embossed designs.

CONFECTIONERY-BAG   PAPER—See CANDY-BAG PAPER.

CONFORMANCE—An affirmative indication or judgment that a product has met the requirements of a specification, contract, or regulation.

CONICAL REFINER—A refiner that uses plug and shell type tackle. A conical refiner which as a rotor included angle of 20° or less from the centerline is often called a Jordan, and is used primarily for cutting fibers. A refiner with a rotor included angle of 90° from the centerline would be called a disc refiner, not a conical refiner. See DISC REFINER; JORDAN; REFINER.

CONIFER—A cone-bearing tree or shrub so called because the fruit of the tree is a cone, as in the pines and firs. The wood is also termed softwood (q.v.).

CONSISTENCY—The percentage by weight of ovendry fibrous material in a stock or stock suspension. It is sometimes called density, solids, or concentration. At a given ratio of chemical to pulp, an increase in pulp consistency increases the concentration of chemicals in solution, and therefore the reaction rate, which makes consistency an important bleaching variable. Bleaching at various consistencies is arbitrarily classified as: low consistency (LC), up to 4%; medium consistency (MC), 10–16%; high consistency (HC), 30% plus.

CONSTRUCTION PAPER—(1) A school paper used for cut-outs, crayon drawings, watercoloring, finger painting, etc. It is usually made from mechanical pulps in basis weights ranging from 40 to 80 pounds (24 x 36 inches – 500). Such paper is also referred to as school poster. (2) Any of a variety of strong, heavyweight, single-ply, laminated, and/or plastic coated papers used as tarpaulins, wet concrete coverings, and various other building construction applications.

CONTACT ANGLE—The angle at which the surface of a liquid meets the surface of a solid or of another liquid. The angle is measured within the liquid so that, for example, the contact angle of water on glass is small, whereas that of water on paraffin or of mercury on glass is large. This is a quantity of interest in the consideration of the penetration of liquids into paper.

CONTACT BED—(1) An artificial bed of coarse material providing extensive surface area for biological growth in a watertight basin. Wastewater exposure to the surface may be accomplished by cycling or by continuous flow through controlled inlet and outlet. (2) An early type of wastewater filter consisting of a bed of coarse broken stone or similar inert material placed in a watertight tank or basin which can be completely filled with wastewater and then emptied. Operation consists of filling, allowing the contents to remain for a short time, draining, and then allowing the bed to rest. The cycle is then repeated. A precursor to the trickling filter.

CONTACT COEFFICIENT—The heat transfer coefficient between the outer surface of a dryer cylinder and the paper web on the cylinder. Normal units are Btu/(h-ft^2-°F) or Watts/(m^2-°K).

CONTACT STABILIZATION PROCESS—A modification of the activated sludge process in which raw wastewater is aerated with a high concentration of activated sludge for a short period, usually less than 60 minutes to obtain BOD removal by absorption. The solids are subsequently removed by sedimentation and transferred to a stabilization tank where aeration is continued further to oxidize and condition them before their reintroduction to the raw wastewater flow.

CONTAINER—A term in the paperboard industry which refers generally to a paperboard box or receptacle. This is usually the outer protection used in packing goods for shipment, in contrast to a folding carton or rigid setup box for individual items or small bulk packaging. See also BAG; CANISTER; CARTON; FIBER CAN; FIBER DRUM; FOLDING PAPER BOX; REGULAR SLOTTED CONTAINER; SETUP BOXES; SHIPPING CONTAINER; SHIPPING SACK; SLIDE BOX.

CONTAINER BOARD—The component materials used in the fabrication of corrugated board and solid fiber combined board: linerboard (q.v.); corrugating medium (q.v.); chipboard (q.v.).

CONTAINER LINER—A creased fiberboard sheet inserted as a sleeve in a container, covering all side walls, to provide extra strength.

CONTAMINANTS—A general papermaking term applicable to extraneous and, usually, harmful matter in pulp or nonfibrous raw materials or in both. The term is more specifically applied to adhesives, wet-strength resins, inks, dirt, coatings, asphalt, plastics, rubber, etc., found in recyclable recovered papers. Also called debris, prohibitives, pernicious contraries. See also ANIONIC CONTAMINANTS; DEBRIS; DISSOLVED CONTAMINANTS.

CONTAMINATED CONDENSATE—Part of the condensate from a multiple effect evaporator (q.v.) that is derived by condensing the steam evaporated in one effect (or stage) in an earlier, lower temperature effect (or stage). This steam is contaminated with low molecular weight organics which volatilize from black liquor (q.v.) during evaporation.

CONTINUOUS—Without interruption of handling or processing, as contrasted to batch operation. Examples are continuous digesters, continuous coating, etc.

CONTINUOUS LAID DANDY ROLL—See SPIRAL LAID DANDY ROLL.

CONTINUOUS MEASUREMENT—The monitoring of a property or a reaction or process on a continuous basis to maintain quality control over the reaction or process.

CONTINUOUS PROCESS—Used to describe pulping, bleaching, and treatment systems where raw material is fed continuously into a vessel (such as chips into a continuous digester) and processes material (pulp) is drawn out of the other end. Chemicals may be added and/or removed at various points along the process.

CONTINUOUS PULPER—A pulper that extracts stock on a continuous, steady-state basis. Typically, the dry furnish components and water are fed discontinuously, or at rates or intervals controlled by operators or automatic controllers, but at relatively short intervals, because true continuous feeding of these components is not practical.

CONTINUOUS TONE—A pictorial image that has not been screened. Tonal gradations are continuous from light to dark, without dots.

CONTRAST RATIO—The ratio of the diffuse reflectance of a sheet when backed by a black body to that of the sheet when backed by a white body. There are several contrast ratios in use, the differences between them being the differences in reflectance of the white backing body. See OPACITY; PRINTING OPACITY; TAPPI OPACITY.

CONTROL CHART—A chart with a central line and upper and lower control limits on which values of some statistical measure for a series of samples or subgroups are plotted.

CONTROLLER TUNING—Adjustment of a controller to respond to the dynamic characteristics of its controlled process.

CONTROL ROLLS—Very small rolls of writing paper which are used on bank proof machines in the processing of bank checks. Control rolls are used in connection with compartment rolls (q.v.).

CONVERSION—See CONVERTING.

CONVERSION COATING—The application of one or more coatings to one or both sides of a paper or paperboard as an operation separate from papermaking. This is sometimes called off-machine coating.

CONVERTER—A plant that manufactures paper products, such as papeteries, envelopes, bags, containers, coated paper, gummed paper. Such operations are frequently carried out by primary paper and board manufacturers as well as by independent companies not engaged in primary manufacturing. See CONVERSION.

CONVERTING—(1) The processes or operations applied to paper, tissue, or board after the normal base sheet has been produced on the paper machine. Various consumer products such as facial tissue, toilet tissue, towels, bags, boxes, or envelopes require converting operations to convert a base sheet to a usable consumer product. Embossing, sheeting, supercalendering, or off-line coating systems are typical converting processes. Some coating operations are in-line on the paper machine and are then considered to be an integral part of the papermaking process. (2) The transformation of half stuff (q.v.) into paper. (3) In nonwovens, the process of modifying nonwoven roll goods by cutting, slitting, folding, application of materials, assembly, and packaging.

CONVERTING MILL—A nonintegrated paper or board mill, i.e., one which does not produce its own pulp.

CONVERTING PAPER—Any paper that may be converted by a separate operation to produce a paper of different characteristics or to produce a product quite distinct from the original paper. Thus, kraft paper is made and sold to be converted into asphalt paper, waxed paper, certain gummed tape, or paper bags; printing paper is converted into envelopes, etc. See also CONVERTING.

COOK—A term used to describe the pulping process. During a cook, fibrous raw materials (chips, straw, or sawdust) are combined with chemicals and heated under pressure for a period of time. The duration of a cook depends on the type of product being manufactured. Unbleached pulps require a shorter cook, and thus less pulping, than bleachable pulps. The most common use of the term cooking is in connection with batch pulping.

COOKED HAM WRAPPER—The inside absorbent sheet, usually crinkled, for a ham wrapper.

COOKERY PARCHMENT—A grade of vegetable parchment paper used in the cooking of foods, such as meats, fowls, fish, vegetables, in which the paper is wrapped about the food to be cooked to retain the juices and to enable several foods to be cooked in the same container. It is also used for lining baking dishes to keep the contents from coming in contact with the dish.

COOKIE—A narrow roll on the ends of a winding set that will be discarded. Also called a butt roll (q.v.).

COOKING LIQUOR—The solution of chemicals used in pulping is collectively called cooking liquor. There are other terms for cooking liquor, depending on the stage of pulping. For example, fresh kraft cooking liquor is called white liquor. After pulping, it is black liquor. During the chemical recovery process and just before causticizing, it is called green liquor. The color at each stage is often used to name the liquor.

COPIER—A machine by which reproductions are made directly from graphic material by xerographic or other methods. See XEROGRAPHY.

COP PAPER—A slack-sized paper, generally made of sulfite or kraft pulp, used in the manufacture of small paper tubes for the textile industry. It is so made that it will roll well on an arbor and paste quickly; the pulp is sufficiently hydrated or otherwise treated to be strong when wet or subjected to steaming. Any coloring materials used in this paper should stand moisture without bleeding.

COPPER ENGRAVING—See DIE STAMPING.

COPPER NUMBER—The number of grams of copper reduced from the cupric to the cuprous state by 100 grams of pulp or paper under specified conditions. Copper number indicates the relative number of reducing groups in the pulp or paper and is used as a measure of its chemical quality and stability.

COP TUBE PAPER—See COP PAPER.

COPYBOOK PAPER—Tablet paper in bound form.

COPYING PAPER—Tablet paper in bound form.

COPYING TISSUE—Any of a group of tissue paper which are slack sized, soft in texture, and of considerable strength, for use in letter press work or for special manifold purposes, such as railroad waybill copying.

COPY PAPER—A high-grade paper, usually uncoated, used for xerographic, ink jet, laser, and other forms of office and home printers. The base stock varies from 100% bleached chemical wood pulp or recycled fiber to mixed rag content papers. Surface smoothness and uniformity are important. Thermal stability at typical fuser roller temperatures is important, particularly with respect to curl, which can cause jam-ups in the paper feed process if not controlled. Normal US dimensions are 8 1/2 inch by 11 inch, with 8 1/2 inch by 14 inch paper also available. Normal ISO paper size is A4, similar to the US 8 1/2 by 11.

CORD—Usually, a pile of pulpwood 8 feet long, 4 feet wide, and 4 feet high, containing 128 cubic feet. See CUNIT.

CORD-REINFORCED PAPER—See REIN-FORCED PAPER.

CORE BOARD—A paperboard for use in making cores upon which paper, textiles, etc., may be wound. It may be made from various grades of paperboard, depending upon the end-use requirements. It is cut into strips for spiral or convolute windings. The board has a surface adaptable for pasting and the resulting core should be crush-resistant. Resistance to splitting and uniformity of caliper are important characteristics.

CORE BUNGS—See CORE PLUGS.

CORE CURL—See ROLL SET.

CORE INSERTS—See CORE PLUGS.

CORE-IRON—Nominal 3 inch iron cores have an internal diameter of 3.068 ± 0.030 inches with an external diameter of about 3.5 inches. They are usually fitted with 3/4 inch keyways. Steel cores can be produced with bore accuracies within 0.005 inch and can wind rolls up to 120 inches wide. Iron cores are usually butt welded while newer designs are fabricated like tubing and drawn over mandrels.

CORE PAPER—A fourdrinier or cylinder sheet usually 165 to 185 pounds in basis weight (24 x 36 inches – 500), customarily made from chemical woodpulp broke papers, sometimes with a proportion of screenings or waste papers, hard-sized and having a low moisture content. It is cut into strips for spiral or convolute winding into cores.

CORE PLUGS—Metal, wood, plastic or composite devices which are fitted into both ends of a fiber core of a finished roll to reinforce the core and prevent collapse or crushing during transport. Also known as bungs or inserts.

CORES-FIBER—See FIBER CORES.

CORESHAFT—A coreshaft is a mandrel designed to provide support for a core and/or a roll of paper. The diameter of a coreshaft can vary from as little as 1 inch to over 30 inches. The coreshaft can be designed to expand and grip the bore of fiber, plastic, or iron cores, and can also be designed for minimal deflection, and for centerwind torque. They are used on reels, rereelers, winders, and many types of converting equipment that make toweling, toilet tissue, or other roll products.

CORE STOCK—Paperboard made for use in the manufacture of cores to be used in rolls of paper, etc. The furnish ranges from screenings to special qualities of virgin pulp, depending upon size of the core and its intended use. See CORE BOARD.

CORE WASTE—The paper which remains on the core after the roll has been unwound.

CORE WOOD—See JUVENILE WOOD.

81

CORK PAPER—(1) Paper printed in such a manner as to resemble cork. (2) Extremely thin, sliced cork laminated to paper. These papers are used chiefly in cigarette tips. (3) A heavy rope or manila paper, coated with ground cork and adhesive. It is used for packing glass, fragile articles, etc. (4) A heavyweight sheet, basis 60 to 375 pounds (24 x 36 inches – 500) made from a furnish of 10 to 25% ground cork, the balance chemical pulp. It is saturated with a mixture of glue and glycerin, then die cut into gaskets.

CORRECTIVE ACTION—The implementation of solutions resulting in the reduction or elimination of an identified problem.

CORRESPONDENCE CARDS—Generally, any heavyweight paper falling into the bristol category, which is used for personal correspondence. Such paper is characterized by a smooth finish, sized for pen and ink writing.

CORRESPONDENCE PAPER—A general term applicable to writing or typewriting papers designed for business and social correspondence, including bonds, ledgers and writings. See also PAPETERIE PAPER.

CORROSION-EROSION—Successive events of the erosion of an otherwise protective corrosion product followed by some amount of corrosion product followed by some amount of corrosion before a protective film is re-formed and the cycle repeats itself.

CORROSION FATIGUE—The process by which a metal fractures prematurely under conditions of simultaneous corrosion and repeated cyclic stress loading, at lower stress levels or fewer cycles than would be required in the absence of the corrosive environment.

CORROSION INHIBITOR—A chemical substance (or combination of substances) which, when present in the proper concentration and form in the environment, prevents or reduces the rate of corrosion.

CORRUGATED BOARD—The structure formed by bonding one or more sheets of fluted corrugating medium to one or more flat facings of linerboard. When this consists of a single facing, it becomes single-face (wrapping material). If bonded on both sides, it becomes double-faced or single wall corrugated board. Similarly, double wall and triple wall corrugated boards are also produced. Corrugated board is most commonly made in four flute sizes, designated A, B, C, and E-flute, although it is also produced in other sizes as well on a limited basis. See also CORRUGATING MEDIUM; DOUBLE FACED CORRUGATED BOARD; DOUBLE WALL CORRUGATED BOARD; FLUTE; LINERBOARD; SINGLE FACED CORRUGATED BOARD; TRIPLE WALL CORRUGATED BOARD.

CORRUGATED CLIPPINGS—See DOUBLE LINED KRAFT.

CORRUGATED COMBINED BOARD—See CORRUGATED BOARD, DOUBLE FACED CORRUGATED BOARD.

CORRUGATED CONTAINER BOARD—See CONTAINER BOARD.

CORRUGATED DOUBLE SORTED—Dry, baled, double sorted corrugated containers, generated from supermarkets or industrial/commercial facilities or both, having liners of test liner, jute, or kraft. Material has been specially sorted to be free of chip, offshore corrugated, plastic, and wax. (Institute of Scrap Recycling Industries, scrap specifications for 1994.)

CORRUGATED ROLL—See SINGLE FACED ROLL.

CORRUGATED SHEET—A sheet of corrugated board used for many purposes where protection, separation, or support is required. It is made of double faced or double wall corrugated board. See DOUBLE FACED CORRUGATED BOARD; DOUBLE WALL CORRUGATED BOARD.

CORRUGATED SHIPPING CONTAINER BOARD—See CONTAINER BOARD.

CORRUGATED WRAPPING—A product of the corrugating machine used for protection of fragile articles. It is made by passing a corrugating medium through fluting rolls of the machine without being pasted to a facing sheet or bonded to a single facing. It may be joined to form a tube for bottles or light bulbs or to be used as a wrapping; it is light, rigid in one direction, and serves as a cushion.

CORRUGATING—The imparting of a wavelike shape to a paper or a board. It is carried out on a corrugating machine by moistening or steaming a roll of corrugating medium prior to passing it between two metal rolls cut with alternate ridges and grooves which are geared to run in complement to each other. This impresses permanent parallel flutes in the paper at right angles to the machine direction.

CORRUGATING MATERIAL—See CORRUGATING MEDIUM.

CORRUGATING MEDIUM—A paperboard used by corrugating plants to form the corrugated or fluted member in making corrugated combined board, corrugated wrapping, and the like. It is usually made from chemical or semichemical woodpulps, straw, or reclaimed paperstock on cylinder or fourdrinier machines. Also called fluting (q.v.).

CORRUGATING MEDIUM TEST—See CONCORA TEST.

CORRUGATION—See FLUTE.

CORRUGATIONS (DEFECT)—A defect, sometimes described as "accordion pleats," which develop due to interlayer movement during a winding or unwinding of a roll. See also FLUTE.

COTTON-BATTING PAPER—A lightweight blue wrapping paper used for wrapping absorbent cotton and cotton batting.

COTTON CONTENT—Percentage of cotton fiber in pulp or paper.

COTTON FIBER CONTENT PAPER—Paper that contains 25% or more cellulose fibers derived from lint cotton, cotton linters, and cotton or linen cloth cuttings. Sometimes flax is used in place of linen cuttings. The term is used interchangeably with rag content and cotton content papers. See RAG CONTENT PAPER.

COTTON LINTERS—The short fibers adhering to cottonseed after the operation of ginning (seed removal and cleaning). These fibers are cut from the seed in a series of passes through cutting blades, and are therefore referred to as "first-cut linters," "second-cut linters," "mill run," etc. Linters are used in the manufacture of cotton fiber content paper and cellulose derivatives.

COTTON PLANT—A plant of the genus *Gossypium,* which yields fiber useful for the manufacture of durable and permanent fine papers and cellulose derivatives. The boll of the cotton plant is a capsule that bursts open when ripe, allowing the seed and attached lint (hairs) to be easily picked. The cotton fiber is removed from the seed by the ginning process. See also COTTON FIBER CONTENT PAPER; COTTON LINTERS; HULL FIBER.

COTTON SAMPLING PAPER—A heavyweight kraft sheet usually in basis 90 pounds (24 x 36 inches – 500) with a smooth finish. It is used to wrap and ship samples from various individual bales of cotton.

COUCH JACKET—A tubular woven thick wool felt which is shrunk onto the top couch roll (q.v.). It is almost obsolete, having been replaced by the suction couch roll (q.v.). See FELT.

COUCH MARK—Mark showing the pattern of the holes in the suction couch roll (q.v.). Shadow mark is a synonym.

COUCH PIT BROKE—See WET BROKE.

COUCH PIT PULPER—An UTM broke pulper (q.v.) located in the couch pit. The sheet is very weak at that point, so normally a simple propeller type agitator with an extraction plate behind it is sufficient for this application.

COUCH ROLL—A paper machine roll primarily involved in dewatering and picking off, or couching, the newly formed paper web from the fabric on which it was formed and partially dewatered. The web is then transferred to the press felt for further dewatering. (1) On a cylinder machine, the couch roll runs against the top of the sheet-forming cylinder mold to produce a pressure nip through which the pick-up felt passes. The sheet formed on the cylinder passes up and through this nip for further dewatering by the nip pressure, is picked off the wire mesh surface of the cylinder mold by the pickup felt, and may be joined by additional plies picked up in turn from other cylinder molds as it is transported by the felt to the wet presses. (2) On a fourdrinier machine, either a suction couch roll or a pressure couch is used. The suction couch roll consists of a heavy metal shell drilled with many small holes through which a suction box inside this shell can apply a high vacuum for rapid removal of water from the sheet as it is carried by the fabric over this roll immediately prior to its transfer from the fabric to a felt for passage through the wet presses. The pressure couch roll consists of a pair of rolls forming a pressure nip through which the fabric and partially dewatered sheet pass for further water removal by pressure immediately prior to transfer of the sheet from the fabric to the press felt. The two rolls involved are termed top couch roll and bottom couch roll.

COUNT—(1) In the paper industry, the actual number of sheets of paperboard of a given size, weight, and caliper required to make a bundle of 50 pounds. (2) The number of sheets which make up a standard unit of a particular kind of paper. Thus, the weight, size, and count of book paper would be, e.g., 50 pounds (basis weight), 25 x 38 (size, inches), and 500 (count), usually written as 50 pounds (25 x 38 – 500).

COUNT CHART (C CHART)—A control chart for evaluating the stability of a process in terms of the count of events of a given classification occurring in a sample.

COUNTER BOARD—See SHOE BOARD.

COUNTER CHECK PAPER—A bond paper used for printing checks, deposit slips, etc., used within banking offices. Such paper is normally made in a 16 to 24 basis weight range (17 x 22 inches – 500) and is devoid of safety features. See also CHECK PAPER; SAFETY PAPER.

COUNTERCURRENT WASHING—A water-saving system for washing pulp in a multistage bleach plant. The filtrate or white water from a later stage is used to dilute stock and wash pulp entering or leaving a preceding stage. Often, recycled filtrates from acidic and alkaline treatments are kept separate throughout the bleach plant, but some bleach plants use full countercurrent washing.

COUNTERIONS—Ions that congregate at charged surfaces of opposite sign as a result of electrostatic attraction. See ELECTRIC DOUBLE LAYER.

COUNTER ROLLS—(1) Rolls of wrapping paper, usually 9 inches in diameter and 12, 15, 18, 24, 30, 36, and 40 inches wide, wound on wooden plugs for use in horizontal holders on store counters. Larger rolls of greater widths and diameters for use in vertical holders are called stand rolls. (2) A term applied to side runs of standard newsprint which have been slit to various widths and rewound for use as wrapping paper.

COUNTERROTATING DISC REFINER—A refiner with two rotating elements, both equipped with refining tackle rotating in the opposite direction. Wood chips or high-consistency pulp is fed through the spokes of one of the rotating discs. Loading is accomplished by positioning one of the rotating assemblies. The relative disc speed is considerably higher, compared, to a single disc refiner.

COUNTER SHEETS—Paper similar in quality to that used for counter rolls, with the exception that it is sheeted into various standard sizes and packed in bundles of standard weights.

COUNT-PER-UNIT CHART (U CHART)—A control chart for evaluating the stability of a process in terms of the average count of events of a given classification per unit of occurrence in a sample.

COUPON PAPER—A grade of bond or book paper used generally in the form of tickets. The basis weight range is from 13 to 20 pounds (17 x 22 inches – 500). Many of these papers now possess the characteristics of safety paper. See also TRADING STAMP PAPER.

COVER-CAP BOARD—A sanitary food board used for specialty caps for milk bottles. After application, the cap covers the lip of the bottle. The board is made of chemical woodpulp on a cylinder or fourdrinier machine. It is used in various colors and is usually waxed. It is normally about 28 points in thickness. See also BOTTLECAP BOARD; HOOD-CAP BOARD.

COVERED ROLL—A roll consisting of a metal body covered with a layer of an elastomeric (rubber) composition or a plastic composite possibly containing a reinforcing fiber. Many different types of natural and synthetic elastomers with different physical and chemical properties are available, and the choice depends on the function of the roll and the operating conditions in the paper machine. The covering may simply protect the metal body from corrosion or provide a cushion for the paper web, forming fabric or felt passing over the roll. On rolls in critical positions on the machine, such as press rolls (q.v.) and pressure rolls, properties of the roll cover composition, including hardness, temperature stability, and thickness, must be carefully designed to produce the required pressure in the operating nip. The cover hardness of paper machine rolls is measured with the Pusey & Jones Plastometer. Hardness values vary from 0 to 1 P&J (rock hard) through 15–40 P&J (medium hard) to 150–250 P&J (extremely soft), determined by the function of the roll and the desired nip action. To measure compositions harder than 0–1 P&J, the Shore D Durometer (q.v.) is used. It is a penetration device and is very useful for measuring rock hard elastomeric and plastic materials. A cover of 0–1 P&J will measure about 85 Shore D. Adhesion to the metal body and the strength and abrasion resistance of the covering must be sufficient to provide satisfactory operating life. Resistance of the covering to the operating chemical environment is an important design consideration. See also COVER ROLL HARDNESS.

COVER PAPER—(1) Any of a wide variety of fairly heavy plain or embellished papers which are converted into covers for books, catalogs, brochures, pamphlets, and the like. (2) A specific coated or uncoated grade made from chemical woodpulps, and/or cotton pulps in basis weights ranging from 40 to 130 pounds (20 x 26 inches – 500) and used as in (1) above. This grade is characterized by good folding qualities, printability, and durability.

CPPA—The Canadian Pulp and Paper Association.

CRACKED EDGE—(1) A broken edge in a paper web usually extending into the web for a short distance only. (2) The term is also applied to a similar defect in metal fourdrinier wires, now obsolete. Cracked edges are not a typical defect in fourdrinier fabrics.

CRACKER-BOX DIVIDER—A chemical or mechanical woodpulp sheet, waxed paper, vegetable parchment paper, or a light board, the weight depending upon the type of product packaged and the size of the container.

CRACKER-BOX LINER—A liner similar to a cereal-box liner. In some cases, depending upon the product, it may be greaseproof and moistureproof, or a waxed paper may be used.

CRACKER-CADDY BOARD—A paperboard used for making boxes in which crackers and cookies are shipped. It may be specially treated to give protection against moisture. The board is strong, odorless, has good bending qualities, and is resistant to rough handling.

CRACKING—(1) Separation of the coating layer or the formation of fissures in the coating during printing or converting processes. (2) Formation of fissures in the crease when a sheet of any paper is folded.

CRACKLE—See RATTLE.

CRASH FINISH—A finish on paper resembling a coarse linen finish, produced by the use of a heavy crash linen cloth in the plating operation or by embossing. Paper so finished is used for a variety of purposes such as cover, papeterie, greeting card. This finish is sometimes referred to as homespun. See LINEN FINISH.

CRATE LINERS—(1) Paperboard used for protection inside of crates. It is a chipboard made with a soft finish to give a cushion between the contents and crate. (2) Papers used to protect the contents of crates from outside contamination. For ordinary merchandise a kraft paper of basis weight 25 pounds (24 x 36 inches – 500) or heavier is generally used. Thus, a laundry-crate liner is made of 25 to 30 pound kraft in a variety of colors. Other liners such as those used in crates for the shipping of celery, lettuce, or other vegetables, or any other product shipped wet, may be made from vegetable parchment, waxed, or waterproof papers. Colored crate liners should have excellent fastness to water bleeding or color transfer to the crate contents under pressure.

CRATERS—Small pits in coated paper, caused by the breaking of air bubbles in the coating.

CRAYON PAPER—A relatively rough-surfaced school paper used for crayon drawing. It is normally manufactured from mechanical pulp in basis weights ranging from 80 to 100 pounds (25 x 38 inches – 500). See also CONSTRUCTION PAPER.

CREASABILITY—The ability of a sheet material to be creased or folded without the appearance in the zone of folding of cracks, sharp lines of bending failure, splitting away of surface coating, or other unsightly manifestations of fractures. This property should be carefully distinguished from brittleness, as the latter property involves a small degree of bending.

CREASE RETENTION—The characteristic of a paper (wrapping, envelope, etc.) to remain in a folded position after being creased by mechanical equipment and without the aid of any adhesive. This property should not be confused with creasability (q.v.).

CREASING STRENGTH—The property of a sheet to retain its tensile strength after folding and creasing under a specified load.

CREEP—The dimensional change with time of the material under constant load following the initial "instantaneous" elastic rapid deformation. Because of the practical difficulty of separating the instantaneous elastic deformation from the delayed deformation, creep is often taken as the total deformation including the "instantaneous" deformation. The phenomenon embraces the partial recovery toward the initial dimensions when the force is removed. Creep at room temperature is sometimes called cold flow.

CREPED—Descriptive of a crinkly paper property produced by crowding a sheet of paper on a roll by means of a doctor, producing thereby an effect simulating crepe. Wet crepe refers to the effect achieved by doctoring a practically dried (50 to 70% solids) sheet, then completing the drying with the sheet in a creped condition. Semi-creped or primary creped or water-creped paper is produced on the paper machine, or as a converting operation, the paper being moistened and passed over a roll equipped with a doctor. Extremely high percentages of crepe may be secured by this process. Modifications of it permit cross-directional creping and diagonal creping, sometimes called an all-directional stretch. Dry creping is a process in which a dry sheet (e.g., 94% solids) is removed from a Yankee dryer by a doctor blade.

CREPED DUPLEX KRAFT PAPER—A duplex sheet composed of two layers of creped kraft laminated to each other with asphalt or other material which is used by nurseries for wrapping roots of nursery shrubs, for wrapping metal parts for export, and as liners for shipping cases. The product may be made by using machine-creped kraft then laminating, or laminating flat kraft and creping in a secondary operation.

CREPED KRAFT PAPER—A bleached or un-bleached kraft paper in various basis weights and with various percentages of stretch, used for wrapping purposes, for bag and barrel liners, and for other converted paper products. It may be creped on the paper machine or in a secondary operation.

CREPED WADDING—Tissue which has been creped and is in roll form prior to converting into a consumer product. See CELLULOSE WADDING.

CREPED WATERPROOF KRAFT PAPER— See CREPED DUPLEX KRAFT PAPER.

CREPE FINISH—(1) A finish produced by embossing or by using creped paper in the plater book in the place of fabrics. (2) See CREPE PAPER.

CREPE PAPER—(1) A general term descriptive of paper made with an effect simulating crepe. See CREPED. (2) MG tissue, water creped (or secondary creped) with the sheet wet in a water solution. Additives such as dyes, sizing adhesives and flame-resisting agents may be present. It has a wide range of uses, including decorating and craft work. (3) See also CREPED KRAFT PAPER.

CREPE RATIO—Cellulose wadding: Crepe is figured on the following bases: 100 times the difference in speed between the Yankee dryer and the winding reels divided by the speed of the reels; the dryer speed divided by the reel speed; or the percentage of stretch in the sheet, using as a basis a given length of the creped wadding. Water crepe and semicrepe: In these grades, the difference in speed of the two ends of the machine is divided by the speed of the Yankee dryer in the case of water crepe or the speed of the creping roll in semicrepe. This percentage is figured by a method exactly the reverse of that used in the case of wadding, i.e., a one-foot sample of uncreped stock would be shortened by creping in the percentages as

listed. Another mode of expressing this ratio is as follows: Crepe ratio 2 means that an inch of the creped paper will stretch or extend to 2 inches when completely pulled out without breaking. This may also be expressed as a percentage, i.e., if a strip is stretched until it is flat and it increases in length from 4 to 6 inches, it is said to have 33 1/3 crepe. Still another method is by the increase in weight over the raw stock, e.g., if a 20-pound sheet, when creped, weighs 30 pounds, it has 50% crepe.

CREPING TISSUE—A dense, well-formed, strong Yankee machine tissue paper, having a high MG finish, and suitable for the manufacture of water-creped paper in a separate converting operation.

CRESCENT FORMER—A gap former (q.v.) in which the headbox jet is injected between a moving felt and a fabric, then wraps a large solid roll in which the felt runs next to the roll and the fabric runs on the outside forcing the stock to drain through the fabric. Transfer boxes ensure the sheet stays in contact with the felt, as it is conveyed to the press and/or dryer. The crescent former is used in tissue-making and the dryer is normally a Yankee (q.v.).

CREVICE CORROSION—Localized corrosion of a metal surface at or immediately adjacent to an area that is shielded from full exposure to the environment, because of close proximity between the metal and another material. Crevice corrosion is observed in mechanical fits (e.g., nuts and bolts, rolled joints) and under deposits or debris.

CRIB SHEET—A pliable, waterproof sheet with a high wet strength. It is used in the nursery to protect the mattress pad under the sheet.

CRILL—Very small bits of fiber, sometimes still attached to the fiber surface, which are largely stripped off during refining. Crill is differentiated from fines (q.v.), in that crill may be much smaller, and highly hydrated, often to the point that it is gel-like.

CRIMP—(1) To crease or break the grain in a sheet of paper so that it will lie flat, as to crimp the binding edge of sheets for looseleaf binders. (2) To crepe.

CRIMP DAMAGE—Indentations in the bottom end of a wrapped roll caused by the crimped folds of the wrapper.

CRIMPER—(1) A device for folding roll wrapper stock over the end of a roll during an automatic roll wrapping operation. (2) A pair of narrow embossing wheels that may have knurled outer diameters that nip multiple plies of tissue in the shear slitter section of a tissue combining operation so that the multiple plies act as one.

CRINKLED—Creped (q.v.).

CRITICAL SPEED—In winding, when a rotating beam such as a winder drum rotates at its natural frequency, that frequency is termed the critical speed. At the critical speed, the amplitude of vibration may be large.

CROCKING—Rubbing off a dye or pigment from the surface of paper and paperboard.

CROSS CUTTER—Obsolete term. See GUILLOTINE TRIMMER.

CROSS DIRECTION—The direction of the paper at right angles to the machine direction (q.v.).

CROSS LAID—The process of laying down a carded web on a moving belt in a back-and-forth motion to produce a nonwoven with a bi-directional fiber orientation.

CROSS LAMINATED—Laminated with some layers of material at right angles to the remaining layers with respect to the grain direction or strongest direction of the sheet. See PARALLEL LAMINATED.

CROSS LINKING—The chemical reaction of a difunctional molecule with each of two molecules of a polymer to form a higher molecular weight material. This structural change of the polymer produces profound changes in its physical properties.

CROSS MACHINE DIRECTION—The direction of the paper at right angles to the machine direction (q.v.).

CROSS-SECTION PAPER—See CHART PAPER; PROFILE PAPER.

CROWN—See ROLL CROWN.

CROWN COMPENSATING ROLL—See VARIABLE CROWN ROLL.

CROWN CURVE—A plot of the crown or camber of a roll developed by recording the roll diameter measurements made with a roll caliper over about 20 equal segments of a roll face. The objective is to simulate a 70° sine curve. However, in some cases, the roll may be ground straight or with a special compound crown.

CROWN FILLER—A hydrated calcium sulfate ($CaSO_4$ $2H_2O$) prepared by the interaction of calcium chloride and sodium sulfate. It is used particularly in high-grade paperies, where a high white color or delicate tint is desired. It is also known as pearl hardening. This has been replaced largely by calcined calcium sulfate.

CRUDE TALL OIL—A dark brown mixture of fatty acids, rosin acids, and neutral materials liberated by the acidulation of soap skimmings. The fatty acids are a mixture of oleic acid and linoleic acids with lesser amounts of saturated and unsaturated fatty acids. The rosin is composed of resin acids similar to those found in gum and wood rosin. The neutral materials are composed mostly of polycyclic hydrocarbons, sterols, and other high-molecular weight alcohols.

CRUSH—See SHEET CRUSHING.

CRUSHED—Having the formation broken by sheet crushing (q.v.).

CRUSHED CORE—A paper roll core which has been crushed.

CRUSHED FINISH—(1) A mottled effect intentionally or unintentionally produced by crushing the paper at the press section of the paper

88

machine so that the paper has a lumpy formation and a mottled finish. (2) A coarse ripple finish applied by a plater press or by embossing.

CRUSHED NEWS—An obsolete term, it referred to old newspapers that had been baled indiscriminately (not piled flat in the baler) as distinguished from flat or overissue news.

CRUSHED ROLL—A roll of paper which has been flattened through pressure or by dropping the roll.

CRUSHING—See SHEET CRUSHING.

CRUSHING STRENGTH—See COMPRESSION (STRENGTH) RESISTANCE.

CRYSTALLINITY—When applied to paper (cellulose) fibers, crystallinity generally refers to the degree of order of the macromolecular cellulose component. Cellulose can exist in allomorphic states. Native celluloses posesses a form labeled Cellulose I, and regenerated cellulose exhibits a new form labeled Cellulose II. Other forms have been proposed. Cellulose I appears to be composed of two distinct forms labeled I alpha and I beta.

CSF—Canadian Standard Freeness. See FREENESS.

CTMP—Chemithermomechanical pulp.

CTO—See CRUDE TALL OIL.

CUAM OR CUPRAM—Cuprammonium hydroxide (q.v.).

CUBICAL DEBRIS—Short fiber bundles with low surface area. There are the primary cause of linting on offset printing. The most suitable tools to remove cubical debris from the pulp are the hydrocyclones and centricleaners.

CUENE OR CUPRIENE—Cupriethylenediamine hydroxide (q.v.).

CULLED BROKE—Dry broke (q.v.) that has been slabbed from a reel or from rewound rolls. Culled broke can be repulped in a dry end broke

pulper, an off-machine broke pulper, or a furnish pulper.

CULTURAL PAPERS—A term applied to papers such as writing and printing used for cultural purposes.

CUNIT—A term used in the measurement of pulpwood—i.e., 100 cubic feet of solid wood, bark excluded. A cunit may be obtained from 2/3 to 1-1/3 cords of wood depending on size, piling, and bark.

CUP BOARD—A paperboard made on either a fourdrinier or cylinder machine and used to manufacture cups of a nested style, constructed with a tapered or sloping sidewall. The caliper ranges from 0.0065 to 0.025 inch (90 to 350 pounds, 24 x 36 inches – 500). The board is usually manufactured of bleached chemical pulp and is hardsized. Folding, beading, and crimping characteristics are important. Cups fabricated from this board are used for hot and cold drinks and in the packing of moist, liquid, and oily foods. See SPECIAL FOOD BOARD.

CUPBOARD LINING—A grade of bleached paper in white and light colors used for lining the shelves of cupboards. See SHELF PAPER.

CUP PAPER—A type of long-fibered bleached kraft or sulfite paper suitable for making cups. The paper is hard sized with rosin and is coated with paraffin or the complete cup is dipped in wax, the weight of the coating depending upon the purpose for which the cup is intended. It has the strength characteristics necessary for crimping, folding, and beading.

CUPRAMMONIUM HYDROXIDE—Often abbreviated cuam or cupram. A solution of cupric hydroxide in aqueous ammonium hydroxide which is capable of dissolving cellulose when the concentrations of copper and ammonium are within certain limits. The solution is also called Schweizer's reagent. The viscosity of a solution of cellulose in cuprammonium is often used as a quality test for pulp strength. The concentrations of copper and ammonia in solvents used for determination of viscosity are specified in standard methods. See VISCOSITY.

CUPRAMMONIUM VISCOSITY—The viscosity of a solution of cellulose or pulp in cuprammonium hydroxide under specified conditions of temperature, cellulose concentration, and solvent composition. It is used as a measure of the average molecular weight of the dissolved cellulose.

CUPRIETHYLENEDIAMINE HYDROXIDE—Often abbreviated cuene or cuprien. A solution of cupric hydroxide in aqueous ethylenediamine which is capable of dissolving cellulose when the concentrations of copper and ethylenediamine are within certain limits. The viscosity of a solution of cellulose in cupriethylenediamine is often used as a quality test for pulp strength. The concentrations of copper and ethylenediamine in solvents used for determination of viscosity are specified in standard methods. See VISCOSITY.

CUPRIETHYLENEDIAMINE VISCOSITY—The viscosity of a solution of cellulose or pulp in cupriethylenediamine hydroxide under specified conditions of temperature, cellulose concentration, and solvent composition. It is used as a measure of the average molecular weight of the dissolved cellulose.

CURING—The process of polymerizing or crosslinking binders which ordinarily occurs at elevated temperatures.

CURING BOX LINER—A vegetable parchment paper, which may be plain or crinkled, in basis weights of 30 to 40 pounds (24 x 36 inches – 500), which is used to line a pickling box (holding about 600 pounds of meat with various ingredients for the pickling process).

CURL—(1) Develops when the two sides of paper do not expand or contract by an equal amount when the paper is exposed to a change in the relative humidity level of the ambient air, such that the paper moisture content changes. The difference in hygroreactivity of the paper between the two sides can be caused by differences in fiber orientation, fines or filler content, or chemical composition. Wetting or drying only one side of the paper can also cause curl. Curl can be introduced by drying the two sides of the paper at a different rate after moisture has been introduced, such as by a size press or coating operation. Generally, the paper will curl toward the side dried last. Curl around an axis parallel to the cross machine direction of the paper is CD curl. (2) Viscoelastic curl is produced when paper or paperboard is wound tightly into a roll and allowed to stand even for a short time interval. The paper near the core will take a permanent set or curl (roll set), which is objectionable when the roll is unwound in subsequent finishing operations such as sheeting. Curl breaking devices which wrap the paper around a small diameter roller or over a sharp edge during unwinding (curl breakers) can remove this source of curl during converting operations.

CURRENCY PAPER—Paper used for printing paper currency, bonds, and other government securities. It may contain distinctive features to protect against counterfeiting. Significant properties are adaptability to printing by the intaglio process, high tensile strength and folding endurance, and resistance to wear. See BANKNOTE PAPER.

CURRENCY STRAP—See BILL STRAPS.

CURTAIN COATING—Applying a paint or plastic coating to a sheet or board which is moving through a continuous flowing curtain of the coating material. The coater may be of gravity or pressure types. Coating films can be varied by adjusting the machine speed and curtain thickness.

CUSHION BOARD—See CORRUGATED BOARD; INDENTED BOARD.

CUT CARDS—Small sizes of cards and tickets in certain standard sizes and shapes, in distinction from cardboard in large sheets. They are used for personal and business cards and for advertising purposes. See CARDS.

CUTLERY PAPER—A thin, white or brown, antitarnish paper (q.v.).

CUT-SIZE CUTTER—A cut-size cutter is a specialized precision sheeter which cuts, slits, and piles into reams of the smaller sheet sizes such as 8 1/2 x 11 inches or 8 1/2 x 14 inches which were formerly "cut-to-size" from piles of large bed sheets in a guillotine trimmer operation. These machines typically cut and slit four rolls from the unwind, and the cutting element is usually a single flyknife unit with a large diameter pull roll for clip accuracy. In some instances, the ream wrapping operation and carton packaging operation for the cut-size papers is "in-line" and continuous.

CUT SIZE PAPER—In the fine paper field, the smaller sheets of business papers typically in sizes 8 x 10, 8 1/2 x 11, or 8 1/2 x 14 (legal size) inches are known as cut size since they are cut down on a guillotine trimmer or on a rotary cut-size cutter.

CUT SCORED—A method of scoring paperboard in which the outer surface is cut by a scoring knife so that the sheet will fold sharply on the score line. It is frequently used in the manufacture of setup cartons, and the surface of the board and scores will be covered by an overlap.

CUTTER—A machine for cross cutting and slitting a single web or a multiplicity of layers into the desired width or length. See also CUT-SIZE CUTTER; DOUBLE FLYKNIFE CUTTER; DUPLEX CUTTER; FOLIO CUTTER; PRECISION CUTTER.

CUTTER BROKE—Trimmings and the waste made during the cutting operation.

CUTTER DUST—Small particles of fibers with mineral coating, or fibers chipped off during the cutting operation, which may adhere to the edge of the sheet and work their way inside the pile of paper, causing printing difficulties.

CUTTER SET—The number of matched rolls on the unwind station of a cutter that are cut simultaneously. Typically, a roll set can vary from only one roll to as many as 10 rolls, depending on the capacity of the cutter design.

CUT TO REGISTER—The cutting of a watermarked paper so that the design falls in a given position in each sheet. See LOCALIZED WATERMARK.

CYANO PAPER—A blueprint paper.

CYCLONE—See CENTRIFUGAL CLEANER.

CYCLONE EVAPORATOR—A device used to contact black liquor (q.v.) with hot recovery boiler flue gases in order to evaporate water from the black liquor.

CYCLOSTYLE PROCESS—A process for making duplicate copies in which a stencil is made by writing or drawing with a pen having at its end a small wheel which makes minute punctures in the stencil paper. Duplicate copies are made by transfer of ink from a small roller through the stencil to the underlaying paper.

CYLINDER BOARD—Any board made on a cylinder machine.

CYLINDER BRISTOLS—This term is applied to any bristol made on a cylinder machine. See INDEX BRISTOLS; MILL BRISTOL.

CYLINDER COUCH ROLL—In a cylinder machine, this roll runs against the sheet-forming mold to produce a pressure nip through which the making felt or mold felt passes. This roll "couches" or picks off the sheet formed by the cylinder mold and causes it to adhere to the bottom of the felt, which transports it to the press section. The making felt may pass over other cylinder molds to pick up additional plies.

CYLINDER DRIED—Dried by passing over internally heated iron rolls. Also termed machine dried. See also BARBER DRYING.

CYLINDER KRAFT LINER—See KRAFT LINERBOARD.

CYLINDER MACHINE—One of the principal types of papermaking machines, characterized by the use of wire or fabric covered cylinders or molds, on which a web is formed. These cylinders may be partially immersed and rotated in vats containing a dilute stock suspension or may be equipped with a headbox or other apparatus for distributing the fibers. The pulp fibers are formed into a sheet on the mold as the water drains through, leaving the fibers on the cylinder face. The wet sheet is couched off the cylinder onto a felt, which is held against the cylinder by a couch roll. A cylinder machine may consist of one or several cylinders, each supplied with the same or with different kinds of stock. In the case of a multi-cylinder machine, the webs are successively couched one upon the other before entering the press section. This permits wide latitude in thickness or weight of the finished sheet, as well as in the kind of stock used for the different layers of the sheet. The press section and the dry end of the machine perform the same functions as those of other types of machines. Due to the slower speeds and heavier basis weights, the press sections may have "open draws" and the dryer sections may have a "stacked" configuration. See DRAW; DRYER SECTION.

CYLINDER MOLD—(1) The complete piece of equipment that rotates in the cylinder vat. It is composed of the shaft, the supporting frame, and the wire or fabric covering. As the mold rotates, the pulp fibers cling to the wire and are carried upward to a contact with the felt. Suction, derived in the vat by the difference in level of the stock inside and outside the mold, regulates the flow and, together with the speed and character of the stock, governs the thickness of the web that is formed. (2) The rotating element in a vacuum filter washer (q.v.).

CYLINDER PAPER—See MOLD-MADE PAPER.

CYLINDERS—DRYER CYLINDERS.

D

DAF CLARIFIER—Dissolved air flotation clarifier (q.v.).

DAMPENING SYSTEM—In offset printing (q.v.), the mechanism that transfers dampening solution to the offset plate during printing.

DAMPERS (DAMPING ROLLS)—See SWEAT DRYER; WATER-COOLED SPRING ROLL.

DAMPING STRETCH—The change in the dimensions of a sheet of paper when it is dampened or moistened.

DAMP STREAKS—Crushed or blackened streaks running in the machine direction. See BLACKENING; CRUSHING.

DANCER ROLL—(1) A compliantly supported web support roll whose position is used as an input to a web tension control system. (2) A roll in a web process whose specific function is to absorb the dynamics of tension transients. See SPRING ROLL.

DANDY—See DANDY ROLL.

DANDY MARK—See WATERMARKING DANDY ROLL.

DANDY PICK—See DANDY ROLL PICK.

DANDY ROLL—A large diameter skeleton roll covered with wire cloth, supported above the forming fabric and allowed to exert a certain pressure on the web at a point where the consistency is in the 2–5% range. It is driven at the same or slightly greater speed than the forming fabric to redisperse the stock remaining to be formed into a sheet, thereby improving its uniformity or formation (q.v.).

DANDY ROLL PICK—Mark caused by the dandy roll picking up fibers from a sheet. These fibers may be subsequently removed or they will interfere with dandy roll operations, printing, etc. See WATERMARKING DANDY ROLL.

DATABASE—Compilation of data in computer storage capable of being shared by multiple information applications. Data can be compared, queried, shared, and reported in any manner desired without disruption to the basic information.

D-BAR SPREADER—A multiple slit web spreading device consisting of a flexible metal strip depressed into a moving web. This adjustable device is oriented perpendicular to the web path across the full width of the slit web just prior to the first winding drum so as to affect the local steering a single or multiple slit webs. This bar has multiple points of adjustment normally located at uniform spacing along the bar.

DBH—A forestry term referring to the diameter of a tree, breast-high.

D BLEACHING STAGE—See CHLORINE DIOXIDE BLEACHING STAGE (D).

DCS—See DISTRIBUTED CONTROL SYSTEM.

DEADBAND—(1) The small band spanning a control setpoint within which the controller can neither sense nor react. (2) The narrow band of regulation between motoring and regeneration of a DC motor.

DEAD BEATEN—See GREASY.

DEADENING FELT—A dry felt used by the construction industry in walls and floors to deaden sounds and to keep out drafts. It is made on a cylinder or fourdrinier machine and may contain from 50 to 70% of roofing rags and from 30 to 50% of news or mixed papers, rags are being replaced to a certain extent by defibrated wood; it is lightly calendered. These felts are made, usually, in three weights of approximately 38, 50 and 75 pounds per roll of 50 square yards (412, 542, and 814 grams per square meter), although lighter and heavier weights may be made on demand. The caliper of the two heavier weights will range from 55 to 60 points (1.4 –1.5 mm) for the lighter and from 82 to 87 points (2–2.2 mm) for the heavier weight. These felts are firm, pliable, durable, free from lumps, and possess a smooth surface. See also CARPETFELT; DRY FELT; LINOLEUM LINING.

DEAD FINISH—A smooth finish without glare.

DEAD SPOTS—Low-finished areas in a highly finished paper.

DEAD TIME—The time delay between an imposed disturbance applied to the input of a process and the resultant change in output. Also referred to as transportation lag.

DEAD WHITE—A neutral white, i.e., one without a perceptible tint.

DEAERATION—(1) The act of removing air from stock, usually by means of a chemical defoamer or by means of a mechanical deaerator. Mechanical deaerators usually spray stock against an impingement plate under a partial vacuum, such that the stock boils, and the entrained and dissolved air is removed. Centrifugal cleaners are often used to spray the stock into the vacuum chamber. (2) Boiler feedwater contains dissolved oxygen and other dissolved gases which can result in corrosion of the boiler and steam lines. Deaeration is the process of reducing the amount of dissolved oxygen, usually by using steam as a purge gas. Dissolved oxygen can also be reduced by using oxygen-scavenging chemicals or applying a vacuum to the feedwater.

DEBARKING—The operation of removing bark from pulpwood prior to shipping, screening, etc. This is carried out by means of a knife (disc), drum, abrasion, hydraulic barker, or by chemical means (rarely used commercially).

DEBENTURE PAPER—Bond paper used in the form of an official document. The usual furnish is cotton fiber which may also be mixed with bleached chemical woodpulps. Good strength and permanence are desirable and significant properties. See BOND PAPER.

DEBRIS—Material not desired in a stock slurry. The designation debris is sensitive to end use requirements—what is not desired for one end use may be highly desired, or at least tolerated, for other end uses. Debris may be naturally occurring, such as shives or chop, ray cells, or bark, or it may be the result of external contamination during processing, or due to additions during converting or end use for recycled fiber.

stock thickening (handwritten)

DEBRIS REMOVAL EFFICIENCY—The debris removed divided by the debris fed, usually in reference to screens or cleaners. Many other definitions and methods of calculation of debris removal efficiency are commonly used; for a complete discussion, see TAPPI Technical Information Sheet TIS 0605-04. Note, it is very important to specify the method of debris analysis when calculating debris removal efficiency.

DECAL—See DECALCOMANIA.

DECALCOMANIA—A process of transferring printed designs to porcelain, wood, glass, marble, etc. It consists usually in gumming the paper or other film bearing the colored picture onto the object and then removing the paper with warm water, the colored picture remaining. Often shortened to decal.

DECALCOMANIA PAPER—An absorbent paper made of cotton fiber mixed with chemical woodpulps or of chemical woodpulps alone, having a smooth, uniform finish and formation with a good wet strength. It is usually made without sizing and in a light-natural color. The basis weights range from 50 to 90 pounds (25 x 38 inches – 500). The base paper is coated with a solution of gum arabic and starch in water (the decalcomania solution). The finished paper is of two types: simplex or single-absorbent paper stock coated with the decalcomania solution; duplex or double-heavy backing paper on which is laminated a very high grade of thin or tissue paper, on which in turn is coated the decalcomania solution which is to receive the printed impression. The heavy backing paper serves to give support to the tissue sheet as it goes through the press and in the placing of the printed design in its proper position on the object. It is used in the manufacture of ceramic or mineral transfers, for curved surfaces, and for very fine lettering. See TRANSFER PAPER.

DECELERATION OFFSET—The change in the cross direction (CD) web steering forces during process deceleration causing a CD displacement of a web in a process such as a winder.

This offset normally occurs with a change in web traction at that speed, such as a collapse of an air film.

DECIDUOUS—A term applied to a tree which loses its leaves annually. Except for tamarack (or larch) and cypress, these are usually broadleaf or hardwood trees.

"deckering" = thicker (handwritten)

DECKER—A stock thickening device that consists of a large, hollow drum covered with a fine filter media, and mounted in a vat. During operation, the stock is introduced into the vat, and it forms a mat on the fine filter media, while most of the water penetrates the filter media, and exits out the ends of the drum. A decker dewaters and thickens pulp without washing it; a washer dewaters and washes a pulp using showers. See also GRAVITY DECKER; VACUUM DRUM FILTER.

DECKLE—(1) In handmade papermaking, the removable, rectangular wooden frame that forms the raised edge to the wire cloth of the mold and holds the stock suspension on the wire. (2) On a fourdrinier papermaking machine, the arrangements on the side of the forming fabric which keep the stock suspension from flowing over the edges of the fabric. The stationary arrangement is a mechanical device for holding a thin and flexible strip of rubber or equivalent material on top of the fabric and just inside the fabric width. The strip restricts the pond or sheet to a chosen width during the period of sheet formation, and therefore varies in its length on different machines. The ruler or strip is made so that its contact with the fabric may be vertical or at an angle and, in some cases, the strip is "showered." As a general rule, deckle rulers are mounted in a fixed position and do not oscillate with shaking fourdriniers. For this reason, pressure of the stationary rubber blade against the traveling fabric varies with the stock, length of ruler, etc. The moving arrangements are a pair of deckle straps that are endless and lie on the fabric at its edges while moving with it. There are two large pulleys at each end to guide and "return" the moving deckle. They retain the pond on a shaking fourdrinier and, in following the fabric minimize "slap" or edge ridging that may occur with the deckle rulers. By traveling at the speed of the fabric they also eliminate the tendency for the stock to roll and "ball up" at the fabric's edge. As machine speeds have increased, the deckle straps are no longer used, but have been replaced by deckle boards, and on lightweight sheets by deckle showers only. Some deckle boards have been replaced by edge curlers. (3) On cylinder machines, the canvas webbing wound around the cylinders at their ends to control the width of the sheet. (4) A term indicating the width of the web formed on the machine.

DECKLE EDGE—The untrimmed feather edge of a sheet of paper formed where the pulp flows against the deckle. This edge may also be produced by means of a jet of water or of air. Generally speaking, handmade paper has four deckle edges and machine-made paper has two; however, by the use of a certain patented procedure, a machine-made paper may be manufactured with four simulated deckle edges. An "imitation" deckle edge is one which is produced on a dry sheet of paper by such means as tearing, cutting with a knife which will give a deckle-edge effect, sand blasting, and sawing.

DECKLE-EDGED BOARD—A board, the edges of which are rough and thin as they come off the machine without trimming. It indicates that the board has been made the full width of the machine. However, the machine can be so equipped that several rolls, each with two deckle edges, can be made at the same time. It is employed mostly for making tubes of the convolute type used in the textile industry.

DECKLE EDGE PAPER—A term most often applied to sheeted fine writings and papeteries having deckle edges on one or more sides. See DECKLE EDGE.

DECKLE STAINED PAPER—Deckle edge paper that is colored or stained along the deckle edges.

DECKLE STRAP—See DECKLE.

DECORATED BLOTTING—A desk-blotting or related grade printed embossed or otherwise embellished for improved appearance characteristics.

DECORATED BOARD—Board that has been embossed, lined, or printed with a design, e.g., oak-grained jute board.

DECORATED BUILDING PAPER—See DECORATED SHEATHING PAPER.

DECORATED COVER PAPER—A kind of cover paper that has been decorated with a design produced by embossing or some other process. The paper may be coated or uncoated, and it may be decorated during or after manufacture.

DECORATED SHEATHING PAPER—A sheathing paper (q.v.) generally gray in color, to one side of which has been applied a design in a combination of colors by printing and suitable for a wall decoration. It is used throughout the southern United States to cover unplastered walls by pasting or tacking, either to the studs or to the wall itself.

DECORATIVE—Having a special design, printed or otherwise produced, intended for decoration.

DECORATIVE LAMINATE—A laminated structure made by heating and pressing together an assembly comprising of fibrous sheets (core sheets) impregnated with a thermosetting resin, such as phenol-formaldehyde, melamine-formaldehyde, or urea-formaldehyde, and a decorative or "print" sheet containing an impregnating resin such as melamine-formaldehyde. In some cases, the print sheet is itself covered by an overlay paper (q.v.), containing an impregnating resin that becomes clear and transparent during the laminating operation. During the laminating operation heat and pressure cure, the impregnating resins and the whole assembly is consolidated into a unitary article. See also PAPER-BASE LAMINATE.

DEED PAPER—Bond paper used in documents such as deeds. The usual furnish is cotton fiber which may also be blended with bleached chemical woodpulps. The paper is generally surface sized. Permanence and durability are significant properties.

DEEP-ETCH OFFSET—A kind of offset lithographic printing in which the image is chemically etched to the order of one ten-thousandth of an inch below the surface of the plate. Its merit is that the plate will stand longer runs than other types of lithographic plates. This process has also been known as offset-gravure (a misnomer) and lithogravure (not recommended).

DEFECT—A product's nonfulfillment of an intended requirement or reasonable expectation for use.

DEFECT DETECTION—The visual, electronic, or mechanical observation of defects such as holes or spots in a paper web or sheet, usually accomplished by a statistical sampling of the product when done off machine.

DEFECTIVE MILL SPLICE—See BAD SPLICE.

DEFECTIVE ROLL ENDS OR EDGES—This type of defect is usually noted as a rough or irregular roll end or edge instead of a smoother, dust free surface. It is usually due to improper winder operation or maintenance with erratic web tension changes, defective slitter knife maintenance or setup, misalignment problems, and/or crushed edges due to improper handling by a clamp tractor during the storage or shipping process.

DEFECTIVE SLITTER EDGE—See DEFECTIVE ROLL ENDS OR EDGES.

DEFECTS—Defects in a paper web, roll, sheet, skid, or package, can result from the raw materials, the papermaking process, the finishing process or during its end use such as printing. Defects can also be the result of transit, handling, or storage. Typical papermaking defects are out of tolerance basis weight, color, caliper, or tensile strength. Typical finishing defects are baggy rolls, corrugations, telescoping, bursts, splices or slitter dust—usually the result of the roll winding process. Typical shipping defects are crimp damage, crushed cores, water damage, and edge damage. Other converting processes also contribute to mechanical defects in paper, which cause problems in subsequent converting operations, such as printing.

DEFIBERING INDEX—An arbitrarily defined and now obsolete term used to express the degree to which paper or broke has been defibered into individual fibers. The currently preferred term is flake content (q.v.).

DEFIBERIZING—See FIBERIZING.

DEFIBRATED PULP—See DEFIBRATED WOOD.

DEFIBRATED WOOD—Pulp produced from wood chips in a refiner, where the wood chips are preheated in a pressure vessel at temperatures over the lignin glass transition point (165°C to 185°C).

DEFIBRATOR—Equipment used for converting wood pulp usually in roll form, to a loose fibrous fluff that ordinarily is used as dry laid paper, diapers, and incontinence products.

DEFLAKER—A high-speed mixer or agitator, or rotor/stator type device through which a fiber-water slurry is pumped to break up any fiber lumps or bits of undefibered paper, and to obtain complete separation of individual fibers. A deflaker typically defibers flakes by fiber-to-fiber rubbing, or hydraulic shear, rather than by close bar-to-bar clearance.

DEFLECTION—The deformation in the direction of the applied load when a corrugated shipping container undergoes the box compression test (q.v.).

DEFLECTION CONTROLLED ROLL—See VARIABLE CROWN ROLL.

DEFLOCCULATION—The dispersion of large bundles of fibers or filler particles into fine agglomerates in aqueous suspension.

DEFLOUR—To remove fines (flour) from pulp by screening or other means. See FINES.

DEFOAMER—A surface-active agent which inhibits the formation of foam or acts on foam or entrapped air to cause the bubbles to break and allow the air to escape. It is usually added in small amounts at the brownstock (q.v.) washers. Also called antifoam agent.

DEGRADATION—In general chemical use, the breakdown of a complex compound to smaller fragments. Specifically for cellulose, the breakdown of the polymer chain, usually by hydrolysis or oxidation. The term degradation is usually applied to changes in chemical structure.

DEGREE OF DEFIBERING—An arbitrarily defined and now obsolete term used to express the degree to which paper or broke has been defibered into individual fibers. The currently preferred term is flake content (q.v.).

DEGREE OF POLYMERIZATION [CELLULOSE]—Degree of polymerization of cellulose represents the number of polymerized anhydroglucose units in the chain molecule. The average degree of polymerization of cellulose represents the average number of polymerized anhydroglucose units per individual chain molecule in a given system. The type of average obtained depends on the method used for the determination. The two most common values are the number average and the weight average.

DEINKED-PAPER STOCK—The pulp resulting from the deinking of recovered paper.

DEINKING—A process for removing most of the ink, varying amounts of the filler and most of the remaining extraneous materials from printed recovered paper. The result is a pulp that can be used alone or with varying percentages of woodpulp in the manufacture of new paper, including printing, writing and office papers, as well as tissue, toweling, and news.

DEINK WASHING—See WASHING (2).

DELAMINATION—The separation of the plies through failure of the bonding. In a pasted or laminated paper product, the separation of the plies through failure of the adhesive.

DELICATESSEN PAPER—A paper used by grocery stores, meat markets, and delicatessen stores as an inner wrap for meats and for soft foods to retain the moisture in the food and to prevent the outer wrapper from becoming water- or grease-soaked. This paper is commonly made from bleached chemical woodpulp and may be given a dry paraffin wax treatment of about 10 to 20% of the weight of the paper (in addition to the ordinary sizing materials). The basis weights range from 20 to 30 pounds (24 x 36 inches – 500). Significant properties include white color, low porosity, uniformity of formation, and waxing in addition to good sizing and resistance to oil penetration. Greaseproof paper and vegetable parchment are also used for this purpose.

DELIGNIFICATION—The process of removing lignin from wood or other cellulosic material by means of chemicals, leaving a residue consisting primarily of cellulose and hemicelluloses.

DEMINERALIZATION—In water treatment, demineralization is the removal of dissolved salts from feedwater. It is an extension of ion exchange to replace dissolved anions and cations by ions that form water.

DENIER—A method of categorizing fibers, filaments and yarns on a grams weight per unit length of 9000 meters basis.

DENSITY—Weight per unit volume. In English units it is expressed as pounds per cubic foot, and in metric units it is expressed as kilograms per cubic meter. See APPARENT DENSITY.

DENSITY ANALYZER—A device that acquires process data, executes necessary computations, and displays the density changes of a winding or unwinding roll during a winding process.

DEOXYGENATION—The removal of oxygen from a sample. Usually refers to water samples.

DEPARTMENT-STORE TISSUE—See WRAPPING TISSUE.

DEPITCHING—The removal of pitch from wood through the action of microbes or enzymes. Lipases are used commercially with mechanical pulps, and pitch-degrading fungi are used commercially with wood chips.

DESIGN OF EXPERIMENTS—A branch of applied statistics dealing with planning, conducting, analyzing, and interpreting controlled tests to evaluate the factors that control the value of a parameter or group of parameters.

DESIGN PAPER—Any paper with a special design marked therein. This type of paper is used for making various types of bags and also as an all-purpose wrapper in department, drug, and notion stores. It is made in basis weights of 25 to 35 pounds (24 x 36 inches – 500). The paper usually has an MG finish and is made in an assortment of colors. Visual properties are most significant.

DESIGN PRINTING—A process of immersing a sheet of paper in a color solution and subsequently passing it between design-marked rollers, so that the design stands out heavily colored or printed against the lighter background of the same color. Gift wrapping paper may be made in this manner; also much safety paper (q.v.).

DESK-TOP PUBLISHING—Use of a computer and graphics program to produce on-screen images that are then converted to hard copy by means of a computer driven printer. No manual layout or paste-up of images and text is required.

DESTRUCTIVELY DISTILLED PINE OIL—See PINE OIL.

DESTRUCTIVELY DISTILLED WOOD TURPENTINE—See TURPENTINE.

DETAIL DRAWING PAPER—A sketching paper used by artists, engineers, draftsmen, etc., for preparing preliminary drawings. It is usually made from chemical wood, cotton or jute pulps in white or cream shades, in basis weights of 60 to 100 pounds (24 x 36 inches – 500). It is characterized by good strength, durability, and erasability.

DETRASHER—Represents a broad class of equipment used to remove gross debris from stock.

DEVELOPING PAPER—A general term for all photographic developing-out papers. They are treated with an emulsion which is sensitive to light but which requires the use of chemicals to make the image visible and permanent.

DEWATER—(1) To extract a portion of the water present in a sludge or slurry. (2) To drain or remove water from an enclosure. A riverbed may be dewatered so that a dam can be built in the dry; a structure may be dewatered so that it can be inspected or repaired.

DEWATERING AID—A flocculant added to stock to increase drainage of water.

DEWAXED WEIGHT—The basis weight of a base paper after wax has been extracted.

DEWAXING—The operation of removing wax or paraffin from a waxed paper.

DEW POINT CORROSION—The action of corrosive condensates at temperatures below the dew point (at which the condensates vaporize and condense at the same rate).

DEXTRIN—A carbohydrate produced from starch by hydrolysis with acids, enzymes, or dry heat. It is a white or yellowish white powder soluble in water or alcohol, and used as an adhesive in envelopes, gummed papers, tapes, etc.

DIAGRAM PAPER—See CHART PAPER.

DIALDEHYDE STARCH—Organic compounds formed by the oxidation of starch. Among the uses of these compounds are the insolubilization of casein and other protein adhesives, improved wet and dry strength of paper, and improved wet rub resistance of coating binders.

DIALYZING VEGETABLE PARCHMENT PAPER—A special vegetable parchment used as a membrane in the dialyzing process.

DIAPER COVER-STOCK—The outside wrapper portion of a diaper that contacts the baby.

DIAPHRAGM PAPER—Obsolete term. An asbestos paper used for filtering purposes.

DIATOMACEOUS EARTH—A filter medium used for filtration of effluents from secondary and tertiary treatments, particularly when a very high grade of water for reuse in certain industrial purposes is required; also used as an absorbent for oils and oily emulsions in some wastewater treatment designs.

DIATOMACEOUS SILICA—An amorphous silica formed from the residues of aquatic plants known as diatoms. It is used as a dulling or flattening agent in coating and as a filler in paper. Also known as diatomaceous earth, diatomite, infusorial earth, and kieselguhr.

DIAZOTYPE BASE STOCK—A special paper designed for light-sensitive diazotype "whiteprint" coatings. It is usually made of chemical wood and/or cotton pulps in basis weights ranging from 17 to 24 pounds (17 x 22 inches – 500) and is characterized by heat stability, chemical purity, good physical strength, cleanliness, and brightness.

DIAZOTYPE PAPER—A paper coated with light-sensitive diazo compounds and used in certain office and engineering machines of the "direct-print" type. Also called whiteprint paper.

DICHROIC—Exhibition of two different colors, dependent on view angle or concentration.

DICHROMATE OXYGEN CONSUMED (DOC)—See CHEMICAL OXYGEN DEMAND.

DID BAGS—Disposable inflatable dunnage (q.v.) used during paper roll transport to secure loads.

DIE CUT—The process by which paper or board is cut or stamped out to a specified shape or size by means of a steel die.

DIE-CUTTING—The process of using sharp steel dies to cut labels, boxes, and other particular shapes from printed work.

DIE EMBOSSING—A supplement to other printing processes whereby a brass or steel die, having its design cut intaglio (q.v.), is used, either hot or cold, on a powerful press to impress the design in relief form into the paper or other substrate.

DIELECTRIC CONSTANT—See SPECIFIC INDUCTIVE CAPACITY.

DIELECTRIC LOSS—See POWER FACTOR.

DIELECTRIC PAPER—A paper substantially free from metallic or other impurities which are capable of conducting electricity; it is used as a dielectric material.

DIELECTRIC STRENGTH—That property of a material which resists the passage of electrical spark discharge. Specifically, it is the potential difference (in volts) at which a spark passes through a specimen of specified thickness under specified conditions; it is usually expressed as a voltage gradient in volts per mil thickness. This property should not be confused with dielectric constant (specific inductive capacity).

DIE STAMPING—An intaglio process (q.v.) for the production of letterheads, cards, etc. by printing from lettering or other designs engraved into copper or steel. Ink is smeared over the surface of the die, the surface is wiped, and the ink remaining in the design is printed under heavy pressure, which also partially embosses the paper. The term copper and/or steel engraving covers this process also.

DIE-WIPING PAPER—(1) A paper used in the printing trade for wiping the surface of printing plates. It is a well-formed sheet made from unbleached or semibleached sulfite or kraft or mechanical pulp, or mixtures of these, ranging in basis weights from 25 to 60 pounds (24 x 36 inches – 500). It usually has a high finish (water finish, machine glazed, or supercalendered) and a smooth surface. It is supplied in rolls of various widths and diameters. Significant properties include sizing, strength, finish (freedom from fuzz or lint), and the absence of abrasive particles. (2) A relatively absorbent paper made of mechanical and chemical woodpulps in a basis weight of 25 to 60 pounds (24 x 36 inches – 500) which is machine creped and lightly calendered. Often shortened to wiping paper.

DIFFERENTIAL PRESSURE—In paper drying, a common term for the pressure difference between steam at the entrance of a dryer cylinder and steam at the exit.

DIFFUSION-TRANSFER BASE STOCK—A paper having excellent formation, a high degree of wet strength and a very smooth surface for the effective application of a silver halide-gelatin emulsion. It is made of a very pure stock free from iron, copper, and sulfur and is resistant to yellowing when exposed to a caustic solution. It may be highly fluorescent. It is converted into copying papers for office reproduction.

DIGESTED SLUDGE—Sludge digester under either aerobic or anaerobic conditions until the volatile content has been reduced to the point at which the solids are relatively nonputrescible and inoffensive.

DIGESTER—A batch or continuous vessel used for pulping fibrous raw materials to remove lignin and produce pulp. The vessels are fabricated from metals that will withstand high temperatures, pressures, and corrosive chemicals. The size of digesters has increased with larger capacity mills. It is possible to have a single continuous digester that will process as much as 10 or more older batch digesters. See PULPING.

DIGESTING—See DIGESTER; PULPING.

DILATANCY—The phenomenon observed when the viscosity of a material increases with an increase in the applied rate of shear. Such behavior is common in high solids suspensions and may cause difficulty in pumping such materials or in their transfer in a coating operation. There is always an apparent volume increase with dilatancy. Dilatancy can be the cause of blade coating scratches and streaks.

DILUTION ZONE—A section at the bottom of a bleaching or storage tower where dilution water is added to reduce pulp consistency to allow easy removal from the tower and pumping.

DIMENSIONAL STABILITY—That property of a sheet of paper that relates to the constancy of its dimensions, especially as they are affected by changes in moisture content, with compressive or tensile stresses, or with time under stable ambient conditions. See CREEP; DRIED-IN-STRAIN; HYGROEXPANSIVITY.

DIMETHYL SULFIDE ($CH_3)_2S$—A colorless, flammable liquid compound with a disagreeable odor. Found in the relief gases in kraft pulpmaking. The compound is very poisonous.

DIMETHYL SULFOXIDE ($CH_3)_2SO$—A waterlike, highly polar, water-miscible, hygroscopic organic liquid commonly referred to as DMSO. It is a by-product of sulfite pulping and is used for various industrial purposes as a solvent, reaction medium, and chemical reactant. It also has been used experimentally for certain medical applications.

DIN—Abbreviation for Deutsche Industrie Norm(en), a set of industrial standards, which preceded the International Standards Organization (ISO).

DIOXIN—A family of toxic polychlorinated hydrocarbon compounds containing a particular benzene or furan ring structure. The most toxic of the family are 2, 3, 7, 8-tetrachlorodibenzo-p-dioxin (TCDD) and 2, 3, 7, 8-tetrachlorobenzo-p-furan (TCDF). The term "dioxin" refers to the entire family of related molecules. See also FURAN.

DIP—Deinked pulp.

DIPPING—Dip dyeing. See TUB COLORING.

DIRECT DYES—A class of aniline dyes, also called substantive dyes, which have a high affinity for cellulose. They are characterized by a high level of light fastness.

DIRECT ENTRY GRADES—See PULP SUBSTITUTES.

DIRECTION (MACHINE)—See CROSS DIRECTION (CD); MACHINE DIRECTION (MD).

DIRECT LITHOGRAPHY—A planographic printing process in which the paper (or cloth, etc.) to be printed comes into direct contact with the inked image area of the printing stone or metal plate in a printing impression. The design to be printed is drawn or transferred to the surface of the stone or metal plate with a special crayon or developed by photographic procedures. The resulting surface of the stone or metal plate is ink receptive in the image area while water receptive in the blank area. The selective wetting by water and ink on these two surfaces when presented in turn to water and ink by appropriate application roller systems permits inking of the imaged area.

DIRECTORY PAPER—A lightweight printing paper designed for the printing of telephone directories, catalogs, and similar products. It is usually made from mechanical, chemical wood, or reclaimed pulps in basis weights ranging from 18 to 28 pounds (24 x 36 inches – 500). It is characterized by good printability, high opacity, and moderate physical strength.

DIRECT PROCESS PAPER—A paper used for direct process reproduction (frequently known as the diazo process) which is made from bleached chemical woodpulps, although some may contain 25 to 50% cotton. The paper must have a very uniform formation, freedom from impurities (especially iron), extremely hard sizing, low pH (ca. 4.5), and a high finish. It must also have good fold, tear, opacity, and bright-

ness. The paper is tub sized (with starch) and is calendered before and after sizing. It is made in basis weights of 17, 20.5, 24, and 32 pounds (17 x 22 inches – 500) and in heavier specially pasted sheets. A certain degree of wet strength is desirable, although the wet tensile strength and wet rub resistance are not as important as in blueprint papers.

DIRECT TYPE DUPLICATOR—See SPIRIT DUPLICATION.

DIRT—Any foreign matter embedded in a sheet of paper, paperboard, or pulp, and which has a marked contrasting color to the rest of the material when viewed by reflected or transmitted light. In paper, it is generally determined by reference to a standard dirt chart.

DISC FILTER—A stock thickening device that consists of multiple hollow discs, covered with a fine filter media, and mounted on a shaft in a vat. In operation, the stock is introduced into the vat, and it forms a mat on the fine filter media surface, while most of the water flows through the filter media, and is removed through the hollow shaft. Also called a vacuum disc filter (q.v.).

DISC REFINER—A refiner (q.v.) whose working elements consist of one or more matched pairs of discs having a pattern of ribs machined into their faces and arranged so that one disc of the pair is rotated. The other disc is usually stationary but may be driven in the opposite direction of rotation. Precision controls are provided for adjusting the clearance between the disc faces. The discs are enclosed in a case arranged so that a suspension of papermaking stock can be pumped in and caused to flow radially from the center out, or vice versa, between the rapidly moving ribbed surfaces of the discs, thus resulting in refining. The refining action on the fiber material is dependent upon such variables as pressure between the two discs, the exact pattern of ribs on the discs, peripheral speed, and consistency of the pulp suspension.

DISCHARGE PERMIT—Permit issued by state or federal government which regulates the dis-

disintegrator

charge of water or air emissions from a commercial facility.

DISHBOARD—See PLATEBOARD.

DISHED ROLL—A roll wound with a progressive edge misalignment, which results in a convex shape on one roll edge and a concave shape on the other.

DISHING—(1) See WAVY EDGES. (2) A condition of paper in piles, in which the sides are higher than the center of the pile.

DISINTEGRATION RESISTANCE—The resistance of pulp or paper stock to complete dispersal in water, using a standard propeller-type disintegrator under specified conditions.

DISPERSE VISCOSITY—The viscosity of pulp dispersed in a suitable solvent, such as cupriethylenediamine, and measured by the falling-ball or viscosity pipet procedure.

DISPERSING AGENT—Any material added to a suspended medium to promote the separation of the individual, extremely fine particles of solids, which are usually of colloidal size. Typical applications of dispersing agents include their use in grinding of pigments for fine enamels and dispersing of certain water-insoluble dyes to secure uniform dyeing. The term is often interchangeable with emulsifying agent or emulsifier.

DISPERSION—A process for dispersing contaminants and fiber flakes in paper and board by submitting a ca.30% consistency slurry of fiber in water to elevated temperatures and pressures in a vessel for 5 to 10 minutes, then diluting to ca.12% consistency and ejecting through a single disc refiner. The process does not remove the contaminants but disperses them so particles are not readily visible.

DISPLACEMENT BATCH PULPING—See EXTENDED DELIGNIFICATION.

DISPLACEMENT BLEACHING—Multistage bleaching performed in one or two towers. Pulp is pushed through a tower in plug flow; chemi-

cals and wash water are injected to treat the pulp, displace the solutions of the previous stages which, in turn, are displaced out of the pulp by subsequent stages. This form of bleaching has the advantages of rapid reaction rates, savings in heat, power, and chemicals, and the reduction of the number of towers required for a bleaching sequence.

DISPLAY BOARDS—Generally thick paperboards, pasted or unpasted, used for printed advertising display.

DISPLAY PAPER—See SEAMLESS DISPLAY PAPER.

DISSOLVED AIR FLOTATION CLARIFIER—A water purifying device in which flocculating and coagulating chemicals are added to incoming raw wastewater, along with water containing air which has been dissolved under high pressure. As the pressurized water is depressurized to atmospheric pressure, the dissolved air forms tiny bubbles, which collect the flocculated suspended solids and float them to the surface of the water. The floated solids are then skimmed off, producing concentrated suspended solids, and relatively clean effluent. Also called a DAF clarifier.

DISSOLVED OXYGEN—Amount of oxygen, expressed in parts per million, dissolved in water.

DISSOLVED SOLIDS—The solids that remain in a water sample after passage through a 0.5-micron glass fiber filter pad. It is a measure of the solids that are dissolved in the water.

DISSOLVING PULP—A special grade of chemical pulp usually made from wood or cotton linters for use in the manufacture of regenerated cellulose (e.g., viscose rayon and cellophane) or cellulose derivatives such as acetate, nitrate, etc.

DISSOLVING TANK—A tank set below a recovery boiler which receives the flow of molten inorganic pulping chemicals and dissolves them in a flow of weak wash.

DISTRIBUTED CONTROL SYSTEM (DCS)— A network of control components distributed throughout a process (or group of processes), sharing their control functions by means of a local area network. This concept offers versatility of computing size, location, and capability, allowing networking within and between systems. A DCS refers to a wide range of hardware, software, and control strategies, including automated startup and grade changes on the paper machine.

DISTRIBUTOR ROLL—Used in certain types of headboxes to even out flow irregularities to create turbulence and keep the fibers deflocculated. Distributor rolls are usually made of highly polished stainless steel and drilled with large holes to make an open area as large as 50%. Some rolls are covered with a rock hard (0–1 P&J) (q.v.) elastomeric cover, which is then drilled to produce the open area. These are also referred to as rectifier rolls, perforated rolls, or holey rolls.

DITHERING—In digital imaging, the process of inserting a pixel between adjacent pixels which has the average tonal value of the adjacent pixel. Dithering smooths images.

DITHIONITE—See SODIUM DITHIONITE.

DIVIDERS—Sheets of boards or paper used to separate the layers of candy, biscuits, etc., when put up in fancy boxes. See CHOCOLATE DIVIDERS AND LAYER BOARD; PADS.

DKL—Double kraft lined.

DLK—Double lined kraft.

DMSO—Dimethyl sulfoxide.

DO—See DISSOLVED OXYGEN.

DOC—Dichromate oxygen demand. See CHEMICAL OXYGEN DEMAND.

DOCTOR—A thin plate or scraper of plastic, metal, or other hard substance placed along the entire length of a roll or cylinder to keep it free from paper, pulp, size, etc., and thus maintain a smooth, clean surface. See also CREPED.

DOCTOR BROKE—Paper that accumulates on the press doctors through breaks in the web.

DOCTOR DUST—Dust that accumulates on dryer and calender doctors. From the doctors, it may be attracted to the paper and be pressed into it.

DOCTOR MARKS—Ridges made by doctors (q.v.) on press rolls which, in turn, mark the paper.

DOCTOR RIDGES—See DOCTOR MARKS.

DOCTOR ROLL—See TAKE-OFF ROLL.

DOCTORS—See PRESS ROLL DOCTORS.

DOCUMENT PARCHMENT—(1) A paper made to resemble animal parchment and used for diplomas, commissions, acts of Congress, and treaties where animal parchment was formerly used. It is made from high-quality linen and cotton fibers on a fourdrinier machine; it may or may not be engine sized with rosin but is surface sized with the highest quality animal glue or with a special tub sizing. The basis weights are 48, 56, 72 and 88 pounds (17 x 22 inches – 1000). The paper should possess excellent durability and permanency. (2) A vegetable parchment paper used for diplomas and documents.

DODGERS—Colored or white newsprint, or groundwood poster used in small sizes and bearing advertising matter for hand distribution.

DOFFER—The last or output roll of a carding machine.

DOG EAR—An oversized corner of a sheet of paper, formed when a sheet having a corner turned under is guillotine trimmed with other sheets in a pile; when the folded corner is then unfolded, it extends beyond the trimmed size of the sheet.

DOG HAIRS—Protruding fibers on the surface, usually of coated paper. Dog hairs are generally longer than fuzz (q.v.).

DOILY PAPER STOCK—See LACE PAPER.

DOMINANT WAVELENGTH—A colorimetric quantity used to designate hue. It is the wavelength of the spectrum color which must be added to (or subtracted from) the illuminant to make the illuminant's hue match that of the specimen when viewed under the same illuminant. Dominant wavelength is one of the three quantities used in the CIE specification of color. See CIE COLOR; COLOR.

DOMINO PAPER—An early kind of wallpaper decorated with a small repeated design (square), picture designs, or even, "marbleized," also with figures and grotesques, originally printed from blocks and colored by hand.

DOPE—A solution of a cellulose ester (such as cellulose acetate or nitrate) or a cellulose ether (such as ethyl cellulose or benzyl cellulose) in a volatile solvent (such as acetone, amyl acetate, etc.), which may be used as a coating material. See also LACQUER.

DOS—Disk operating system. See IBM.

DOT GAIN—Enlargement of printed dot size relative to plate dot size, which causes darker colors and reduced detail.

DOT-MATRIX PRINTER—A computer printer which uses tiny dots to create images. Earlier dot-matrix printers produced crude images; later dot-matrix printers produce near-letter-quality printing.

DOTS PER INCH—Also called DPI, dots per inch is a measure of resolution of a screen or printing plate.

DOUBLE CALENDERED—Paper which has been run through the supercalenders twice, generally coated paper.

DOUBLE COATED—(1) A term applied to a paper or board which has been coated twice on the same side with the same or different materials. The term is also used (incorrectly) to designate a paper or board coated on both (C 2 S) sides. (2) A term applied to a paper or board with a heavy coating (but not necessarily two coatings).

DOUBLE DECKLE—Having a deckle on both edges (of machine-made papers).

DOUBLE DISC "DD" REFINER—A refiner where the rotating disc is located between two stationary discs. Both sides of the rotating disc are equipped with refining tackle (refiner plates). Each side of the rotating disc interacts with one of the stationary discs. The rotating disc floats between the two stationary discs and positions itself in the center to balance the thrust load. The refiner is used for low-consistency, stock preparation applications.

DOUBLE-DUTY SISAL TAPE—See GUMMED SISAL TAPE.

DOUBLE FACED CORRUGATED BOARD—The board most commonly used for corrugated containers and many other purposes. It is made on a corrugating machine wherein a corrugating medium after fluting is faced on each side, usually with linerboard. Also called single wall corrugated board (q.v.).

DOUBLE-FACED PAPER—Paper having a different color on each side. See also DUPLEX.

DOUBLE FELTED PRESS—Any press nip (q.v.) where the sheet is sandwiched between two press fabrics or felts (q.v.), each felt backed by a roll or shoe support.

DOUBLE FLYKNIFE CUTTER—Uses two synchronized rotary knives to cut webs to length as compared to a conventional cutter with one rotary knife and a bed knife to perform the scissor action. Double flyknife cutters can handle heavier basis weights and knives will run longer and produce cleaner cuts than single flyknife cutters.

DOUBLE KRAFT LINED—See DOUBLE LINED KRAFT.

DOUBLE LINED KRAFT—The converting waste product when corrugating medium and linerboard are converted into multi-wall corrugated board. Also called double kraft lined (DKL) or corrugated clippings.

DOUBLE MANILA-LINED CHIPBOARD—Double manila-lined board. See COMBINATION BOARD.

DOUBLE SIZING—A method of tub sizing, in which the paper is sized in the usual manner, then dried by passing over paper-machine dryers, after which the sizing operation is repeated. Obsolete as a commercial practice.

DOUBLE STRENGTH CORRUGATED BOARD—Term replaced by double wall corrugated board (q.v.).

DOUBLE THICK COVER PAPER—A heavy coated paper made of two sheets of regular weight cover paper pasted together.

DOUBLE VAT LINED—See VAT LINED.

DOUBLE WALL CORRUGATED BOARD—A board made by combining two single face (q.v.) webs with an additional facing. The three facings and two corrugated members thus make a single board which possesses greater strength than double faced corrugated board made from the same materials. See also TRIPLE WALL CORRUGATED BOARD.

DOUBLE WHITE PATENT–COATED BOARD—A paperboard vat lined on both sides. See PATENT COATED.

DOUBLING—The unintentional printing of two images slightly out of register. It is particularly harmful in halftones where it increases tone and color values. It can be caused by rippling or premature contact of paper with the offset blanket or by sheets slipping or stretching. Doubling can also be caused by cylinder misalignment and other mechanical press conditions.

DOUGHNUT-BAG PAPER—A paper which is sometimes supercalendered, used for the manufacture of doughnut bags. Greaseproofness is a desirable characteristics.

DOWNFLOW TOWER—A bleaching vessel in which pulp enters from the top and is removed at the bottom. Retention time may be varied by changing the level or consistency of pulp in the tower.

DOWNSTREAM—Reference to the direction of a process or flow of paper. In the case of winding, downstream points to the discharge side of the winding rolls.

DP—Degree of polymerization.

DPI—Dots per inch (q.v.).

DRAFTING PAPER—See DRAWING PAPER.

DRAG SPOTS—Irregular thin streaks or lumps caused by agglomerations of stock adhering to the slice. Such an agglomeration reduces the flow of stock at that point, causing a thin streak and, when it breaks loose, causes a lump.

DRAINAGE BOX—See DRAINAGE SHOE.

DRAINAGE FACTOR—The slope of the lineal graph resulting from plotting the quantity of stock in a sheet machine against the drainage time, expressed as seconds required per gram of pulp to drain under standard conditions.

DRAINAGE SHOE (DRAINAGE BOX, BLADED BOX)—A drainage device located on the underside of a forming fabric on a flat fourdrinier or inside one of the forming fabrics on a multi-wire former. A drainage shoe usually refers to a stationary box that has a solid surface, usually curved. The fabric conforms to the surface and its direction changes as it goes over the surface. Drainage occurs when water is forced into the backing fabric (on a multi-wire machine) by centrifugal force.

A drainage box is a device, similar to a drainage shoe, in which the surface has an open area, usually slots which are formed as part of the cover. The drainage box can have either a flat or curved shape, and it can be at atmospheric pressure or under vacuum.

A bladed box, also similar to a drainage shoe, is a device in which the slotted open area is the space between the blades. The major difference between a drainage box and a bladed box is that the blades can be removed individually.

DRAINAGE TIME—The number of seconds required for a charge of stock to form a mat in a laboratory sheet-making machine, under standard conditions. It gives a measure of the drainage rate of the stock.

DRAINER STOCK—Pulp or paper stock kept in a drainer (a chest with bottom so arranged as to permit the escape of water but not of fibers).

DRAPERS' CAPS—Very thin brown papers, which are machine glazed, used for wrapping small articles.

DRAW—(1) The tension applied to the paper between sections of a paper machine, such as the press section and dryer section. (2) The difference in speed between sections of a paper machine. It is calculated by substracting the speed of the upstream section (closest to the headbox) from the downstream section (closest to the reel). (3) A machine-direction gap the paper web must pass through where it is not supported by a fabric. Also called an open draw. See also SHEET TRANSFER; WEB DRAW; WEB TENSION.

DRAW CONTROL—A control device or system to control the draw or relative draw between any two stations. This is used in web processes where the web stiffness is very low.

DRAW DOWN—A method of surface application for testing purposes utilizing a wire-wound rod coater (q.v.), a machined metal block, or other instrument to wipe an excess of coating color (q.v.) or ink from a sheet of paper thus leaving a metered quantity on the surface.

DRAWER-LINING PAPER—See SHELF PAPER.

DRAWING BOARD—A paperboard used for crayon or water-color drawings. It is made of woodpulp and reclaimed paper stock of sufficient thickness to withstand bending. The board is sized, has a good texture, and is finished without gloss.

DRAWING PAPER—(1) Paper used for pen or pencil drawings by artists, architects, and draftsmen. It is usually a machine-made paper in North America, but in some countries most of it is handmade. There are a number of papers in this class or type that have certain properties emphasized to fit a special need. Thus, there are architect's, art, charcoal, crayon, detail, drafting, manila, matte art, rope, school, and vellum drawing papers. Chemical woodpulps, cotton, or mixtures of these fibers are used in the manufacture of drawing paper in basis weights of 80 and 100 pounds (24 x 36 inches – 500). This paper has a good writing surface for pencil, good erasability, and a dull or low finish. (2) A grade of paper containing about 75% mechanical pulp with the balance unbleached chemical woodpulp. The sheet is principally used by schoolchildren for sketching, crayon, or watercolor work. It is hard sized and has a "toothy" surface. The usual basis weights are 56, 64, and 72 pounds (24 x 36 inches – 500). It is usually supplied in manila and gray colors.

DREGS—Fine particulate in raw green liquor (q.v.) derived from the carbon in recovery boiler smelt and lime mud particles suspended in weak wash.

DREGS WASHER—A piece of equipment to receive dregs from the underflow of a green liquor (q.v.) clarifier and to wash the sodium salts out of the dregs with fresh water.

DRIED-IN STRAIN—The portion of the potential strain that is retained in paper because of tension or the restraint of shrinkage during drying. The magnitude of the dried-in strain decreases with time; this decrease is accelerated by exposure to high humidity or by wetting. Other terms sometimes used in reference to this property are dried-in stress, built-in strain or stress, and frozen-in strain or stress.

DRIED-IN STRESS—See DRIED-IN STRAIN.

DRILL METHOD (MOISTURE)—A method of sampling (1) baled pulp in sheets, (2) roll pulp, (3) baled shredded pulp having 50% moisture or less and (4) hydraulic laps having 64% moisture or less. The samples to be tested for moisture are taken by boring or drilling into each hole with a tool which cuts a disc about 4 inches in diameter.

DRINKING-CUP PAPER—See CUP PAPER; CUP BOARD.

DRINKING STRAW PAPER—A paper made from strong, bleached chemical woodpulp, in basis weights from 26 to 41 pounds (24 x 36 inches – 500), which is used for the manufacture of spirally wound tubes for drinking purposes. The tube is paraffined in the conversion process and simulates a hollow stem or stalk. Essential properties are strength (particularly tensile strength), stiffness, freedom from dirt specks, and uniformity of caliper; the finish, sizing, density, and hardness requirements vary with the conversion process.

DRIOGRAPHY—A planographic printing process not requiring a dampening system.

DRIVE SIDE—The side of a machine where the connecting drives for the rotating paper machine elements are located. It is the side opposite the aisle, or operating side of the machine, and is normally not as accessible to operating personnel.

DROPLEG—The pipe attached to the side of the cylinder washer that creates a vacuum inside the washer drum by functioning as a syphon.

DROP MARKS—Marks in or on paper caused by water dropping onto it while on the forming fabric.

DROP-OFF—(1) A dropping off of the couched sheet from the underside of the bottom felt on a cylinder machine. (2) A small side sheet made when the order does not fill the machine trim.

DROP OUT—A form of relief printing where the image is the uninked area within a wider inked area.

DROP TESTS—Procedures for determining the ability of containers and other packages to withstand impact in free fall drops.

DRUG BOND WRAPPING—See DRUG WRAPPING.

DRUG WRAPPING PAPER—A wrapping paper made from bleached chemical woodpulp in basis weights of from 30 to 40 pounds (24 x 36 inches – 500) with a low finish and in a variety of colors, which may be printed, embossed, or decorated by means of distinctive press marks or by marking felts. It is specified in rolls 9 to 36 inches in width and 9 inches in diameter. It is used for wrapping purposes in drugstores and fancy-goods stores. Cleanliness is especially important.

DRUM—See FIBER DRUM.

DRUM DRYERS—See DRYER CYLINDERS.

DRUM FILTER—See VACUUM DRUM FILTER.

DRUM GROOVING—The specific pattern of grooving machined into the cylindrical surface of a winding drum. The function of the grooving is for traction improvement and/or for air film reduction.

DRUMHEAD MANILA—A strong rope paper which derives its name from the fact that it is used for musical drumheads. It is often used for other purposes because of its exceptional strength and durability. The basis weight is about 140 pounds (24 x 36 inches – 500).

DRUM LINER—A quality of linerboard used in the fabrication of fiber drums.

DRUM SPEED DIFFERENTIAL—The difference in the surface speed of the drums of a two drum winder, normally expressed as a percentage of winding speed.

DRUM TEST—Procedure, using a revolving drum, to determine the ability of a shipping container to withstand a variety of shocks and impact stresses simulating those to be expected in actual handling or shipping or to determine the relative ability of packages to protect their contents under such conditions.

DRY BROKE—Broke produced at the dry end of the paper machine, or is culled from the reel or from rewound rolls. See BROKE.

DRY COATING—The process of applying an adhesive to paper by means of a roll or brush and immediately running it through a box in which finely divided pigment is suspended. See HOT-MELT COATING.

DRY CREPING—See CREPING TISSUE.

DRY END—The mill term for the dryer section of the paper machine, consisting mainly of the dryers, calenders, reels, and slitters.

DRY END PULPER—A UTM broke pulper (q.v.) located under the dry end of a paper or board machine. Dry end pulpers often handle stock from a variety of broke-generating locations near the dry end of the paper or board machine, including the various dryer sections, the calender, the winder and, sometimes, the rewinder.

DRYER BARS—Bars located on the inner diameter of a steam-heated dryer cylinder which enhance the condensate heat transfer coefficient. These bars are oriented in the cross-machine direction. They are spaced around the circumference of the cylinder to induce a sloshing in the condensate at the natural frequency of the pool of water between the bars.

DRYER CANS—See DRYER CYLINDERS.

DRYER CYLINDERS—A series of steam-heated metal cylinders, 30 to 84 inches (762 to 2134 mm) in diameter, varying in number from 20 to 130 or more (newer paper machines have some 20 to 50 cylinders), and arranged in sections. The dryers within a section have a common fabric. The cylinders are driven by interconnected gears or by electric drives on cylinders or fabric rolls with the non-driven rolls turned by the fabric. The temperature of the cylinders, their number, and their speed determine the capacity of the paper machine. Dryer cylinders are also known as dryer cans and drum dryers. See also BABY DRYER; DRYER SECTION; FOURDRINIER MACHINE; PLATEN DRYER; TUNNEL DRYER; YANKEE DRYER; YANKEE MACHINE.

DRYER FABRIC—The machine clothing (q.v.) that conveys and/or holds the sheet to the dryer cylinder and removes moisture by the evaporative process. It is usually an open mesh structure installed via an "on-machine" seam. Because the dryer section is exceptionally hot and humid, dryer fabrics are made of polyester and of patented or proprietary fibers and filaments that resist hydrolysis. These fabrics are usually made in a layered construction to increase their stiffness, durability, and tension-holding capability. Shaped strands such as "flat warp" are used to increase contact area and fabric stability. The sheet side of the dryer fabric can be made to have a high contact area to improve drying and improve the ability of the fabric to "hold the sheet," e.g., single-tier dryer configurations (q.v.). The air-carrying and air-pumping characteristics of dryer fabrics are important considerations for sheet stability in some dryer configurations. Dryer fabrics are generally named for the part of the dryer section they clothe. Common names are first top, first bottom, second top, second bottom, etc. Dryer fabrics are also called dryer screens.

DRYER FABRIC MARKS—Although unusual, some dryer fabrics can cause a faint mark or impression in the sheet. Usually, this impression is much less intense than a felt mark. Dryer fabric seams can also cause a mark in the sheet. See FELT MARK.

DRYER FELT—See DRYER FABRIC.

DRYER FELT MARKS—See DRYER FABRIC MARKS.

DRYER HOOD—A covering over the whole or part of the dryer section of the paper machine or coating unit which serves to collect the hot moisture-laden air from the drying process and exhaust it through suitable fans and duct work. It is designed to control air flow for uniform and rapid drying. Modern units include energy recovery systems.

DRYER PICK MARKS—Marks caused by the plucking out of small particles of the sheet or coating which adhere to a dryer roll.

DRYERS—See DRYER CYLINDERS.

DRYER SCREEN—See DRYER FABRIC.

DRYER SECTION—That part of the papermaking machine which follows the press section (q.v.) and dries the sheet by evaporation from 40 to 50% solids to about 95% solids, usually with steam heated dryer cylinders (q.v.) or "dryer cans." The dryer cylinders can be arranged in one of the following configurations:

Single-tier

Two-tier

Stacked

Serpentine (Unorun)

Single-tier: With this configuration, the steam-heated dryer cylinders are located in the same plane. The sheet contacts about 230° of each dryer cylinder, with the same side of the sheet contacting subsequent dryer cylinders. The sheet is supported continuously by a dryer fabric which limits and causes a more uniform sheet shrinkage. To provide uniform drying through the sheet thickness, single-tier sections alternate in contacting the opposite side of the sheet until the sheet is dry.

Two-tier: A configuration of steam-heated dryer cylinders in which the sheet travels down the machine in a serpentine path through two rows of dryer cylinders stacked on top of each other. The top row of dryer cylinders is offset such that the centerline of the top dryer cylinder is between two bottom dryer cylinders. The sheet travels up and over the first top dryer cylinder, down and under the first bottom dryer cylinder,

then up and over the second top dryer cylinder, etc. Pocket dryer fabric rolls are added so that top and bottom dryer fabrics can be added to hold the sheet against the top dryer cylinders. The dryer section has subsections to allow for draw (q.v.) control, as the sheet shrinks when it is dried and elongates as it is stretched. Each subsection has separate top and bottom felts, and has a separate drive for draw control. In the serpentine configuration, pockets (q.v.) are formed by the sheet, the dryer cylinder surface, and the dryer fabric. These pockets become humid and retard drying unless pocket ventilation (q.v.) is practiced.

Stacked dryer: A configuration of steam-heated dryer cylinders used in drying selected board grades. Here the dryer cylinders are directly over each other, usually three or four high. The sheet is threaded so that it serpentines up one stack, crosses over to a second stack on top, and serpentines down the second stack crossing over the third stack on the bottom. This process continues until the sheet is dry. Because dryer fabrics cannot be run in this configuration, a stacked dryer section is generally used on heavy weight grades where sheet tension can be used to bring the sheet into contact with the dryer cylinder.

Serpentine dryer (single-felted two-tier): A two-tier dryer configuration in which a single fabric travels down the machine on a serpentine path through the two tiers of dryer cylinders. Normally, the web is between the fabric and dryer on the top cylinders and on the outside of the fabric on the bottom cylinders. This geometry is used in the wet-end dryers to support the wet weak sheet. This type of section is sometimes called a Unorun.

DRY FELT—An absorbent sheet of felted fibers of vegetable or animal origin, or mixture thereof, suitable for use in the manufacture of bituminous saturated felt, roll roofing, siding, shingles, etc. This material is also designated as organic or rag felt.

DRY FINISH—(1) A process in which paper or paperboard is calendered without the application of surface moisture. (2) A descriptive term applied to paper, particularly wrapping paper, and paperboard processed in this manner and characterized by having an unglazed, fairly rough surface. Dry-finished boards have a relatively low density and are normally used in the manufacture of setup cartons.

DRY-FINISH BUTCHERS WRAP—A well-sized, dry-finished wrapping paper, made from mechanical and/or chemical woodpulp, especially adapted to the wrapping of meats in over-the-counter trade. It is ordinarily specified in a basis weight of 35 to 50 pounds (24 x 36 inches – 500) and is sold in standard size counter rolls and sheets. It is strong and resistant to penetration by meat fluids. See also BLOODPROOF PAPER.

DRY-FINISH SCREENINGS—A paper which has the same finish as dry-finish fiber, but which is made of screenings.

DRY-FINISH SULFITE WRAPPING—A typical wrapping paper colored or white, made wholly or principally from sulfite pulp, in weights up to 80 pounds (24 x 36 inches – 500). It has an unglazed surface.

DRYING—The process of evaporating a fluid, usually water, from a nonwoven or other material.

DRYING CRACKS—A cracking of the surface of a coating due to excessively high rates of evaporation of the moisture in the coating.

DRYING PAPER—See INTERLEAVING BLOTTING.

DRYING RESTRAINT—Applying forces in the plane of the web to prevent it from shrinking as it dries.

DRY-LAID—The process of forming a nonwoven by either carding or an air laying method.

DRY LAKE—See COLOR LAKE.

DRY LAP—Pulp that has been formed into laps and dried, usually by contact with heated dryer drums, or by contact with heated air.

DRY LIMESTONE PROCESS—A method of controlling air pollution caused by sulfur oxides. The polluted gases are exposed to limestone which combines with oxides of sulfur to form manageable residues.

DRY-MOUNTING TISSUE—A thin paper (approximately 0.001 inch in thickness) treated on both sides with a thermoplastic adhesive coating for a total thickness of about 0.002 inch. It must be of uniform thickness and free from any matter capable of detrimentally affecting a photographic image. It is flexible when cold and becomes adhesive under the application of heat and pressure, retaining its adhesiveness upon becoming cold. It does not become soft or sticky upon standing under normal conditions of temperature and humidity.

DRY PRINTING—Any paper which dries rapidly after printing. Any papermaking materials may be used. It is essential, however, to have a surface and ink receptivity such as to ensure rapid drying of quick-drying ink.

DRY PROOFING PAPER—A paper used by a printer on a proofing press. The type as setup is printed on the proofing paper and checked for mistakes. Usually a cheap paper having quick-drying properties owing to surface, texture, etc., is used. Newsprint is frequently employed for this purpose. See also GALLEY-PROOF PAPER.

DRY RUB RESISTANCE—The resistance of the dry surface of coated or uncoated paper or paperboard to disruption of the surface when subjected to rubbing or scuffing. Often used as a measure of ink durability. See also WET RUB.

DRY-WAXED PAPER—Paper which has been passed between rolls, one of which revolves in a bath of molten paraffin or other wax and subsequently through squeeze rolls, so that the major portion of the wax is driven into the interior of the sheet and the paper feels "dry." The wax may also be applied by means of transfer rolls. The usual weight of the paper before waxing is 18 to 50 pounds (24 x 36 inches – 500); such a paper after waxing will carry 5 to

10 pounds of wax. See also WAXING PAPER; WET-WAXED PAPER.

DUAL COMPARTMENT BOXES—See HYDROFOILS.

DULL-COATED PRINTING PAPER—A special coated printing paper characterized by a dull, flat surface with minimum gloss. This type of surface is usually achieved by incorporating calcium carbonate or blanc fixe into the surface coating formulations. See DULL FINISH.

DULL FINISH—A finish with a low gloss. With respect to coated book paper, a finish with a glare test less than 55%.

DULL-GLAZED ART PAPER—A type of dull-coated printing paper designed for art reproductions and other high-quality printing jobs.

DULL SLITTER—A slitter that produces an unacceptable slit quality because of excessive wear or because of poor geometry.

DUMMY—A set of blank sheets made up to show in advance the size, shape, form, and general style and plan of a contemplated piece of printing, such as a book.

DUMP CHEST—A chest which receives stock only periodically, usually from a batch pulper or other batch processing device, or multiples thereof, and either discharges stock continuously to downstream process equipment, or discharges stock periodically to a surge chest or a blend chest.

DUNNAGE—Material used to fill voids after the product is loaded for shipment by rail or truck. See also DUNNAGE BAG.

DUNNAGE BAG—A bag manufactured of heavy polyethylene coated kraft paper (or entirely of rubber), used to hold rail car contents in place during transit. The bag is inflated with air as the rail car is loaded, then deflated during unloading.

DUNNAGE PAPER—See CAR LINER.

DUOTONE FINISH—A finish having two depths of color (visual brightness) of the same hue, produced by unequal pressure in the finishing operation.

DUPLEX—Having two or more plies or sheets. See also DUPLEX COLOR; DUPLEX FINISH; DUPLEX TEXTURE; and other entries beginning with DUPLEX.

DUPLEX ASPHALT PAPER—A waterproof paper consisting of two sheets of paper, of either similar or dissimilar composition, which have been caused to adhere by a film of asphalt or bituminous material. The weight of such paper is usually given as the weight of the sheets and of the asphalt used, e.g., 30.30.30 indicates that two 30-pound sheets have been laminated with 30 pounds of asphalt per ream of 500 sheets.

DUPLEX BLOTTING PAPER—A combined blotting paper made by pasting together two sheets of blotting distinctly different in color or made on a cylinder machine by combining plies of different color. It is absorbent on both sides and differs in this respect from enameled blotting, which is absorbent only on one side.

DUPLEX BOARD—A general term for a board made of two different stocks or colors on a cylinder machine, or a combination cylinder-fourdrinier machine, or a fourdrinier machine equipped to receive two different pulp furnishes. It is a combination board limited to two stocks or colors.

DUPLEX BOXBOARD—See DUPLEX BOARD.

DUPLEX BRISTOL—A bristol with a different color on each side.

DUPLEX COATED BRISTOL—A solid-center bristol base, coated with a bright or deep color on one side and with a tint or white on the reverse side. The bristol base is usually made of softwood and hardwood chemical pulps, approximately 125 pounds per 500 sheets (22 1/2 x 28 1/2 inches). It is used for advertising postcards, folders, covers, and other forms of direct mail publicity.

DUPLEX COATER—Obsolete term replaced by simultaneous two-sided coating (q.v.).

DUPLEX COLORS—Colors that are different on the two sides of the sheet, as made by laminating sheets of different colors, staining one side and not the other, or coating the sides with different colors.

DUPLEX CUTTER—A specialty cutter (sheeter) designed to accommodate dual layboys and piling sections so two different sheet sizes can be cut from the same unwinding webs.

DUPLEX ENAMEL BOOK—A coated two side book paper made with a bright or deep color on one side and a harmonizing tint or white on the reverse, with either a high finish or a dull coating; it is usually made in 22 1/2 x 35 inch and 35 x 45 inch sizes with a basis weight of 80 pounds (25 x 38 inches – 500). It is used for colorful direct mail advertising produced with odd folds, trims, or die cutting so that both sides of the paper are visible.

DUPLEX FINISH—A finish that is different on the two sides.

DUPLEX FOIL BACKING—An MG paper made of bleached chemical woodpulp in basis weights of 25 to 35 pounds (24 x 36 inches – 500). It has a high, smooth finish on the wire side and an antique finish on the felt side. The paper is soft, has good tensile strength and good formation. Wet strength and/or drying may be required to resist water-based adhesives.

DUPLEX MACHINE FINISH (MF) LITHO—A paper used for lithography (q.v.), one side of which has a rough surface and the other a smooth machine finish. This differs from MG litho in that it is dried without the use of a Yankee dryer.

DUPLEX MILL WRAPPERS—Heavy, water-finished wrapping papers duplex as to color. They are used for the protection of rolls or bundles of paper, textiles, etc.

DUPLEX OFFSET BLOTTING PAPER—A combined blotting paper made by pasting a sheet of white offset paper with one of blotting (which may be variously colored). The offset paper side is largely used for color printing. This blotting is normally made in basis weights of 100 to 140 pounds (19 x 24 inches – 500).

DUPLEX PAPER—Any paper showing different colors, textures, or finishes on the two sides of the sheet.

DUPLEX PHOTOGRAPHIC FILM PAPER—A paper usually made from chemical woodpulp on a combination fourdrinier-cylinder machine. It has a high tensile strength and is free from material which would affect the photographic film. One side is black and the other side is generally white or colored.

DUPLEX SHEATHING—A single sheet of sheathing paper (q.v.), which is stained on one side only or which is stained with a different color on each side.

DUPLEX STAINLESS STEELS—Duplex stainless steels contain >18% chromium and <8% Ni. They have a mixed crystal structure, part austenite and part ferrite. They have approximately twice the yield strength and fatigue strength of the austenitic stainless steels and are highly resistant to stress-corrosion cracking.

DUPLEX SUPER—Any supercalendered paper which has a higher finish on one side than on the other. It is made by running the paper through the supercalender so that one side does not come into contact with the chilled cast-iron rolls.

DUPLEX TEXTURE—Texture produced by laminating two papers or boards with different textures, by lining a board with two kinds of stock, or by using two different stocks on a cylinder or cylinder-fourdrinier machine.

DUPLEX VARNISHING LITHO—A paper used for lithographic printing where the paper is printed and varnished on one side and the other side is left rough to facilitate the pasting operation. This paper is made from bleached chemical woodpulps processed and made to resist penetration of the inks and varnish. It is usually calendered to give a higher finish on one side than the other. The usual basis weights are 50 to 60 pounds (25 x 38 inches – 500).

DUPLEX WALLPAPER—(1) A duplex sheet in which the colored sheet made with fast-to-light dyes is produced on a fourdrinier machine and is united with a backing sheet produced on a single-cylinder machine. (2) A duplex sheet of hanging paper made on a cylinder machine with one side of the sheet smooth to take adhesive and the other side rough as in oatmeal paper (q.v.).

DUPLEX WINDER—A winder with two powered coreshafts and one supporting drum to provide surface winding torque. Some models provide controllable centerwind torque to each winding roll at the same time that surface winding is provided by the driven winder drum. All types of duplex winders stagger the rolls on each side of the winder drum so they are separated during the winding process.

DUPLICATING NOTE PAPER—Tablet or writing paper converted into a form in which part or all of the sheets are carbonized so that writing may be duplicated by inserting blank sheets.

DUPLICATING PAPER—Paper used either as masters or copy sheets in the aniline-ink or hectograph process of reproduction. This process involves two distinctly different types of machines. The spirit type (also called the liquid or direct process type) and the gelatin type. In the spirit machines, the master paper is generally a smooth, level 60 pound (25 x 38 inches – 500) coated book grade on which the original copy is typewritten using a special ribbon or carbon paper containing a heavy aniline pigment coating. This prepared master is placed in a spirit duplicator and copies are made there from by contacting it with spirit-dampened copy paper [a chemical wood pulp sheet with a high finish usually made in 16 and 20 pound weights (17 x 22 inches – 500)]. In the gelatin process, the duplicating machine comprises a gelatin surface (either a pan of gelatin or a gelatin-covered fabric). The copy to be duplicated is prepared on a master sheet (an ordinary chemical woodpulp bond paper is commonly used) using a hectograph typewriter ribbon, carbon paper, ink, or indelible pencil. This master copy is pressed against the gelatin surface depositing the hectograph ink thereon. Copies are then made, using the same type of paper as noted above for the spirit process, by simply pressing the copy paper into intimate contact with the gelatin surface.

DUPLICATING TISSUE—A tissue paper used for producing wet copies of manuscripts. It is made on a fourdrinier machine in basis weights of 10 to 15 pounds (24 x 36 inches – 480). It has a fairly well-closed formation and contains some sizing. Modern duplicating methods have largely displaced the use of this paper.

DUPLICATOR—A machine or device, largely for office use, that reproduces graphic materials by spirit, stencil, lithographic, or other processes. Many modern duplicators are small lithographic presses. Duplicators have been supplanted by xerographic copiers and computer printers.

DUPLICATOR PAPER—See DUPLICATING PAPER.

DURABILITY—The degree to which a paper retains its original qualities under continual usage. This is not to be confused with permanence which is the degree to which a paper resists chemical action which may result from impurities in the paper itself or agents from the surrounding air. See PERMANENCE.

DUST—The debris found on the surface or edges of a sheet or web of paper. This may adversely affect the printing process. See also SLITTER DUST.

DUSTING—(1) A condition encountered in some papers where fine particles of filler, fibers, or coating material leave the sheet during the finishing, converting, or printing operation. When the dusting is of appreciable quantity, it interferes with the operation, especially in printing.

See also CHALKING. (2) Small paper particles created by paper cutting, known variously as slitter dust (q.v.), cutter dust (q.v.) or trimmer dust, for its source.

DUSTING PAPER—A soft, absorbent, nonabrasive sheet made from chemical pulps in a wide range of basis weights, which is treated with a furniture or polishing oil and used for dusting or polishing purposes.

DUST pH—The pH of an aqueous solution of recovery boiler precipitator dust. Often used as an approximate indicator of the dust chemical composition, which is useful in assessing black liquor combustion.

DUTCH PAPER—A deckle edged paper made in Holland.

DYESTUFFS—See ACID DYES; BASIC DYES; COLOR LAKE; DIRECT DYES; PIGMENT.

DYNAMITE PAPER—See BLASTING PAPER; DYNAMITE-SHELL PAPER.

DYNAMITE-SHELL PAPER—A well-formed, smooth, highly finished sulfite or kraft sheet, usually in basis weights of 60 and 90 pounds (24 x 36 inches – 500) to be converted into tubes for packaging of dynamite, powder, etc. Tensile and tearing strength and moisture resistance, as well as finish and uniformity, are important characteristics.

E

EARLYWOOD—The portion of an annual ring produced during the early part of the growing season (in the spring in temperate zones); the inner portion of the annual ring. It is usually lighter (less dense) than the summerwood.

EASY-BLEACHING PULP—A term applied to a thoroughly cooked pulp, containing less lignin, usually of lighter color than regular unbleached pulp, which can be bleached with a minimum of bleaching agent.

E BLEACHING STAGE—See ALKALINE EXTRACTION STAGE (E).

ECF—See ELEMENTAL CHLORINE FREE.

ECONOMIZER—The economizer is usually the last heat transfer section of a boiler wherein the boiler feedwater is heated while cooling the combustion gases to their exhaust temperature. Most economizers are designed such that the feedwater will not begin to boil, although some are designed specifically for some boiling to occur near the outlet.

ECT—Edge crush test. See EDGEWISE CRUSH RESISTANCE.

EDGE CRACK—A small tear or discontinuity at the edge of a web. When the web is under tension, this can cause local stress concentrations that can encourage partial or total failure of that web.

EDGE CRUSH TEST—Also called short column test (q.v.). See EDGEWISE CRUSH RESISTANCE.

EDGE GUIDE—A device which senses the offset or sideways movement of a traveling web and counteracts this undesirable movement by continuously adjusting a guide roll to maintain a controlled position of the web so roll edges are straight and smooth during a winding process. Edge guides are also used to guide wires, fabrics, or felts on papermaking machines.

EDGE PROTECTOR—Hard fiber, heavy board, or corrugated board used to protect the ends of rolls from being damaged during handling, transit, and storage.

EDGE NAIL STRENGTH—The maximum resistance of a nail to lateral movement through a fiberboard.

EDGE-TEARING RESISTANCE—The resistance by paper to the onset of tearing at the edge of a sheet. It is commonly judged by sensing the resistance when a tear is started from the edge by applying, by means of the thumbs and index fingers, of a twist and a force. It appears to depend on the extensibility and the tensile strength of the paper. See TEARING RESISTANCE.

EDGEWISE COMPRESSION STRENGTH— (1) For paperboard, this property is measured in a compression mode imposed parallel to the plane of the paper. This property can be measured by the ring crush test (q.v.) and short span compression test (q.v.) (2) For corrugated board, see EDGEWISE CRUSH RESISTANCE.

EDGEWISE CRUSH RESISTANCE—The maximum compressive force in the direction of the flutes which a plane, rectangular test piece of corrugated board, standing on its edge, can withstand without failure. The test for this property is known as the edge crush test (ECT) or short column test.

EF—English finish (q.v.).

E.F.—English finish.

EFFECTIVE ALKALI—The sum of the sodium hydroxide concentration and half of the sodium sulfide concentration in white liquor expressed as equivalent concentrations of sodium oxide, Na_2O.

EFFLUENT—(1) A discharge of pollutants into the environment, partially or completely treated or in its natural state. Generally used in regard to discharges into waters. (2) Washing filtrate from a bleaching stage or any process water that has been chemically or physically altered by its use in or passage through the bleach plant.

EFFLUENT TREATMENT—The treatment applied to discharge from any plant operation. Usually involves primary and secondary treatment (q.v.).

E-FLUTE—See FLUTE.

EGG-CARTON BOARD—A folding boxboard used for making cartons to package eggs for local shipment. A wide variety of grades of boxboard is used which have the common qualities of high stiffness, water resistance, and good

bending. The board is die-cut and folded to form a separate compartment for each egg.

EGG-CASE BOARD—See EGG-CASE FLATS.

EGG-CASE FILLER BOARD—An egg-case board used for the honeycomb or cell partitions which separate the eggs as packed in a layer. The board is cut into strips and fed into a die-cutting and forming machine. The strips interlock at right angles forming the cells.

EGG-CASE FLATS—An egg-case board used to separate one layer of eggs from another when placed in an eggcase. These are plain sheets cut to the required size and are sometimes indented to add to the cushioning effect. They also include molded flats (see MOULDED PULP PRODUCTS) made of over-issue news or virgin mechanical pulp and formed on a vacuum cylinder mold, the wire surface of which is indented to impart a double-cupped shape to each flat. One cupped flat is placed between the layers and two cupped flats are placed back to back in the bottom of each section of the case and two on top.

EGGSHELL BOOK PAPER—A book paper having a eggshell finish (q.v.). The basis weight ranges from 45 to 80 pounds (25 x 38 inches – 500).

EGGSHELL FINISH—A relatively rough finish given to paper, so called because it resembles the surface texture of an eggshell. It is produced by the use of special felts which mark the paper as it enters the dryers with relatively round hills and valleys not definitely aligned with the grain direction. It is generally considered to be rougher than an MF finish and smoother than an antique finish. See ANTIQUE FINISH for further discussion.

EIS—An acronym for Environmental Impact Statement (q.v.).

ELASTICITY—That property of a material which enables it to undergo deformation and to recover its original dimensions after removal of the deforming stress. Elasticity is determined more by ability to recover initial shape than by capacity to be deformed or extended. It should not be confused with extensibility or stretch (q.v.).

ELASTIC MODULUS—The ratio of stress (nominal) to corresponding strain below the proportional limit of a material. It is expressed in force per unit area based on the average initial cross sectional area. Also known as Young's Modulus.

ELASTOMER—Synthetic polymer with rubber-like characteristics. Used for gaskets, roll covers, and sometimes as tank linings. See COVERED ROLL.

ELECTION BRISTOL—See CAMPAIGN BRISTOL.

ELECTRIC CABLE PAPER—See CABLE PAPER.

ELECTRIC DOUBLE LAYER—A physical entity that exists when oppositely charged ions (counterions, q.v.) firmly attach themselves to a charged surface that is fixed, and thus form a double layer with some of the counterions remaining mobile beyond the shear plane of the liquid. By definition, the remaining surface potential at the shear plane is called zeta potential. See ELECTROKINETIC CHARGE.

ELECTRICAL-CABLE FILLING PAPER—See CABLE PAPER.

ELECTRICAL CONDUCTIVITY—The conductance of unit area of a sheet to an electric current flowing through the sheet. This is to be differentiated from the more general definition of conductivity, which is the conductance of a unit cube of the material. This property is strongly influenced by the presence of foreign conducting particles, a matter of great importance in condenser papers and in other papers used in the electrical industry. Electrical insulating papers must have very low conductivity after treatment and must be as free as possible of foreign conducting particles.

ELECTRICAL FIBER—See ELECTRICAL INSULATION FIBER. The term is also applied to other papers and boards which are used for electrical insulating purposes.

ELECTRICAL INSULATING MATERIALS—See INSULATING MATERIALS for a classification of these materials.

ELECTRICAL INSULATING PAPER—See CABLE PAPER; CAPACITOR PAPER; CONDENSER PAPER; ELECTRICAL INSULATION FIBER; INSULATING TISSUE; LAYER INSULATION PAPER; TRANSFORMER BOARD.

ELECTRICAL INSULATION FIBER—A tough, flexible grade of vulcanized fiber (q.v.) having high dielectric, tensile, and bending strength. It is primarily intended for electrical insulation applications and others involving difficult bending or forming operations. It is made in thicknesses from 0.004 to 0.125 inch in colors ranging from gray to a bluish gray.

ELECTRICAL PRESSBOARD—See TRANSFORMER BOARD.

ELECTROCONDUCTIVE POLYMER—A polymeric material capable of carrying an electric current.

ELECTROKINETIC CHARGE—When papermaking particles having at least one very small dimension, such as fiber fines, filler particles, or rosin size particles are suspended in water they take on an electrostatic charge (usually negative) at their surface. This causes a rearrangement of dissolved anoins and cations in the vicinity of the particle surface. Some of the ions (counterions) are bound tightly to the particle surface and travel with the particle when it moves. Other ions are less tightly bound and do not travel with the particle. The potential at the shear boundary between a moving particle with its strongly bound counterions and the other, less strongly bound ions in the surrounding water is called the zeta potential. It is usually expressed in millivolts (mV). The entire system is called the electric double layer (q.v.).

Phenomena which involve the movement (kinetic) of charged particles (electro) are referred to as electrokinetic phenomena and the associated charge is often called the electrokinetic charge.

Methods for measuring the zeta potential where suspended particles and their suspending medium are made to move relative to one another are referred to as electrokinetic effects. Examples which are applied in the paper industry are microelectroporesis, streaming potential, and streaming current. See MICROELECTRO-PHORESIS, STREAMING POTENTIAL, COLLOID TITRATION, STREAMING CURRENT, and ELECTRIC DOUBLE LAYER.

ELECTROKINETIC CHARGE MEASUREMENT—See MOBILITY.

ELECTROKINETIC CHARGE TITRATION—Oppositely charged polyelectrolytes neutralize one another in an approximately stoichiometric manner. This principle is used to titrate quantitatively the available charge of paper furnishes using microelectrophoresis, streaming potential, streaming current, or colloid titration to identify the zero charge endpoint. See CATIONIC DEMAND; COLLOID TITRATION; MICROELECTROPHORESIS; STREAMING CURRENT.

ELECTROLYTIC CAPACITOR PAPER—See CAPACITOR PAPER.

ELECTROLYTIC PAPER—An asbestos paper which is used as a diaphragm in certain electrolytic processes.

ELECTROPHOTOGRAPHY—See XEROGRAPHY.

ELECTROSTATIC COPY BASE STOCK—A bond or writing paper designed for coating with zinc oxide formulations to produce copy papers used in Electrofax type office reproduction machines.

ELECTROSTATIC COPY PAPER—See ELECTROSTATIC COPY BASE STOCK.

ELECTROSTATIC PRECIPITATOR—An air pollution control device that removes particulate matter by imparting an electrical charge to particles in a gas stream (e.g., flue gases) for mechanical collection on an electrode.

ELECTROSTATIC PRINTING—A reproduction process in which the image on the paper or other surface is formed by applying an electrical charge to the toner or pigment particles, bringing them into contact with an oppositely charged image area on the surface and treating the electrostatically held image to bind it to the surface.

ELEMENTAL CHLORINE FREE—A term used to describe a pulp bleached without gaseous chlorine.

ELONGATION AT RUPTURE—See STRETCH.

ELUTRIATION—The process of treating sludge by flushing fresh water through it, resulting in sludge that will dewater more advantageously.

EMBOSSED—See EMBOSSING; EMBOSSING CALENDER.

EMBOSSED BLOTTING—Any blotting paper which has been treated after manufacture in an embosser or plater in order to impress on the surface, either on one or both sides, a figured design.

EMBOSSED COVER PAPER—Any cover paper with an embossed surface created by passing the sheet through a pair of matched patterned steel rolls or through a combination of a pattern roll and a smooth backing roll.

EMBOSSED GLASSINE—Glassine paper which has been decorated with a continuous formal design by embossing rolls. See GLASSINE PAPER.

EMBOSSER—A machine that embosses a paper or board web. See EMBOSSING; EMBOSSING CALENDER.

EMBOSSING—A process which converts a smooth surfaced web to a decorative surface by replicating the design on the rolls which form the nip of the embosser. Embossing machines can consist of two steel rolls (with contrasting male and female patterns), one etched steel roll and a mating paper/wood filled roll which is impressed with the reverse pattern of the steel roll, or one etched steel roll mated to a resilient surfaced roll which produces more of a two-sided product. This type of embossing is known as "free running" whereas the matched rolls require specific surface ratios and gearing to maintain repetitive roll surfaces. See also DIE EMBOSSING; EMBOSSING CALENDER.

EMBOSSING CALENDER—An embossing calender (embosser) replicates a pattern on the substrate, which is imparted by the driven engraved steel (male) roll, which can be mated to either: (1) a smooth surfaced elastomeric roll which merely presses the web against the engraved roll, or (2) to a roll that will present the inverse pattern of the male roll. This female roll can be steel or it can be a paper/wool filled roll which has been impressed with the inverse pattern of the male roll. Case (1) is called a "free running" pattern and only the engraved roll is driven. A micro-engraved "matte finisher" uses this method. In case (2), the dynamic circumferences of the rolls must match or must be exact multiples of a 1:1 ratio. Both rolls are geared together so the inverted patterns are continuously matched. See SOFT-NIP CALENDER.

EMERGENCY STOP—See E-STOP.

EMERY—A kind of abrasive (pulverized corundum) used in a surface-coating mixture with an adhesive in the manufacture of abrasive paper.

EMERY CARD CLOTHING—A sandpaper-like material used to cover some stationary top cards when processing man-made fibers.

EMERY PAPER—An abrasive paper (q.v.), the base stock of which is a kraft sheet in basis weight of 65 to 70 pounds (24 x 36 inches –

480), coated with emery powder with a glue or resin bond, which is used in sheets and in narrow rolls of polishing operations on metals.

EMISSION—Any discharge from a process; can be a gas (vapor), liquid, or solid.

EMISSION FACTOR—The average amount of a pollutant emitted from each type of polluting source in relation to a specific amount of material processed. For example, an emission factor for a blast furnace (used to make iron) would be a number of pounds of particulates per ton of raw materials.

EMISSION STANDARD—The maximum amount of a pollutant legally permitted to be discharged from a single source, either mobile or stationary.

EMULSION—A system consisting of finely divided liquid dispersed within another liquid.

ENAMEL—A term applied to a glossy coated paper.

ENAMELED—Originally a term applied to supercalendered coated papers in which the coating pigment was largely satin white or blanc fixe. Today the term is used generally for any coated paper.

ENAMELED BLOTTING PAPER—A blotting paper to which a coated book paper or card has been pasted, thus giving one side a smooth, hard surface suitable for printing and lithographing.

ENAMELED BOARD—See COATED BOARD.

ENAMELED BOOK PAPER—See BOOK PAPER (COATED).

ENAMELED CARD—See COATED BOARD.

ENAMELED PAPER—See COATED PAPER.

ENAMELED POSTCARD—A postcard board single- or double-coated on one or both sides, used for souvenir postal cards and similar work where a high finish is desired. It is often linen finished on one side or on both sides.

ENCLOSED POND APPLICATOR—A device used to apply wet film in some designs of metering size presses. It consists of two rotating rods which seal the pond and control the amount of wet film applied plus a hydrofoil to prevent turbulence in the enclosed pond.

END BANDS—Heavy paper used to protect the ends of paper rolls for shipment. See EDGE PROTECTOR.

END-LEAF PAPER—A white, colored, or ornamental paper used for binding a book's contents to its cover. It is usually made of chemical woodpulp in basis weights of 50 to 80 pounds (25 x 38 inches – 500), and is characterized by high tearing and folding strength, and ability to paste smoothly to the book cover. It is also referred to as end leaves, end sheets, end paper, and fancy end.

ENDLESS FELT—See PRESS FELT.

END SHEET—The outer sheet of a bound signature of pages, which is printed on one side and glued to the cover on the other side. See also END-LEAF PAPER.

ENGINE SIZING—See BEATER SIZING.

ENGLISH FINISH—A finish between that of machine-finished and supercalendered in degree of smoothness. For printing papers with basis weight of 45 pounds (25 x 38 inches – 500), it has a bulk of 650 to 750 pages per inch. It is the smoothest of machine finish papers and best adapted to halftone printing. See FINISH; SUPERCALENDERED FINISH.

ENGRAVED ROLL—A steel roll which is acid-etched with a specific pattern to treat paper or board surfaces for a decorative effect. The patterns are mechanically imprinted on a rotary die or master roll by artisans and transferred to the engraved roll by means of a mechanical device which utilizes the master die to expose portions of the steel roll to acid etching. The steel roll surface matches the master die pattern and is normally protected by an overlay of chromium plating.

ENGRAVERS BRISTOL—A high-grade bristol, the quality and finish of which are especially suitable for plate engraving. Important properties are color, finish, and resiliency.

ENGRAVERS PROVING PAPER—A paper used by the manufacturers of halftone plates to prepare a printed proof of the plate. The paper is a high-grade coated paper, carefully made to give a perfect surface for printing. The paper has a very uniform glossy surface. The usual basis weight is 80 to 100 pounds (22 x 28 inches – 500).

ENVELOPE-LINING TISSUE—A tissue paper used for increasing the opacity of an envelope and for decorative purposes. It is made on a fourdrinier machine, in basis weights of 10, or more generally, 15 pounds (24 x 36 inches – 480). It may be beater or surface colored and may be further decorated by printing.

ENVELOPE MANILA—A fourdrinier MF paper made in manila color of chemical woodpulps, usually with some percentage of mechanical woodpulp.

ENVELOPE PAPER—A general term for papers used in the manufacture of envelopes. Because of the wide variations in use requirements, envelope papers vary in weight, appearance, and finish to such an extent that many kinds of paper may be employed. See COMMERCIAL WOVEN ENVELOPE; ENVELOPE MANILA; ENVELOPE PAPER, KRAFT; GLASSINE PAPER; JUTE ENVELOPE PAPER; MANILA WRAPPING; ROPE MANILA PAPER.

ENVELOPE PAPER, KRAFT—A fourdrinier MF or MG paper made of unbleached, semibleached, or full bleached sulfate pulp, used in the manufacture of envelopes when strength is a primary requirement. The basis weights range normally from 16 to 44 pounds (17 x 22 inches – 500). Other desirable properties include smooth fold, strength at crease, good printability, and lack of tendency to curl or cockle.

ENVELOPES—Converted paper products designed for wrapping or enclosing material to be stored or mailed. Envelopes are made in many forms from many different grades of paper and some typical items are: bank by mail, commercials, outlook, window, airmail, self-seal, postage saver, booklet, clasp or snap, photo mailer, two compartment, coin, open end, open side.

ENVELOPE SHAPE—The usual shape of envelopes, as distinguished form that of bags.

ENVELOPE STUFFER—A paper or card, blank or printed, such as a small circular, which is designed for enclosure with or without a letter in an envelope.

ENVIRONMENTAL IMPACT STATEMENT—A document prepared by a Federal agency on the environmental impact of its proposals for legislation and other major actions significantly affecting the quality of the human environment. Environmental impact statements are used as tools for decision making and are required by the National Environmental Policy Act.

ENVIRONMENTAL POLLUTION ABATEMENT—The act or process of minimizing or eliminating the effects on the environment of municipal or industrial air, water, and land pollutants.

ENZYME—A protein capable of catalyzing a chemical reaction. Examples of enzymes are xylanase (which cleaves linkages in xylan) or ligninase (which cleaves linkages in lignin). Related enzymes are used in bleaching pre-treatment. Almost all biological reaction are enzyme-catalyzed.

ENZYME BLEACH BOOSTING—See BIOBLEACHING.

ENZYME BLEACHING—See BIOBLEACHING.

ENZYME DEINKING—The use of enzymes (usually microbial cellulases or lipases) to facilitate ink removal in paper recycling.

Eo BLEACHING STAGE—See OXIDATIVE EXTRACTION STAGE (Eo).

EQUILIBRIUM RELATIVE HUMIDITY—The ambient relative humidity, at a given temperature, at which exposure of paper will not result in moisture absorption or desorption.

EQUIVALENT WEIGHT—The weight of a given paper of any size (usually expressed as the weight of a ream) in terms of some other size; equivalent weights are in direct proportion to the areas of the single sheets. Thus, a 50-pound paper, 25 x 38 inches, is equivalent to a 74-pound paper, 32 x 44 inches. See also BASIS WEIGHT.

ERASABILITY—That property of a sheet which is concerned with the ease of removing typed or written characters and impressions or both from the sheet by mechanical erasure, the cleanliness or the amount of abrading on the erased portion, and the suitability of the erased portion for reuse.

ERASABLE PARCHMENT BOND—A special type of vegetable parchment paper designed for use as a typewriter paper where printability and ease of erasability are required.

EROSION—The progressive loss of material from a solid surface due to mechanical interaction between that surface and a fluid, a multi-component fluid, or particles carried by a fluid.

ERROR SIGNAL—The difference between the process measured output value and the setpoint.

ESPARTO—A coarse grass grown chiefly in southern Spain and Northern Africa, containing short fibers which are usually extracted by alkaline pulping processes. Esparto pulp is most often used in the production of book papers. Esparto is also known as alfa, esparto grass, and Spanish grass.

ESPARTO PAPER—Paper made from esparto pulp mixed with chemical woodpulp. Esparto pulp tends to provide a more uniform formation and a bulk for a given weight usually in excess of that obtainable in chemical wood papers.

E-STOP—Short for emergency stop. This is a maximum deceleration of a machine (e.g., a winder) to protect an operator, product, or machine from further damage after an accident.

ETCHING—(1) An intaglio process (q.v.) of pictorial production wherein the image is either scratched (dry-point) or chemically etched through a protective resist, forming a pattern of lines and dots below the surface of the plate (usually copper or zinc). Printing is accomplished by smearing the plate with heavy ink, which is then wiped off, leaving the surface clean (or nearly so). Each resulting proof is individual in its effect. Editions are limited, even when the plates are steel-faced. (2) The chemical engraving part of a photochemical process by means of which printing plates are produced from photographic intermediates.

ETCHING PAPER—A high-grade strong drawing paper with a smooth surface. It is the same as steel-plate paper. Good erasability and a good surface are significant properties. The basis weight is 120 pounds (22 x 28 inches – 500). Much of this paper is made according to specifications.

EUCALYPTUS—A large genus of short-fibered broad leaved wood species, originally found in Australia but cultivated in many parts of the world. Eucalypt pulps are primarily used for printing and writing paper grades.

EUTROPHICATION—The process of nutrient enrichment of a water course. This process can occur naturally or be induced by human activities.

EVAPORATOR—In the paper industry, an indirect contact heat exchanger, usually cylindrical, used to evaporate water from an aqueous solution by heat transfer from a hotter fluid.

EVAPORATOR ECONOMY—An expression of the efficiency of a multiple effect evaporator (q.v.), which is the ratio of the total water evaporated to the steam supplied for evaporation.

EXCELSIOR TISSUE—Sideruns or other paper which are put through a shredder to be cut into strips about 1/8 of an inch wide. The basis weights may vary from 10 to 18 pounds (24 x 36 inches – 500). The excelsior should be fluffy and free from dust after being cut; it is used for packing fragile articles. See also WAXED TISSUE EXCELSIOR.

EXCELSIOR WRAPPER—A kraft or sulfite sheet usually specified in basis weights of 30 pounds (24 x 36 inches – 500) and heavier, used for wrapping wood excelsior, for making pads for protection of various commodities, especially furniture, in transit. A strong sheet, capable of high production on converting machines, is an important qualification.

EXCESS SHRINKAGE—Cross-direction shrinkage of the web between successive impressions in printing, sufficient to cause misregister.

EXCITATION PURITY—A colorimetric quantity used to designate depth of color. It is one of the three quantities used in the CIE specification of color and is a ratio of two distances in the CIE trichromatic diagram. Excitation purity is mathematically related to colorimetric purity.

EXERCISE-BOOK PAPER—Flat writings, usually ruled, for use with pen or pencil.

EXPANDABLE CORE SHAFT—A specially designed shaft on which cores are mounted, and that locates the cores in a winding process. The cores are held in place by some mechanical, pneumatic, or hydraulic expansion means.

EXPANDABLE PAPER—See EXTENSIBLE PAPER.

EXPLODED FIBERS—Wood fibers prepared by subjecting chips to a high-pressure steam treatment for a few seconds, followed by a quick release of the pressure, thus causing the fibers to separate or "explode." These are used in the manufacture of various types of boards.

EXPRESS FIBER—See MILL WRAPPER.

EXPRESS PAPER—(1) See MILL WRAPPER. (2) A term sometimes used to designate a heavy weight paper having an extremely high actual density (approaching a specific gravity of unity) and made to caliper specifications. It is used in the production of sheet plastics, the paper being impregnated and laminated under pressure.

EXTENDED DELIGNIFICATION—A modification of the kraft pulping process. Instead of putting all the chemicals into the digester at the beginning of the cook, various concentrations of black and white liquor are exchanged during the cook. This allows lower target kappa numbers with higher yield and strength than normal kraft pulp. Commercially, extended delignification is called modified continuous cooking or displacement batch pulping.

EXTENSIBILITY—The capacity for extension (extension being the increase in length per unit length). Rubber affords a good example of great extensibility. See STRETCH.

EXTENSIBLE PAPER—A smooth-appearing stretchable paper with high-energy absorption properties. A controlled amount of stretch to meet specifications may be imparted in a number of ways, either on or off the paper machine by methods generally differing from those used to produce creped papers. Extensible papers, made in a variety of basis weights and grades, are used for multiwall sacks, packaging, converting, laminating, wrapping, etc.

EXTRA HIGH BULK BOOK PAPER—A book paper which, under 35 pounds pressure per square inch, bulks 344 pages or less to one inch for a basis weight of 45 pounds (25 x 38 inches – 500). Other weights are in proportion. See also BULKING BOOK.

EXTRA STRONG—Having strength characteristics greater than those usually attributed to an identified grade of paper.

EXTRACTION—See ALKALINE EXTRACTION; OXIDATIVE EXTRACTION.

EXTRACTIVES—A general term for non-fibrous materials found in woody plants, not an integral part of the cell wall structure, which can be removed by neutral solvents such as ether, benzene, alcohol, and hot or cold water. These materials, 3 to 10% of the wood, are volatile oils, turpenoids, fatty acids, resin acids, esters, alcohols, and carbohydrates other than cellulose and hemicellulose.

EXTRACTOR COUCH ROLL—On cylinder machines this roll runs against a wire mesh roll to remove water from the sheet prior to the suction drum roll (q.v.). Extractor couch rolls are usually covered with a medium soft (90–100 P&J) (q.v.) elastomeric covering.

EXTRANEOUS MATERIALS—Non-cell-wall substances in wood that are resistant to extraction and may include pectins, starches or inorganic material.

EXTRAS—An additional number of sheets in a ream package.

EXTRUDER—A unit that forces a material usually at elevated temperature under pressure through an orifice to form a continuous tube, rod, filament, film, or coating. Usually, but not necessarily, associated with the processing of plastic materials.

EXTRUSION COATING—A coating applied by means of extrusion and lamination onto the surface of the material to be coated. Coatings of the extrusion type are normally hot melt polymers applied at elevated temperature, usually associated with plastics.

F

FABRIC—See DRYER FABRIC; FORMING FABRIC; PRESS FABRIC.

FABRIC MARK—The impression left in the paper by the machine fabric or dandy roll of the paper machine. The term is also applied to the laid lines in handmade papers, although these are more usually termed wire lines.

FABRIC PRESS—Similar to a plain roll press (q.v.) except for its clothing. A multilayer weaved fabric belt passes through the press nip sandwiched between the rubber covered roll and the felt or fabric. This compression-resistant inner fabric aids in water removal by venting the press fabric or felt. In greater use in Europe than elsewhere.

FABRICS AND FELTS—These are the materials that support the web being made on tissue, paper, and board machines. Although frequently referred to as wires, wet felts, and dryer felts in the forming section, press section, and dryer section, respectively, current usage favors the use of forming fabrics, press felts, and dryer fabrics.

FABRIC SIDE—That side of a sheet of paper which was formed in contact with the fabric of the paper machine during the process of manufacture. See also FELT SIDE.

FABRIC SPOT—A spot in a paper web caused by an imperfection in the forming fabric.

FABRIC WRAP—See WRAP.

FACIAL TISSUE—(1) A name given a class of soft absorbent papers in the sanitary tissue group. Originally used for removal of creams, oil, etc., from the skin, it is now used in large volume for packaged facial tissue, toilet paper, paper napkins, professional towels, industrial wipes, and for hospital items. It is made of bleached sulfite or sulfate pulp, sometimes mixed with bleached mechanical pulp, on a single-cylinder or fourdrinier Yankee machine, with creping on a Yankee dryer at low moisture content, the finished crepe ratio being 10 to 25%. It varies in weight from 8.5 to 13 pounds (24 x 36 – 500) after creping. Desirable characteristics are softness, strength, and freedom from lint. (2) Facial tissue stock in sheets usually packed for resale.

FACING—A sheet of linerboard used as the flat outer surface of corrugated board.

FACING PAPER—Lightweight paper, such as fancy cover, book, and manila. It is pasted on various thicknesses of base stock or filler board to produce picture mounts, photomounts, and other boards requiring plain or fancy covering.

FACSIMILE—A device which combines a scanner, a printer, and a telephone. In practice, a telephone connection is made between facsimile machines, the printed material is scanned and transmitted page by page, with the receiving facsimile machine printing as the material is transmitted. Also called telecopier.

FACULTATIVE ORGANISMS—Organisms that can function in an aerobic or anaerobic environment. See ANAEROBIC BIOLOGICAL TREATMENT.

FADING—(1) A gradual change in color of paper. It is usually applied to the change produced by light. (2) See also FUGITIVE COLORS; YELLOWING.

FANCY END—See END-LEAF PAPER.

FANCY GIFT-WRAPPING PAPER—See GIFT WRAPPING PAPER.

FANFOLD—See FORM BOND; REGISTER BOND.

FAN FOLD—See ACCORDION FOLD.

FANNING—A method of sorting and inspecting paper, in which the operator "fans" the corners and edges of the sheets on the pile and either removes those sheets which are damaged or makes a foldover tear for easy identification and later removal. This paper may or may not be counted simultaneously.

FAN PUMP—The pump which feeds stock to the paper machine headbox or former.

FAST COLOR—A color resistant to the action of external agents, such as light, acids, and alkalies.

FASTNESS—That property of a paper, pigment, or dyestuff that renders it resistant to change in color. Depending upon its use, a paper or paperboard may be required to show good resistance (fastness) to change in color after exposure to destructive influences such as light, acids, alkalies, bleaching agents, or water usually under specified test conditions.

FAST WHITE—See BARIUM SULFATE.

FATIGUE—In corrosion, the phenomenon leading to fracture under repeated or fluctuating stresses having a maximum value less than the tensile strength of the material. Fatigue cracks are progressive, beginning as minute cracks that grow under the action of the fluctuating stress.

FATIGUE FAILURE—The failure resulting from a number of repetitions of loan (or strain) in contrast to creep, which is the deformation caused by the continuous application of load for an extended period of time.

FATTY ACID—A glyceride compound (more specifically a triacylglycerol compound) derived from many different carboxylic acids.

FAX—Short for facsimile (q.v.) or facsimile machine.

FEATHER EDGE—(1) A thin, rough edge, like a deckle edge (q.v.). (2) (of papers) Made with thickness tapering from that of the body sheet to the edge. Such edges are used on fireworks papers, for example, so that, when formed into a tube, the outer edge will paste down smoothly.

FEATHER-EDGED BOARD—A term used to denote deckle edged boards. See DECKLE EDGED BOARD.

FEATHERING—A print defect characterized by ragged edges of the printed lines.

FEATHERWEIGHT—(1) Having light weight. (2) Having very light weight in proportion to bulk.

FEATHERWEIGHT BOOK PAPER—Paper used principally in the manufacture of novels, especially where good bulk is required for a given number of sheets. The paper is made with an antique finish. The principal basis weights are from 50 to 80 pounds (25 x 38 inches – 500).

FEATHERWEIGHT COATED PAPER—See CATALOG PAPER.

FECULOSE—An acetylated starchy product produced by treating starch with glacial acetic acid. Unlike normal starch, it gives a clear solution in hot water and is useful as a size in paper.

FEEDBACK CONTROL—A control method that samples the output of the process variable and compares it with the setpoint to determine the amount of process deviation from setpoint. The controller then sends a corrective signal and continuous feedback/comparison/adjustment provides automatic control. Correction takes place after the fact.

FEEDFORWARD CONTROL—A control method that anticipates process change before the change takes place, and attempts to make process correction before the change occurs. It requires a very accurate algorithm model of the process. Works well with processes having long dead time characteristics.

FEEDFORWARD-FEEDBACK CONTROL—A combination of feedback control and feedforward control, providing the after-the-fact adjustment benefit of feedback with the anticipatory benefit of feedforward. Shortcomings of feedforward are compensated by feedback catch-up control adjustments.

FEEL—The impression obtained by touching and handling paper to judge its finish and general quality.

FELT—Felt is ordinarily produced by a fulling process where woolen fibers migrate to their root end because of the action of an alkali, water and agitation on the wool scales. The term felt is sometimes associated with needle-punched materials such as papermakers felts and other nonwoven products. See also DRYER

FABRIC; MACHINE CLOTHING; PRESS FELT.

FELT CARRYING ROLL—See FORMING FABRIC, FELT, AND DRYER FABRIC CARRYING ROLLS.

FELT CONDITIONING—The cleaning, flushing, and mechanical decompaction of the press fabric by various forms of showers that apply water and chemicals to the fabric or felt. Used in conjuction with felt suction boxes.

FELT DRIVE—A method of driving the dryer section where only a few of the rolls are directly connected to a drive motor. The rest of the dryer cylinders and felt rolls are driven by the dryer felt, rather than through interconnected gearing. See also FELT ROLL.

FELT FINISH—(1) A finish applied to paper at the wet press by the use of felts of peculiar weave, such as corduroy, instead of plain weaves. See also WET-END FINISH. (2) A finish produced on a Yankee machine (q.v.) by the felt which holds the paper in contact with the dryer. The felts carry special markings which are pressed into the paper. (3) A finish applied to dry paper by the use of felts and mechanical pressure in a plater.

FELT HAIRS—Hairs from the fabric of felts, that appear in the surface of paper.

FELTING—The migration of wool fibers by agitation in the presence of water, an alkali, or soap and warm temperatures to produce a fabric-like or nonwoven material. A small amount of natural or manmade fibers may sometimes be used, but this normally will not exceed 40% of the blend when using fine wool fibers.

FELT MARK—(1) A mark or pattern on paper or paperboard produced by the impression of the press felt. (2) A mark or thin spot in paper or paperboard caused by a dirty felt or by a dirty spot on the felt.

FELT MARKING—(1) A design produced by special patterns woven directly into the felt. The effect lies between that of the embossing or plater method and the dandy-roll or stamping-roll method. (2) May also be an undesired marking of the sheet by the pattern in the surface of felts.

FELT PAPER—See DEADENING FELT; DRY FELT.

FELT ROLL—Used to convey the press felts or dryer felts through the machine. Felt rolls are usually covered with a rock hard elastomeric or plastic composition.

FELT ROOFING—A term sometimes used in reference to sheet roofing in the form of heavy mineralized, wide selvage edge, smooth roll roofing, or universal base sheet. See ASPHALT FELT; TARRED FELT.

FELTS AND FABRICS—See FABRICS AND FELTS.

FELTS, DEADENING—See DEADENING FELT.

FELT SIDE—That side of the paper web that has not been in contact with the forming fabric (q.v.) during manufacture. It is the top side of the sheet.

FELTS, SATURATING—See DRY FELT.

FERTILIZER BAG PAPER—See SHIPPING SACK KRAFT PAPER.

FESTOON—A method of in-line storing of paper during printing, which uses a separable set of rollers.

FESTOON DRYING—A method of air drying. The paper is hung in a single continuous web, in short festoons or loops, on traveling poles or slats moving though a drying chamber in which the temperature and humidity are controlled. Obsolete as a commercial practice but may still be in use for specialty or handmade paper.

FIBER—A thread-like body or filament, many times longer than its diameter. Paper pulps are composed of fibers, usually of vegetable origin, but sometimes animal, mineral, or synthetic, for special types of papers.

FIBERBOARD—(1) A general term to describe a board made from chemical woodpulp, wastepapers, other waste materials, or a combination of such materials, with or without the addition of chemicals. Its principal uses are for luggage, containers, electrical products, and shoes. See SHOE BOARD. (2) A term sometimes used to designate container boards in general, as well as trunk board and other products of this character. (3) Often a designation for vulcanized fiber.

FIBERBOARD SHEATHING—See BUILDING BOARD; INSULATING BOARD.

FIBERBOARD SHIPPING CONTAINER BOARD—See CONTAINER BOARD.

FIBER BUNDLES—See SHIVES.

FIBER CAN—A cylindrical receptacle generally with a capacity of ten gallons or less, which is made from paperboard or combinations of fiberboard, paper, and metal foils for the sidewalls and sheet metal ends. These are frequently referred to as composite cans and may be designed for either dry or liquid products. See CANISTER; FIBER DRUM.

FIBER COMPOSITION—The percentages of different types of fibers in pulp or a sheet of paper. It is determined by the anatomy of different fibers and by various staining reactions and the application of weighting factors.

FIBER CONTAINER—See CONTAINER.

FIBER CORES—Paper tubes, commonly called fiber cores, are fabricated from laminations of narrow fiberboard which is spirally wound onto an appropriately sized mandrel at an angle to the axis and reinforced with an adhesive between plies. Typical shipping rolls in the paper industry use fiber cores at about 3 inches (76 mm) I.D. x 4 1/8 inches (105 mm) to 4 1/2 inches (115 mm) O.D. with about 20 plies of 30 point premium grade kraft board. Cores are available in a variety of sizes using paperboard

from 20 pt. (0.5 mm) to 40 pt. (1.0 mm) in thickness that typically contains from 20 to 40 plies. Lengths can range up to 10 feet (3.1 m), and they can be fitted with metal caps to provide crush resistance and improved torque capacity.

FIBER CUT—A cut which occurs in a paper web at the location where a fiber larger than a normal papermaking fiber is contained in the web and goes through a calender or press nip.

FIBER CUTTING—Shortening of fiber length in papermaking pulps. The primary purpose being improved sheet formation.

FIBER DRUM—A cylindrical, convolute, or spiral-wound container made in various sizes from paperboard or from combinations of paperboard with paper, plastic films, plastic coatings, or metal foil, with fiber, metal, or wooden ends. These containers may be designed for packaging fluid or dry products. See CANISTER; CONTAINER; FIBER CAN.

FIBER FRACTIONATION—See FRACTIONATION.

FIBERIZING—A process for the reduction of fiber aggregates to individual fibers from such sources as wood chips, pulp sheets, dry broke, reclaimed paper stock, and the like. The reduction is usually accomplished by mechanical equipment such as disc mills, conical refiners, deflakers, and vortex pulpers. The raw materials may or may not have been softened by water, steam, or by chemical treatment prior to the mechanical action.

FIBER OPTICS—Flexible transparent fiber devices used for either image or data transmission. They are relatively small and lightweight, and provide immunity to electromagnetic interference, radio frequency interference, crosstalk, ground loops, short circuits, and signal leakage. Has capability for high-density signal transmission.

FIBER-REINFORCED PLASTICS (FRP)—Composite materials made in their final shape from thermosetting resin reinforced by a fiber matrix (usually glass).

FIBER SUPPORT INDEX (FSI)—FSI is a useful characterization of a forming fabric's ability to support a sheet that is moderately machine-direction oriented. The index indicates the degree of fiber embedment between the strands of the fabric permitted by a particular fabric.

FIBER WATERPROOF PAPER OR BOARD—Waterproof paper or board that is laminated or made from chemically parchmentized pulp; used for tags subject to contact with oil or grease, dry cleaning, or laundering.

FIBRILLATION—In refining wood pulp fibers, the loosening of threadlike elements from the fiber wall to provide greater surface for forming fiber-to-fiber bonds. Also called brushing out.

FIBROUS ROLL—See FILLED ROLL.

FILE-BACK PAPER OR BOARD—Document manila, usually 8 x 13 inches, used in offices for fastening papers together.

FILE-COPY TISSUES—See MANIFOLD PAPER.

FILE FOLDER—See FOLDER; FOLDER STOCK.

FILL—The maximum trimmed width of paper or paperboard that can be made on a given paper machine.

FILLED BOARD—Board (made on a cylinder machine or a multi-ply former) containing inner plies of a different stock from that of the outer or liner plies. Because this applies to virtually every grade of cylinder board, the term is seldom used today. It is more common to specify when a board is not filled by the prefix "solid." See SOLID BLEACHED SULFATE.

FILLED BRISTOL—A bristol board, made on a cylinder machine or a multi-ply former, with the center of different fiber content from the liners. Many mill bristols are of this type. It is used mainly for printing purposes.

FILLED ROLL—A term applied to a calender roll or embossing roll which is composed of laminations of paper, cotton, wool, or composite fibers fitted to a steel supporting shaft and highly compressed to provide a resilient surface with a hardness which could vary from 70 to 93 Shore Durometer (q.v.). Filling pressures in the roll filling press could vary from 8000 to 13,000 pounds per square inch. Filled rolls are also called fibrous rolls or bowls.

FILLER BOARD—(1) The inner plies of a combination board. (2) A chipboard made for solid fiber container manufacture to be used as the inner member in the process of pasting the three or more layers which make up the container board (filler chip ranges from 0.012 to 0.051 of an inch in thickness and from 36 to 180 pounds per 1000 square feet in weight). (3) The inner members used in manufacture of wallboard. (4) The center or middle part of a homogeneous board where this portion differs in any respect from the outside surfaces or liners.

FILLER CLAY—A kaolin or other type of clay which is added to the papermaking furnish prior to sheet formation to increase its opacity, brightness, and printing smoothness. See CLAY; KAOLIN.

FILLER, FILLER PLY—(1) A material, generally nonfibrous, added to the fiber furnish of paper. See LOADING. (2) The center ply or plies in a multilayer board. It is usually a lower grade of stock than the top liner (q.v.) or back liner (q.v.). In a three layer sheet, it is sometimes called the base sheet.

FILLER PAPER—A tablet or bond-type paper usually made from bleached chemical woodpulps in white Sub. 16, 20 and 24 (17 x 22 inches – 500) for 3-ring notebooks, spiral bound books, and similar items requiring marginally punched holes. See NOTEBOOK PAPER.

FILLET CLOTHING—The clothing for rolls found on cards and certain air-lay machines. The clothing is constructed by punching a staple-like wire through a base supporting media such a multi-layer fabric. The wire often has a bend called a knee part way up the wire.

FILM COATED—See PIGMENTED SURFACE SIZE.

FILM SPLIT/FILM SPLITTING—A physical phenomenon which occurs when a viscous fluid leaves a diverging nip, such as at the exit of a size press. The film is slit or separated into parts that adhere to the roll and to the sheet surface. Film splitting is a component of size press pickup; it also can cause a surface quality defect know as film split pattern (q.v.).

FILM SPLIT PATTERN—A surface quality defect of paper from film splitting. It may appear as either a pattern of lines or as blotches or "orange peel" (q.v.). A sheet with film split pattern will have areas of differential ink absorption, leading to a blotchy or mottled printed surface appearance.

FILM TRANSFER SIZE PRESS—See METERING SIZE PRESS.

FILTER—A fabric, sheet, or medium for separating two or more materials usually for the purpose of cleaning or separation.

FILTER PAPER—A porous, unsized paper for filtering solid particles from liquids or gases. It is made from cotton fiber or chemical woodpulp or both, in basis weights from 15 to 200 pounds (20 x 20 inches – 500). Important properties are uniformity of formation, moderate strength when wet, high retention of particulate matter, high filtering rate and, usually, high chemical purity. For most purposes, the pore size is carefully controlled, since this determines the speed of filtration and the size of particles removed from the fluid (liquid or gas). For analytical filter paper, the ash content should be as low as possible. Industrial filter papers are used on filter presses and machines in the chemical and allied industries for the removal of foreign particles or clarifying solutions. Those industrial filter papers with a basis weight of 40 pounds or higher are interchangeably known and used as blotting papers, in that they are used because of their absorptive quality rather than for their filtering quality. See also ANALYTICAL FILTER PAPER.

FILTRATION—(1) The process of passing a liquid through a filtering medium (which may consist of granular material, such as sand, magnetite, or diatomaceous earth, finely woven cloth, unglazed porcelain, nonwovens, or specially prepared paper) for the removal of suspended or colloidal matter. (2) It is also the medium by which successive layers of fibers are deposited stepwise onto the woven cloth, known as a forming fabric (q.v.) from the suspension deposited by the headbox.

FINE PAPERS—A broad category of papers made from bleached chemical, mechanical, and cotton pulps and/or from recycled pulps, which are used for printing and writing purposes. See PRINTING AND WRITING PAPERS.

FINENESS—A term often used to express the relative diameter of fibers.

FINENESS OF GRIND—A measure of the dispersion or pigment particle size in an ink.

FINES—(1) Very short pulp fibers or fiber fragments and ray cells. These may pass through the fabric during sheet formation. (2) Pigments and other nonfibrous additives that may pass through the fabric during sheet formation. (3) Small wood particles in chips. Sometimes also referred to as sawdust or wood flour.

FINGER PAINT PAPER—A sized, smooth, coated paper made in heavier basis weights for a child's finger painting medium.

FINISH—(1) A process or treatment which may be chemical, mechanical, or high energy, that improves the performance or appearance of materials. (2) The surface property of a sheet determined by its surface contour, gloss, and appearance. It is usually determined by inspection. (3) A term used to designate the density of paperboard, specifically boxboard.

FINISH BROKE—The wastepaper resulting from the various finishing operations.

FINISHING PAPER—An abrasive paper used for hand sanding, usually with a base of fourdrinier chemical woodpulp paper, in a basis weight of 40 pounds (24 x 36 inches – 480), which is coated with aluminumoxide, silicon carbide, garnet, or flint (quartz), using an animal glue as the abrasive bond. This paper is tough and flexible.

FINISHING (PAPER)—The finishing process modifies, by mechanical means, the surface characteristics, the size, or the shape of webs of material by operations such as calendering, embossing, winding, sheeting, trimming, packaging, handling, storing, and shipping and receiving. The calendering or embossing process (finishing) can be done during or after the papermaking process is completed. In line soft-nip calendering can be done on-machine, but is still considered to be a finishing process.

FINISHING STACK—A calender located at the dry end of the paper machine, whereas a breaker stack (q.v.) is usually located mid-machine or prior to an on-machine coater or size press. See CALENDER.

FINISH POINTS—The total thickness in points (thousandths of an inch) of 50 pounds of paperboard cut to a standard sheet size. Two standards for sheet size exist: 25 x 40 and 26 x 38 inches. The 25 x 40-inch size is quite commonly used, and the other on limited occasions.

FINISH VARIATION—Marked change in surface smoothness or gloss within a roll or between rolls on either the felt or wire side.

FINISH WASTE—See FINISH BROKE.

FIRECRACKER PAPER—A paper made to give sheets with deckle edges from which strips can be cut counter to the machine direction. In wrapping the firecrackers, the deckle edge serves to terminate the roll without leaving a sharply cut edge.

FIREPROOF—Fire-resistant. The sheet may be composed of inorganic fibers or of organic fibers rendered partially fireproof (flameproof) by treatment with suitable chemicals. Such paper may char but will not burn with a flame.

FIREPROOF CREPE—Any creped paper treated with chemicals to resist flame and used for decorative purposes in public places where fire is a hazard. It is made in a variety of colors.

FIREPROOF PAPER—Any paper that is treated with chemicals so that it will not support combustion. It is not actually fireproof but is self-extinguishing and will not propagate a flame. Fire-resistant is generally considered a more accurate term since no paper is fireproof unless it is made entirely of inorganic fibers such as asbestos, ceramic, or glass.

FIREWORKS PAPER—Bright-colored papers used in the manufacture of fireworks. This is a mechanical and chemical woodpulp paper made in light weights and sized to take paste. There are no special strength requirements.

FIRST ORDER SYSTEM—One whose output is modeled by a first order differential equation. Typical example is a storage tank.

FIRST PASS RETENTION—The process of retaining all stock additives [chemicals, fillers, cellulosic fines (q.v.)] in the forming paper web on the first pass of the headbox stock over the paper machine forming section. First pass retention affects the runnability of the paper machine and the quality of produced paper. See ASA; RETENTION.

FIRST PRESS—See PRESS SECTION.

FIRST-STATE BIOCHEMICAL OXYGEN DEMAND—That part of oxygen demand associated with biochemical oxidation of carbonaceous, as distinct from nitrogenous, material. Usually, the greater part, if not all, of the carbonaceous material is oxidized before the second stage, or substantial oxidation of the nitrogenous material, takes place. Nearly always, at least a portion of the carbonaceous

129

material is oxidized before oxidation of nitrogenous material even starts. See also ULTIMATE BIOCHEMICAL OXYGEN DEMAND.

FISCAL MANUSCRIPT COVER—See MANUSCRIPT COVER.

FISH EYES—(1) Small, round, glazed or transparent spots resulting from slime, undefibered portions of stock, or foreign materials, which are crushed in calendering the sheet. (2) Round transparent spots in the surface of coated paper or board, which may be caused by excess defoamer.

FISH PAPER—See ELECTRICAL INSULATION FIBER.

FISH WRAPPER—A vegetable parchment paper or a waxed paper which possesses some wet tensile strength and does not impart an odor to the fish wrapped therein.

FLAG—(1) A small piece of paper or similar material placed in a roll so that it extends beyond the end to denote a splice or a defect. (2) To insert a marker to denote a splice or a defect.

FLAKE CONTENT—A measure of the residual undefibered paper flakes after some degree of pulping or deflaking or refining of broke or recovered fiber. The related terms DEFIBERING INDEX, DEGREE OF DEFIBERING and PULPING INDEX are historic, poorly defined, and made obsolete by the current preferred term flake content (q.v.).

FLAKING—The separating of the coating material from a coated sheet in the form of flakes.

FLAMEPROOF—Flame-resistant. See FIREPROOF.

FLAME RESISTANCE—The resistance of treated paper or paperboard to the spread of flame when ignited.

FLASH DRIED PULP—Pulp that has been dried by hot air in crumbled form.

FLAT BOX—A high vacuum suction box (q.v.).

FLAT BUNDLE—A method of packing sheets of paper, either flat or soft folded, so as not to crease the individual sheets. The designated quantity, either by weight or count, is usually wrapped with heavy kraft paper and securely tied with rope or twine.

FLAT CRUSH RESISTANCE—The resistance of the flutes in single-faced or single wall (double-faced) corrugated board to a crushing force applied perpendicularly to the surface of the board, under prescribed conditions. See CONCORA TEST.

FLAT DRINKING-CUP STOCK—See CUP PAPER.

FLAT FINISH—A smooth finish free from glare.

FLAT KRAFT—See EXTENSIBLE PAPER; FLAT PAPER OR FLAT KRAFT.

FLAT NEWS—Newsprint cut into sheets.

FLAT PAPER OR FLAT KRAFT—(1) Paper which comes from the mill in flat sheets, without fold or crease. The term is applied especially to writing papers in packages of flat sheets of standard sizes and finishes. (2) Standard multiwall or bag paper grades as distinguished from extensible paper (q.v.).

FLAT REAM—A ream of paper which is shipped flat, i.e., not folded.

FLAT ROLLS—Rolls of paper that have been flattened through dropping or crushed by a pressure on the side of the roll.

FLATS—See FLAT WRITINGS.

FLAT WALLET—See RED WALLET.

FLAT WRAPPERS—A general term applied to wrapping paper which has been put up in bundles of flat sheets instead of folded.

FLAT WRITINGS—(1) Usually school flats and tablet paper. (2) Writing paper with a smooth, flat finish as distinguished from a bond finish.

FLAX—The bast fiber of the flax plant *(Linum usitatissimum)* has been the source of linen for several millennia. Linen rags, cuttings, threads, etc., have long been used in papermaking. More recently the straw from flax cultivated for seed has been used for the manufacture of cigarette paper and similar papers.

FLAX BOARD—A fibrous insulation board manufactured largely from flax straw pulp.

FLEXIBLE BLADE COATING—See BLADE COATING.

FLEXIBLE COVER—A cover paper designed to be flexible and used for book covers, pocketbooks, etc.

FLEXIBLE FIBER—A grade of vulcanized fiber (q.v.) made soft and resilient by incorporating a plasticizer to make it suitable for gaskets, packing, and similar nonelectrical applications. It is made in thicknesses from 1/64 to 7/16 inch usually in red, black, or gray.

FLEXIBLE PACKAGING—The use of a material, such as an adhesive or coating, to convert one or more flexible webs to prepare a composite material for the packaging of food or other items to provide functional and synergistic benefits unattainable with a single, unmodified web.

FLEXIBLE PACKAGING LAMINATING—The process of combining flexible plastic films, paper, foil, or other webs with an adhesive material to form composite structures that possess certain aesthetic, convenient, automatic handling, and product protection characteristics.

FLEXOGRAPHIC PRINTING—A rotary letterpress printing process in which the ink used consists of a dye or a pigment or both and generally a binder in a rapidly evaporating solvent vehicle. The printing press is much simpler than conventional machines, because only two rolls maybe required, one running partially sub-merged in the fountain, the other transferring the ink to the plate. The rolls and plates are ordinarily made of rubber. This process is widely used for printing on many different media including various grades of paper and paperboard, laminations, film, foil, etc. The process was formerly called aniline printing.

FLEXURAL RESISTANCE—The resistance of paperboard, including corrugated board, to deflection when a specimen is supported at the ends and load is applied in the middle of the span. This property is of interest in connection with bookbinding boards and rigid insulation boards.

FLINT BACKING PAPER—A base paper for the manufacture of abrasive paper (q.v.).

FLINT-GLAZED BOX-COVER PAPER—Coated box-cover paper with a high finish once obtained by flint glazing (q.v.).

FLINT GLAZING—A method of imparting a hard, brilliant polish to paper, more especially to coated papers, by means of rubbing with a smooth stone or stone burnisher on a flint glazing machine. This is an obsolete method of paper finishing that has been replaced by supercalendering, soft-nip calendering, gloss calendering, machine calendering.

FLINT PAPER—An abrasive paper (q.v.) in which the base paper is a kraft sheet in basis weights of 50 to 60 pounds (24 x 36 inches – 480), coated with flint (quartz) with a glue bond. This is the common hardware-store type of sandpaper for hand sanding. Aluminum oxide flint papers are also popular hardware store items.

FLINTS—See FLINT GLAZING; FLINT-GLAZED BOX-COVER PAPER.

FLOATING DRYER—A term applied to the first dryer, after the doctor, on some semicrepe tissue machines. It runs at a slower speed than the dryer rolls immediately following, permitting a higher crepe ratio than would be possible without it.

131

FLOC—An aggregate of particles which may be present in a liquid or vapor.

FLOCKING TISSUE—Highly finished tissue, usually with a basis weight of 10 and 12 pounds (24 x 36 inches – 480), which is used in textile manufacture to protect dress and other thin materials which are printed. The high finish prevents the tissue from linting onto the fabric with which it is used. This paper is also designed interleaving tissue.

FLOCK PAPER—A kind of wallpaper or cover paper prepared by being sized, either over the whole surface or over special parts constituting the pattern only, and then powdered over with flock (powdered wool, cotton, or rayon), which is specially dyed. It was originally intended to imitate tapestry and Italian velvet brocades.

FLONG—This is synonymous with stereotype dry mat and is the term generally used in foreign countries. In the United States, mat is the term generally used in the trade, although flong is referred to in some patents. See MATRIX BOARD (1).

FLOORING FELTS—See CARPET FELT; DEADENING FELT; SHEATHING PAPER.

FLOORING PAPER—See CARPET FELT; DEADENING FELT; SHEATHING PAPER.

FLORISTS BOXBOARD—A paperboard used to make boxes for the packaging of flowers. It is a combination board, hard sized and waterproof, usually 0.016 to 0.048 of an inch in thickness, with a mist gray (or other color) vat liner on one side and a green-colored vat liner on the other.

FLORISTS CREPE PAPER—A paper used for wrapping the outside of flower pots and for other decorative purposes. It is made of bleached or unbleached chemical pulp, usually creped and colored. Basis weight is about 30 pounds (24 x 36 inches – 500) and when colored, the dyestuffs are fast to water, i.e., do not bleed.

FLORISTS PARCHMENT—A vegetable parchment specially prepared to protect flowers in transit. It is distinguished by the fact that it is soft, yet durable and moistureproof. The usual basis weight for this purpose is 27 pounds (24 x 36 inches – 500).

FLORISTS TISSUE—Usually a colored waxed tissue used by florists as a flower wrap.

FLOTATION—In the paper industry, a process for separating printing inks and minerals from printed recovered paper by sparging air through the pulp slurry which has been treated with soaps or synthetic surfactants. When properly chosen, the surfactants cause preferential wetting of solid ink particles and certain types of oil, while other non-ink-based contaminants are not wetted. The former are carried to the surface by the air bubbles and thus floated away and separated from the wetted particles.

FLOTATION DEINKING—The process of separating ink, usually with some mineral fillers, from a slurry of recycled fibers, by the froth flotation process. In this process, the slurry is mixed in a high shear environment with air, surfactants, and special collecting chemicals. The ink, and some of the minerals, are preferentially attached to the air bubbles and float to the upper surface of the slurry, where the resulting scum is removed from the slurry in concentrated form. Also called froth flotation.

FLOUR—(1) A term applied to the fine fibers or fiber fragments of a pulp. They are also known as fines. (2) Wood flour (q.v.).

FLOUR-SACK PAPER—A paper used for the manufacture of flour sacks, which is made from a variety of raw materials including rope, sulfate, and sulfite pulps, either as such or in combination. These papers may be made from natural colored, semibleached, or fully bleached pulps; they may be uncoated or coated, and they are supplied in a number of finishes, such as plain, embossed, or supercalendered and embossed. Among the types of paper which maybe used are white enameled coated manila rope paper, kraft paper, or a composition rope and kraft paper, which are blue lined and are used for single wall bags. These papers, generally manufactured on a cylinder machine, are made in basis weights ranging from 60 to 100 pounds (24 x 36 inches – 500), the weight not including the coating material subsequently applied. An uncoated bleached white sulfate paper is also used, which may be unlined or blue lined. Double wall paper flour sacks are usually made with an outer wall of 50-pound uncoated bleached kraft paper with an inner wall of blue or natural kraft, ranging in basis weights from 30 to 50 pounds (24 x 36 inches – 500). See also SHIPPING SACK KRAFT PAPER.

FLOWBOX—British term for headbox (q.v.).

FLOWER-POT PAPER—A stiff, smooth, lined paper which may be sized or waxed and is used in the manufacture of paper flower pots.

FLOWER-POT COVERING PAPER—See FLORISTS CREPE PAPER.

FLOWMETER—In wastewater treatment, a meter that indicates the rate at which wastewater flows through the plant.

FLOW-ON COATING—Flow-on coating consists of flowing a suspension of pigment and adhesive directly onto the wet web as it is being formed on a fourdrinier papermaking machine.

FLUE GAS—Gaseous mixture of the products of combustion of any fuel along with other gases associated with the specific unit process. Flue gas for power boilers consists of combustion products of whatever fuel is being fired. Flue gas for a recovery boiler (q.v.) is the products of combustion of the organic portion of black liquor. Flue gas for a lime kiln (q.v.) includes products of combustion of the fuel plus the carbon dioxide from lime mud (q.v.) calcination.

FLUFF—(1) Dust from paper which gathers on the rolls or doctors of a paper machine or on the rolls of a printing press. (2) A type of pulp, characterized by high bulk density, low fines, and good absorbency, often used for diapers, and other air-laid (q.v.) products.

FLUFF PULP—A chemical, mechanical, or combination chemical-mechanical pulp, usually bleached, used as an absorbent medium in disposable diapers, bedpads, and hygienic personal products. Also known as "fluffing" or "comminution" pulp.

FLUIDITY—The reciprocal of viscosity. The unit of fluidity is the rhe.

FLUIDIZED BED BOILER—Combustion air is used to keep a bed of inert material, usually sand, in motion to aid combustion. In a circulating fluidized bed, the inert material is circulated with the combustion gases, separated, and returned to the bed. In a bubbling fluidized bed, the inert material circulation is only in the lower portion of the boiler; separation of the inerts takes place above the bed and the inerts fall back into the bed. A circulating unit is better for less volatile fuels, while a bubbling bed is better with higher volatility fuels even if they are wet.

FLUIDIZERS—Chemical compounds which decrease the viscosity and reduce the thixotropy of pigment dispersions both in the initial dispersion, and in the pigmented coating color.

FLUORESCENCE—(1) That property of substances that causes them to emit radiation as the immediate result of, and only during, the absorption of incident radiant energy of different wavelengths. (2) The radiation emitted in this process. The fluorescent radiation usually has a longer wavelength than the exciting radiation. When irradiated by ultraviolet light, the fluorescent radiation is often visible. This phenomenon is the basis of "optical brighteners," substances which are added to paper so that, when illuminated by daylight or other light containing ultraviolet, the paper fluoresces and appears brighter.

FLUORESCENT DYES—Synthetic dyestuffs used for enhancing the brightness of uncolored paper. These materials absorb ultraviolet light and re-emit it at a higher wavelength (usually in the violet end of the spectrum) thus increasing the "blueness" and "brightness" of the paper as seen by the eye.

FLUORESCENT PAPER—A paper containing a fluorescent material which absorbs ultraviolet "black" light and re-emits it in the visible spectrum. Such papers glow brilliantly in various hues (depending on the fluorescent agent) when illuminated by ultraviolet energy. Bluish fluorescing colorless dyes or pigments are commonly employed in white papers to enhance brightness when viewed in daylight which contains some ultraviolet energy. This phenomenon is often referred to as optical bleaching. See also FLUORESCENT WHITE.

FLUORESCENT WHITE—(1) A term descriptive of fluorescent dyes or pigments colorless as applied to "white" paper or paperboard but which increase the brightness by absorbing the ultraviolet energy of daylight and re-emitting it as visible light. (2) "White" paper or paperboard containing such fluorescent material.

FLUTE—The geometric configuration formed by one of the undulations of the corrugated medium in corrugated board. The exact dimensions of the flute will vary slightly, depending on corrugating roll contour, material characteristics, converting equipment, and technique. The three common types in conventional corrugated board used in shipping containers and boxes are A-, B- and C-Flute approximately 3/16, 3/32, and 5/32 inch high, respectively [not including the thickness of the liner(s)]. The number of flutes per foot are approximately 34–36, 47–50, 39–42 for A, B, and C flute, respectively. E-flute and F-flute, approximately 1/16 inch high and 3/64 inch high, respectively, and spaced about 88 to 100 flutes per foot, are used mainly in corrugated board for folding cartons.

FLUTED—See CORRUGATED BOARD.

FLUTED PAPER—Paper with a corrugated appearance; e.g., paper used on the inside of light bulb containers or in baking cups.

FLUTING—See CORRUGATING MEDIUM; FLUTE.

FLY ASH—All solids, including ash, charred paper, cinders, dust, soot, or other partially incinerated matter, that are carried in a flue gas stream.

FLY ROLL—A roll used to support or change direction of a web of material on a processing machine.

FLYING DUTCHMAN—A name formerly applied to the Yankee machine (q.v.).

FLYING PASTER—A high-speed automatic device for splicing rolls of paper to permit continuous operation of converting and printing equipment.

FLYING SPLICE—A machine feature which cuts and splices a web onto a new unwind or windup roll without stopping.

FLYING SPLICE UNWIND—A special unwind stand designed to splice the tail of a new roll to the tail of an expiring roll while maintaining control of the web. The flying splice is used for continuous operations, such as an off-machine coater. Flying splice unwinds may also be used on supercalenders and winders to improve time cycles.

FLYKNIFE—The rotating knife on a cutter or sheeter. Some converting machines also incorporate rotary cut-off knives called flyknives.

FLYLEAF—(1) See END SHEET. (2) In printing, any leaf which is free from printing but which is a part of one of the printed signatures. See also END-LEAF PAPER.

FLYPAPER—(1) The base paper for flypaper. It is a well-formed kraft or sulfite sheet of a basis weight of about 60 pounds (24 x 36 inches – 500), prepared from well-beaten stock. (2) Sticky flypaper is such a base paper coated on one or both sides with a special glue preparation for the entanglement of flies. Poisonous flypaper is a water-leaf sheet impregnated with a toxic substance, together with sugar or honey; the paper is kept moist in a tray.

FOAM BONDING—The process of bonding nonwovens using a froth of binder and air.

FOAM MARKS—Marks caused by foam in the stock as it comes onto the paper-machine wire. Such marks may also occur in coated paper. Also called foam spots.

FOIL CUPSTOCK—See FOIL LAMINATE.

FOIL LAMINATE—The lamination of a thin sheet of aluminum foil to paper to enhance barrier properties and increase resistance to the penetration of oxygen and/or moisture. Used in the manufacture of cigarette soft packs. Sometimes called foil cupstock.

FOIL-MOUNTING BOARD—A paperboard which is specifically designed for foil laminating work. Essential characteristics are an even formation, smoothness, and uniform finish.

FOIL-MOUNTING PAPER—A paper for mounting metallic foils for protective and decorative purposes. Smoothness, uniform caliper, and freedom from alkali are important characteristics. A wide variety of grades of foil-mounting paper is employed depending on end-use requirements and laminating processes. Basis weights range from lightweight tissue up to 35 pounds (24 x 36 inches – 500).

FOILS—See HYDROFOILS.

FOLDED—(Of reams of writing papers) Folded in half, quired, or interfolded; not flat. A folded sheet is also termed fly. See HARD FOLD; INTERLAPPED; SOFT FOLD; QUIRE.

FOLDED FOURDRINIER—See MULTIPLE FOURDRINIERS.

FOLDED NEWS—Unsold newspaper collected from newsstands. This recovered paper grade is usually included in the category of overissue news, which consists of unused, or overrun, regular newspapers printed on newsprint, baled or securely tied in bundles. The term is no longer in common use.

FOLDED WRITINGS—Writing paper folded to a definite size. (Rare).

FOLDER—(1) A circular or other piece of direct advertising material which is folded before mailing. (2) A heavyweight sheet, folded once and used for filing purposes. (3) A grade of boxboard suitable for scoring and folding. See BENDER.

FOLDER STOCK—A board or bristol used for the manufacture of folders for business filing. It is commonly made of woodpulp and reclaimed paper stock, though some grades are made from rope or jute stock. The basis weights vary from 100 to 225 pounds (24 x 36 inches – 500); common thicknesses are 0.008, 0.011, and 0.014 of an inch. It may be surface sized or treated to give greater wearing qualities. Significant characteristics include tearing resistance, stiffness, folding and noncurling properties, and uniform high finish without mottle.

FOLDING—See BENDER.

FOLDING BOX—See FOLDING PAPER BOX.

FOLDING BOXBOARD—A paperboard suitable for the manufacture of folding cartons which can be made from a large variety of raw materials on either a cylinder machine or a fourdrinier machine. It possesses strength qualities that permit scoring and folding, and has variable surface properties depending upon the printing requirements. This classification includes such products as clay-coated boxboard, white patent coated news, manila lined news, and fourdrinier bleached kraft board. See BOXBOARD; PAPERBOARD.

FOLDING BRISTOL—A paperboard of the mill bristol type but with longer and more flexible fibers to enable the board to be folded. Important qualities include folding ability, color, finish, and printing properties.

FOLDING CARTON—See FOLDING PAPER BOX.

FOLDING ENAMEL—A coated book paper with extra strength, which is particularly suitable for folding.

FOLDING ENDURANCE—The number of folds under specified conditions in a standard instrument which a paper will withstand before failure. See also LOG FOLD.

FOLDING PAPER BOX—A container (other than solid fiber or corrugated shipping container) which is the product of a cutting and creasing (die-cutting) operation on relatively lightweight folding boxboard. The carton is capable of being folded flat for shipment by the fabricator in contrast to the setup box which is sent out already formed. The carton is usually formed up, filled, and closed by the user. It is produced in many styles, shapes, and sizes, of which the four-sided style with flap closure at ends or top and bottom is the most important.

FOLDING STOCK—Paper made from a strong long-fibered chemical woodpulp sometimes mixed with rag pulp to be coated with a coating mixture that is plastic enough not to crack when folded and opened. This paper is made in a wide variety of basis weights.

FOLDING STRENGTH—See FOLDING ENDURANCE.

FOLDING TRANSLUCENTS—Clay-coated bristols possessing high finish and folding qualities.

FOLIO—A ream or sheet in its full size. When used in connection with books, folio means that the sheet has been folded once, producing four pages. See OCTAVO; QUARTO.

FOLIO CUTTER—Somewhat similar to a cut-size cutter except that it is designed to cut folio size, which is generally two times the size of cut-size sheets. For example, a typical sheet size is 17 x 22 inches, which is twice the size of an 8 1/2 x 11 inches sheet cut on a cut-size cutter (q.v.).

FOODBOARD—See SPECIAL FOODBOARD.

FOOD-SHOP PAPER—See DELICATESSEN PAPER.

FOOD WRAPPERS—Any paper which is suitable for wrapping food. Cleanliness is an important property. It may be a dry-finish butchers, dry-finish grocers, glassine, greaseproof, or vegetable parchment sheet. Waxed, coated, lacquered, and laminated papers of many kinds including a wide variety of combinations of paper, metal foil, and synthetic plastic film is also used to meet special end-use requirements.

FOOLSCAP—A size of writing paper 13 x 16 inches. The name comes from a similar size of paper, 13 1/2 x 17 inches originally made in the United Kingdom, which had a watermark of a jester's head with cap and bells.

FOOTAGE—The length or area of a roll of paper expressed in linear or square measure.

FORCED-DRAFT AIR—See FORCED-DRAFT FAN.

FORCED-DRAFT BOILER—A forced-draft boiler is one in which all of the combustion air and combustion gases are pushed through the boiler by raising the pressure of the inlet air sufficiently.

FORCED-DRAFT FAN—The fan used to force combustion air into the furnace of a boiler.

FOREIGN PARTICLES—Particles other than fiber, embedded in the pulp, appearing opaque when viewed by transmitted light, and having an area not less than 0.02 square millimeters.

FOREST GENETICS—The basic science dealing with causes of differences and similarities among trees related by descent. Genetics takes into account the influence of the genes (units of inheritance) and environment on tree growth and utility. See FOREST TREE IMPROVEMENT.

FOREST RESIDUES—Comprises chips, particles, and fibers arising as by-products of logging operations including culled material, slash, limbs, saplings, etc. Also includes other secondary forest material not usually defined as logging residues, such as tops, branches, standing saplings, and cull trees.

FOREST TREE IMPROVEMENT—The applied science of systematic genetic improvement of a species or a population. Employed are such techniques as selection, hybridization (combining parents of unlike genetic makeup), and mutations (the change in structure or number of basic units of inheritance). The term is sometimes also applied to the improvement of trees or stands via non-genetic means such as the use of chemical growth stimulators.

FORKED NEEDLE—A needle-punch needle ordinarily having a "U" shaped tip that is capable of producing velour and patterned effects.

FORMAMIDINE SULFINIC ACID (FAS)—$H_2NC(NH)SO_2H$. A powerful reductive bleaching agent, commonly used as a dye stripper and brightener for recycled fiber. FAS may be applied in the pulper or as a post-bleaching stage. FAS is also used to bleach colored broke and, in Europe, as a strength preservative in ozone bleaching.

FORMATION—(1) A property determined by the degree of uniformity of distribution of the solid components of the sheet with special reference to the fibers. It is usually judged by the visual appearance of the sheet when viewed by transmitted light. This property is very important, not only because of its influence on the appearance of the sheet, but because it influences the values and uniformity of values of nearly all other properties. (2) The process of forming a sheet.

FORM BOARD—A paperboard used to make models or shapes, such as window-display figures for clothing. It is made of reclaimed paper stock and may contain some rag pulp and is 12 points or more in thickness, depending upon the size and character of the form. It is pliable and waterproof. The board is usually painted or colored after the form is made.

FORM BOND—A lightweight commodity paper designed primarily for printed business forms. It is usually made from chemical wood and/or mechanical pulps. Important product qualities include good perforating, folding, punching, and manifolding properties. The most common end use for this grade is a carbon-interleaved multi-part computer printout paper which is marginally punched, cross-perforated, and fanfolded. See REGISTER BOARD.

FORMER—The forming section (q.v.) of a papermaking machine.

FORMING BOARD—The first drainage element under the fabric on a fourdrinier or multi-wire former is known as the forming board or making board. The positioning and open area of this stationary bladed structure is critical to papermaking. Typically, the headbox jet lands on the first blade in such a manner that a small percentage of the jet is immediately doctored (scraped) away.

FORMING FABRIC—An endless belt woven of plastic or metal for use on the forming section (q.v.) on which belt the fibers are felted into pulp, board, and paper. The fabrics are woven "flat" and joined (seamed) by mechanical means or they are woven "endless" and come off the loom like a tube and do not need to be joined. When woven flat and joined, the "warp strands" are in the machine direction while the "weft strands" are in the cross direction. When woven "endless," the "warp strands" are in the cross machine direction and the "weft strands" are in the machine direction. The weave configurations are constantly being expanded to fit the various drainage and sheet requirements in the industry. The fineness of the fabric is approximated by the mesh count (i.e., the number of warp strands per inch and the number of weft strands per inch). Layered constructions offer the option of using fine yarns with a high fiber support index (q.v.) on the sheet side and a more rugged construction on the other side for greater stiffness and wear resistance. The synthetic yarns are manufactured from specially designed resins or blends of resins or copolymers.

FORMING FABRIC CARRYING ROLL—See FORMING FABRIC, FELT AND DRYER FABRIC CARRYING ROLLS.

FORMING FABRIC DRIVE ROLL—See WIRE TURNING ROLL.

FORMING FABRIC, FELT AND DRYER FABRIC CARRYING ROLLS—On flat fourdriniers and twin wire machines, a number of rolls are used to support and convey forming fabrics and felts through various parts of the machines. They are usually covered with a hard elastomeric composition. The names of the rolls define their function, such as guide roll (q.v.) and stretch roll (q.v.)

FORMING FABRIC TURNING ROLL—This roll is located below the suction couch roll (q.v.) and turns the forming fabric back under the table towards the breast roll (q.v.). It is sometimes called a wire drive roll, and it shares the power load required to drive the forming fabric with the suction couch roll (q.v.). It may be covered to increase its coefficient of friction.

FORMING ROLL—These rolls play an essential part in the headbox operation of twin wire formers. The forming roll is located at the jet headbox delivery point where the two forming fabrics converge between this roll and a backing roll. There are many types in use, depending on the design of the particular twin wire machine. In general, these rolls consist of a solid or drilled shell and may be covered with a metal or fabric screen. Some designs contain an internal suction box.

FORMING SECTION (FORMER)—That part of the paper machine which transforms the headbox slurry of fibers (usually about 0.5 to 1.0% solids and 99.0 to 99.5% water) into a web of about 25% solids and 75% water. The main functions of the forming section are water removal, fiber orientation, and achieving a uniform web. The most common type of formers are: fourdrinier (or flat fourdrinier), twin wire, top wire, multi-ply, cylinder, Inverform, crescent, and suction breast roll.

FORTIFIED SIZE—A chemically modified rosin acid used in place of, or in combination with, rosin size for producing improved water resistance in paper and board.

FORWARD CLEANER—A hydrocyclone for removing contaminants with specific gravities greater than one from a slurry of fibers in water. The feed enters the cone tangentially at the top with the heavy contaminants being thrown centrifugally to the outer wall and exiting at the bottom of the cone. The cleaned pulp normally discharges out of the top center of the cone. One design operates to discharge the accepts from the bottom of the cleaner. This application was the only one practiced in the paper industry until about 1970. Prior to that time, forward cleaners were commonly referred to by the more generic term centrifugal cleaner (q.v.). See also REVERSE CLEANER.

FOUNTAIN SOLUTION—In offset printing, the solution of water and other chemicals that dampen the plate in the non-image areas.

FOUR COLOR PRINTING—Use of four inks, cyan, magenta, yellow, and black, to generate all colors. In practice, a plate for each color is developed by color separation. The combination of the ink layers creates the full color spectrum.

FOURDRINIER BOARD—Board made from a fourdrinier machine (q.v.).

FOURDRINIER BRISTOLS—This term includes index bristols and sometimes mill bristols. The stock may be rag or chemical woodpulps in varying proportions. It is used principally for index, record, business, and commercial cards, social announcements, invitations, specialties, etc. See INDEX BRISTOL; MILL BRISTOL.

FOURDRINIER FABRIC—See FORMING FABRIC; TWIN WIRE FORMER.

FOURDRINIER KRAFT LINERBOARD—Kraft linerboard (q.v.) formed on a fourdrinier paper machine.

FOURDRINIER MACHINE—The fourdrinier machine, named after its sponsor, with its modifications and the cylinder machine (q.v.) comprise the machines normally employed in the manufacture of all grades of paper and board up to the 1970s. The fourdrinier machine, for descriptive purposes, may be divided into four sections, the wet end, the press section, the dryer section, and the calender section. In the wet end, the pulp or stock (at a consistency or concentration of 0.2 to 1%, depending upon the grade and weight of paper to be manufactured) flows from a headbox through a slice (q.v.) onto a moving endless fabric belt called the fourdrinier fabric or fabric, of brass, bronze, stainless steel, or plastic. The fabric runs over a breast roll under or adjacent to the headbox, over a series of tubes or table rolls, or more recently, drainage blades which maintain the working surface of the fabric in a plane and aid in water removal. The tubes or rolls create a vacuum on the downstream side of the nip. Similarly, the drainage blades create a vacuum on the downstream side where the fabric leaves the blade surface, but also performs the function of a doctor blade on the upstream side. The fabric then passes over a series of suction boxes, over the bottom couch roll (or suction couch roll) which drives the fabric and then down and back over various guide rolls and a stretch roll to the breast roll. The second section, the press section, usually consists of two or more presses, the function of which is to mechanically remove further excess water from the sheet and to equalize the surface characteristics of the felt and fabric sides of the sheet. The wet web of paper, which is transferred from the fabric to the felt at the couch roll, is carried through the presses on the felts, the texture and character of the felts varying according to the grade of paper being made. The third section, the dryer section, consists of steam-heated cylinders, and the paper is held close to the dryers by means of the dryer fabric. As the paper passes from one dryer to the next, first the felt side and then the fabric side comes in contact with the heated surface of the dryer. As the paper enters the dryer section, approximately 40% dry, the bulk of the water is evaporated in this section. Moisture removal may be facilitated by blowing hot air onto the sheet and in between the dryers in order to carry away the water vapor. Within the dryer section and at a point at least 50% along the drying curve, a breaker stack is sometimes used for imparting finish and to facilitate drying. This equipment usually comprises a pair of chilled iron and/or rubber surfaced rolls. There may also be a size press located within the dryer section, or more properly, at a point where the paper moisture content is approximately 5%. The fourth section of the machine is known as the calender section. It consists of from one to three calender stacks with a reel device for winding the paper into a roll as it leaves the paper machine. The purpose of the calender stacks is to finish the paper, i.e., the paper is smoothed and the desired finish, thickness or gloss is imparted to the sheet. Water, starch or other solutions, wax emulsions, etc., may be applied for additional finish. The reel winds the finished paper into a roll, which for further finishing can be taken either to a rewinder or, as in the case of some machines, the rewinder on the machine produces finished rolls directly from the machine reel. The wet end, the press section, the several dryer sections, the calender stacks, and the reel are so driven that proper tension is maintained in the web of paper despite its machine direction elongation, or cross machine direction shrinkage during its passage through the machine. The overall speed of the machine is determined by the grade and weight of paper being manufactured. There are two modifications of the fourdrinier in use: the Harper and the so-called Yankee or MG machine which, in principle, are similar to the fourdrinier machine. Some machines contain one or more coating stations to impart smoothness, gloss, and color to one side (C1S) or two sides (C2S) of the paper. Some machines incorporate on-machine finishing, typically with soft nip calenders. Papers varying in weight from lightweight tissue to heavy paperboard are made on the fourdrinier machine. See also MULTIPLE FOURDRINIERS.

FOURDRINIER YANKEE MACHINE—See YANKEE MACHINE.

FOURDRINIER WIRE—See FORMING FABRIC.

FOURTH PRESS—See PRESS SECTION.

FOXED, FOXY—Containing foxing (q.v.).

FOXING—Stains, specks, or spots in paper, e.g., prints or books, mostly caused by mold or mildew.

FRACTIONATION—A separation of fibers on the basis of some physical characteristic, such as fiber length, width, specific surface area, freeness, degree of refining, and flexibility. Most washers, thickeners, screens, and cleaners in the pulp mill, stock preparation, or recovered fibers processing areas fractionate to some degree. The extent of fractionation can be adjusted substantially by adjusting the operating parameters of these unit operations. Sometimes, fractionation is deliberately induced, to produce multiple streams containing fibers of differing characteristics, which may be treated or used separately.

FRAME BUNDLE—A method of packing paper in which a solid or latticed wooden top and bottom are placed on the flat sides of the bundle and the three are banded or tied together in both directions.

FREENESS—The rate at which water drains from a stock suspension through a wire mesh screen or a perforated plate. It is also known as slowness or wetness, according to the type of instrument used in its measurement and the method of reporting results. When measured by the Canadian Standard Freeness (CSF) test, the CSF freeness is reported as the volume (in milliliters) of water flowing through the side orifice of the tester under test method conditions.

Riegler Schopper - Freeness 262

FREE SHEET—(1) In paper manufacture, a term denoting paper made from stock having rapid drainage on a paper machine. (2) Paper free of mechanical woodpulp.

FREE STOCK—A pulp suspension (stock) having rapid drainage on a paper machine.

FRENCH FOLD—A sheet of paper printed on one side only and so folded as to expose the printing.

FRENCH FOLIO—A lightweight writing paper used for second sheets and for taking printers'

proofs. It is also used for overlays and underlays in make ready in a printing plant. It is usually made from bleached chemical woodpulp in white and various colors. It is smoother than most manifold papers and has no specific strength specifications. This paper is made in basis weights ranging from 18 to 24 pounds (25 x 38 inches – 500) or from 7 to 10 pounds (17 x 22 inches – 500).

FRENCH WRITING PAPER—See FRENCH FOLIO.

FRESH CHIPS—Chips that go straight to pulping rather than to storage prior to pulping.

FRESH LIME—Lime (q.v.) derived from calcining limestone by driving off carbon dioxide from calcium carbonate at high temperature. Also referred to as purchased lime or makeup lime.

FRICTION CALENDER—Obsolete term. See CALENDER; GLOSS CALENDER; HOT CALENDER.

FRICTION GLAZED—A term applied to paper which has a very high finish, secured by passing the sheet through chilled iron rolls revolving at different peripheral speeds. Now obsolete, this process was used largely in finishing coated box-lining papers, waterproof papers, bronzed, and silver papers, etc. See CALENDER; GLOSS CALENDER; HOT CALENDER.

FRISKET PAPER—A manila-colored wrapping paper used for friskets on printing presses.

FRONT DRUM—The winding drum that is closest to the winder operator when approaching the winder from the discharge side of a two drum-winder.

FRONT SIDE—The operating side of a paper machine; the side away from the drive (back side).

FROSTED KRAFT PAPER—A machine-glazed or machine finish kraft wrapping paper, decorated with white or colored pulp which is distributed on the paper web while on the fourdrinier wire. The added pulp blends with the fibers, giving a frosted design. See also CLOUD FINISH.

FROTH FLOTATION—See FLOTATION DE-INKING.

FROTH SPOTS—See FOAM MARKS.

FROZEN-FOODS PAPERS—A type of highly moisture- and water-vapor resistant papers used for inner liners in frozen-foods packaging. They are usually glassine or bleached chemical woodpulp papers specially treated for high water-vapor resistance; waxed papers and plain, coated, or waxed vegetable parchments are also used. They are pliable so as to resist cracking under the low temperatures employed in quick-freezing and storage of foods. Properties required are stripping quality, strength and flexibility, resistance to penetration of liquids and vapors, high wet tensile strength, and purity.

FROZEN-IN STRAIN—See DRIED-IN STRAIN.

FROZEN-IN STRESS—See DRIED-IN STRAIN.

FRP—See FIBER-REINFORCED PLASTICS.

FRUIT-BAG PAPER—See TWISTING PAPER.

FRUIT WRAPS—Lightweight MF or MG papers used to wrap or pack fruit. They are made in white and colors, in a variety of furnishes to meet specific requirements. Some grades are given a special mineral oil or other treatment to prevent "scalding," retard decay, or decrease shrinkage of the fruit. This treatment must not impart odor and must not have any harmful effect on the taste of the fruit. Basis weights range from 10 pounds (24 x 36 inches – 480) to 21 pounds (24 x 36 inches – 500). Fruit wraps may be plain or printed and are supplied in various sizes.

FSI—Fiber support index.

FUEL-SACK PAPER—See CHARCOAL KRAFT PAPER.

FUGITIVE COLORS—Colors not fast to light. However, the term is also applied to colors which are readily destroyed by acids, alkalies, or bleaching agents.

FULL CREPED—See CREPING.

FUME—Very fine particulate found in recovery boiler flue gas (q.v.) derived from volatilization of sodium compounds during the combustion of black liquor (q.v.).

FUNCTIONAL COATING—A coating applied to paper which has a specific function, other than increased smoothness. Functional coatings are often intended to interact chemically with the paper, the environment, or another substrate later, compared to normal coatings, which are generally non-reactive. Often, functional coatings require special or separate treatment during broke handling. Examples include coatings containing micro encapsulated dyes for carbonless paper, light sensitive emulsions for photographic paper, silicone for release paper, or thermal-sensitive coatings for thermal printing. See also COATING.

FUNGUS (PLURAL FUNGI)—Fungi are a large group of microorganisms characterized by having chitin in their cell walls and possessing true nuclei (eukaryotes). They play a major role in decomposing organic materials in nature.

FURAN (TCDF)—A chlorinated aromatic organic compound, chemical name 2, 3, 7, 8 - tetrachlorodibenzofuran (TCDF). One of the most toxic of the chlorinated furans and used as a benchmark for furan toxicity.

FURNISH—(1) The ingredients contained in a slurry for making wet-laid products. (2) The mixture of various materials that are blended in the stock suspension from which paper or board is made. The chief constituents are the fibrous material (pulp), sizing materials, wet-strength or other additives, fillers, and dyes.

FURNISH PULPER—A pulper used primarily to form a slurry from wet lap pulp, dry lap pulp, or recovered fiber. While broke, often, is added to the furnish pulper as well, a pulper used primarily to form a slurry from broke would more properly be called a broke pulper (q.v.).

FURNITURE BOGUS PAPER—A furniture wrapping paper made in light weights, such as 40 pounds (24 x 36 inches – 500) for protective stuffing and also in 70 to 90 pound weights for protective pads and wraps.

FURNITURE PAPER—See FURNITURE-WRAPPING PAPER.

FURNITURE POLISHING PAPER—See DUSTING PAPER.

FURNITURE-WRAPPING PAPER—A common heavyweight wrapping paper, generally made of screenings, reclaimed paper stock, or kraft, which is used to protect furniture against damage in transit and while in storage.

FUSEE PAPER—A strong sheet with feather deckles (similar to cartridge paper), usually red, in basis weights of 80, 100, and 500 pounds (24 x 36 inches – 500). It is used in the manufacture of fuses for railway and track signal purposes.

FUSIBLE INTERLINING—The inside liner material applied to suits, jackets and other apparel using thermally meltable powders, polymers, and adhesives which previously had been applied to one surface of the nonwoven material.

FUZZ—(1) That property which causes a sheet to exhibit fibrous projections on its surface to develop such fibrous projections in use. (These two interpretations of the property might more descriptively be called "fuzziness" and "fuzzability," respectively.) (2) Fibers projecting from the surface of a sheet of paper.

G

GALACTAN—A hemicellulose with a backbone of galactose monomeric units. The most common galactan in wood is the beta-D-1→4-linked galactan which occurs in compression wood. This polymer with some arabinose substitution at C-6 of galactose also occurs associated with pectic substances in the middle lamella. See ARABINOGALACTAN.

GALACTOMANNAN—See MANNOGALACTAN.

GALLEY-PROOF PAPER—Sheets of paper having sufficient width to take proofs from type standing in galleys. The nature of the paper is not important as long as it will take a clean impression of the type. See also DRY PROOFING PAPER.

GALVANIC CORROSION—Accelerated corrosion of a metal because of electrical contact with a more noble metal or nonmetallic conductor (e.g., carbon, mill-scale) in an electrolyte.

GALVANIZED—Galvanized steel is formed by dipping steel in molten zinc. This leaves an outer layer of zinc over a series of zinc-iron alloys on the surface of the steel. Galvanized steel resists atmospheric corrosion.

GALVANIZED APPEARANCE—An uneven sheet surface appearance resembling galvanized metal caused by variations in smoothness or gloss.

GAMMA-CELLULOSE—That portion of cellulosic material that dissolves in the alkaline solution under the conditions of the alpha-cellulose determination, and remains in solution on neutralization of the alkaline solution in contrast to the beta-cellulose fraction. See BETA-CELLULOSE.

GAMPI—A shrub *(Wikstroemia canescens)* of the family *Thymelaeaceae* which grows wild in the mountain forests of central and southern Japan. The bast fiber of the inner bark is used in papermaking.

GAP FORMER—A type of twin wire former in which the headbox jet is injected directly between two converging fabrics.

GARBAGE-BAG PAPER—(1) A kraft or sulfite sheet, in basis weights ranging from 35 pounds (24 x 36 inches – 500) upward, which is coated or impregnated with paraffin, wax, or oil, or a duplex asphalt paper, used for making bags to line garbage pails. (2) Also a sheet of shipping sack kraft, having good wet strength, used in the construction of outdoor garbage and trash bags.

GARMENT BAG PAPER—A paper made from kraft, sulfite, or red rope pulp, in basis weights of 25 pounds (24 x 36 inches – 500) and higher, which is used for making large bags to hold suits, dresses, and coats. It may be bleached, unbleached, or colored. For storage purposes, it may be impregnated with mothproofing agents.

GARNET PAPER—An abrasive paper (q.v.) used for handwork in the wet or dry sanding of easily abraded surfaces. The base paper is a kraft sheet in basis weights of 40, 70, or 90 pounds (24 x 36 inches – 480), which is coated with garnet using a varnish-resin adhesive.

GARNETT—A machine similar to a carding machine which was designed for reclaiming fiber from waste, roving, yarn, and fabric, but which sometimes is used to manufacture nonwoven webs.

GAS CYCLONE—See CENTRIFUGAL CLEANER.

GASIFICATION—The partial combustion of a liquid or solid fuel with less than stoichiometric air to produce a combustible gas. The product gas may be burned in a second-stage combustion or used as a fuel for a combustion turbine or engine.

GASIFIER—A device used to volatize or partially burn a solid or liquid fuel in order to produce a fuel gas. See also GASIFICATION.

GASKET BOARD—A paperboard which is subsequently treated with chemicals and cut into gaskets. It is usually a chipboard or woodpulp board without special finish, but which has absorbent properties.

GAS-PHASE BLEACHING—Treatment of a pulp that has been shredded and fluffed into loose fiber aggregates at a high consistency with a gaseous reagent to provide maximum mass transfer from gas phase to fibers.

GAS SCRUBBERS—(1) Devices in which gas scrubbing is performed. (2) Equipment for absorption of gaseous pollutants (e.g., chlorine and chlorine dioxide) emitted during pulp bleaching either from process equipment or bleach plant washers. See also GAS SCRUBBING.

GAS SCRUBBING—The removal of pollutants from a gas stream by contacting it with a liquid or a solid. Liquid scrubbing is more common, but dry scrubbing is also performed.

GATE ROLL COATER—A gate roll size press (q.v.) used to apply pigmented coating.

GATE ROLL SIZE PRESS—A design of metering size press that consists of six rolls, i.e., three rolls on each side of the web. Surface sizing material is delivered by a header to a pond formed by the two outer "gate" rolls. These rolls are of different hardness and run at different speeds to control the amount of material metered through the nip. The metered film of surface sizing material is transferred to the third "applicator" roll, which applies it to the sheet surface in the size press nip. A gate roll size press treats both sides of the sheet simultaneously.

GATF—The Graphic Arts Technical Foundation which is concerned with such matters as research, development, and education in the graphic arts.

GE BRIGHTNESS—A directional brightness measurement utilizing essentially parallel beams of light to illuminate the paper surface at an angle of 45°. A brightness measurement more commonly used within the paper industry in the United States as compared to Canada, Europe, and South America where Elrepho or ISO brightness is used.

GEL—A system composed of colloidal particles that form a jelly-like structure.

GELATIN—A colorless, odorless albuminous material extracted from animal bones, hides, etc. It is used as a high-purity alternative for glue in paper coating and sizing.

GELATIN DUPLICATING PROCESS PAPER—See DUPLICATING PAPER.

GELATIN DUPLICATOR—A reproduction machine or process in which copies on smooth finish writing paper are "pulled" from a gelatin film containing hectograph ink images imparted thereto from a previously prepared master sheet.

GELATIN PAPER—See DUPLICATING PAPER; GELATIN-PRINTING PAPER.

GELATIN-PLATE PAPER—A smooth paper used in printing from a prepared gelatin plate. Uncoated or coated paper may be used, but it is well sized and has good tearing and surface strength so that the surface will not pick (q.v.).

GELATIN-PRINTING PAPER—A paper used for gelatin printing and having the same general characteristics as gelatin-plate paper (q.v.).

GENERATING BANK—The section of the boiler where a portion of the boiling heat transfer and feedwater evaporation occurs.

GENERATOR—In electrical systems, a generator is an electromagnetic machine that transforms mechanical energy into electrical energy. Alternating current is produced by rotating a magnetic field close to conductors wound on a magnetic stator. The electromotive forces developed in the stator result in a voltage and current flow. In a direct current generator, the magnetic field is stationary and the conductors are rotated.

GENUINE VEGETABLE PARCHMENT—See VEGETABLE PARCHMENT PAPER.

GENUINE WATERMARK—See WATERMARK.

GEOMETRY—In paper optical properties, a term referring to the arrangement of the illuminating system, the specimen, and the observer or light-measuring device.

GIFT WRAPPING PAPER—Plain gift wrapping paper is usually a 10-lb. tissue (24 x 36 inches – 500) made from chemical woodpulps, although heavier weights are occasionally used, and is put up in folds or rolls for resale. It may be white or colored. Fancy or decorated gift wrapping paper is a sheet of good quality paper, decorated or embossed, or printed in one or more colors by any one of several printing processes. It may be any basis weight from 10 lb. (24 x 36 inches – 500) up, and is put up in either folds or rolls for resale. It is used to dress up a gift and enhance its eye appeal.

GILLING—Glazing by means of a "gill" machine, named after the inventor.

GLARE—See GLOSS.

GLASS FIBER FELT—A mat made from glass fibers bound together with a synthetic resin.

GLASSINE-LINED BOARD—A paperboard to which a glassine paper has been laminated.

GLASSINE PAPER—A supercalendered, smooth, dense, transparent or semitransparent paper manufactured primarily from chemical woodpulps, which have been beaten to secure a high degree of hydration of the stock. This paper is grease resistant, and has a high resistance to the passage of air and many essential oil vapors used as food flavoring and, when waxed, lacquered, or laminated, is practically impervious to the transmission of moisture vapor. It is made in white and various colors; opaque glassines are produced by the addition of fillers. The basis weights may range from 12 to 90 pounds, the ordinary range being from 15 to 40 pounds (24 x 36 inches – 500). Glassine paper is used as a protective wrapper for all kinds of foodstuffs, tobacco products, chemicals, and metal parts, as well as for many purposes where its transparent feature is useful. For these purposes it is often converted into bags, envelopes, printed wraps, fluted cups, etc.; it is also used for lining boxes, cartons and as windows in window envelopes. It is also called glazed greaseproof paper. The German name is Pergamyn.

GLASS PAPER—(1) Paper made from glass fibers. (2) A term sometimes applied to abrasive papers.

GLASS TRANSITION POINT—For lignin, the temperature at which the thermoplastic lignin in the wood softens but remains in a glassy state.

GLAZED—Having a high gloss or polish, formed on the surface of the paper by methods such as gloss calendering, calendering, plating, or drying on a Yankee dryer. See MACHINE GLAZED.

GLAZED CASING—Papers which are highly burnished on both sides.

GLAZED COATED BOOK PAPER—Any coated book paper having a high supercalendered or glossy brush-finished or similar surface.

GLAZED COATED COVER PAPER—A kind of coated cover paper which possesses a highly polished surface.

GLAZED GREASEPROOF PAPER—See GLASSINE PAPER.

GLAZED PAPER—See GLAZED.

GLAZING—The operation of producing glazed papers or boards. See FRICTION GLAZED; GLAZED; MACHINE GLAZED.

GLOSS—The angular selectivity of reflectance of surface-reflected light responsible for the degree to which reflected highlights or images of objects may be seen as superimposed on the surface. Gloss depends on the kind of illumi-nation, the angles of its incidence and reflection, the kind of use of the paper, and the relative position of the paper and the observer.

GLOSS AGENTS—Additives that improve the optical smoothness of the surface of a coated sheet to improve its ability to reflect incident light.

GLOSS CALENDER—A calender that uses a highly polished and heated roll to replicate a coated web surface by means of nip pressure to cause thermoplastic flow of the coated surface. A gloss calender is normally used for heavy basis weight coated paper or board. See HOT CALENDER.

GLOSS CALENDER ROLL—The roll presses the paper web against the heated metal roll in a gloss calender (q.v.) to impart high levels of gloss on heavy basis weight coated paper or board. Gloss calender rolls are usually covered with a medium hard elastomeric covering. If both sides of the web are to be finished, a roll is used on each side of the cylinder.

GLOSSMETER—An instrument that measures specular gloss in a laboratory or, for example, "in-line" on a supercalender or a gloss calender.

GLUCOMANNAN—A hemicellulose with a straight chain backbone composed of glucose and mannose monomeric units. The predominant hemicellulose of softwoods is an O-acetylgalac-toglucomannan in which the glucomannan chain is occasionally substituted by acetyl or galactose groups. Typical values for the ratios of the components mannose, glucose, acetyl and galactose are 3:1:0.24:0.15, respectively. Unsubstituted glucomannans occur in hardwoods.

GLUE—Organic colloids of complex protein structure obtained from animal materials such as bones and hides in the meat packing and tanning industries. It is used for gumming, tub sizing, and as a general adhesive. It also serves as a coating adhesive for specialty products. The term is sometimes loosely used in a general sense synonymous with adhesive (q.v.).

GLUE-COATED PAPER—A coated paper in which glue is used as an adhesive for the coating material.

GLYCERIN PAPER—A paper which has been impregnated with glycerin. It may be used for wrapping products which are to be protected from the moisture of the air; it is also used as a base for oilcloth and paper drapes and for making gaskets.

GMELINA—A fast-growing hardwood species, *Gmelina arborea,* native to India but cultivated in other parts of the world for use in paper-making pulp.

GOLD ANNOUNCEMENTS—A gold-colored writing paper used for advertising purposes. It is made from chemical woodpulps in basis weights ranging from 13 to 24 pounds (17 x 22 inches – 500).

GOLDBEATERS TISSUE—An unbleached tissue paper which has a hard surface free from lint and is used as an interleaving between sheets of goldleaf.

GOLDEN BROWN—A shade of paper, usually kraft, used for wrapping paper, gummed tape, and kraft envelopes. The base stock may be unbleached or semi-bleached, and is usually colored with a small amount of dye.

GOLD-MAILING PAPER—(1) A gold-colored chemical-mechanical woodpulp sheet in basis weights of 25 and 35 pounds (20 x 26 inches – 500), which is moderately hard sized. It is used in hand addressing machines for labeling newspapers and magazines. (2) See GOLD ANNOUNCEMENTS.

GOLD PAPER—Metallic bronze-coated paper. There are many kinds, ranging from lightweights used in the paper box industry to heavy bristols. See BRONZE PAPER; METALLIC COATING.

GOLD ROTOGRAVURE PAPER—A metallic bronze pigment coated paper sized to print well in the rotogravure process.

GRAB SAMPLE—A sample of pulp or paper for testing or analysis taken at random.

GRADE—(1) A class or level of quality of a paper or pulp which is ranked, or distinguished from other papers or pulps, on the basis of its use, appearance, quality, manufacturing history, raw materials, or a combination of these factors. Some grades have been officially identified and described; others are commonly recognized but lack official definition. (2) With reference to one particular quality, one item (q.v.) differing from another only in size, weight, or grain; e.g., an offset book paper cut grain long is not the same grade as the same paper cut grain short.

GRAIN—The machine direction of paper.

GRAIN DIRECTION—See MACHINE DIRECTION.

GRAINED BOARD—Any paperboard used for certain types of fancy setup boxes upon which a design resembling wood graining has been printed or embossed.

GRAINED PAPER—An embossed or decorated paper with a surface to imitate various grains, such as wood, marble, alligator, and Spanish leather. It is usually made in cover or box-cover weights.

GRAININESS—In printing, the effect produced by a random pattern of light and dark specks or grains in halftones and solids. The grains can be due to roughening of the edges of halftone dots, random specks of ink between dots, discontinuous ink films on halftone dots and solids, specular reflections off inked fibers in the surface of the paper, etc. Graininess is caused by fine grains which cannot be easily resolved by the eye and should not be confused with mottle—large blotches due to uneven absorption or formation in the paper, or with wire patterns—which have regular rather than random distribution. See also GRAINY.

GRAINING—The result of printing various designs on paper or board to simulate various wood grains, marble, etc.

GRAINLESS PLATE—A lithographic printing plate that has not been roughened by graining. Many so-called grainless plates have a slight chemical or mechanical roughening of their surface.

GRAINY—Small variations in the surface appearance of paper or board, resulting from any of a variety of causes, such as impressions of wires or felts, irregular distribution of color, and uneven shrinkage in drying.

GRAINY EDGES—A grainy condition extending for varying distances in from the edge of the sheet, rougher than the rest of the sheet.

GRAMMAGE—The metric equivalent basis weight of paper or paperboard expressed in grams per square meter instead of pounds per ream. See also BASIS WEIGHT.

GRANITE—Lightly colored or tinted, and containing a small percentage (usually less than one percent) of heavily dyed, fairly long fibers of a different color. See MOTTLED.

GRANITE NOTE—A mottled writing paper cut to note size and used for social correspondence. It is made by adding a fraction of a percent of heavily dyed long fibers to a furnish of white pulp. Blue, red, and black are the most common colors for mottling fibers.

GRANITE ROLL—A press roll consisting of a thick walled granite roll shell machined from quarried blocks of granite and fitted to supporting internal structures. Widely used as the top press roll (q.v.) on machines running light weight sheets, such as newsprint, because of excellent peel or release of the wet sheet from the roll surface. Granite rolls are now challenged by rolls covered with elastomeric compositions or ceramic coatings.

GRANULAR CARD—A carding machine for processing man-made fibers in which the principal carding elements are composed of an emery type sandpaper.

GRAPH PAPER—See CHART PAPER.

GRAPHITE PAPER—A paper made from stock which has been treated with colloidal graphite or which has been coated by spraying, painting, or dipping with an aqueous paste of colloidal graphite. The paper is usually gray to gray black. The addition of graphite increases the opacity of the sheet, renders it less sensitive to color changes by sunlight, and gives the paper the properties of lubricity and electrical conductivity. Graphite may be added to paper of almost any basis weight.

GRATE BOILER—The solid fuels are distributed on a stationary or moving grate for combustion. Part of the combustion air is fed under the grate to promote combustion. Gaseous, liquid, or solid fuels such as pulverized coal may be fired in burners above the grate.

GRAVITY DECKER—(1) See VALVELESS DECKERS. (2) A stock thickening device that consists of a large, hollow drum covered with a fine filter media and mounted in a vat. During operation, the stock is introduced into the vat, and it forms a mat on the fine filter media, while most of the water penetrates the filter media, and exits from the ends of the drum. Gravity deckers differ from vacuum drum filters (q.v.) in that gravity deckers use only the head differential between the vat level and the white water level inside the drum to thicken the stock.

GRAVITY SCALPING SCREEN—A stock thickening or washing device that consists of a fine filter media mounted at an angle to the horizontal, and which is fed low-consistency stock. The fibers roll or slide down the filter media, and most of the white water (q.v.) passes through the filter media. Homemade versions of a gravity scalping screen are often called sidehill screens (q.v.).

GRAVURE PAPER—See ROTOGRAVURE PAPER.

GRAVURE PRINTING—See INTAGLIO PRINTING. The term gravure is also used in one form of offset lithography. See DEEP-ETCH OFFSET.

GRAY EXPRESS—See MILL WRAPPER.

GRAY ROSIN SHEATHING PAPER—Same as red rosin sheathing paper (q.v.), except for the color.

GRAY SCALE—A range of standard gray tones, from white to black, which can be used to compare the tonal gradation of the original copy.

GREASE—Transparent spots in the coating.

GREASEPROOF BOARD—Any paperboard upon which there has been pasted a paper that is greaseproof, such as glassine, or a board that has been treated to render it grease- and oil-resistant.

GREASEPROOFNESS—Ability to resist the passage of greases and oils. This property is important in papers used in packaging greasy and oily substances.

GREASEPROOF PAPER—(1) A protective wrapping paper made from chemical wood-pulps which are highly hydrated in order that the resulting paper will be resistant to oil and grease. The basis weights range from 20 to 40 pounds (24 x 36 inches – 500). This paper is used extensively for wrapping greasy food products. (2) A descriptive term for any paper which has been treated or coated to render it resistant to grease or oils.

GREASE SPOTS—(1) Dirt spots in paper caused by oil or grease. (2) Portions of paper containing less pulp than the rest of the sheet, caused by a grease spot on the wire, which inhibits proper sheet formation.

GREASY—A term applied to a pulp which has been refined to a very low freeness and has a characteristic slippery or greasy feel. Such a pulp is sometimes referred to as dead beaten.

GREAT NORTHERN GRINDER—See MAGAZINE GRINDER.

GREEN—The naturally occurring moisture content of wood. May also refer to incompletely dried or seasoned wood.

GREENHOUSE EFFECT—A hypothesis that predicts the warming of the earth's surface due to the buildup of carbon dioxide in the atmosphere. This increased atmospheric carbon dioxide reduces the amount of infrared radiation (heat) from the earth which is radiated into space.

GREEN LIQUOR—Aqueous solution of sodium salts produced in the sulfate process by dissolving recovery boiler smelt in weak wash. See also RAW GREEN LIQUOR.

GREEN LIQUOR CLARIFIER—A clarifier used to remove dregs from raw green liquor (q.v.).

GREETING CARD BRISTOL—A bristol selected for color, finish, or other special characteristics and used for the manufacture of greeting cards. It may be an index bristol, a mill bristol, or a wedding bristol. Important properties are color, finish, rigidity, and sizing. See BRISTOLS.

GREETING CARD PARCHMENT—(1) A translucent greaseproof or genuine vegetable parchment paper resembling animal parchment, and used for printing certain types of greeting cards. See VEGETABLE PARCHMENT. (2) An imitation parchment paper made from long chemical wood fibers with a wild formation, and often with certain chemical additives. It is designed to be an imitation of genuine vegetable parchment, and it is used largely for greeting card manufacture.

GREETING CARD STOCK—(1) A heavyweight paper or card stock, made either in a solid sheet or lighter weights pasted together. A wide range of colors, from pastel shades to very brilliant ones, in a great variety of fancy finishes. It is used as the name implies. See BRISTOLS. (2) Any paper used in the manufacture of greeting cards. See also PAPETERIE PAPERS.

GRINDER—In mechanical pulping, a machine for producing mechanical pulp, groundwood. It consists of a rotating pulpstone against which debarked logs are pressed and reduced to pulp.

Various models are; the packet grinder, the chain grinder, the magazine grinder and the Roberts grinder.

GRINDING—In mechanical pulping, a method of preparing mechanical pulp from debarked logs, where the temperature of the process is controlled at 65°C to 70°C by showers on the stone under atmospheric conditions. See also CHAIN GRINDING.

GRINDING PAPER—See ABRASIVE PAPERS.

GRINDSTONE—See PULPSTONE.

GRIPPER EDGE—The leading edge of a sheet held by the grippers as the sheet is transported through the press.

GRIT—(1) Hard particles in any component of the sheet furnish or coating color. In specifications for filler or coating pigments, grit is usually expressed as the percent remaining on a 325-mesh screen. (2) The abrasive particles in a pulpstone which are responsible for the grinding action of the stone. The size, shape and the hardness of these particles affect the quality of the groundwood produced. (3) Coarse particles separated in the slaker classifier from the slurry of lime and green liquor. Generally coarser than lime mud (q.v.) particles and very unreactive with the chemical constituents of green liquor (q.v.).

GROOVED ROLL—A roll in which the cover is cut with grooves to provide receptacles for the water expelled in the press nip. The helically cut grooves are typically 3.2 mm (0.125 in.) in depth, 0.5 mm (0.02 in) wide on 3.2 mm (0.125 in.) centers (i.e., 8 grooves per inch). Grooved rolls are used in transverse press designs because they vent the felt and reduce hydraulic resistance to flow in the press nip for enhanced water removal versus a solid roll. They consist of an undrilled metal body and may be operated under high pressure. The grooved roll in the dryer section reduces the volume of air forced in and can be useful in controlling sheet flutter. Covers must be relatively hard (less than 10 P&J) to prevent the grooves from closing

under high pressure. Some mills use stainless steel grooved rolls to avoid this problem. See GROOVED ROLL PRESS.

GROOVED ROLL PRESS—A press in which one of the rolls supporting the press fabric is a grooved roll (q.v.).

GROOVING–CHEVRON—See CHEVRON–GROOVING.

GROUND LIMESTONE—See CALCIUM CARBONATE.

GROUND WATER—The water contained in the pores of the ground below the surface of the earth. After a certain depth saturation, a continuous interconnected system of water, is achieved.

GROUNDWOOD BOOK PAPERS—See GROUNDWOOD PRINTING PAPERS.

GROUNDWOOD FREE—Containing no mechanical woodpulp. In practice, a paper found to contain less than 5% of mechanical pulp, by microscopic staining techniques, is considered groundwood free. See FREE SHEET.

GROUNDWOOD (GWD)—A mechanical pulp produced by pressing debarked logs sideways against a rotating pulpstone in the presence of water and reducing the wood to a mass of relatively short fibers and fines.

GROUNDWOOD PAPERS—Papers other than newsprint, made with substantial proportions of mechanical pulp, and used for printing or converting. See GROUNDWOOD PRINTING PAPERS.

GROUNDWOOD PRINTING PAPERS—Low cost printing papers made primarily from mechanical pulps; they are also referred to as groundwood specialties. Such papers are characterized by relatively high bulk-to-weight ratios, high opacity, and high-speed printability. They are made in a wide range of basis weights from 18 to 100 pounds (25 x 38 inches – 500 and 24 x 36 inches – 500).

GROUNDWOOD PULP—See GROUND-WOOD.

GROWTH RING—See ANNUAL RING.

GUAR GUM—A polysaccharide, mainly manno-galactan, derived from the seed endosperm of the guar plant grown in India and the United States and used as a beater or wet end additive primarily for improving strength properties. It may also be used as a surface sizing agent. See MANNOGALACTAN.

GUIDE EDGE—Side of paper corresponding to side guide on sheet-fed press feed board.

GUIDE PALM—A stationary levered arm resting against the moving machine clothing. When the clothing moves away from it, a signal is sent to the guide roll to move in a direction causing the machine clothing to move back into it, until the desired position is achieved.

GUIDE ROLL—A forming fabric, press felt or dryer felt carrying roll whose purpose is to keep machine clothing centered on the machine. The guide roll is usually pivoted on one side of the machine (e.g., backside) and is free to travel in the machine direction on the other side (front side). The position of the guide roll is determined by the position of the machine clothing and the guide palm (q.v.).

GUIDE SHEET—Paper used to check the guides between two presses.

GUILLOTINE—See GUILLOTINE TRIMMER.

GUILLOTINE TRIMMER—A machine that trims one side of a pile of paper by means of a powered knife bar. The pile of paper is transported into position on a film of air and positioned accurately by a backstop before the operator actuates the knife which moves vertically to trim one edge at a time or to split into smaller piles or reams of paper. This finishing operation has been superseded by precision cutters such as cut-size cutters (q.v.) used by many mills.

GUM—(1) A substance of high molecular weight, usually with colloidal properties, which produces a gel, or viscous suspension or solution, at low solids content in an appropriate solvent or swelling agent. (2) More commonly, a plant polysaccharide or derivative thereof, which is dispersible in water to produce a viscous mixture or solution. See GUAR GUM; KARAYA GUM; LOCUST BEAN GUM.

GUMMED CLOTH TAPES—Tapes, which may be clay filled, fiber filled, or cloth combined with paper, used by manufacturers of corrugated and solid fiber shipping containers for the corner stay (called manufacturer's joint). Generally one corner is taped.

GUMMED CORRUGATOR TAPE—See GUMMED REINFORCED PAPER TAPE.

GUMMED FLAT PAPERS—Strong, hard-sized MF, English, supercalendered or coated papers which have been gummed for use as gummed labels, embossed seals, drug labels, and other applications. They may be white, colored, or metallic. Standard sizes when sheeted are 17 x 22 and 20 x 25, with 500 sheets to the ream. Different gummings are required depending upon the surface to which the paper is to adhere.

GUMMED PAPER—Any paper coated on one side with adhesive gum, the adhesive being a dextrin, fish or animal glue, or resin, or a blend of any of these.

GUMMED REINFORCED PAPER TAPE—A tape made of two plies of kraft, reinforced with fibers or threads such as sisal, rayon, nylon, glass, combined with asphalt or other laminants and coated on one side with water-activated adhesive. It is commonly used to seal cartons.

GUMMED SEALING TAPE—A kraft paper, usually in basis weights of 35, 60, or 90 pounds (24 x 36 inches – 500), which is coated on one side with a water-activated adhesive and slit to rolls in various widths and prescribed yardage. It is used largely for sealing packages, bundles, and cartons. A certain amount of lightweight sealing tape is made from sulfite or kraft papers in white and various colors; this is used by retail stores where appearance is important.

GUMMED SISAL TAPE—See GUMMED REINFORCED PAPER TAPE; REINFORCED PAPER (1).

GUMMED STAY—Gummed sealing tape used in the construction of setup paper boxes for staying the corners. It is heavier than the normal sealing tape, the basis weights of the kraft paper used being 90, 100, and 120 pounds (24 x 36 inches – 500). This tape may be precreased in the center to facilitate application. See BOX STAY TAPE.

GUMMED TAPES—Tapes, which may be clay filled, fiber filled, or cloth combined with paper, used by manufacturers of corrugated and solid fiber shipping containers for the corner stay (called manufacturer's joint). Generally one corner is taped. See TAPE.

GUMMED VENEER TAPE—A gummed paper tape used in the manufacture of veneered wood to hold the edges together during the process of gluing to the core. It is sandpapered off the wood in the finishing process. The usual basis weights are 35, 40, 50, and 60 pounds (24 x 36 inches – 500).

GUMMED WATER-RESISTANT TAPE—A gummed tape usually made from water-resistant, asphalt-laminated paper and a special adhesive which is activated by a solvent. This tape should have a water resistance of twenty-four hours minimum. It is adapted for sealing frozen food packages and also for export packaging.

GUMMING—The operation of applying a gum or adhesive to a sheet of paper.

GUMMING PAPER—(1) For labels, a strong, hard-sized paper made of chemical wood pulp (usually bleached) in rolls for gumming purposes. It may be white or colored and should be free from lint. The usual basis weight is 45 pounds (24 x 36 inches – 500). (2) For sealing tape, see GUMMED SEALING TAPE.

GUM ROSIN—See ROSIN.

GUM SPIRITS—See TURPENTINE.

GUM TURPENTINE—See TURPENTINE.

GUN WADDING—A felt or chipboard used in loading shotgun shells.

GUSSET—The reverse folds in the sides or bottoms of some styles of bags or shipping sacks.

GUSSET STOCK—A heavy natural kraft or red wallet stock laminated to russet or khaki cambric. It is used in the manufacture of expanding envelopes and cases.

GUTTER—The unprinted space at the edge or between print images.

GWD—Groundwood (q.v.).

GYMNOSPERMS—Plants whose seeds are not enclosed in an ovary. The common trees of this type are conebearing. See ANGIOSPERMS.

GYPSUM—The hydrous form of calcium sulfate occurring in nature having the chemical formula of $CaSO_4 \cdot 2H_2O$. Sometimes this material is also referred to as pearl filler, puritan filler, or terra alba. The pigment is used as a filler in paper, especially in building boards. See CALCIUM SULFATE.

GYPSUM BOARD—A rigid noncombustible building board composed essentially of gypsum and reinforced on the surfaces by a covering of paper or other fibrous material firmly bonded to the gypsum core. The board is manufactured in 1/4, 5/16, 3/8, 1/2 and 5/8-inch thicknesses, and in weights from 900 to 2500 pounds per 1000 square feet. It is designed to be used for walls, ceilings, and partitions and affords a surface suitable to receive decoration without plaster. Gypsum board is also manufactured with a wood-grained appearance and with aluminum foil backing. Also commonly known as plasterboard, gypsum wallboard, one of a class of wallboards.

151

GYPSUM LATH—A special grade of gypsum board used to replace wood laths. The outer layer of paperboard is unsized to be receptive to the plaster. Available in sheets 16 x 48, 3/8 or 1/2 inches in thickness.

GYPSUM SHEATHING—A special form of gypsum board (q.v.) used under the siding on exteriors. The paperboard is asphalt- or oil-treated to make it water resistant. The core is also treated to make it water resistant.

H

HAIR CUT—A cut similar to a fiber cut (q.v.) caused by the pressure of a felt hair or human hair in the web.

HALF-FINE METALLIC PAPERS—Papers produced by laying patches of thin copper or aluminum alloys, about five inches square, on an adhesive-coated sheet. The patches are applied to overlap both each other and the edges of the base sheet to form a continuous metallic surface. The extending patch edges are brushed free, forming clean straight edges.

HALF PLATE PAPER—A machine-made paper of fine and soft texture used for woodcuts.

HALFSTUFF—(1) Rags after cooking, washing, defibering, and bleaching into pulp ready to be charged into the beater. After beating it is called whole stuff or simply stuff. (2) Any partially refined fiber.

HALFTONE—(1) A printing plate (usually copper, zinc or photopolymeric) produced by the photoengraving process. It is a reproduction of a photograph, drawing, painting, print, or other object having a gradation of tones, the surface of the plate consisting of dots of various sizes uniformly placed and being capable of rendering not only the highlights and shadows of a picture but all the gradations between these. (2) An impression or print made from a halftone plate by the letterpress, intaglio, or planographic process.

HALFTONE BLOTTING PAPER—A blotting paper which has been subjected to a smoothing treatment on one or both surfaces in order that it will print readily without disturbing the surface fibers. It retains its blotting characteristics while at the same time it acquires some of the properties of a printing paper. It is primarily a blotting paper with a smoothed surface that is suitable for printing from coarse deep-etched halftones. The basis weights are 100, 120, and 140 pounds (19 x 24 inches – 500).

HALFTONE PAPER—A printing paper characterized by a high finish, and suitable for halftone printing.

HALFTONE SCREEN—A sheet of glass with finely ruled cross lines or, now more commonly, a film sheet with a square pattern of fine vignetted dots, by means of which a variable-tone image can be photographically converted to one made up of fine dots uniformly spaced center-to-center but varying in size. Special types of halftone screens produce images made up of fine lines of varying width instead of dots.

HAM JACKETS—A plain or crinkled grease-proof paper or parchment, in basis weights of 30 to 40 pounds (24 x 36 inches – 500), used to protect the inside of a boiler from pitting during the process of cooking ham.

HAM WRAPPER—(1) A protective paper for wrapping hams, which may be a vegetable parchment, a greaseproof paper, or a "sulfite" sheet, usually in basis weights of 40 pounds (24 x 36 inches – 500) or heavier. The paper may be grease resistant and suitable for printing. (2) An absorbent sheet, the chief property of which is its ability to absorb grease. See COOKED HAM WRAPPER.

HANDBILL PAPER—Any paper for printing advertisements to be handed out to the public.

HANDLE—The impressions of touch and sound received when a sheet of paper is handled. Handle includes such properties as feel and rattle.

HANDLING AND TRANSIT WASTE—See PAPER WASTE.

HANDMADE FELT—A press felt (q.v.) that gives the paper surface the appearance of handmade paper.

HANDMADE FINISH—A plate or embossed finish giving the effect of a handmade paper.

HANDMADE PAPERS—(1) Small sheets of paper made by hand for felting or specialty purposes. Low-consistency pulp stock is placed into the mold, or the mold is dipped into a vat of stock. On removing water through the wire screen at the bottom of the mold, the sheet is formed. It is pressed and dried manually. (2) Paper made by hand molds in single sheets, having rough or deckle edges on four sides. The mold, of the required size, is dipped into a vat containing the stock and is lifted with a particular motion, forming the sheet. It is sometimes called deckle edged paper.

HAND MOLD—See MOLD (4) and HANDMADE PAPER.

HANDSHEET—A sheet made from a suspension of fibers in water, with or without the addition of sizing, loading, or coloring agents, in an operation whereby each sheet is formed separately by draining the pulp suspension on a stationary sheet mold. It is generally used for testing the physical properties of the pulp or its combinations with other materials.

HANGING PAPER—The raw stock used in the manufacture of wallpaper. The converter usually clay coats the sheet and then prints it or, in the heavier weights, embosses it. It is usually manufactured with a substantial portion of mechanical woodpulp, the balance being unbleached or bleached chemical woodpulp. However, some grades contain no mechanical woodpulp. The sheet is hard sized, has a "toothy" surface to enable the coating color to adhere to the sheet, is uniform in surface so that the design will print uniformly, and is especially adapted to hold deep embossing. The commonly used weights are 38, 42, 50, 58, 66, 74, 82, and 98 pounds (24 x 36 inches – 480), which

are respectively referred to in the trade as 9, 10, 12, 14, 16, 18, 20, and 24 ounces. The ounce nomenclature presumably arose out of the weight in ounces of a roll of wallpaper of a given length and width, but it no longer represents the weight of any standard size roll of wallpaper and is a purely arbitrary designation. See also TILE STOCK.

HANGING RAW STOCK—See HANGING PAPER.

HARD BEATING—A term applied to pulp which must be given a long treatment in a beater or refiner to develop the required papermaking properties. See also HARD STOCK.

HARD BLEACHING PULP—Imprecise term denoting a pulp that is difficult to bleach because of dark color and/or high lignin content, resulting in higher bleaching chemical usage in the bleach plant.

HARDBOARD—A panel manufactured primarily from interfelted lignocellulosic fibers which are consolidated under heat and pressure in a hot-press to a density of 31 pounds per cubic foot or greater. Other materials may be added to improve certain properties, such as stiffness, hardness, finishing properties, resistance to abrasion, and moisture, as well as to increase strength, durability, and utility.

HARD COOK—A batch cook or period of time in a continuous digester when the target pulping conditions are not maintained and the result is undercooking. Low temperature, too short cooking time, and insufficient chemical will result in a hard cook. The pulp produced is called a hard pulp. In some mills this is also called a raw cook. The kappa number of a hard cook is higher than the target.

HARD COPY—A readable printout or readout on paper, as provided by a teleprinter, computer printer, etc.

HARD FIBER—A general term used to include stiff boards of a dense nature. See ELECTRICAL INSULATION FIBER; VULCANIZED FIBER.

HARD FOLD—A method of preparing large sheets of paper for shipment, which consists of folding a comparatively small number of sheets by hand and then compressing the fold by means of a round stick. Upon opening the paper, a definite crease remains. See also SOFT FOLD.

HARD HARDWOODS—The hard and dense species of hardwoods, as differentiated from the softer species. It includes such pulpwood species as yellow birch, beech, the hard maples (sugar and black), and the oaks. See SOFT HARDWOODS.

HARDNESS—(1) That property of a sheet which resists indentation by objects of specified size, shape, and hardness. Other definitions of hardness might be deduced from subjective judgment of hardness in handling and in use, e.g., a paper which is stiff, one which produces a strong rattle, or one which feels hard when crumpled in the hand, may very easily be thought of as being hard. (2) When applied to pulp, a term usually referring to the degree of cooking, a hard pulp resulting from milder than normal digesting conditions. See HARD COOK. (3) For roll cover hardness, see PUSEY & JONES PLASTOMETER; ROLL COVER HARDNESS; SHORE DUROMETER. (4) For roll hardness, see SCHMIDT HARDNESS TESTER.

HARDNESS TESTER—See PUSEY AND JONES PLASTOMETER.

HARDNESS TESTER (PAPER)—An instrument used in finishing operations to check the cross deckle hardness profiles of wound rolls (shipping rolls) of paper. The Rhometer or Schmidt testers are hand held and generally used for quality control. The "backtenders friend" is an on-line dynamic instrument used for the measurement and control of hardness on the reel of a paper machine. See RHOMETER; ROLL HARDNESS.

HARD PAPER—(1) Usually a kraft paper, although other stocks may be used, which is impregnated with synthetic thermosetting resins. It is used in the form of plates, pipes, and molded pieces for electrical insulating purposes. (2) A paper with a hard, smooth surface, mostly writing paper, which, because of its sizing, is harder to print than ordinary book paper.

HARD PULP—Pulp resulting from a hard cook, i.e., one which is mildly cooked or undercooked. See HARD COOK.

HARD SIZED—Sized to give a high degree of water resistance.

HARD-SIZED BOOK PAPER—A book paper which has been hard sized for special purposes. The term applies only to the sizing characteristics of the paper.

HARD-SIZED NEWS—An inaccurate term (now practically obsolete) used to describe a class of papers made with approximately the same proportions of mechanical and chemical woodpulp as standard newsprint, but differing from it in many respects, including hard-sizing.

HARD STOCK—A term applied to pulps produced from rags, rope, or jute. See also HARD BEATING.

HARD WHITE SHAVINGS—See SHAVINGS.

HARDWOOD—Wood obtained from a class of trees known as angiosperms, such as birch, maple, oak, gum, eucalyptus, and poplar. These trees are characterized by broad leaves and are usually deciduous in the temperate zones.

HARDWOOD PULP—Any pulp made from a hardwood or mixture of hardwoods.

HARD WRINKLES—See WRINKLES.

HARD WRITING FINISH—A term applied to carbon paper (q.v.).

HARPER MACHINE—A type of fourdrinier machine in which the machine forming fabric travels away from the presses, the headbox being placed between the breast roll and the first press, and the wet sheet being couched from the fabric by a pickup felt and carried back above the fabric to the first press. It is generally used in the manufacture of lightweight papers.

HAZARDOUS AIR POLLUTANT—According to law, a pollutant to which no ambient air quality standard is applicable and that may cause or contribute to an increase in mortality or in serious illness. For example, asbestos, beryllium, and mercury have been declared hazardous air pollutants.

HAZARDOUS MATERIALS—Materials that present a hazard to human or animal life; includes toxic materials, flammable materials and explosives. Materials are often classified as hazardous or acutely hazardous; the differentiation is made on the basis of how quickly and at what dose they cause death or other serious harm.

H BLEACHING STAGE—See HYPOCHLORITE BLEACHING STAGE (H).

HD STORAGE—High-density storage (q.v.).

HEADBOX—(1) On fourdrinier machines: (a) air-padded type. A large flow control chamber which receives dilute paper stock or furnish from the stock preparation system, and by means of baffles and other flow evening devices, spreads the flow evenly to the full width of the paper machine and provides delivery of stock to the fourdrinier fabric uniformly across its full width. The height of the liquid in an open headbox or the air pressure in a closed headbox provide the requisite speed of flow of the stock onto the fourdrinier fabric. On high-speed machines the headbox is always enclosed. (b) On many fourdrinier machines, especially those with a top-wire former, and on all twin-wire machines, the headbox is a nozzle with contiguous small diameter tubes or comparable differently shaped turbulence inducing elements in one or more banks across the width of the box in its front end followed by a converging open section the width of the machine in which the flows from the turbulence inducing elements combine into a single flow the width of the

machine and the height of the slice (q.v.) opening. (2) On cylinder machines: a flow regulating device which controls the volume of stock flowing to the screens and mixing boxes before the vats.

HEADBOX SLICE—See SLICE.

HEARTH HEAT RELEASE RATE—Ratio of the total heating value input to a recovery boiler (q.v.) furnace to the plan area of the hearth or furnace floor. Often used as design guideline and a measure of the boiler loading.

HEARTWOOD—The central portion of a tree root or branch in which all cells have died although the tree is still living. It is usually darker colored than the surrounding sapwood due to extractives deposited there, and has a lower moisture content.

HEAT-AFFECTED ZONE—In welding and corrosion, the area adjacent to a weld, where the welding thermal cycle has caused microstructural changes in the metal (which may adversely affect mechanical and/or corrosion properties).

HEAT EXCHANGER—A tube-filled apparatus in which fluid inside the tubes is heated or cooled by fluid outside the tubes.

HEATING VALUE—A measure of the chemical energy of a fuel that can be released as thermal energy during combustion.

HEAT RECOVERY STEAM GENERATOR—Also referred to as a waste heat boiler, a heat recovery steam generator is designed to raise steam from the hot exhaust gases of a gas turbine. It may be equipped with supplementary firing with fuel to increase the combustion gas temperature, thereby increasing the steam temperature and quantity.

HEAT-SEALING PAPER—Any paper, the surface of which has been coated so that it becomes adhesive when heated. (1) Heat-sealing or self-sealing paraffin-waxed paper is bread wrapping or other grade shaving enough surface wax to permit sealing so that the wrapper is effectively held together and sealed against moisture loss or regain. (2) Heat-sealing grades of varnished or lacquered papers contain resins or other thermo-adhesive material.

HEAT SET INKS—Inks which require heat to dry.

HEAT TRANSFER PRINTING—A textile decoration method based on the transfer, using heat, of dyes from paper previously printed by any one of the principal processes, most commonly gravure.

HEAVY METAL IONS—Metal ions such as iron, copper, aluminum, chromium, and manganese that have an adverse effect on brightness development and brightness stability. The adverse effects of these heavy metal ions during brightening can be minimized through the use of a chelating agent or acid treatment. There are also heavy metal ions which are beneficial for bleaching. Magnesium is one of these. See ACID TREATMENT; CHELATION STAGE (Q).

HEAVY SPAR—See BARYTES.

HEAVY WEIGHTS—(1) Papers made in weights above the middle range basis weight which are usual for the grade. (2) In cover paper, the term may refer to the double thick or pasted weights.

HECTOGRAPH—A duplicating process utilizing a gelatin pad that receives copy of a specially prepared image using a special type of aniline ink and then transfers it to a suitable paper. The term is sometimes used to refer to the spirit process.

HECTOGRAPH PAPER—A smooth-surfaced (or coated with glue-glycerol composition) noncurling paper used for making transfer copies with hectographic inks from a gelatin surface. The paper should be relatively nonabsorbent and free from loose surface fibers. See also DUPLICATING PAPER.

HELD COOK—See SOFT COOK.

HELIOGRAPH PAPER—A photographic printing paper developed by exposure to sunlight.

HELIOGRAVURE PAPER—See HELIOGRAPH PAPER.

HEMICELLULASES—Enzymes that hydrolyze hemicelluloses to the component sugars. Hemicellulases are multiple enzymes each of which hyrolyzes specific linkages; thus complete hydrolysis of a given hemicellulose to component sugars often requires several enzymes, even though single enzymes—xylanases and mannases—can extensively depolymerize the backbone polymers.

HEMICELLULOSE—A broad class of polysaccharides which are associated with cellulose in plant cell walls. Hemicelluloses are of lower molecular weight than cellulose, typically 20 to 40 kg/mol. Hemicelluloses are generally branched or substituted with other sugars or uronic acids and usually contain more than one type of monomeric sugar. Hemicelluloses of wood are primarily composed of five "wood sugars": glucose, mannose, galactose, xylose, and arabinose.

HEMP—(*Cannabis sativa*). A plant grown in nearly all the temperate countries of the world. It furnishes a bast fiber, obtained by a retting process, which is used for rope and textiles. Some of the fiber enters the paper industry as waste material. The term hemp has also come to be used in a generic sense as fiber and is then preceded by an adjective, for example, manila hemp (see ABACA), sisal hemp (see SISAL).

HEMP PAPER—See ROPE PAPER.

HERBARIUM PAPER—A lightweight bristol or cardboard, usually cut 11.5 by 16 inches, on which pressed plants are mounted.

H FACTOR—A measure of the amount of chemical and heat energy applied during chemical pulping. The higher the H factor, the lower the yield and resulting kappa number. Cooks made with the same chips at the same H factor should have close to identical pulp yield and kappa numbers.

HICKIES—Plain or doughnut-shaped spots in printed matter, especially in solids. In lithography, plain white spots are caused by paper dust or pick-outs that adhere to the offset blanket, become saturated with moisture, and refuse to transfer ink. Doughnut-shaped white spots are caused by particles of ink skin or other ink-receptive material usually adhering to the printing plate that depress the offset blanket and cause it to bear off the plate in the surrounding area.

HIGH BULK BOOK PAPER—A book paper which, under 35 pounds pressure, bulks from 440 to 344 pages to one inch for a basis weight of 45 pounds (25 x 38 inches – 500). Other weights are in proportion.

HIGH CONSISTENCY BLEACHING—See HIGH DENSITY (CONSISTENCY) BLEACHING.

HIGH DENSITY CLEANERS—See COARSE CLEANERS.

HIGH DENSITY (CONSISTENCY) BLEACHING—Bleaching done at consistencies greater than 18%. Above 25%, the fibers are usually treated with a bleaching agent in the gaseous state.

HIGH DENSITY STORAGE—(1) The storage tower used for the storage of pulp within the 8 to 15% consistency range. (2) The storage tower used for brownstock storage before and bleached stock after the bleach plant for intermittent pulp storage.

HIGH DENSITY PUMP—A pump specially designed for pumping high-consistency pulp stock.

HIGH IMPULSE PRESS—Any press having exceptionally wide nips and high nip loads. Shoe presses (q.v.) and jumbo roll presses are two basic configurations.

HIGH INTENSITY MIXER—Active mixer producing a fluidized fiber suspension, specifically designed for mixing liquids and gases with medium consistency stock.

HIGH PERFORMANCE CONTAINER-BOARD—Linerboard and corrugating medium produced at a higher edge compression strength to basis weight ratio than standard grades.

HIGH PERFORMANCE LINERBOARD—See HIGH TEST OR HIGH PERFORMANCE LINERBOARD.

HIGH SHEAR MIXER—Mixer using a high shear mixing action for the mixing of gaseous and liquid bleaching chemicals (e.g., chlorine dioxide solution, gaseous chlorine, gaseous oxygen, hydrogen peroxide). This mixer provides greater mixing efficiency than the older peg mixers and the in-line static mixers. See HIGH INTENSITY MIXER.

HIGH TEST LINER—See HIGH TEST OR HIGH PERFORMANCE LINERBOARD.

HIGH TEST OR HIGH PERFORMANCE LINERBOARD—Linerboard made with edgewise compression strength equivalent to that of conventional linerboard of higher basis weight. See LINERBOARD.

HIGH VELOCITY CONVECTION DRUM DRYER—A machine used for drying coated paper consisting of a large drum enclosed by an evaporating hood.

HIGH YIELD PULP—A papermaking pulp where none or only a small percentage of the lignin is removed during the pulping process. See MECHANICAL PULP.

HIGH-YIELD PULP BLEACHING—A brightening treatment with hydrosulfite or hydrogen peroxide in which the lignin and carbohydrates are not degraded, thus preserving the yield advantage of the pulping process. It is usually referred to as brightening to distinguish it from lignin-removing processes. See BRIGHTENING (MECHANICAL PULPS).

HINGED LEDGER—(1) A ledger paper with a flexible section made into the sheet by removing some of the fiber on the paper machine, ordinarily by suction. The hinge is about 11/4 inches wide and is located 5/8 of an inch from the binding edge of the sheet. The paper is thinner at the hinge, permitting loose-leaf sheets to lie flat when used in a binder. (2) A ledger sheet which has a linen hinge, made by pasting a strip of linen to one side of the sheet.

HISTOGRAM—A graphic summary of variation in a data set.

HOLDOUT—The extent to which a paper or board surface resists penetration by aqueous or nonaqueous fluids. Where the fluid involved is water or water vapor, this property is usually termed sizing (q.v.). Nonaqueous fluids of concern include printing inks, lacquers, and various oils or waxes. Too little holdout reduces gloss, but too much holdout contributes to set off. See OFFSET PRINTING.

HOLE—In papermaking, an opening in a paper sheet, caused by slime, stock lump, coating splash, or other causes.

HOLEY ROLL—See DISTRIBUTOR ROLL.

HOLLANDER—The original name given to the beater (q.v.).

HOLLOW ROLL—See BARRELING.

HOOD-CAP PAPER OR BOARD—A paper or paperboard used for making a coverall cap for milk bottles. It may be made from chemical or mechanical woodpulp on a fourdrinier or cylinder machine. It is about 9 points in thickness.

HORIZONTAL POROSITY—See LATERAL POROSITY.

HOSE-WRAP PAPER—A flat, crinkled or creped single-ply or duplex kraft paper used for wrapping individual coils of garden hose or for bundling individual coils which may have been wrapped previously. It may also be waterproofed.

HOSIERY INSERT PAPER—A paper used as an insert in packaging women's hose. It is normally made with a high short fiber pulp con-

tent on a fourdrinier machine in basis weights of from 40 to 50 pounds (25 x 38 inches – 500). The chief requirements are stiffness and bulk to keep the hose in place and a surface sufficiently smooth to avoid causing snags in the hose.

HOSIERY PAPER—Wrapping or tissue paper used in connection with the packaging of hosiery. See also HOSIERY INSERT PAPER.

HOT CALENDER—Machine calenders and supercalenders fitted with heated rolls to improve the efficiency of the calendering process. A gloss calender is, in effect, a hot calender. The heated roll can be steam heated, electrically heated, or heated by gas-fired infrared systems. See GLOSS CALENDER.

HOT CHLORINATION—A modified chlorination stage incorporating higher temperature, usually used in conjunction with higher levels of chlorine dioxide substitution for chlorine.

HOT EMBOSSING—The operation of embossing with heated rolls or plates.

HOT GRINDING—See STONE GROUNDWOOD (CONVENTIONAL).

HOT-MELT ADHESIVE—A material which melts readily when heated and then solidifies rapidly upon cooling to form a firm bond.

HOT-MELT COATING—A method of applying molten wax or plastic materials to a base stock without solvent or other carrier, using a roll, knife, casting, or extrusion method. This process usually gives high gloss and is frequently employed for the application of barrier materials to paper and board.

HOT PRESS—Any press in which sheet (water) temperature is increased to reduce water viscosity and increase water removal. This is accomplished indirectly by applying steam showers to the sheet or directly by heating a roll in contact with the sheet.

HOT PRESSED—Originally applied to a process of applying pressure and heat to paper; now applied to papers which have been finished by plate glazing.

HOT ROLLING—Obsolete term. See GLOSS CALENDER; HOT CALENDER.

HOT SMASHING—A term applied to a printing process whose purpose is to crush flat an area in a rough sheet of paper so that a halftone or other impression can be printed on the smoothed surface. A brass die is attached to an electrically heated base and, with a heavy impression, the paper can be given a very smooth surface. Also called coining.

HOT STAMPING—A foil embossing print technique which uses a heated, raised-image die to create the image in foil on the print surface. The piece of stock to be hot stamped is positioned under the die. An unused piece of adhesive-coated foil (thin leaf of gold, other metal or pigment) is placed under the die. The die comes down on the foil, and the image area is transferred onto the surface of the stock. Used for lettering or decorating book bindings, leather articles, etc.

HOT WAXED BOARD—See WAXED BOARD.

HOUSE SHEATING PAPER—A sheet consisting of about 21-pound dry felt (21 pounds per 480 square feet), saturated with an asphalt to the extent of approximately 135% of its own weight and then coated on both sides with asphalt. It is then dusted with talc. The finished product weights about 50 to 60 pounds per 200 square feet. Some heavy papers are waxed. It is used as a highly waterproof sheathing paper.

HUE—That attribute of colors which permits them to be classed as reddish, greenish, bluish, yellowish, purplish, etc. See COLOR.

HULL FIBER—(1) The pulp obtained upon digesting the hulls of cotton seeds after they have been crushed. See also COTTON LINTERS. (2) A fiber such as that from coconut.

HUMIDIFIER—(1) A device to add moisture to paper in webs or sheets. (2) A device to add moisture to a room.

HUMIDITY—The water vapor present in air. See ABSOLUTE HUMIDITY; RELATIVE HUMIDITY.

HUNTER L,A,B, SCALES—Opponent color scale in which L = lightness, +a = redness, -a = greenness, +b = yellowness, -b = blueness (similar to, but numerically different from Adams coordinates and CIELAB L*, a*, b*).

HYBRID FORMER—See TWIN WIRE FORMERS.

HYDRATION—(1) In the physical sense, the condition of materials containing water of adsorption. (2) In papermaking, the treatments, essentially mechanical refining, which increases the amount of water held by the fibers. Increased hydration results in slower drainage rate and rather profoundly influences sheet properties, especially increased physical strength and decreased opacity. (3) The pulp characteristics resulting from the above treatment.

HYDRAULIC CYCLONE—See CENTRIFUGAL CLEANER.

HYDROCELLULOSES—A general term applied to the water-insoluble products of the hydrolysis of cellulose. The average degree of polymerization (DP) and the DP distribution depend on the nature of the hydrolysis and on the original cellulose. Hydrocelluloses generally have a higher degree of crystallinity than the cellulose from which they were derived, because the less crystalline regions of cellulose are more easily hydrolyzed.

HYDROCHLORIC ACID—Hydrogen chloride or hydrochloric acid, HCl is often found in flue gases in low concentrations when chlorine is present in the fuel in some form.

HYDROCYCLONE—See CENTRIFUGAL CLEANER.

HYDROENTANGLING—A method of bonding nonwovens which mechanically entangles the fibers using a multitude of fine water jets.

HYDROFOILS—Stationary drainage devices that exert a suction and doctoring action. A hydrofoil is a stationary doctor blade shaped to form an angle with the moving fabric at the trailing edge of the blade. When liquid is present, a vacuum is generated in this "divergent nip." This vacuum dewaters the paper slurry. Critical parameters include doctoring efficiency, the machine direction length of the blade, machine direction spacing between blades, and the divergent angle between the fabric and the blade. Typically, several hydrofoil blades are mounted on a single support structure. When the support structure is enclosed, vacuum can be applied to augment the vacuum in the divergent nip. This vacuum also acts on the slurry between the blades. When the hydrofoil blades are flat (zero divergent angle), the device is called a high or a low vacuum box. A box large enough to have a low vacuum section and a high vacuum section is called a dual compartment box. The major difference between flat-bladed vacuum boxes and suction boxes (q.v.) is the construction of the surface in contact with the fabric. Flat-bladed vacuum boxes (low or high) are generally constructed of removable blades like hydrofoils. Suction boxes generally have a one piece top with slots machined into the top. Alternatively, the suction boxes can have a drilled pattern. Open area is a critical parameter of flat-bladed vacuum boxes and suction boxes.

HYDROGEN-ION CONCENTRATION—The concentration of hydrogen ions (H^+) in an aqueous solution. It is a measure of the active acidity or alkalinity, and is expressed as the number of moles (1.0078 gram) of (H^+) per liter of solution. It may also be expressed in terms of pH (q.v.).

HYDROGEN PEROXIDE—H_2O_2. A chemical with oxidizing properties used to bleach pulp in an alkaline solution, frequently as one of a group of pulp bleaching agents. The active bleaching agent is the perhydroxyl ion OOH. Hydrogen peroxide and peroxides are particularly sensitive to catalytic decomposition by metallic ions. For bleaching, the metal ions are normally removed by pretreatment of the pulp with sequestering agents (q.v.). Magnesium sulfate and sodium silicate are added to control peroxide decomposition. Magnesium sulfate is especially advantageous in preventing strength loss during bleaching of chemical pulps with peroxide.

HYDROGEN PEROXIDE BLEACHING STAGE (P)—Peroxide (H_2O_2) brightening stages are used in mechanical pulp mills and in kraft pulp mills. Peroxide has been used as a replacement for chlorine dioxide brightening stages in applications where very low adsorbable organically bound halogen (AOX) or total chlorine compound free (TCF) pulps are required. Typical peroxide stage conditions are temperatures of 80°C to 90°C, pH over 10.8 and retention times over 3 hours. It is important to maintain a high concentration of peroxide through the stage to reach the desired brightness.

HYDROPHILIC—Readily wetted by water.

HYDROPHOBIC—Water-repellent; not wetted by water.

HYDROSULFITE—See SODIUM DITHIONITE.

HYDROTROPIC PULPING—A pulping process in which the cellulosic raw material is digested with an aqueous solution of a hydrotropic substance, i.e., one which has the property of markedly increasing the solubility of materials which ordinarily are but slightly soluble in water (e.g., sodium *m*-xylene sulfonate) for the removal of lignin.

HYGROEXPANSIVITY—The change in dimension of paper that results from a change in the ambient relative humidity. It is commonly expressed as a percentage and is usually several times higher for the cross direction than for the machine direction. This property is of great importance in applications where the dimensions of paper sheets and cards or construction board (wallboard, acoustical tile, etc.) are critical.

HYGROSCOPICITY—Ability to absorb water vapor from the surrounding atmosphere. Paper is strongly hygroscopic relative to most materials. Its hygroscopicity is measured by the change in moisture content when the relative humidity of the atmosphere is changed, and this change in moisture content affects its physical properties.

HYMNAL PAPER—A strong opaque book paper used for printing hymnals. It is made from bleached chemical woodpulp in the lighter basis weights.

HYPOCHLORITE BLEACHING STAGE (H)—An oxidative bleaching stage, often in the third stage of a bleaching sequence, where the active component is sodium hypochlorite or calcium hypochlorite in an alkaline medium. The brightening is achieved by destructive oxidation of the lignin, and the continuous presence of the alkali leads to the solution of the reaction products, thus opening deeper layers of lignin in the fiber to further attack. Hypochlorite degradation of cellulose is minimized by maintaining a high pH. Because it is not lignin-specific, hypochlorite can achieve a limited brightness in one stage without substantial damage to the pulp strength. It is the original bleaching agent used on woodpulps, initially in one batch stage and then in multistages from which modern sequences have developed. Typical conditions during a hypochlorite bleach are: temperature 35–40°C; time 2 hours; consistency 10%; and pH usually 9.5 to 11.0.

HYPOCHLORITE BRIGHTENING STAGE—A mild hypochlorite treatment, used occasionally for hardwood mechanical pulps where a significant brightness gain can be made with minimal loss of pulp yield and strength.

HYPOCHLORITES—$NaOCl$ or $Ca(OCl)_2$. Oxidative bleaching reagents prepared by reacting chlorine with either $NaOH$ to produce $NaOCl$ or with $Ca(OH)_2$ to produce $Ca(OCl)_2$. Both compounds are soluble in water, but start to decompose below pH10. Consequently, their manufacture and storage require the maintenance of highly alkaline conditions.

HYPOCHLOROUS ACID (HOCl)—A weak acid existing as one of the components of a chlorine water solution at equilibrium. Its concentration is pH-dependent, and is greatest at pH 4–6 where almost all chlorine is in the form of chloride ion and hypochlorous acid. It is a strong bleaching agent, but it reacts so rapidly that it is difficult to control, and is detrimental to pulp viscosity and strength. The formation of HOCl in the range pH 2–4 is now thought to be the primary cause of pulp degradation (viscosity loss) in chlorination.

HYPO NUMBER—A chlorine number (q.v.) test in which a reaction of a pulp sample with acidified sodium hypochlorite under carefully defined conditions indicates the chlorine demand of the pulp.

HYPO REACTOR—A device in which hypochlorite is produced by chlorine being absorbed into either caustic ($NaOH$) or milk-of-lime ($Ca(OH)_2$).

HYSTERESIS—Difference in the value of a property of a substance depending on whether a given value of a related condition or variable is approached from a higher or lower level. For example, paper conditioned at 50% relative humidity will have a greater moisture content when this environment is approached from a higher relative humidity than from a lower relative humidity.

I

IBM—International Business Machines, manufacturer of computer and business information systems. IBM developed the DOS (disk operating system) used by many computer manufacturers for personal computers (PCs). IBM was the first major developer of the PC.

ICE-BLANKET PAPER—Sheets of vegetable parchment paper, cut to sizes to cover cakes of ice, as a protection against heat.

ICE BOARD—See ICE PAPER.

ICE-CREAM BOARD—A paperboard used for packaging ice cream. It is made of chemical woodpulp, commonly from 0.016 to 0.022 of an inch in thickness. It is subsequently waxed, coated or otherwise treated resulting in a clean moisture-resistant board. See SPECIAL FOOD BOARD.

ICE-CREAM BRICK WRAPPER—A waxed paper or vegetable parchment sheet used for wrapping ice-cream bricks.

ICE PAPER—A well-sized paper which is coated with dextrin, gum arabic, or other adhesive and a salt, such as zinc sulfate, barium chloride, or sodium acetate, which forms crystals upon drying and thus gives the paper a frosted appearance.

IDLER ROLL—A web support roll that has no external drive mechanism. Through the application of web tension and available friction, the web attempts to accelerate and decelerate and rotate the roll at the web process speed.

IDLING LOAD—See CIRCULATING LOAD.

IGT TEST—IGT is the Institute for Graphic Techniques, Amsterdam, Holland. A test used to measure the ink (oil) pick resistance of paper. See SURFACE STRENGTH.

ILLUSTRATED LETTER PAPER—A bond paper so treated, as by coating on one side, as to make it suitable for fine illustrations on the one side and for ordinary typewriting or writing on the other.

ILLUSTRATION BOARD—A pasted board used principally for ink and water color. A typical drawing paper is pasted on both sides of the board (usually a filled pulp-lined board or a pasted board). Usual properties of drawing paper, such as finish and sizing, are essential, but hardsizing and good erasing quality are most important. The finished board should be as free as possible from warping. The basis weight is 150 pounds (17 x 22 inches – 500); the thickness is 0.0325 of an inch.

IMITATION ART PAPER—Highly finished printing paper prepared by the addition of a high percentage of china clay to the pulp. It has a water finish, giving it a surface, opacity, and absorbency suitable for printing halftones. The distinction between art paper and imitation art paper is that in the former the clay is coated on the surface, whereas in the latter it is mixed with the fiber.

IMITATION GREASEPROOF PAPER—A term applied to a heavily sized paper made from well-beaten bleached or unbleached chemical woodpulp. This paper possesses some resistance to moisture and blood penetration but has less grease-resisting qualities than greaseproof paper (q.v.).

IMITATION HANDMADE—A term used to indicate a machine-made paper that has been processed to appear as if it had been made by hand. It is often applied to plate or embossed finishes.

IMITATION HANDMADE PRINTING—See IMITATION HANDMADE.

IMITATION HANDMADE WRITINGS—See IMITATION HANDMADE.

IMITATION JAPANESE PAPER—A paper made of rope fibers to imitate in strength and appearance Japanese vellum. It is used for insulating purposes in armatures.

IMITATION JAPANESE VELLUM—A strong printing paper of wild formation made to imitate the appearance of Japanese vellum.

IMITATION KRAFT PAPER—A paper sometime used as a substitute for kraft paper where strength and durability are not necessary. It is commonly made of mechanical woodpulp, unbleached sulfite, or from wastepapers, and is colored brown to give the appearance of kraft.

IMITATION LEATHER—See ARTIFICIAL LEATHER PAPER.

IMITATION PARCHMENT—A single-process all chemical woodpulp sheet, originally so called to designate it as an imitation of vegetable parchment (but not an imitation of animal parchment). It now bears practically no resemblance to vegetable parchment paper, nor does it possess any of the qualities of that sheet. At present the term is used to describe a wrapping paper made of unbleached or semibleached long-fibered chemical woodpulps in a basis weight of about 30 pounds (24 x 36 inches – 500); when made of bleached pulps, it is more hydrated and has some greaseproof characteristics. It is employed in the grocery and provision trades. See GREETING CARD PARCHMENT.

IMITATION PRESSBOARD—A heavily calendered, cylinder machine board made of chemical and mechanical pulp and used for notebook covers, etc. It is usually paste laminated into thick board. It simulates transformer board (q.v.) but lacks its high density and dielectric characteristics. It is made in a variety of colors and has a characteristic mottled surface. See also INDEX PRESSBOARD.

IMPACT PAPER—See CARBONLESS PAPER.

IMPACT RESISTANCE—See IMPACT TEST FOR SHIPPING CONTAINERS.

IMPACT TEST FOR SHIPPING CONTAINERS—A method for measuring the resistance of fiberboard containers to impact from successive blows of increasing force, caused by carrying the loaded box on a dolly down an inclined plane to a bumper.

IMPERVIOUS—Resistant to the penetration of moisture, grease, oil or chemicals.

IMPREGNATING PAPERS—See ABSORBENT PAPERS; DRY FELT; SATURATING PAPERS.

IMPREGNATION—(1) For chemical pulping to be uniform, chemicals and heat must reach all parts of the fibrous raw materials such as chips. Impregnation is the process of moving the chemicals and heat into the material through pressure treatment or allowing enough time for diffusion to occur. If done before the cook is started, it is called pre-impregnation. (2) A method of bonding nonwovens by saturation.

IMPREGNATOR—In mechanical pulping, a compression device for compressing wood chips before impregnating them with a chemical solution.

IMPRESS—The different hand and mechanical methods of applying reading matter, illustrations, decorations, or rulings to paper.

IMPRESSED WATERMARK—See WATERMARK.

IMPRESSION—(1) The compression of paper and backing materials needed to transfer ink from one surface to another, as plate to paper, plate to blanket, or blanket to paper, usually expressed as thousandths of an inch beyond that needed to produce first contact between two printing cylinders. (2) A printed copy.

IMPRESSION PAPER—See DUPLICATING PAPER; MIMEOGRAPH PAPER; PROOF OR PROOFING PAPER.

IMPULSE—The product of the average force and the time during which it acts. This product is equal to the change in momentum produced by the force. See IMPULSE TO RUPTURE.

IMPULSE DRYING—A method that utilizes a press with a hot roll to dewater paper. The hot roll typically would have a surface temperature greater than 150°C, and directly contacts the paper. This method extends dewatering above that achieved with a press.

IMPULSE TO RUPTURE—A test employed in evaluating the strength of paper. It is expressed as the integrated product of the force (to cause rupture) and the time interval over which the force acts on the specimen.

INCH-HOUR—A term indicating the rate of operation over a period of time of one or more paper machines taking into account machine width but not speed. The figure is arrived at by multiplying inches of machine width by the number of hours in the period.

INCLINED PRESS—See PRESS SECTION.

IN-CONTROL PROCESS—A process in which the statistical measure being evaluated is in a state of statistical control (i.e., the variations among the observed sampling results can be attributed to a constant system of common causes). See also OUT-OF-CONTROL PROCESS.

INDENTED—A term used to indicate a paper or paperboard with raised knobs developed or formed into the sheet in the primary process. It is produced with large or small indentations or knobs without puncturing the sheet, producing a soft, bulky sheet of paper. Such material is especially suitable for packing or wrapping purposes when it is desired to prevent jarring of the articles so treated.

INDENTED BOARD—A paperboard used for wrapping purposes, carpet linings, and wherever a soft cushion is required. It is usually made of reclaimed paper stock. In the primary process, the sheet passes through rollers with raised knobs and corresponding hollows which impart indentations without puncturing. Large or small designs are used depending on requirements. The caliper before indenting varies from 0.012 to 0.023 of an inch.

INDENTED PACKING PAPER—See INDENTED BOARD.

INDEX—See INDEX BRISTOLS.

INDEX BOARD—A general term applied to various types of boards used principally for index records. See also INDEX BRISTOLS.

INDEX BRISTOLS—A class of heavyweight papers used for index cards, and the like. They are usually made from chemical wood and/or cotton pulps in solid or two-ply pasted form, in white and colors. Basis weights range from 90 to 220 pounds (25.5 x 30.5 inches – 500) but pasted items are made as high as 440 pounds.

INDEX CARD—A record card, cut from index bristol or heavy ledger; usual sizes are 3 x 5, 4 x 6, or 5 x 8 inches, in different weights. See BRISTOLS; INDEX BRISTOLS.

INDEX PRESSBOARD—A general term applied to those boards used for the manufacture of filing materials, guides, folders, etc., and for book covers, such as composition books and receipt books. This class may be divided into genuine and imitation pressboards, the difference between the two being in the density, rigidity, and finish. The furnish is generally chemical woodpulp, although some grades may contain a small percentage of rags. For the most part, they are made on a cylinder machine, although some are made by the so-called wet process. Genuine pressboard is finished by means of a glaze roll, which gives a high density and a highly polished surface. Imitation pressboard (q.v.) is finished by means of calender rolls.

INDIA BIBLE PAPER—See BIBLE PAPER.

INDIA OXFORD PAPER—See BIBLE PAPER.

INDIA PAPER—See BIBLE PAPER.

INDIA PROOF PAPER—A paper of straw color, extremely soft and absorbent, unsized, which can readily conform to the surface pattern of a printing plate to absorb ink without smearing, thus giving a true proof impression. The term is also loosely used to include all illustrations printed on India paper.

INDIA TINT—Light buff color.

INDIA TRANSFER PAPER—See INDIA PROOF PAPER.

INDICATOR PAPER—See TEST PAPERS.

INDUCED-DRAFT BOILER—See INDUCED-DRAFT FAN.

INDUCED-DRAFT FAN—The fan used to draw the flue gases out of the boiler in order to maintain the furnace at an operating pressure just below the surrounding atmospheric pressure.

INDUSTRIAL PAPERS—Papers intended for industrial uses, as opposed to those for cultural or sanitary purposes.

INDUSTRIAL WASTES—The liquid wastes from industrial processes as distinct from domestic or sanitary wastes.

INDUSTRIAL WIPES—Paper towels that are especially made for industrial cleaning and wiping uses. Capacity to absorb oil and water, high wet-strength properties, lint freeness, and pliability are important characteristics. Some are single-ply and others are made from two or more plies of a special facial-type stock.

INERTIA COMPENSATION—An automatic adjustment of the gain of a winder/unwind drive controller to avoid sluggish performance at large roll diameters and instability at small roll diameters.

INFLUENTS—Streams that flow into a treatment process.

INFORMATION RETRIEVAL—The searching of past and present information sources, generally accomplished by computer, with the purpose of retrieving information on a given subject. See also INFORMATION STORAGE AND RETRIEVAL.

INFORMATION STORAGE AND RETRIEVAL—The collection and storage, in a computer, of data from multiple sources using a specialized database program and incorporating means to access and report the data with sorting, logging, and graphical capabilities.

INFRARED (IR) DRYER—A noncontact drying method that uses infrared radiation to heat the paper. Most common usage is in coating drying.

INFRARED SPECTROSCOPY—The science dealing with the spectral analysis of compounds using radiation in the infrared region (780 to 400,000 nanometers).

INGRAIN—A descriptive term for a mottled or granite appearance in paper.

INGRAIN ART PAPER—A typical art paper with a special finish.

INHERENTLY BONDED (SELF BONDING)—A nonwoven material that is bonded at the instant of manufacture, such as a highly linear film which is broken down into a fibrous state by use of pins, stretching, or flexing.

INHIBITIVE TISSUE—Generally a kraft interleaving tissue impregnated with sodium chromate to prevent water stains on flat sheets of aluminum. See INTERLEAVING PAPER; INTERLEAVING TISSUE.

INITIAL TEARING RESISTANCE—See EDGE TEARING RESISTANCE.

INK ABSORPTION—See ABSORBENCY.

INK HOLDOUT—See HOLDOUT; INK RECEPTIVITY.

INK-JET PRINTER—Computer-driven device that generates microscopic droplets of ink that are shot at the paper surface to create images.

INK-JET PRINTING—A printing process in which tiny droplets of ink are shot onto the paper surface to form the printed image. These droplets are emitted from an orifice on a trajectory determined by the computer to result in the desired image appearing on the substrate.

INK MOTTLE—A non-uniform appearance of the ink film in printed areas, with respect to density, color, or both. It may be caused by varying ink film thickness, by variation in the ink receptivity and absorption of the paper, or by both these causes acting together. See also ORANGE PEEL.

INKOMETER—An instrument for measuring the tack of inks in terms of the force required to split an ink film between rollers with controlled speed, temperature, and ink-film thickness.

INK RECEPTIVITY—See CASTOR-OIL TEST.

INK RUB-OFF—The smudging and spreading of a print due to friction against paper, fingers, clothing, or any other surface. Ink rub-off is particularly objectionable in newspapers, which normally use inks with a mineral-oil base.

INK SET-OFF—The transfer of ink from fresh prints to any other surface: (a) First impression set-off occurs when the paper is printed on the second printing nip while the print from the first impression is still fresh. (b) Opposite page or companion page set-off occurs when printed pages are folded or printed sheets are stacked soon after printing.

INK TRANSFER—The amount of ink film transferred to a receiving surface as the result of a printing impression, expressed as a percentage of ink available.

IN-LINE INJECTOR—A device for injecting chemicals into a stock line under pressure.

IN-LINE MIXER—A device mounted within a process line for mixing chemicals into stock. The energy for mixing can be supplied either by the stock pump (in the case of a static mixer), or an external motor can supply power to a rotating impeller, in which case, it is called a high shear or a dynamic mixer. See MIXER.

INNER BACKING FABRIC—See TWIN WIRE FORMING FABRICS.

INNER BARK—See PHLOEM.

INNER CONVEYING FABRIC—See TWIN WIRE FORMING FABRICS.

INNERFRAME STOCK—A paperboard, usually 0.012 of an inch thickness and fully bleached, used for support and as the inner ply of cigarette boxes (flip-top).

INNERSOLE BOARD—See SHOE BOARD.

INOCULUM—Living microbial seed material used to initiate growth of the microbe in or on a new substrate. For example, inoculum consisting of spores is used to introduce a pitch-reduction fungus into wood chips.

INSECT REPELLENT PAPER—More properly an insect resistant paper. A paper which has been treated for protection against insect penetration. There are a variety of insecticides which can be applied to the paper, but the most common group of chemicals are pyrethrins and piperonylbutoxide. These are frequently used in the manufacture of cartons or multiwall paper bags where protection is necessary against boring insects.

INSERT—(1) A term applied to a single sheet of the same or different quality inserted in a magazine, newspaper, or book. When the insert is of better quality than the overall stock, as for instance when it is coated, it enhances the display features thereof. (2) A thin filler or frame of paperboard or wadding used to take up space or separate articles within a package. (3) A pad of the same material as the box which is dropped into the gap between the inner flaps when all flaps do not meet. (4) See ENVELOPE STUFFER.

INSPECTION—The measurement, examination, testing or gauging of one or more characteristics of a product and comparing the results with specified requirements to determine whether conformity is achieved.

INSULATING BLANKET—A product used in building construction. It is a flexible material composed of mineral or vegetable fibers loosely felted together so as to contain a maximum amount of entrapped or dead air. It is generally covered on one or both surfaces with kraft paper or foil, which is cemented to the fibrous mat. It is available in a number of thicknesses depending upon the desired amount of insulation.

INSULATING BOARD—A type of board composed of some fibrous material, such as wood or other vegetable fiber, sized throughout, and felted or pressed together in such a way as to contain a large quantity of entrapped or "dead" air. It is made either by cementing together several thin layers or forming a nonlaminated layer of the required thickness. It is used in plain or decorative finishes for interior walls and ceilings in thicknesses of 0.5 and 1 inch (in some cases up to 3 inches) and also as a water-repellent finish for house sheathing. Desirable properties are: low thermal conductivity, moisture resistance, fire resistance, permanency, vermin and insect resistance, and structural strength. No single material combines all these properties but all should be permanent and should be treated to resist moisture absorption. See STRUCTURAL FIBER INSULATING BOARD.

INSULATING LATH—Insulating board in sheets 18 x 48 inches with long edges ship-lapped and all edges beveled. It is used as a base for a plaster coat. This material is furnished in 1/2- and 1-inch thicknesses and is made from mechanically shredded wood fibers.

INSULATING MATERIALS—Insulating materials include the following classes:

Heat-insulating materials

Felt

Refrigerator paper

Floor felts

Roofing felts

Floor lining or deadening felts...plain or indented

Insulating crepe-wadding blankets

Asbestos

Sheathing

Red, gray, or blue rosin sheathing

Paper-lined felted fibrous products

Structural fiber insulating board

Building board

Lath (for plaster base)

Roof-insulation board

Interior boards (factory finished)

Interior-finish board

Panel board (or tileboard)

Sheathing

Interior boards (flame-resistant-finished surface)

Interior-finish board

Panel board (or tile board)

Electrical insulation materials

Untreated

Coil papers (those papers often described as layer insulation; used as insulation between wire layers and as wrap around insulation

Cable papers (or turn insulation..conductor wrap)

Slot papers (slot and commutator segment insulation)

Crepe papers (for forming purposes, taping of coils, leads and insulting pads)

Capacitor tissue (capacitor dielectrics)

Electrolytic condenser paper (relatively porous paper used as a spacer between foils in electrolytic condensers)

Transformer board (sometimes called pressboard and used as layer insulation in transformers, and as formed parts, separator, and mechanical supports in various electrical applications)

Treated

Impregnated (with resins, varnishes, waxes, and dielectric liquids)

Coated (with resins, varnishes, waxes, etc.)

Chemically or mechanically pretreated

Vulcanized fiber (a cotton paper treated with zinc chloride) in various grades called electrical insulation fiber, hard fiber, hermetic fiber, bone fiber

Mechanically hydrated (paper prepared from pulp subjected to considerable mechanical hydration)

Sound-insulating materials

Acoustical board

INSULATION PAPER—A paper, usually kraft, used in the manufacture of insulation batts. Open-face batts are made with a layer of paper on one side only. This sheet, usually 35, 40, or 50 pounds (24 x 36 inches – 500) may be

167

printed or colored or both, and is sometimes laminated to foil. The reverse side is usually coated with asphalt which acts as a vapor barrier and also adheres the sheet to the insulating material (rock wool, mineral wool, fiberglass, etc.); hence, it is known as the vapor barrier sheet. Paper-enclosed batts employ the same vapor barrier sheet on one side and a lighter weight sheet on the other. This weighs from 20 to 30 pounds (24 x 36 inches – 500) and it is usually flame resistant. It is called the breather sheet.

INSULATING SHEATHING—A general term indicating any type of sheathing that provides insulation. See ASPHALT SHEATHING PAPER; RED ROSIN SHEATHING PAPER; SHEATHING PAPER.

INSULATING TISSUE—Thin paper used for insulating radio wire, magnet wire, transformer wire, telephone cable wire, and for use under the rubber in lamp-cord wire. Generally, it ranges in thicknesses from 0.0005 to 0.003 of an inch and in basis weight from 5 to 30 pounds (24 x 36 inches – 500). It is customarily made from manila fiber, chemical woodpulp, or mixtures of the two and normally is fairly open, porous, tough, and long-fibered. See CABLE-PAPER; CONDENSER PAPER.

INTAGLIO—(1) Engraving incised or cut into the surface of wood or metal, as distinguished from engraving in relief. See INTAGLIO PRINTING. (2) A type of watermarking with a dandy roll. It is also called shadow watermark. See WATERMARK.

INTAGLIO PAPER—Any paper suitable for intaglio printing. See ROTOGRAVURE PAPER.

INTAGLIO PRINTING—Printing from plates in which the image is intaglio, or sunken below the surface, as distinguished from printing by letterpress (raised image) or planography (flat plane). Included in this class of printing are rotogravure, sheet-fed gravure, photogravure, offset gravure, copper and steel engraving, etching, stipple engraving, aquatint, and mezzotint. The image consists of regular (screen) or irregular (grain) etched depressions or lines and dots engraved by hand. These are filled with ink, and the surface is wiped or scraped clean by a doctor blade, wiping paper, or counter rotating polymeric wiping cylinder. This feature gives gravure prints an unusually wide range of tonal expression, ranging from full, velvety depths to purest highlights. Steel line engraved intaglio is traditionally associated with security printing, as for currency, stock certificates, automobile titles, and birth certificates.

INTEGRATED MILL—A paper or board mill that produces substantially all its own pulp. A partially integrated mill is one that produces some but not all of its pulp.

INTENSE WRITING FINISH—A term applied to carbon paper (q.v.).

INTERFACING—Material which is applied to the inner surface of apparel to add resilience, stability, weight, and warmth.

INTERFOLDED TISSUE—(1) Waxed tissue that has been folded and cut so that one sheet at a time dispenses from a carton. Used primarily for food handling in commercial food establishments. (2) Toilet tissue that has been folded and cut so that one sheet at a time dispenses from a specially mounted wall dispenser. (3) Facial tissue that has been folded and cut so that one sheet at a time dispenses from a carton.

INTERGRANULAR CORROSION—Preferential corrosion at, or adjacent to, the grain boundaries of a metal or alloy.

INTERIOR BOARDS—Various building boards such as gypsum board and insulating fiberboard, used for interior construction.

INTERIOR PACKING—A term inclusive of pads, partitions, liners, etc., used inside a shipping container to separate or give added protection to the contents. It may be made of single faced, double faced, or double wall corrugated board or solid fiberboard cut to sizes and shaped to specifications; or it may be cellulose wadding or indented board (q.v.).

168

INTERIOR WRAPS—Materials used to protect individual items against penetration by water, other liquids, gases, and the like.

INTERLAPPED—A method of packing paper whereby one lift is placed so that it half covers another lift; the exposed half of the lower lift is soft-folded over the edge of the top lift and the other half of the top lift is soft-folded over the pile in order to form a package.

INTERLAYER SLIPPAGE—The circumferential or lateral displacement of the internal layers of a wound roll.

INTERLEAVING BLOTTING—A thin blotting paper to be inserted between printed sheets to prevent offsetting or blotting and between leaves of books for the purpose of absorbing ink from freshly written manuscripts. The usual basis weight is 20 pounds (19 x 24 inches – 500).

INTERLEAVING PAPER—(1) A paper (usually of tissue weight) which is placed in front of illustrations in books or between two or more engravings, etchings, sheets of cellulosic films, etc. (2) A paper which is inserted between sheets as they come off the printing press to prevent offset. (3) Thin blottings used in diaries. See INTERLEAVING BLOTTING. (4) See STEAK INTERLEAVING PAPER. (5) A paper made from mechanical pulp which is used in the printing of textiles. (6) See STEEL INTERLEAVING PAPER.

INTERLEAVING TISSUE—A tissue, used for separating, or protective purposes, in a variety of grades. See also FLOCKING TISSUE; INHIBITIVE TISSUE; INTERLEAVING PAPER.

INTERMEDIATE BLACK LIQUOR—Black liquor at any intermediate stage of evaporation. Frequently refers to the black liquor coming out of the multiple effect evaporators (q.v.) and entering the concentrator (q.v.).

INTERNAL BOND—(1) The force with which fibers are bonded to each other within a sheet of paper. (2) The force with which fibers are bonded to each other within a sheet of paperboard or the force which plies are bonded together in a multi-ply sheet of paperboard. See BONDING STRENGTH.

INTERNAL REVENUE STAMP PAPER—A special paper manufactured to specifications of the United States government for printing internal revenue stamps. It is made of bleached chemical wood fiber in a basis weight of 42.5 pounds (24 x 36 inches – 500). The paper is either white or blue, the color of which is fast to light and water. It is printed on an intaglio press.

INTERNAL SIZING—The process of sizing paper by the application of sizing materials in the beater, or to the furnish prior to sheet formation, as distinguished from surface sizing or tub sizing. It usually refers to the use of rosin size and alum, but other sizing agents may be used.

INTERNAL TEARING RESISTANCE—The force in grams required to tear a single sheet of paper after the tear has been started. It should not be confused with initial tear or edge tear. It is normally tested on an Elmendorf tester.

INTERNATIONAL SYSTEM OF UNITS (SI UNITS)—A system of units based on a set of 7 dimensionally independent units (meter, kilogram, second, ampere, Kelvin, mole, candela), 2 supplementary units (radian and steradian), and derived units formed by combining base units, supplementary units and other derived units. SI units are now used in all scientific work.

INTERWEAVING—In winding, the overlap of two adjacent webs in a set as they wind. The result is two rolls that are stuck together.

INTRINSIC VISCOSITY—An empirical test for evaluating the degree of degradation of cellulose. It is used as a control procedure in the manufacture of rayon and other cellulose-based products such as films, lacquers, and plastics. Technically, it is equal to the specific viscosity divided by the concentration when the concentration approaches zero.

INVERFORM—A papermaking device used to manufacture single or multi-ply grades of paper and paperboard. The stock flows from a headbox to a bottom fabric (similar to a fourdrinier) and is then joined by a top fabric so that water removal from the stock is accomplished through the top fabric as well as the bottom fabric. In multi-ply operation the bottom fabric and formed web continue under subsequent headboxes where additional plies are laid down. Each headbox is followed by another top fabric. For each additional ply thus laid down, virtually all of the water removal is upward through the top fabric. The machine is capable of producing high basis weight paper.

INVERSION—An atmospheric condition where a layer of cool air is trapped by a layer of warm air so that it cannot rise. Inversions spread polluted air horizontally rather than vertically so that contaminating substances cannot be widely dispersed. An inversion of several days can cause an air pollution episode.

INVOICE PAPER—(1) See BILLING-MACHINE PAPER. (2) Any paper used for billing purposes.

ION—An atom or a group of atoms having either a positive or negative charge due to having lost or gained one or more electrons.

ION EXCHANGE—Salts dissolved in water dissociate to form ions. Ion exchange is the process of using an ion exchange medium to exchange unwanted ions for less harmful ions in feedwater. Ion exchange media can replace both negatively charged and positively charged ions.

ION-EXCHANGE PAPER—A paper having the property of selectively absorbing either positive or negative ions, and used for the separation of ions or for removal of certain ions from solution. The paper may be made from modified cellulose fibers, from noncellulosic fibers having acidic or basic properties, or as a sheet containing cellulose fibers and ion-exchange resins.

IRIDESCENCE—See BRONZING OF INK.

IRIDESCENT PAPER—See MOTHER-OF-PEARL PAPER.

IRON, ACID-SOLUBLE—The portion of the iron present in paper that is soluble in hydrochloric acid, and thus is considered to be potentially chemically reactive.

IRON CORES—See CORE-IRON.

IRON FREE ALUM—An aluminum sulfate which is essentially free from iron as required in the sizing of photographic and other specialty papers.

IRON SPECKS—Fragments of iron or rust in the sheet.

ISBN—International Standard Book Number, a 10 digit number used to identify the publisher, country of origin, and title of a book,

ISO—International Organization for Standardization which sets standards and test methods for many areas of international commerce.

ISO 9000 SERIES STANDARDS—A set of international standards on quality management and quality assurance developed to help companies effectively document the quality system elements to be implemented to maintain an efficient quality system.

ISRI—The Institute of Scrap Recycling Industries, Inc. is the national trade association that represents processors, brokers, and consumers of scrap metals, paper, plastics, glass, rubber, and textiles, as well as suppliers of equipment and services to the industry.

ITEM—One grade of paper made in one size, weight, grain, finish, and color.

ITSC—Impact test for shipping containers.

IVORY—A cream-white color.

IVORY BRISTOL—A heavy, uncoated, semi-translucent, smooth-finished paper designed primarily for printing or engraving of business cards. It is usually made from chemical wood and/or cotton pulps in basis weights of 70 to 90 pounds (22.5 x 28.5 inches – 500). Ivory bristol is referred to as ivory board in countries other than the United States.

J

JACKET—(1) See COUCH JACKET. (2) The detachable outer cover of a bound book.

JACQUARD BOARD—Paperboard made from specialty kraft fiber stock (some grades use a percentage of new rags) suitable for cutting or perforating. This board is used on jacquard looms and has exceptional punching qualities, as each card is perforated with a great many holes. The card should also lie flat and show minimum change in dimensions under extreme atmospheric conditions.

JACQUARD PAPER—Paper made specifically for use on the paper tape jacquard machine. (Commonly known in Europe as "Verdol Paper.")

JAPAN ART PAPER—A paper used for artists' proofs or engravings. It may be classed as an art parchment. It is a paper with long fiber as made in Japan or imitations of it. Significant properties are those of art paper, *viz.*, formation and surface—especially for the so-called Japanese vellum.

JAPANESE COPYING PAPER—Specially thin and strong papers made in Japan from long fibers, such as mitsumata and paper mulberry, and used for copying books. These papers are largely handmade, the fibers pulped by hand and the sheets made on molds of bamboo or hair. The length of the fiber gives a paper of exceptional wearing qualities, the fibers pulling apart and not tearing.

JAPANESE DECORATING PAPER—Any native Japanese paper which is suitable for decorating by water colors.

JAPANESE PAPER—See JAPAN ART PAPER.

JAPANESE PARCHMENT PAPER—See JAPANESE VELLUM.

JAPANESE VELLUM—A thick paper, made in Japan, of native fibers, which are characterized by their length. The formation is very cloudy and the paper is very tough and durable. The color is usually a cream or natural color. It is finished with a good surface and is suitable for certificates or for other purposes where a tough durable paper is necessary.

JAPAN PAPERS—A special type of paper with irregular formation, imitating the old imperial vellum, which gives the surface a beautiful, mottled effect. Besides this surface effect, the papers are characteristically long fibered and strong. The basis weights range from 50 to 150 pounds (25 x 36 inches – 500). These papers are used for offset printing, novelties, greeting cards, and the like.

JAPON—A French imitation of Japanese vellum.

JCP—The Joint Committee on Printing (Congress of the United States). Its Committee on Specifications develops specifications for papers used in printing and binding by the Federal Government.

JELLY PROTECTORS—Heavy waxed bond or vegetable parchment paper cut to desired sizes and used for covering jelly, or other preserves, when in jars.

JET DECKLE—A deckle edge made on the fourdrinier fabric or cylinder wire by means of a jet of water or air.

JEWELERS BRISTOL—See JEWELRY-CARD BRISTOL.

JEWELERS TISSUE—An antitarnish tissue used for wrapping silver and other articles. It is made from cotton fiber and/or chemical woodpulp. It is usually white in color and is frequently fine ribbed.

JEWELRY-CARD BRISTOL—A card manufactured usually from mill bristol and used for the display of jewelry. It usually has a light pink color and has antitarnish qualities. See BRISTOLS.

J-LINE—A measurement of nip-induced inter-layer slippage during roll winding or unwinding.

JOB LOT—(1) Paper produced in excess of a specific order. (2) A discontinued line of paper. (3) Paper rejected because of a defect or nonconformance to specifications. (4) A nonstandard paper because of specification changes. This category of paper is normally sold at a reduced rate as "off quality" or it is utilized as "broke" and reprocessed.

JOG—(1) A control on a machine (like a winder) which provides a momentary forward movement of the web. (2) A vibratory operation on a pile of paper intended to orient individual sheets so the pile is straight-sided and smooth.

JORDAN—A refiner (q.v.) or conical refiner (q.v.) whose working elements consist of a conical plug rotating in a matching conical shell. The outside of the plug and the inside of the shell are furnished with knives or bars commonly called tackle (q.v.). In operation, the rotating conical plug is pushed into the shell to press against the shell knives or bars and gives macerating action on the fibrous material in water suspension that is passed between them. Stock is usually introduced into the small end of the jordan and withdrawn from the large end though it may also be pumped through in the other direction. A jordan is distinguished from other conical refiners by its very low rotor included angle, 20° or less.

JOULE—A unit of work equal to the work done by a force of one newton acting through a distance of one meter. A newton is that force which gives a mass of one thousand grams an acceleration of one meter per second per second.

JUMBO ROLL—Any large reel of paper or large shipping roll. A jumbo reel could produce up to four shipping rolls and could exceed 120 inches in diameter. A large shipping roll of fine paper can be 50 inches in diameter. Three and

3.5-meter wide printing presses require jumbo rolls from 118 to 138 inches wide.

JUNIOR CARTON—Generally speaking, a smaller version of a regular carton. Whereas a regular carton is generally considered to be of such size as to hold from about 120 to 150 pounds of paper, a junior carton normally contains about 40 to 50 pounds. Junior cartons are most commonly associated with the packing of writing and printing papers, in small "office sizes," such as 8 x 11 inches. Three junior cartons are usually considered to be equivalent to one regular carton for pricing or other purposes. See CARTON.

JUNK BOX—A pulper detrashing device that consists of a simple collection box or container, usually located near the bottom of a pulper. Heavy debris, such as nuts and bolts or bits of bale wire, accumulate in the trash box until they are isolated from the pulper, and emptied, either manually or automatically.

JUNK TOWER—A pulper detrashing device that accumulates gross debris. It generally consists of a large diameter vertical pipe mounted next to a pulper, and which extends well below the bottom of the pulper tub. It is connected to the pulper tub near the bottom of the pulper tub, such that gross, heavy debris in particular can exit the pulper tub and accumulate inside the junk tower. The accumulated debris is periodically removed, usually with a clam shell grapple, lowered from above.

JUSTIFICATION—Alignment of the ends of printed lines. In left justified text, the left margin is straight, the right margin ragged. Right justified text is the opposite. Text may be both left and right justified, as in newspaper columns.

JUTE—(1) An Indian bast fiber, white jute *(Corchorus capsularis)*, and tossa jute *(C. olitorius),* which is used for the manufacture of coarse sacking and bags (gunny sack). Old gunny and sacking are used as raw materials in papermaking. (2) A term indicating a furnish consisting substantially of paper stock reclaimed from post consumer papers. See JUTE LINERBOARD.

JUTE BAG PAPER—Bag paper made from jute fiber.

JUTE BOARD—(1) A paperboard for use in shipping containers. See JUTE LINERBOARD. (2) A combination board for use in folding box manufacture, made on a cylinder machine wherein one or both of the outer plies are made of kraft or reclaimed kraft paper stock and the remainder of mixed paper stock. Although this is termed jute, it contains no jute fiber. The board must be able to withstand scoring and folding.

JUTE BRISTOL—A bristol in which the furnish contains more than 50% of jute fiber. It is characterized by unusual strength, especially high tearing resistance, and is used for various purposes where durability is essential. See BRISTOLS.

JUTE ENVELOPE PAPER—A strong, opaque paper with good folding qualities, made of jute fiber alone or in combination with kraft, and used for envelopes. It may be machine or water finished.

JUTE LINER—See JUTE LINERBOARD.

JUTE LINERBOARD—A paperboard used chiefly as an outer facing in the manufacture of corrugated or solid fiber shipping containers. It is made primarily of paper stock reclaimed from old corrugated containers, shipping sacks, and the like with or without small additions of virgin kraft pulp. No jute fiber is employed in the furnish. It is usually formed on a multicylinder machine. Nominal grade weights range from 28 to 110 lb/1000 ft^2 and thicknesses 9 to 30 points (0.009 to 0.03 inch). See RECYCLED FIBER LINERBOARD.

JUTE PAPER—Any paper made from jute fiber (*Corchorus capsularis* and/or *C. olitorius)* or burlap waste with various proportions of kraft or sulfite pulp. The basis weight is from 20 to 300 pounds (24 x 36 inches – 500). Jute papers are used extensively for envelopes, folders, tagstock, wrappers, cover stock, bristols, pattern papers, and a variety of specialties; also hydrated lime and cement bags, flour sacks, etc.

JUTE PATTERN—See PATTERN PAPER.

JUTE PULP—Jute pulp is made from reclaimed sacking, burlap, and string by cleaning, chemical cooking with lime or caustic soda, and bleaching. Paper of acceptable strength and durability is made from such pulp.

JUTE WRAPPING—See JUTE PAPER.

JUVENILE WOOD—The core of a coniferous tree, usually the first 10 to 15 growth rings from the pith. Juvenile wood has shorter fiber length, lower specific gravity, and a higher percentage of earlywood fibers than the wood toward the outside of the tree, called mature wood.

K

KAOLIN—A whitish earthy material composed primarily of the clay mineral kaolinite, a form of aluminum silicate. In refined form, kaolin is used in papermaking as a filler, coating component and opacifying agent. See also PAPER CLAY.

KAPPA FACTOR—A ratio of the amount of (equivalent) chlorine applied in the initial chlorination stage to the amount of lignin in the pulp entering the initial chlorination stage. The kappa factor was previously called the active chlorine multiple (ACM). It is the sum of all chlorine compound additions to the first chlorination stage (converted to active chlorine equivalent, i.e., ClO_2 wt% charge x 2.63) divided by the kappa number of the pulp entering the first chlorination stage.

KAPPA NUMBER—A test value that relates linearly to the amount of lignin remaining in pulp after pulping (degree of delignification) and thus an estimate of the bleaching chemical demand. The kappa number is the number of milliliters of standardized 0.1N $KMnO_4$ (potassium permanganate) solution reduced by 1 gram of ovendried pulp under controlled conditions and corrected to consumption of 50% of the permanganate. It is linear on all types and grades of pulp up to yields of 70%. Unbleached bag and paperboard pulps are cooked to a high kappa number, in the range 50 to 90. Pulps that will be bleached are cooked to lower kappa numbers, in the range 12 to 35.

KARAYA GUM—An acetylated polysaccharide obtained from the dried exudation of the *Sterculia urens* tree grown in India. In the deacetylated form, it is used as a fiber deflocculating agent for long-fibered pulps.

K-B BOARD—A paperboard made by a patented process and used for automobile panels and other purposes where waterproofing is an important factor. It is made on a cylinder or a wet machine with a furnish composed of wastepaper stock and an emulsion of asphalt about 15 to 20% in weight. The board is vat lined or subsequently lined with paper or pasted onto plain or embossed paperboard. It is stiff and waterproof.

K-B SHEATHING—A sheathing paper which has been made by the K-B patented process to procure waterproofing. See K-B BOARD; SHEATHING PAPER.

KENAF—An annual plant *(Hibiscus cannabinus)*, originally from the East Indies but now widespread. The fiber can be used for paper pulp and for cordage.

KERNING—The process of adjusting the space between letters to achieve even spacing. This is shown in the overlap of letter combinations such as AW or LT.

KETONE—Any class of organic compounds containing a carbonyl group (>C=O) attached to two organic groups (e.g., acetone, $(CH_3)_2CO)$). This class of compounds has been studied for catalyzing the formation of "activated" oxygen species, which can be very powerful and specific delignification agents.

KEYBOARD PAPER—See MONOTYPE PAPER.

KICK-UP—A term applied to needle-punch needles to indicate the amount that the barbs protrude above the surface of the needle.

KID FINISH—A vellum finish on a soft texture paper. It resembles, in appearance and feel, undressed kid leather. It is similar to a smooth eggshell finish, but it has a finer surface texture. It is used for bristols, weddings, and papeteries.

KID WEDDINGS—Typical wedding paper having a kid finish, i.e., smooth without gloss.

KINETIC FRICTION—The friction between two surfaces, such as cartons, in sliding contact with each other.

KITE PAPER—A 20-pound MG sulfite or kraft paper used for the manufacture of toy kites.

KLASON LIGNIN—A standard method for the quantitative determination of lignin in wood, isolated as the residue after treatment of wood meal with 72% sulfuric acid for a certain period, followed by dilution and cooking. It is the material remaining after 72% H_2SO_4 hydrolysis of resin free pulp expressed as % of ovendried pulp. Up to 20%, or even more, of the lignin may dissolve in the acid, depending on the pulping process and whether the pulp is partly bleached. The amount of dissolved lignin can be measured by ultraviolet spectroscopy (@ 280 or 210 nm), and is called UV lignin. Therefore, total lignin = Klason lignin + UV lignin.

KLISCH—Short for helioklischograph, a device that scans print images, converts the scanned input into electronic signals that drive a diamond stylus to reproduce these images on a gravure cylinder.

KLUDGE—An operating system, particularly computer related, which is assembled out of mismatched parts. Efficiency is usually far from optimum. The adjective form is kludgey, and variant spellings.

KNIFE COATING—A coating process in which a doctor, knife, or a straight edge is employed to spread and control the amount of coating on the paper. See AIR KNIFE COATING; BLADE COATING.

KNIFE EDGE—That edge of the paper which is cut by a knife. It may be guillotine trimmed or cut with a rotary knife. See also SLITTER EDGE.

KNITS—See NITS.

KNOTS—(1) Lumps in paper stock resulting from incompletely cooked or defibered woodpulp, usually as the result of poor cooking liquor penetration due to the presence of a locally dense area in the wood where a tree limb projected from the trunk or another branch. (2) Textile fibers or cloth, twisted into a noticeable lump in the stock.

K NUMBER—See PERMANGANATE NUMBER.

KOZO—See PAPER MULBERRY.

KRAFT—Relating to the sulfate pulping process, the resulting pulp, and the paper or board made therefrom.

KRAFT BAG PAPER—A paper made of sulfate pulp and used in the manufacture of paper bags. It normally has a greater bulk and a rougher surface than the usual kraft wrapping paper. So-called kraft bag papers have been made from sulfite pulp and colored to resemble a true kraft bag paper. The basis weights are those of regular kraft wrapping paper, i.e., 25 pounds (24 x 36 inches – 500) and heavier. See also SHIPPING SACK KRAFT PAPER.

KRAFT BITUMEN PAPER—A kraft paper that has been coated, impregnated, or laminated with asphalt.

KRAFT BOARD—A paperboard in various thicknesses made of kraft pulp. The chief characteristics of this board are its strength and bending qualities. See KRAFT CORRUGATING MEDIUM; KRAFT LINERBOARD.

KRAFT BUTCHERS—See DRY FINISH BUTCHERS WRAP.

KRAFT CORRUGATING MEDIUM—A corrugating medium (q.v.) usually made on a fourdrinier machine from a furnish which is 75% or more virgin kraft pulp.

KRAFT ENVELOPE—See ENVELOPE PAPER, KRAFT.

KRAFT LINER—See KRAFT LINERBOARD.

KRAFT LINERBOARD—A linerboard (q.v.) made on a cylinder or fourdrinier machine from a furnish containing 80% or more virgin kraft woodpulp. Nominal grade weights range from 26 to 90 lb/1000 ft^2 and thicknesses from 9 to 30 points.

KRAFT MANILA—Any paper made from kraft pulp, varying in basis weight from 20 to 120 pounds (24 x 36 inches – 500), which is colored yellow to simulate a manila shade.

KRAFT PAPER—A paper made essentially from woodpulp produced by a modified sulfate pulping process. It is a comparatively coarse paper particularly noted for its strength, and in unbleached grades is used primarily as a wrapper or packaging material. It is usually manufactured on a fourdrinier machine with a regular machine-finished or machine-glazed surface. It can be watermarked, striped, or calendered, and it has an acceptable surface for printing. Its natural unbleached color is brown, but, by the use of semibleached or fully bleached sulfate pulps, it can be produced in lighter shades of brown, cream tints, and white. Kraft paper is most commonly made in basis weights from 25 to 60 pounds (24 x 36 inches – 500) but may be made in weights ranging from 18 to 200 pounds. In addition to its use as a wrapping paper, it is also converted into a wide variety of products such as: grocers' bags, envelopes, gummed sealing tape, asphalted papers, multiwall sacks, tire wraps, butchers wraps, waxed paper, coated paper, cable sheathing, insulating and abrasive papers, as well as all types of specialty bags and sacks. Many paper grades including tissues, printing and fine papers, formerly manufactured from bleached sulfite are now made from bleached kraft.

KRAFT PULP—See KRAFT PULPING.

175

KRAFT PULPING—The alkaline pulping process that uses a combination of sodium hydroxide and sodium sulfide. It is derived from the German word meaning "strong," a fitting term since kraft pulp is the strongest chemical pulp. An alternative term is sulfate pulping.

KRAFT TEST LINER—See KRAFT LINERBOARD.

KRAFT TWISTING—A kraft paper suitable for twisting into paper twine, etc. See TWISTING PAPER.

KRAFT WATERLEAF—See WATERLEAF.

KRAFT WATERPROOF—A kraft wrapping paper which has been treated with paraffin, asphalt, or other material to render the sheet highly resistant to penetration by moisture.

KRAFT WRAPPING—A wrapping sheet of general commercial purposes, made of kraft pulp, MF and MG finish, in a large range of basis weights, usually from 25 to 120 pounds (24 x 36 inches – 500). See KRAFT PAPER.

KUBELKA-MUNK—Authors of the theory of scattering and absorption of radiation in intensively scattering media.

L

LABEL—A separate slip or sheet of paper affixed to the surface of a corrugated shipping container for identification or description.

LABEL CLOTH—A cloth-lined paper used when extra strength is required.

LABEL MANILA PAPER—Any paper of manila color made especially for gumming. The gummed paper should lie flat.

LABEL PAPER—A paper usually made from chemical woodpulp for label printing. It normally has a smooth machine or supercalendered finish for good lithographing and gumming qualities, and when coated on one side, it is referred to as CIS Label. Normal basis weights range from 50 to 80 pounds (25 x 38 inches – 500).

LACE PAPER—A bleached chemical woodpulp paper, hard sized, with good tensile and tearing strength. The basis weight ranges from 30 to 50 pounds (24 x 36 inches – 500). Cleanliness is of great importance. This paper is sold in jumbo rolls to converters, who slit it to the desired widths and use it for the manufacture of paper doilies, box laces, Valentines, etc.

LACQUER—A solution in an organic solvent of a natural or synthetic resin, a cellulose ester, such as cellulose nitrate or cellulose acetate, or a cellulose ether, such as methyl or benzyl cellulose, together with modifying agents, such as plasticizers, resins, waxes, and pigments. The solvent evaporates after application of the lacquer, leaving the dissolved material as a shiny, more or less continuous protective film on the surface of the material so treated. Lacquers are used for coating paper to give them functional qualities such as decreased water-vapor transmission rates, heat-sealing properties, grease resistance, gloss, and decorative effects.

LACQUERED PAPER—A general term for any paper which has been coated on one or both sides with a lacquer or plastic.

LAER—Lowest Achievable Emission Rate, which is achieved by applying all available treatment technology to the emission.

LAGOON—(1) A shallow body of water, as a pond or lake, which usually has a shallow, restricted inlet from the sea. (2) A pond containing raw or partially treated wastewater in which aerobic or anaerobic stabilization occurs.

LAID—The ribbed appearance in writing and printing papers produced by the use of a dandy roll on which the wires are laid side by side instead of being woven transversely.

LAID ANTIQUE—(1) Any paper watermarked with a laid dandy roll. (2) A book or writing paper having an overall laid watermark and an antique finish.

LAID DANDY ROLL—A dandy roll (q.v.) made with wires parallel to the axis of the roll and attached to the frame and kept in position by chain wires evenly spaced and encircling the circumference of the roll. This creates a ladder-like appearance in the sheets of paper. The roll is placed on top of the forming sheet at a point where its consistency is 2 to 5%.

LAID LINES—The closely spaced light lines in laid papers, produced by the laid wires of the mold or dandy roll. Laid lines usually run across the grain of the paper, but spiral laid paper has lines parallel with the grain or in the machine direction.

LAID MOLD—A hand mold in which the cover or sieve is composed of wires laid parallel to each other, in contradistinction to a woven mold formed of wire cloth. It is used to make laid paper.

LAID PAPER—Paper watermarked with a laid dandy roll.

LAID WIRES—The closely spaced wires of a laid dandy roll.

LAID WRITING—A correspondence paper usually made from chemical wood and/or cotton pulps and characterized by an overall laid watermark. It is usually made in a basis weight range from 16 to 24 pounds (17 x 22 inches – 500).

LAKE—See COLOR LAKE.

LAKE PIGMENTS—See COLOR LAKE.

LAMBDA TUNING—A method of controller tuning which allows the coordinated tuning of all the loops in a process plant, so that the speed of response of each loop matches the needs for the uniform manufacture of quality product. The technique allows the user to select a desired speed of response for the given loop. It requires performing a series of process bump tests to identify the nature and speed of the process response, together with any actuator deficiencies. This information allows the determination of both the control algorithm, and the controller settings needed to achieve a loop response at the desired speed in a smooth nonoscillatory manner. The ability to set the speed of response and to achieve such smooth control makes this method suitable for the uniform manufacture of continuous product in a process plant, such as a pulp and paper mill.

LAMINATED—See LAMINATION.

LAMINATED BOARD—Paperboard laminated either by (a) combining two or more plies of board; (b) combining to it on either one or two sides, a paper, plastic film, or other sheet material with specific properties. The adhesive used may be either a water solution of glue, casein, or starch, or a thermoplastic wax or resin composition. The lining may be of such grades of paper as book or hanging, for the general purpose of improving the appearance and the printing surface of the board, or a special barrier material such as greaseproof or glassine, for the purpose of imparting some specific property which could not be built into the board itself.

LAMINATED FELT (COMPOSITE FELT)— In the lamination process, two base fabrics are woven. The base fabrics have batt needled into them, or they are attached to each other by the needling process, usually by applying batt to one or both outside surfaces. A laminated felt is usually a press felt (q.v.).

LAMINATED GLASSINE—Laminated glassine is made of two or more sheets of glassine bonded together with an adhesive. When a wax base adhesive is used, the resulting paper is resistant to the passage of water vapor.

LAMINATED PAPER—A laminated product made of paper only. See LAMINATED; LAMINATING.

LAMINATING—(1) The operation of combining two or more layers of paper or paperboard with an adhesive in such a way as to form a multi-ply paper product, the purpose generally being to increase thickness and rigidity or to impart special properties, for example, moisture- and grease-resistance. An example is the lamination of glassine or greaseproof to paperboard with a thermoplastic wax or resin combination for the baking industry. (2) The operation of combining similar or dissimilar webs for the purpose of obtaining added strength and improved resistance to moisture vapor and grease on the folds and creases in packaging papers; also, improved functional properties such as toughness, pliability on automatic packaging equipment and the like. See LAMINATED GLASSINE; LAMINATION.

LAMINATION—A combination of two or more layers, usually but not necessarily, with the use of an adhesive to provide a material with enhanced or synergistic properties not possessed by any of the single components alone.

LAMPBLACK PAPER—See CARBON-BLACK BAG PAPER.

LAMPSHADE BRISTOL—A well-formed, clean bristol with high oil absorbency, decorated by the converter and used for lampshades. It is usually made with particular attention to translucency and cleanliness after oiling.

LAMPSHADE PAPER—A paper used to cover the frame in lampshades in place of textiles. It is usually made of chemical wood pulps, a heavy vegetable parchment, or laminated glassine. Ability to absorb flame-resistant or retardant materials and materials to give translucency are significant properties.

LAN—The local area network system that links distributed computer terminals and centralized server devices.

LANDFILL—A technology or process which disposes of solid waste through burial. It employs special precautions if done in a state-of-the-art manner. These precautions include ground water protective barriers, leachate collection, daily cover, final cover, and a long-term site utilization plan.

LAPLACE TRANSFORM—A mathematical tool which allows solution of process control dynamic analysis complex equations by means of algebraic equations, at the same time providing direct relationships of inputs and outputs of a process.

LAPPED—Descriptive of a method of packing reams of paper in a bundle by overlapping even portions and then folding over the two ends in a soft fold. This method of packing is now rare.

LAPPING—The operation of extracting water from screened pulp by a wet press and collecting the fibers in sheets dry enough to enable them to be folded or lapped into a stack or bundle.

LAPS—See WET LAP.

LARD PAPER—A greaseproof paper or vegetable parchment used to line lard cartons, lard pails, and lard tubs to prevent the grease from penetrating the board and to protect the lard from absorbing odors from the container. The thickness of the sheet depends upon the type of container used. High purity and lack of odor and taste are important qualities.

LASER—The acronym for light amplification by stimulated emission of radiation, a coherent, narrow-band-width light used in many aspects of graphic arts.

LASER PRINTER—A computer-driven printing device that uses a laser beam to image the digitized text or picture, which is then electrostatically transferred to the paper. The precision of the laser beam produces crisp, sharp printing.

LATENCY—In mechanical pulping, latency is introduced into mechanical pulp during high consistency (over 25%) refining by the twisting action of the rotating disc. The so-created kinks and curls are frozen into the fibers when they are discharged from the high temperature refiner casing into the atmosphere. Latency needs to be removed before screening by exposing the pulp to low consistency at high temperatures (70°C to 80°C) where the fibers will relax and straighten out.

LATERAL POROSITY—Porosity (q.v.) measured in a direction parallel with the plane of the sheet. Sometimes called transverse porosity.

LATEWOOD—The part of an annual ring produced during the latter part of the growing season (in the summer in temperate zones); the outer portion of the annual ring. It is usually denser than earlywood because the cell walls are thicker.

LATEX—A colloidal water dispersion of high polymers from sources related to natural rubber or of synthetic high molecular weight polymers which resemble natural rubber. The term was originally applied to the milky sap obtained from the Hevea tree. Latex is used in paper as an adhesive, in pigment coating, as a barrier coating and as a saturant for specialty papers.

LATEX BOARD—See LATEX-TREATED PAPERS.

LATEX-TREATED PAPERS—Papers manufactured by two major processes. In one, rubber latex is incorporated with the fibers in the beater prior to formation of the sheet. In the other, a preformed web of absorbent fiber is saturated with properly compounded latex. Papers made by the beater process can be produced on any of the regular types of papermaking machinery including cylinder, fourdrinier, and wet machines. Latex-impregnated papers made by the saturating process are manufactured on specially designed equipment consisting essentially of a suitable bath for impregnation, a pair of squeeze rolls for removal of excess latex, and drying equipment. Fibers commonly used consist of rags and chemical woodpulp. Latex-impregnated products range from approximately 0.004 to 0.250 of an inch in thickness. They vary widely in their physical and chemical characteristics. As a class, they are characterized by toughness, folding endurance, flexibility, durability, and resistance to splitting and abrasion in varying degrees as may be necessary to meet end-use requirements. The heavier products find use in the shoe industry as innersoles and midsoles. The mechanical and process industries use them as gaskets. The thinner products are used variously; by the artificial leather industry in coated form as simulated leather; by the pressure-sensitive tape industry as a base for masking, holding, and protective tapes; by the automobile manufacturing industry as antisqueak and facing materials.

LATH—See GYPSUM LATH.

LAUNDER RING—A well around the periphery at the top of an upflow tower. Pulp rising in the tower is moved into the launder ring by rotating rakes. Dilution water is normally added to the launder ring forcing the pulp to fall into a washer vat.

LAUNDRY PAPER—(1) A highly sized, well-formed, chemical woodpulp sheet or a grease-proof paper, which is moisture resistant, used for separating wet wash laundry and preventing various colors of laundry from contacting one another. It is made in basis weights of 30 to 35 pounds (24 x 36 inches – 500) and sold in standard size rolls and sheets. The paper generally has a high wet tensile strength and fast colors. (2) A kraft wrapping paper of the usual type, for outside wrapping of finished laundry. It is usually specified in basis weights of 30 and 35 pounds (24 x 36 inches – 500). It is sometimes colored, printed, or decorated.

LAWBOOK COVER—A descriptive term for cover paper used for lawbooks and pamphlets.

LAWN FINISH—A finish produced with very fine-weave linen cloth on a plater press. It is distinguished from linen finish in that the paper is conditioned to increase the moisture content before plating. The plater book is made up of a zinc plate, a sheet of linen, a sheet of paper, a sheet of linen, and a zinc plate and the number of books made up to the capacity of the plater. It is plated under heavy pressure, giving a smooth but distinct linen surface to the paper; the effect is especially apparent when looking through the paper. It is used particularly with papeteries.

LAYBOY—The mechanical piler on a sheeter or cutter where the cut sheets are overlapped, counted, jogged, and piled onto a skid or platform.

LAYER BOARD—A paperboard used to separate two or more layers in packaging of candies, crackers, etc., and to form nestings therefore. It may be made of any grade of board 0.012 of an inch or more in thickness, which is stiff enough for the purpose. It may be plain or surface-coated with a grease-resistant material. See also CHOCOLATE DIVIDERS AND LAYER BOARD.

LAYER INSULATION PAPER—An unbleached kraft paper with a high dielectric strength which is used between layers of wire in transformers. Depending upon the usage it may be high-density, water-finished paper made from well-hydrated stock or it may be low-density paper from relatively freestock. It must be flexible without cracking and free from metallic or other conducting particles. It must be able to withstand long exposure to elevated temperatures and maybe chemically treated to improve this property. Since power transformers are usually oil-filled, this paper is completely saturated with and immersed in oil. Usual thicknesses are from 0.005 to 0.030 inch.

LAYOUT—(1) The process of arranging text and pictures prior to final design. (2) The arrangement of pictures, text, tables, titles, etc., on a page or other printed work.

LEACHATE—Liquid that has percolated through solid waste or other medias and has extracted dissolved or suspended materials from it.

LEACHING—The process by which soluble materials in the soil, such as nutrients, pesticide chemicals or contaminants, are washed into a lower layer of soil, or into ground waters.

LEAD-PENCIL PAPER—(1) Tablet or writing paper not well sized, suitable for writing on with pencil. (2) A strong paper prepared from chemical woodpulp used for spirally wrapping the "lead" in the manufacture of marking pencils or crayons. The basis weight is in the range of 70 to 80 pounds (24 x 36 inches – 500). It is generally made to a bulk specification.

LEAF FIBER—A fiber from a leaf or leaf stalk, such as New Zealand flax, sisal, pineapple, and manila (abaca).

LEAKAGE AIR—Air that flows into a boiler furnace through various openings because of the lower pressure in the furnace compared to the surrounding atmospheric pressure.

LEATHER OR LEATHERETTE PAPER—An imitation leather made of paper colored like leather, either in the body or on the surface, and then embossed with a grain simulating leather. It is used to cover boxes, as notebook covers, or wherever an imitation leather is desired. See ARTIFICIAL LEATHER PAPER; LATEX-TREATED PAPERS.

LEDGER PAPER—Originally, a smooth, well-sized cotton content writing paper characterized by good tearing resistance, high folding endurance, ruling quality, permanence, durability, etc., and used for manual (i.e., pen and ink) entry account books, ledgers, record books, diaries, and the like. The term is now applicable to a broad variety of chemical woodpulp record-type papers used in mechanical-entry accounting machines, looseleaf and other notebooks, etc. Ledger papers are commonly made in white, buff, and green-tint shades in a basis weight range of 24 to 36 pounds (17 x 22 inches – 500). See also BOOKKEEPING MACHINE PAPER; POSTING LEDGER; STATEMENT LEDGER.

LENS TISSUE—A tissue paper used for wrapping and polishing photographic, optical, and other lenses and for other cleaning purposes requiring high-grade tissue. It is made of long-fibered stock which is free of unbleached and mechanical pulps. The paper has a high degree of softness and is free of abrasiveness, lint, or dusting. It is uncalendered and generally made in weights of 5.5, 8.5, and 16 pounds (24 x 35 inches – 500). Certain papers contain a silicone and others are given a wet-strength treatment for heavy-duty wet cleaning applications.

LETTER-COPYING PAPER—Typical writing paper of manifold grade suitable for use in obtaining several copies of the same letter.

LETTER PAPER—Writing paper cut to proper size for correspondence purposes.

LETTERPRESS PRINTING—A process also known as relief or typographic printing. These are interchangeable terms applied to any printing produced from a raised, or relief surface, as distinct from planographic or intaglio printing. Letterpress printing employs type or plates or any character cast or engraved in relief on metal, wood, rubber, linoleum, etc. The ink is applied to the printing surface below which all nonprinting areas or spaces are recessed (the exact opposite of gravure or intaglio printing). Impressions are made by pressure against a flat area of type or plate (as on a platen press), by pressure of a cylinder rolling across a flat area of type (flat-bed cylinder press printing), or by having a cylindrical plate against which another impression cylinder revolves, carrying a continuous paper web. Rotary printing of flat sheets is also common, this being accomplished by having the printing areas electrotyped, i.e., duplicated in a copper or other metal plate, which is then curved to fit the plate cylinder.

LEVELING CHEST—See SURGE CHEST.

LICKERIN—In nonwovens, a roll clothed with wire, immediately following the feed roll, that opens and delivers fibers to the main cylinder.

LIFT—A quantity of sheets of paper or paperboard which can be readily lifted from one operation to another.

LIFTING—See PICKING.

LIGHT FASTNESS—See FASTNESS.

LIGHTWEIGHT—A term applied to papers made in weights below the normal minimum basis weight of the grade in question.

LIGHTWEIGHT CATALOG AND DIRECTORY PAPER—See CATALOG PAPER.

LIGHTWEIGHT CHIP—A paperboard used principally by the corrugated container manufacturers as a facing for single-faced corrugated rolls, for pads and partitions, and for other types of interior packing. It is made of chipboard, generally 0.006 to 0.012 of an inch in thickness, and has a surface adapted to the adhesive used in corrugated board manufacture.

LIGHT WEIGHT COATED PAPER—Used for publications, light weight coated paper is sometimes referred to as publication grade paper. The paper, normally, is in the 30 to 40 pound basis weight range (25 x 38 inches – 500). Typically, it contains mechanical pulp, and is produced without size press pretreatment, usually with blade coaters. The paper is printed in web form on either rotogravure or web offset presses. The sheets are usually classified as No. 4 and 5 glossies in paper grade reference.

LIGNIN—The noncarbohydrate portion of the cell wall of plant material; it is usually determined as the residue after hydrolysis with strong acid of the plant material, after removal of waxes, tannins, and other extractives. Lignin is amorphous, has high molecular weight, and is predominantly aromatic in structure. The monomeric units are P-hydrocinnamyl alcohols. It is not one compound, but varies in composition with the method of isolation and with the species, age, growing conditions, etc., of the plant. It is more or less completely removed during chemical pulping, but is not removed by mechanical pulping. Bleaching of the pulp further removes or modifies any remaining lignin.

LIGNINASE—See ENZYME.

LIGNIN MODEL COMPOUNDS—Synthesized compounds which are assumed to be reasonable representatives or models of lignin derivative units. Common model compounds are produced from oxidative coupling of coniferyl alcohol or p-coumaryl alcohol. Model compounds are used to study the kinetics of bleaching reactions and to characterize bleaching reactions in order to improve the selectivity, efficiency, and environmental impact of bleaching processes.

LIGNIN SULFONATE—See LIGNOSUL-
FONATE.

LIGNIN (UV)—See KLASON LIGNIN.

LIGNOSULFONATE—A water-soluble product
resulting from the reaction of lignin and sulfite
pulping reagents.

LIME—Calcium oxide (CaO) derived by calcining the calcium carbonate in lime mud (q.v.) or
limestone by driving off carbon dioxide at high
temperature. It is the reactive material added
to green liquor (q.v.) in order to convert sodium
carbonate into sodium hydroxide (caustic) and
to reclaim the calcium as lime mud or calcium
carbonate.

LIME AVAILABILITY—A measure of the fraction of calcium oxide in reburned or fresh lime.

LIME KILN—The unit that converts lime mud
($CaCO_3$) into lime (CaO) by calcination. Most
often a direct-fired, counterflow rotary kiln.

LIME KILN CHAIN—Large gauge chain lengths
hung inside the kiln near the feed end of rotary
lime kilns which act as a regenerative heat exchange medium for transfer of heat from the
hot kiln flue gas to the wet lime mud.

LIME KILN REFRACTORY—The interior lining of lime kilns usually consisting of bricks of
alumina-silica material, which are resistant to
attack by lime and soda compounds and have
good strength and structural integrity at high
temperature. May also be a monolithic structure instead of pre-formed bricks. Also, may
be multi-layered to provide both chemical attack resistance and insulating value.

LIME KILN RINGS—Deposits which form on
the interior surface of rotary lime kilns usually
in a continuous circumferential band. They may
form at any location by a variety of mechanisms, and usually consist primarily of calcium
carbonate somewhat enriched in different locations by calcium sulfate, calcium oxide, or
sodium compounds.

LIME MUD—Slurry or cake of fine particles of,
primarily, calcium carbonate derived from the
reaction of reburned or fresh lime (q.v.) with
green liquor (q.v.).

LINCRUSTA PAPER—A paper used as a substitute for lincrusta (a canvas fabric used for
hangings, ceilings, etc.). It may be made of a
furnish containing mechanical woodpulp or
reclaimed paper stock, but it is generally heavier
and stronger than hanging paper.

LINEAR LAID—A thin writing paper with
watermarked lines to serve as a guide in writing.

LINED BOARD—(1) Mill-lined or laminated:
any board that is lined with paper such as newsprint, book paper, and cover paper, after the
board is made on the board machine and while
it is still in the roll, before being cut into sheets.
(2) Vat-lined: any board made on a cylinder
machine, where the top or bottom liner or both
are of different quality stock from the filler or
center of the board. This term refers to a one-
process sheet, the liner or liners being a part of
the board as it is being formed on the wet end
of the machine. (3) Sheet-lined: board that is
pasted, sheet by sheet, on a sheet-lining machine. Any board can be sheet lined with any
quality of lining paper.

LINEN—Linen fibers are the bast fibers of the flax
plant. In the paper industry, it usually refers to
the linen rags and cuttings received from the
textile industry, for use in the manufacture of
high quality cotton fiber content paper (q.v.)
which term embraces both cotton and linen.
Chemical pulp derived from the flax plant is
called flax pulp.

LINEN-FACED PAPER—Originally, a wrapping
paper or other paper lined or faced, on one or
both sides, with linen, but now a paper with a
linen finish. The original meaning is usually
covered by the terms cloth-lined, faced, or
mounted.

LINEN FINISH—A finish produced by compressing sheets of paper between alternate

sheets of linen cloth so that the pattern of the cloth is impressed upon the surface of the paper or by pressing the continuous web of paper between two endless belts of linen cloth by means of press rolls. A similar effect is obtained by embossing a continuous web of paper with a steel roll which has been knurled or engraved to simulate the surface of linen cloth.

LINEN PAPER—(1) Unless otherwise indicated, paper which has a linen finish. (2) A paper made wholly or in part from linen rags. See also LINEN-FACED PAPER.

LINENS—See LINEN PAPER.

LINER—(1) A creased fiberboard sheet inserted as a sleeve in a container, covering all side walls, to provide extra strength. (2) A heat-sealed plastic sack inserted inside a paper sack to provide barrier protection. (3) Liners of paper in bags of burlap or cotton to prevent leakage. See BAG LINERS; CONTAINER LINER.

LINERBOARD—A paperboard made on a fourdrinier or cylinder machine and used as the facing material in the production of corrugated and solid fiber shipping containers. Linerboard is usually classified according to furnish and method of web formation, as for example fourdrinier kraft linerboard (q.v.), cylinder kraft linerboard, jute linerboard (q.v.).

LINERS—(1) The outside layers (i.e., the vat-lined or sheet-lined surface) of a built up or combination board (q.v.) or combination paperboard (q.v.). (2) Incorrectly used for the outer facings of corrugated board (q.v.) or solid fiberboard. See LINERBOARD. (3) A creased fiberboard sheet inserted as a sleeve in a container, covering all side walls, to provide extra strength. See CONTAINER LINER. (4) Various grades of light- and medium-weight sheets, generally waxed, used as inner or protection wrappers in packaging food such as crackers, cookies, and powdered milk. (5) A heat-sealed plastic sack inserted inside a paper sack.

LINING PAPER—Any paper used as a covering, such as box lining, boxboard lining, and trunk lining. Generally a nonfading, opaque machine-finished paper is required, sized for adhesives, and of sufficient strength to go through the laminating process and subsequent converting operations. Coated-one-side lining papers are also used in a wide range of qualities, colors, and finishes, usually on a mechanical woodpulp base paper and with a range of high finishes from a low calender to a friction-glazed. See BOX COVER PAPER; FACING PAPER.

LINOCUT—Linoleum cut printing, where the unprint area is cut away with a knife or burin. A type of relief printing.

LINOLEUM LINER FELT—See LINOLEUM LINING.

LINOLEUM LINING—An absorbent felt paper, usually weighing three-fourths of a pound or one pound per square yard. Sometimes the felt paper has been treated or impregnated with asphalt. This material is used as an interlining between linoleum and the floor upon which it is to be laid. The lining felt is usually pasted or cemented to the floor, and the linoleum, in turn, to the lining felt. The principal function of this product is to prevent cracking, breaking, or loosening of the linoleum because of expansion or contraction of the floor boards. The lining felt should have a proper degree of absorbency in order to allow setting of the adhesive and should also possess sufficient strength, stretch, and resistance to delamination to fulfill its purpose. See also DEADENING FELT.

LINOLEUM UNDERFELT BOARD—See DEADENING FELT.

LINOLEUM UNDERFELT PAPER—See DEADENING FELT.

LINT—(1) Particles of fibers that separate or "dust off" from paper during manufacturing or converting operations. See DUSTING. (2) The ginned cotton textile fiber is technically known as lint.

LINTERS—See COTTON LINTERS.

183

LINTING—Fibers with low specific surface located on the surface of the paper and adhering to the ink film on the blanket of an offset printing press. Lint-causing materials are short, stiff, and with relatively smooth surfaces that have a low bonding potential.

LINTLESS BLOTTING—See HALFTONE BLOTTING PAPER.

LIPASES—Enzyme that catalyze the hydrolysis of triglycerides or other esters of glycerol, yielding glycerol and the acid residue. For example, the triglyceride triolein is hydrolyzed to one mole of glycerol and three moles of oleic acid.

LISTING CARD BRISTOL—See INDEX BRISTOLS.

LITHO BLANKS—Coated blanks which are made of cardboard middles lined with book paper, which are clay coated on one or both sides, and are subsequently given a high finish by calendering or plating. Litho blanks are practically clay-coated, book-lined blanks, using number 1 or 2 cardboard middles as the coating raw stock. It includes clay-coated litho boards, coated litho boards, coated blanks, coated litho blanks, and clay-coated litho blanks.

LITHO BOARD—See LITHO BLANKS.

LITHO-COATED PAPER—A paper coated on one side with a coating made to withstand the water used in the lithograph process. It is made in a wide-range of basis weights. Good pick strength is essential.

LITHOGRAPH BOOK—See LITHOGRAPH PAPER.

LITHOGRAPH BOX WRAPS—Coated one-side litho paper which has been lithographed and cut to size of paper boxes. It is applied to boxes in place of box-cover paper.

LITHOGRAPHERS' PLATE WIPER—A converted product made from embossed, creped cellulose wadding and used for wiping and cleaning offset lithograph plates. It should be free from abrasive particles, have high alcohol absorbency rate, and good strength when saturated with alcohol.

LITHOGRAPHIC MASKING PAPER—A paper coated on one or two sides with a water-resistant surface and having an orange red color, used to mask plate areas on a lithographic printing plate.

LITHOGRAPHIC PROCESS—See LITHOGRAPHY.

LITHOGRAPH LABEL PAPER—See LABEL PAPER.

LITHOGRAPH PAPER—A paper for use in lithographic printing made of bleached chemical woodpulp alone or in combination with mechanical woodpulp or deinked paper stock. Essential characteristics are: surface cleanliness, a degree of water resistance sufficient to inhibit penetration into the paper of water encountered in the printing process, relative freedom from curl and high pick strength. The paper is made both uncoated and coated in either sheets or rolls and with machine, supercalendered, or duplex finishes. Usual basis weights are: uncoated 45 to 70 pounds, coated 50 to 100 pounds (25 x 38 inches – 500).

LITHOGRAPHY—The original meaning of lithography was based on the affinity of a greasy surface for printer's ink, the ink being also repelled by a damp surface. It involved the preparation of designs on a stone with a special crayon or liquid drawing medium and the production of printed impressions therefrom on a flat-bed press. Today it is divided into two classes, direct lithography and offset lithography (q.v.).

LITHOGRAVURE—See DEEP-ETCH OFFSET.

LITHO PAPER—See LITHOGRAPH PAPER.

LITHOPONE—A mixture of 28–30% zinc sulfide with 72–70% of barium sulfate, resulting from the cross precipitation of barium sulfide and zinc sulfate solutions. The pigment is used as a filler in paper. See ZINC SULFIDE.

LITMUS PAPER—See TEST PAPERS.

LOAD—The total force applied to a given specimen in testing for such properties as compression resistance and tensile strength of paper or paperboard. Load differs from stress; the latter is force per unit of load area. Load, rather than stress, is usually used because of difficulty in determining the loaded (cross-sectional) area of paper.

LOAD CELL—An electronic sensor that measures force. On a winder, load cells under the ends of an undriven roller are often used to measure web tension.

LOADING—(1) The incorporation of finely divided relatively insoluble materials, such as clay and calcium carbonate, in the papermaking composition, usually prior to sheet formation, to modify certain characteristics of the finished sheet, including opacity, texture, printability, finish, weight, etc. (2) Mineral matter, such as clay, used as a filler in paper.

LOAN PAPER—An obsolete term for high grade cotton content writing papers used for certain legal documents.

LOCALIZED WATERMARK—A watermark arranged to appear at definite intervals in a sheet of paper. See also CUT TO REGISTER.

LOCKER PAPER—Flexible sheet materials used for the wrapping and protection of frozen foods and meats during freezing and storage. The wide assortment of grades includes waxed papers and wax laminated combinations. All grade must be resistant in some degree to moisture vapor, grease, and moisture. Pliability at low temperatures, nontoxicity, and lack of odor are important characteristics. High bursting strength and cleanliness are desirable features. See also FROZEN FOODS PAPER.

LOCUST BEAN GUM—A polysaccharide derived from the seed endosperm of the locust bean or carob tree grown in the Mediterranean regions of Africa and Europe; used as a beater or wet-end additive primarily for improving strength properties.

LOFT DRYING—A form of air drying. The wet paper is hung over poles in a drying loft, where the atmospheric conditions are regulated to give the desired rate of drying. Now almost obsolete as a commercial practice, but may still be in use for speciality or handmade papers.

LOG FOLD—The logarithm to the base 10 of the folding endurance (q.v.).

LOGGING—(1) Process of cutting down trees and removing the raw unprocessed products such as tree lengths, saw timber, pulpwood, logs, bolts, and whole tree chips from the forest. (2) Collection of process information and placement of that data in archive storage to allow future retrieval, trending, and comparison.

LOG (PAPER)—A large roll of paper wound at the end of a paper machine on a reel.

LOIN PAPER—A paper used by packing and locker service concerns to prevent loss of moisture from meats. It is made from bleached or semibleached chemical woodpulp in basis weights of 35 to 60 pounds (24 x 36 inches – 500) and treated with 15 to 20% of its weight with petrolatum or white oil. Significant properties include strength, resistance to penetration of oil and water, uniform greaseless surface, and a moderate resistance to transfer of water vapor through the paper.

LONG DIRECTION—See MACHINE DIRECTION.

LONGEVITY—See DURABILITY; PERMANENCE.

LONG FOLD—A term denoting that, if folded lengthwise, the sheet will be folded with the grain. In bristol boards, the term indicates that the grain runs lengthwise.

LONG NIP PRESS—See PRESS SECTION.

LONG STOCK—Stock which, after refining, has a relatively long fiber—i.e., one in which the fibers have not been greatly reduced in length by the refining treatment. Obsolete as a commercial practice.

LOOK-THROUGH—The appearance of paper when viewed by transmitted light, thus disclosing the texture or formation (q.v.) of the sheet.

LOOP DRYING—See FESTOON DRYING.

LOOSE COATING—Coating not firmly bound to the base stock, tending to pick during printing.

LOOSE-LEAF FILLER—Any writing paper used as filler for loose-leaf notebooks. See also FILLER PAPER; LEDGER PAPER; NOTEBOOK PAPER.

LOOSE-LEAF LEDGER PAPER—See LEDGER PAPER.

LOT—A defined quantity of product accumulated under conditions that are considered uniform for sampling purposes.

LOTTERY PAPER—A general term for papers which are used for lottery tickets. Usually, such papers are treated to prevent fraudulent alteration or duplication.

LOUDSPEAKER (RADIO) CONE PAPER—A paper, characterized by its bulk and low finish, which is made of chemical woodpulp or cotton fibers (which may be blended with other types of fibers) on a fourdrinier machine. The stock has a relatively high freeness. The paper may be slack sized to improve its handling properties during the conversion operation. A 15-point all-cotton cone paper has a basis weight of 45 pounds (19 x 24 inches – 500); a 17-point kraft paper has a basis weight of 90 pounds (24 x 36 inches – 480). The paper is used in the manufacture of conoidal diaphragms for radio loudspeakers.

LOW-ANGLE GLOSS—The gloss of paper measured at 75 degrees from the vertical. This angle is chosen as the most nearly duplicating the angle at which an observer judges the gloss using the naked eye. See GLOSS.

LOW CONSISTENCY—Pulp stock in which fiber and insoluble material are 2–5% of the weight of the solution.

LOW COUNT—A number of sheets less than the standard number required in a bundle of paperboard or a ream of paper.

LOW DENSITY—See LOW CONSISTENCY.

LOWERING TABLE—A device that is designed to lower a roll from one elevation to a lower elevation in a linear (normally vertical) path.

LOW FINISH—See ANTIQUE FINISH.

LUMEN—A void region in plant cells enclosed by the cell wall.

LUMINOSITY—See LUMINOUS REFLECTIVITY.

LUMINOUS PAPER—See FLUORESCENT PAPER; PHOSPHORESCENT PAPER.

LUMINOUS REFLECTIVITY—One of the three qualities necessary for specifying a color. It is the ratio of the luminosity of the illuminated specimen to that of a standard reflector when both specimen and standard reflector are illuminated by a specified illuminant under identical conditions and as viewed by a standard observer. Luminous reflectivity can be calculated by a standard procedure, when the spectral reflectivity of the specimen is known. Also, it can be measured directly with a suitable colorimeter.

LUMP—A localized thickened area in paper caused by an agglomeration of fiber or other materials.

LUMPBREAKER ROLL—A large diameter, light weight roll running under low pressure, located above the suction couch roll (q.v.) and used to smooth out the paper web before it leaves the forming fabric. These rolls are covered with a very soft (180–250 P&J) elastomeric covering which may be as thick as 50 mm (2 inches).

LUNCH ROLL PAPER—A wrapping paper, usually a 16-pound basis (24 x 36 inches – 480) waxed to 20 to 24 pounds, cut in 12-inch rolls and sold for household purposes.

LWC—See LIGHT WEIGHT COATED PAPER.

M

M—The Roman symbol for 1000. A thousand sheets.

MACARONI PAPER—An MF or MG paper made from unbleached kraft or sulfite pulp, in basis weights of 25 and 30 pounds (24 x 36 inches – 500). It is made in rolls and sheets in a dark blue color for packaging macaroni. The sheet should have good wrapping qualities.

MACARONI WRAPPING PAPER—See MACARONI PAPER.

MACHINE BROKE—Wet or dry paper resulting from breaks in the paper web during manufacture, or defective paper discarded at the dry end of the machine.

MACHINE CALENDER—A set or "stack" of vertically oriented and possibly steam-heated, chilled-iron rolls utilized to improve the smoothness and/or gloss of the paper surface. A paper machine can incorporate a finishing stack at the dry end of the paper machine as well as an additional "breaker" stack positioned before an on-machine coater. Paper machines may also incorporate soft-nip calenders in lieu of chilled-iron machine calenders. See BREAKER STACK; FINISHING STACK.

MACHINE CHEST—Also called paper machine chest. The last chest before the fan pump in a paper machine approach flow system (q.v.).

MACHINE CLOTHING—The forming fabrics, press felts and dryer fabrics of a paper machine. The term may also include deckle straps (q.v.), aprons (q.v.), jackets (q.v.) and other traveling parts. See also FABRICS AND FELTS.

MACHINE COATED—Paper or board which during the process of manufacture has been coated with a mixture of a mineral pigment and a suitable adhesive by means of a device which is a part of the paper machine. See COATING.

MACHINE-COATED BOARD—See MACHINE COATED.

MACHINE-COATED PAPER—See MACHINE COATED.

MACHINE COATING—The process of applying coating (q.v.) to paper or paperboard with equipment which is an integral part of the paper machine. It is sometimes called on-machine coating and is distinguished from off-machine coating or conversion coating.

MACHINE CREPED—See CREPED.

MACHINE DIRECTION—The direction of paper parallel with the direction of movement on the paper machine (MD). It is also called grain direction, and for roll paper this is the direction the roll unwinds. For sheet paper, the machine direction is the stiffer direction. The direction at right angles to the machine direction is called the cross-machine direction or simply cross direction (CD).

MACHINE DRIED—Dried on the paper machine.

MACHINE FINISH—(1) Any finish obtained on a paper machine. It may be that of the sheet as it leaves the last dryer or as it leaves the calender stack. It may also be a dry or water finish. (2) When used in conjunction with the name of a grade or type of paper, a machine finish of less than the maximum range of smoothness.

MACHINE FINISH BOOK PAPER (MF BOOK)—A printing paper having a medium finish, good opacity, printability, and suitability for book printing. It is made from chemical or mechanical woodpulp, normally in basis weights of 25 to 100 pounds (25 x 38 inches – 500), and must usually conform to specific bulk standard, i.e., pages per inch.

MACHINE FINISH COVER BOOK—A heavy cover paper whose finish is developed on the paper machine calender stack and ranges from smooth to antique. It is also called MF cover, and its name distinguishes it from embossed cover.

MACHINE GLAZED—The finish produced on a Yankee machine, where the paper is pressed against a large steam-heated, highly polished revolving cylinder, which dries the sheet and imparts a highly glazed surface on the side next to the cylinder, leaving the other side rough—i.e., with the texture of the felt used on the machine. See YANKEE MACHINE.

MACHINE GLAZED LITHO—A printing paper made from chemical wood and/or reclaimed pulp on a Yankee machine which gives a high finish on one side and a relatively rough finish on the other. It is commonly made in basis weights of 50 to 60 pounds (25 x 38 inches – 500). Is also referred to as MG litho.

MACHINE-GLAZED POSTER—See MACHINE-GLAZED LITHO.

MACHINE IMPRINTED—Having a design or mark impressed by means of a metal or rubber plate at a point where the paper sheet contains sufficient moisture to be plastic. See WATERMARK.

MACHINE LOADING—The application of loading material to the surface of paper on the paper machine at the size press. It is differentiated from machine coating by the physical characteristic of the material applied and the equipment used. See also WIRE LOADING.

MACHINE-MADE DECKLE-EDGE PAPER—A paper used as a cover or text stock, for announcements, greeting cards, house organs, and advertising pieces, etc. It is similar in texture, surface properties, and deckle edges to handmade paper but manufactured instead on paper machines. The paper is made from cotton fiber and chemical woodpulp. Basis weights range from 65 to 130 (occasionally 180) pounds (25 x 38 inches – 500).

MACHINE MARK STRIPES—A series of stripes on a paper web running in the machine direction resulting from the application of devices to the surface of the sheet while being formed on the forming fabric. These stripes are used for identification purposes, particularly in the enforcement of tariff regulations of some countries.

MACHINE POSTING INDEX—See INDEX BRISTOLS.

MACHINE POSTING LEDGER—See BOOKKEEPING MACHINE PAPER; STATEMENT LEDGER.

MACHINE TREATED BUTCHERS PAPER—A wrapping grade used in the retail trade for packaging meat products. It is manufactured of bleached or semibleached chemical pulps or a combination of these pulps, with a basis weight of approximately 40 pounds (24 x 36 inches – 500). The paper is treated to impart wet strength, dry strength, blood-penetration resistance, and freedom from sticking to the packaged product.

MACHINE WIRE—See FORMING FABRIC.

MACINTOSH—Icon- and mouse-based computer program by Apple Corp. Especially useful for graphics.

MACRO MIXING—Mixing of chemical into pulp on a large scale, greater than the distance of diffusion of chemicals in the time of a typical bleaching stage, say over 2 mm. See also MICRO MIXING.

MAGAZINE-COVER PAPER—See COVER PAPER.

MAGAZINE GRINDER—A machine for producing mechanical pulp, groundwood. It consists of a rotating pulpstone against which debarked logs are pressed and reduced to pulp. The four-feet long debarked logs are fed into two magazines located on both sides of the stone and pressed against the rotating pulpstone by hydraulic cylinders.

MAGAZINE NEWS—See ROTOGRAVURE PAPER.

MAGAZINE PAPER—Any of a wide range of coated and uncoated book-type papers used for magazine or periodical printing. See also COATED MAGAZINE PAPER.

MAGAZINE STOCK—Pulp produced by de-inking printed magazines and books and often used with other pulps for making paper.

MAGNEFITE PROCESS—A two-stage magnesium base sulfite pulping process. The acidity of the first stage is greater than that of the second stage.

MAGNESIUM BASE—See ACID SULFITE PROCESS.

MAGNESIUM BISULFITE—$Mg(HSO_3)_2$. The chemical base for the magnesium-base sulfite process. See ACID SULFITE PROCESS.

MAGNETIC FLOWMETER—An obstructionless flow tube surrounded by a magnetic field. It operates on the principle of an electrical generator where the conductor length is the diameter of the flow tube and the velocity of the fluid cutting the lines of magnetic field generates a voltage signal. Measurement errors can be caused by entrained air and saline solutions, but for the most part it can be used as a measurement device for paper stock.

MAKE AND HOLD ORDER—A given quantity of paper which is made on order and held by the manufacturer awaiting customer shipping instructions.

MAKEREADY TISSUE—A sheet of 9 to 15-pound (24 x 36 inches – 480) manila-colored tissue, well sized, which is used in the print shop for making a form ready for printing.

MAKEREADY WASTE—See PAPER WASTE.

MAKING BOARD—See FORMING BOARD.

MAKING ORDER—Any order that is not supplied from stock and is made to the purchaser's specifications.

MANHOLE—In paper drying, an opening in the head of a dryer that provides access to the inside of the cylinder.

MANIFOLD COPYING PAPER—See MANIFOLD PAPER.

MANIFOLD PAPER—A lightweight writing paper designed primarily for carbon copies of typewritten material. It is normally made from chemical wood and/or cotton pulp in basis weights ranging from 7 to 9 pounds (17 x 22 inches – 500) and in glazed and unglazed finishes. See also ONIONSKIN PAPER.

MANILA—A color and finish similar to that formerly obtained with paper made from manila hemp (rope) stock. Under present usage, the term has no significance as to fiber composition.

MANILA BRISTOL—See ROPE BRISTOL.

MANILA COLOR—A light straw or yellowish color.

MANILA DRAWING PAPER—Drawing paper having a manila color.

MANILA FIBER—See ABACA.

MANILA FOR OILING—See PACKERS OILED MANILA.

MANILA HEMP—See ABACA.

MANILA ROPE PAPER—See ROPE PAPER.

MANILA ROPE SHIPPING SACK PAPER— Cylinder-made paper containing 75% or more

189

of manila rope fibers and used in the construction of single (and sometimes double wall) paper shipping sacks.

MANILA TAG—See TAG BOARD; TAG STOCK.

MANILA WRAPPING—A term applied to a group of manila-colored wrapping papers, made of chemical, or a mixture of chemical and mechanical pulps. They are used for various wrapping and printing purposes and also for envelopes, filing folders, etc.

MANILA WRITING—See RAILROAD MANILA.

MANNASES (GLUCOMANNASE)—Enzymes capable of hydrolyzing Beta-1, 4 mannans or beta-1, 4-glucomannans.

MANNOGALACTAN—A carbohydrate found in the seeds of certain plants which, upon acid hydrolysis, yields monmeric sugars, mannose, and galactose. Mannogalactan (also called galactomannan) is used in papermaking as a beater or wet-end additive to improve sheet formation and fiber bonding. See GUAR GUM.

MANUFACTURER'S JOINT—The body joint or joints of corrugated shipping containers, made by the manufacturer. It may be lapped, butted, or spliced at the option of the customer.

MANUSCRIPT BINDER—See MANUSCRIPT COVER.

MANUSCRIPT COVER—A lightweight cover paper, used for protecting legal papers and manuscripts. It is made of varying proportions of chemical pulps, with or without the inclusion of cotton fibers. The basis weight is 40 pounds (18 x 31 inches – 500). It is also referred to as MSS cover.

MAP PAPER—(1) Any paper used for map printing. Map papers are generally made in basis weights ranging from 16 to 28 pounds (17 x 22 inches – 500) or 40 to 70 pounds (25 x 38 inches – 500) and may have special qualities such as wet strength, water repellency, mildew resis-

tance, abrasion resistance, and luminescence. (2) A cotton content bond paper designed for the printing of nautical charts, military maps, and similar products requiring high durability or permanence, or both. (3) An offset printing paper made from chemical woodpulp and designed especially for road maps, atlases, etc.

MARBLE PAPER—(1) A decorated end-leaf paper used in blank and printed books. (2) An intaglio-printed end-leaf paper comprising a marble-like pattern. (3) An old style wallpaper with the appearance of marble.

MARBLED PAPER—See MARBLE PAPER.

MARGARINE WRAPPER—A paper with exceptionally high grease resistance and wet-strength properties. See BUTTER PAPER.

MARK—The design or impression in paper produced by a watermark or by the use of embossing rollers or marking felts.

MARKET PULP—Wood, cotton or other pulp produced for, and sold on, the open market, as opposed to that which is produced for internal consumption by an integrated pulp and paper mill.

MARKING FELT—A felt which is so made that it imparts a distinctive pattern or mark to a sheet of paper.

MARKING ROLL—See WATERMARKING DANDY ROLL.

MASKING PAPER—Any paper used to block out the surface of a design so that, after using an air brush, the portion of the design covered by the paper is not printed or enameled. It is used principally in the finishing of automobiles.

MASKING TAPE—An impregnated creped unbleached or semibleached kraft sheet to one side of which is applies a pressure-sensitive coating. The adhesive consists essentially of a mixture of rubber solution and resins, and is of a type which does not harden during storage after the treated sheet has been slit and rewound into narrow tape rolls. Masking tape is used mainly for application to surfaces such as automobile bodies, so that, when these are sprayed with lacquer, only the exposed portion of the body will be sprayed. It is therefore necessary that the masking tape, when removed, separates cleanly from the steel or other object to which it is attached without marring the surface or leaving any of the coating on the surface.

MASSTONE—The color of ink in the can.

MAT—See MATRIX BOARD (1).

MAT BOARD—A paperboard lined with a plain or decorated cover paper. It possesses rigidity and is used for mounting specimens or articles.

MATCH-BOOK COVER BOARD—A paperboard used for the outside covers of books of paper matches. It is made of bleached manila-lined board, white patent-coated or clay-coated board, of about 0.016 to 0.020 of an inch in thickness. The board is a good bender with a surface suitable for printing in colors.

MATCH-BOX BOARD—A paperboard used for making boxes in which matches are packed. It is a manila-lined chipboard about 0.020 of an inch in thickness with a surface suitable for printing. It possesses good bending qualities.

MATCH-STEM STOCK—A paperboard, sometimes news filled, about 0.040 of an inch in thickness, and frequently colored. It is used for the shaft of the matches contained in matchbooks. The board is stiff and absorbent for receiving the chemicals necessary in match making.

MATRIX—The mold in which printing plates are cast. See MATRIX BOARD.

MATRIX BOARD—(1) Stereotype dry mat. A board used for making a mold for casting printing plates primarily for newspapers but also for some magazines. It is also used extensively in commercial advertising work. In newspaper and magazine printing, pages are composed of type, halftone, and line engravings. These are locked into a form known as a chase. The stereotype dry mat is placed on the form and pressure applied to form a right reading impression. After drying, which imparts rigidity, hardness and shrinkage to the mat, it is ready for casting the printing plate. In commercial advertising, a stereotype mat is molded and sent to the newspaper where a flat cast plate is made; then this is inserted in the chase during makeup of the page. Dry mats are made from chemical woodpulps, and for some grades, reclaimed paper stock may be used. Clay, talc, or other fillers are included in the furnish. A release coating is added in a finishing operation. The board is made on wet machines from a thin web which is allowed to wind on the press roll until the desired thickness is obtained. The wet laminated board is then cut off the press roll and dried on a rotary drum dryer or a continuous belt dryer. Mats are made in two ranges of thickness: the first is 0.024 to 0.036 inch, and such mats require packing or support, after molding, in the spaces which are not to print from the printing plate; a higher range of thickness is 0.055 to 0.070 inch for packless mats; such mats take the entire depth of impression in molding and are flat on the back so do not require supplementary packing. Important properties of the dry mats are their ability to stretch without fracture and compress during the molding operations, a smooth surface texture which will accurately reproduce detail in the copy, and in newspaper mats shrinkage is of particular importance. The form comprising the newspaper page is standard size but newspapers generally print on newsprint much narrower than the standard form. A shrinkage of 3/8 to 1 1/8 inches across the page is obtained by fiber treatment during manufacture and adjusting the moisture content from 10% up to virtual saturation. (2) Wet mat. This product, now obsolete in the United States, was a laminate formerly used for making a mold for casting printing plates. The laminate was made up by pasting together multiple sheets of tissue made from wood pulp, plus a facing of colored rag tissue to produce a smooth surface. The wet laminate was placed on the form and after molding was dried on the form on steam tables especially made for this purpose. This was time consuming and tied up many forms. Important properties of the wet mat were uniform thickness and density, suitable moisture content, stretch and compressibility when wet, rigidity and hardness after molding and drying.

MATRIX FIBER—A fiber containing two different materials or polymers dispersed throughout the fiber in a uniform manner. Activated carbon fibers and superabsorbent fibers are typical examples. The preferred terminology is matrix fiber.

MAT STOCK—A matte-finished cover stock in various furnishes and colors, which is used for mounting pictures, pamphlet covers, etc.

MATTE ART PAPER—Art paper having a matte finish.

MATTE FINISH—See DULL FINISH.

MATTE FINISHER—An "on-line" or "off-line" micro-embossing unit which replicates a "sandblasted" or dull finish to coated paper. This operation provides a sheet which can have high printed gloss (good ink holdout) with contrasting low paper gloss. Normally tandem, single-nip, matte-finishing units are utilized to treat both sides of the web. See SOFT-NIP CALENDER; TANDEM FINISHER.

MATURE—Seasoned for some time before use.

MATURE WOOD—See JUVENILE WOOD.

MC MIXERS—See MEDIUM CONSISTENCY MIXERS.

MD—An abbreviated term for machine direction (q.v.).

MD CONTROLS—Machine direction measurement and control of paper machine variables such as basis weight and moisture to provide better product uniformity and quality. Occurs perpendicular to cross machine direction (MD) controls.

MEALY—See MOTTLED.

MEAN—The sum of all values divided by the number of values in a data set (commonly referred to as the "average").

MEASURING TAPE PAPER—A thin strip of paper, usually kraft, which is marked with yardage and enclosed with a bolt of cloth to show its length. Such paper is nonstretching, strong, and capable of being printed.

MEAT WRAPPER—See BLOODPROOF PAPER; DRY-FINISH BUTCHERS WRAP.

MECHANICAL AERATOR—A mechanical device for the introduction of atmospheric oxygen into a liquid.

MECHANICAL DECKLE EDGE PAPER—An imitation deckle edge paper produced by subjecting the edge or edges of the dry sheet to mechanical abrasion or other treatment. See also DECKLE EDGE.

MECHANICAL FINISHING—A method of finishing fabrics and nonwovens using mechanical means such as napping, compaction, and embossing.

MECHANICALLY BONDING—A method of bonding nonwovens using mechanical means such as the needle-punch, hydroentangling, and stitch-through methods.

MECHANICAL PAPERS—(1) See COARSE PAPERS. (2) Papers made in major part from mechanical woodpulp.

MECHANICAL PULP—A high-yield pulp produced from wood by a variety of mechanical pulping processes. It is characterized as a woodpulp with high lignin content. It is primarily suitable for printing grade papers, but can be used in other paper grades, like tissue, board, and fluffed pulps. See also ALKALINE PEROXIDE MECHANICAL PULP; BLEACHED CHEMITHERMO-MECHANICAL PULP; CHEMIGROUNDWOOD; CHEMIMECHANICAL PULP; CHEMITHERMOMECHANICAL PULP; GROUNDWOOD (GROUNDWOOD PULP); PRESSURIZED GROUNDWOOD; REFINER GROUNDWOOD; SUPERPRESSURE GROUNDWOOD.

MECHANICAL WOOD PULP—See MECHANICAL PULP.

MEDIAN—The middle value in a data set.

MEDICAL PAPERS—Generally, nonwoven, disposable papers that are used to make surgical drapes, hospital garments and gowns, hospital pillow covers and sheets, and shoe covers.

MEDIEVAL LAID—See ANTIQUE PAPER; LAID PAPER.

MEDIO BOARD—See PHOTOGELATIN BRISTOL.

MEDIUM—See CORRUGATING MEDIUM.

MEDIUM CONSISTENCY (MC) BLEACHING—Bleaching technology based on the favorable mixing properties of high shear medium consistency (MC) mixers. All bleaching stages are carried out at medium consistency (10% to 15%).

MEDIUM CONSISTENCY (MC) MIXERS—Mixers that fluidize pulp at medium consistency (10% to 15%) by applying high shear forces to it in a narrow gap between a fast rotating part and a stationary part. Bleaching chemicals can be dispersed thoroughly into the fluidized pulp. See MIXER.

MEDIUM DENSITY CLEANER—A type of forward cleaner that, typically, operates at consistencies between 1% and 3%, and which, generally, has a reject trap on the apex end of the conical section, such that high specific gravity debris is collected for some period of time, and is discharged continuously. See also CENTRIFUGAL CLEANER; FORWARD CLEANER.

MEDIUM DENSITY FIBERBOARD—A board manufactured from refined lignocellulose fibers with a synthetic resin or other suitable binder which is dry felted and hot pressed to a density of 31 to 50 pounds/cubic foot by a process in which substantially the entire inter-fiber bond is created by the added binder. Other materials may be added to improve certain properties.

MEDIUM FINISH—An intermediate finish, i.e., neither high nor low, nor rough nor smooth. The term maybe used with any type of finish.

MEDIUM WRITING FINISH—See CARBON PAPER.

MELAMINE RESIN ACID COLLOID—A cationic colloidal solution of melamine-formaldehyde resin in dilute acid, used for imparting wet strength by addition to the furnish prior to sheet formation.

MELDED NONWOVEN—A nonwoven consisting of at least some thermoplastic fiber or material that is bonded by application of heat to soften the thermoplastic to form the bond.

MELT BLOWN—A method of forming nonwovens where melted polymer exits a spin-nerette and is attenuated by hot air. The discontinuous lengths of fiber are then collected on a screen or drum to form a continuous nonwoven.

MEMO COVERS—Cover stock especially made for memorandum, bank, and passbooks.

MENU—A list of options on a computer screen.

MENU BRISTOL—A heavyweight paper similar to index bristol commonly used for printing menus.

MERCAPTAN—Any of a class of sulfur containing organic compounds with the type formula RSH, the low boiling members of which have an extremely offensive odor.

MERCERIZATION—The process of treating vegetable fibers with an alkaline reagent, with or without tension, so as to increase their diameter, density, strength, luster, and receptiveness to dyes.

MERCHANT'S BRAND—A line of papers, the names of which are owned by the paper merchant, in contrast to mill brand, where a paper manufacturer is the owner.

193

MERCHANT'S STOCK ORDER—Paper purchased for merchant's warehouse stock.

MESH—In papermaking, the number of warp and weft (shute) wires per inch determine the mesh of a fabric. See FORMING FABRIC.

METACHROMATYPE PAPER—Paper colored with dyes that undergo color changes in physical conditions, such as temperature or relative humidity.

METAL IONS—See HEAVY METAL IONS.

METALLIC BLOTTING—A blotting paper laminated with a coated-one-side metallic paper.

METALLIC COATING—Coating for either a paper or a board, the pigment content of which is a composition of metallic flakes such as aluminum or bronze powder and the vehicle of which is either (a) casein or other aqueous binder or (b) a lacquer. The bronze may be uncolored to give the effect of silver, or colored to give the effect of gold, gunmetal, or other special shades. See FOIL LAMINATE.

METALLIC PAPER—(1) A specially coated paper on which marks may be made with a metal point or stylus (of silver, aluminum, lead, etc.) which cannot be erased. It is used for notebooks and indicator diagrams. The coating may be of zinc white or a mixture of clay, whiting, lime, and barytes. (2) Paper coated with metallic substances to produce the effect of a metallic surface or paper which has been combined with metallic foils. (3) Paper which has been coated by the condensation of vaporized metal while in high vacuum. This process is known as vacuum metalizing. See ALUMINUM PAPER; BRONZE PAPER.

METAL TRANSFER PAPER—See TRANSFER PAPER.

METAMERISM—The property of color in which the perceived color varies, depending on the illuminant.

METER PAPER—A ledger paper usually made from chemical woodpulp, in basis weights from 28 to 40 pounds (17 x 22 inches – 500). It is commonly used with recording instruments, in circular dial form. Good sizing, freedom from lint, and good dimensional stability are important properties.

METERING SIZE PRESS—A size press in which the sizing material is metered and distributed as a film onto the surface of the size press rolls, then transferred to the sheet surface in the size press nip. Types of metering size presses include blade metering size presses, gate roll size presses, and rod metering size presses.

METHANOL—CH_3OH. A reducing agent used to manufacture chlorine dioxide. See CHLORINE DIOXIDE MANUFACTURE.

METRICATION—The process of converting from the English system of weights and measures to the metric system.

METRIC SYSTEM—A system of units based on the meter, kilogram, and second (mks system).

MEZZOTINT—An intaglio process (q.v.) of engraving on copper or steel by first applying a roughened surface with a "rocker," a kind of steel chisel whose edge is set with minute teeth and curved so that the teeth are rocked into the plate at various angles to produce the desired light and shade effects. Nonprinting areas are then smoothed. This is one of the oldest intaglio processes. The plates are printed on a hand press in limited editions.

MF—See MACHINE FINISH.

MG—See MACHINE GLAZED.

MG SULFITE WRAPPING PAPER—A predominantly sulfite sheet made on a Yankee machine in various weights and colors. It has a high-glazed finish on one side and a relatively rough finish on the reverse side.

MIC—See MICROBIALLY INFLUENCED CORROSION.

MICA PAPER—A paper coated with ground mica, usually of 180 mesh or finer, incorporated with an adhesive, such as casein. Dyes may be added for tinting. The coated product may be calendered or embossed and is widely used for greeting folders, box coverings, and other decorative purposes. The base paper is usually of chemical woodpulp in basis weights of 38 to 65 pounds (25 x 38 inches – 500); it should be well sized and of uniform formation and density; the finished weights range from 20 to 30 pounds (20 x 36 inches – 500).

MICR—Short for magnetic ink character recognition, special font characters printed with magnetic ink, designed to be machine read and processed. Numbers on the bottom of checks are MICR.

MICROBIALLY INFLUENCED CORROSION—Corrosion processes initiated by or accelerated by the growth of microorganisms at the metal surface.

MICROCRYSTALLINE WAX—One of a series of waxes obtained from petroleum and differing from paraffin in having smaller crystal size and in containing more branched-chain hydrocarbons. The color may range from white to dark brown or black. In general, the microcrystalline waxes are tougher, more flexible, more adhesive, and have a higher melting point than the paraffin waxes, although different grades will differ widely. Microcrystalline waxes are used for laminating and in blends with paraffin wax for surface treatments having increased sealing strength.

MICROELECTROPHORESIS—A widely used electrokinetic technique for the measurement of the electrokinetic charge of small papermaking stock components. The mobility (q.v.) of microscopic particles [fines (q.v.)] in an electric field is the quantity determined.

MICROMETER—An instrument for measuring thickness of paper and paperboard. See THICKNESS.

MICRO MIXING—The mixing of bleaching chemicals and pulp on a very small scale, of the order of the diffusion distance of chemicals in the time of typical bleaching stage, say, less than 2 mm. See also MACRO MIXING.

MICROPARTICULATE PROCESS—A process involving the sequential addition of a charged (usually cationic) polymer followed by anionic microparticles to form small, dense flocs which produce good paper formation and rapid drainage. The anionic microparticles are believed to form bridges between cationic flocs formed upon polymer addition. Generally, optimum performance is achieved when the furnish zeta potential is close to zero. Either synthetic cationic polymers or cationic starch can be used as the polymer component. Colloidal silica and Bentonite clay are used as microparticles. See FIRST PASS RETENTION.

MICROWAVE PAPERS—Papers designated to be used in microwave cooking. Often they incorporate a "receptor" which allows the sheet to absorb the microwave radiation, thus increasing the temperature in the package to provide for "browning" or, for popcorn bags, higher oil temperatures.

MICR PAPER—Generally, a writing paper suitable for magnetic-ink character recognition printing.

MIDDLE LAMELLA—The intercellular portion of wood structure. This region is very thin (ca. 250 nm) and is composed primarily of lignin and pectic polysaccharides. The term usually includes the cell corners where three or four cells come together. Cell corners are thicker than the portion of the middle lamella between two cell walls and may have a somewhat different chemical composition. The region encompassing the middle lamella and the primary wall of adjacent cells is called the compound middle lamella.

MID-FEATHER AGITATOR—A side entry propeller agitator fitted into a long, narrow mixing tub which has a wall (the mid-feather) dividing the tub into an oblong flow channel. The agitator, typically, occupies the entire width of one channel, and all stock passes through the agitator as it progresses around the oblong flow channel.

MIGRATION—A term often associated with binders to indicate the movement of binder to the surface of the coated paper or the nonwoven due to temperature and drying differential. Binder migration may also occur when printing binder on a nonwoven.

MILK-BOTTLE CAPS—See BOTTLE-CAP BOARD.

MILK-CAN GASKET—A sheet of vegetable parchment paper laid over the top of a milk can, upon which the cover is placed, in order to make a tight seal and protect the milk from contamination, facilitate easy removal of the lid, and reduce clean-up time. It is made in basis weights of 30 to 40 pounds (24 x 36 inches – 500).

MILK-CAP BOARD—See BOTTLE-CAP BOARD.

MILK-CARTON BOARD—A grade of special food board (q.v.) which is strong, tough, usually plastic coated, and capable of being formed into a container for milk, cream, or other beverages. It is made of bleached pulp and normally has a thickness of 16 to 30 points, the actual caliper depending on the type and size of the container.

MILL BLANKS—Mill blanks are cylinder-machine products, generally consisting of top and bottom liner of white stock, vat lined on a filler of mechanical pulp, news, or similar stock. Principal uses are for menus, posters, advertising cards, etc. Standard thicknesses are 3, 4, 5, 6, 8 and 10 ply. Thicknesses such as 12 and 14 ply and heavier are usually made by pasting together two thinner plies, which are white on one side only with a news back. No. 1 mill blanks have liners consisting of bleached and unbleached chemical woodpulps, deinked paper stock, and soft white shavings in varying amounts. The center is usually blank or printed news, though it may consist of mechanical pulp in the better grades. No. 2 mill blanks are, in general, characterized by being of poorer color and quality and not as bright as the No. 1 grade.

The filler may be the same as that of No.1 mill blanks or, as is more usual, it may contain a larger proportion of printed news.

MILL BOARDS—Heavy paperboards used for bookbinding, box making, carriage panels, shoes, etc. They are made from fiber refuse, wastepaper, screenings, and mechanical woodpulp; the better grades may contain some hemp and flax fibers. The sheet is made on a wet machine and is calendered by passing several times through the board calender. They are hard, flat, and non-warping.

MILL BRAND—A line of papers, the names of which are owned by the mill manufacturing them.

MILL BRISTOL—A term applied to a group of bristols, usually made on a cylinder machine. The basis weights range from 90 to 200 pounds (22 x 28 inches – 500). They are normally made for printing purposes. See BRISTOLS; CYLINDER BRISTOLS.

MILL BROKE—See BROKE.

MILL CLARIFIER—See CLARIFIER.

MILL COUNT—The count (q.v.) by the mill. This term is used by merchants, in invoicing a shipment directly to the consumer, to indicate that the shipment has not been re-counted by the merchant.

MILL CUT—Obsolete term. See DOUBLE KNIFE CUTTER; PRECISION CUTTER; SLITTER.

MILL EDGE—The slightly rough edge of untrimmed papers.

MILL HEADS—See MILL WRAPPER.

MILL LINE—A proprietary line of papers. See MILL BRAND.

MILL ROLL—A roll of paper as it comes from the paper machine reel, which is subsequently converted to smaller rolls or to sheets.

MILL SCALE—The black oxide scale on the surface of steel (or related alloys) which have been hot-worked. Hot-rolled steel ("black iron") is covered with mill-scale or "magnetite," a strongly magnetic Fe_3O_4.

MILL SIZE—A term used to denote a rosin size prepared at the mill.

MILL SPLICE—A splice prepared at the paper mill as opposed to a splice prepared at some later stage in the converting of the roll.

MILL WATER—Process water that enters the mill. Raw mill water may be treated before use in the mill to remove undesirable minerals.

MILL WRAPPER—A general term to designate grades of paper used by paper mills for wrapping purposes. The grade depends on the quality of the contents to be wrapped and upon the custom of the mill. (1) Heavyweight paper made of screenings, jute, rope, sulfate, or wastepaper and used for wrapping paper mill runs or large rolls and paper put into bundles. (2) See SCALING PAPERS. (3) Bogus or flexible paperboard for the above purposes.

MILOX PROCESS—See SOLVENT PULPING.

MIMEO BOND—A grade of writing paper used for making copies on stencil duplicating machines. It is normally made of chemical woodpulps in basis weights of 16, 20, and 24 pounds (17 x 22 inches – 500). It is characterized by good opacity, ink absorbency, and freedom from lint, plus good printing, writing, and mimeographing quality.

MIMEOGRAPH—The trademark applied to the stencil duplicator invented by Albert Blake Dick in 1884. A stencil duplicator is an apparatus by which reproductions of typewritten and illustrative material may be produced by the use of a stencil. The principal part of the machine is a perforated revolving cylinder, usually carrying on the inside an inking device; a felt ink-pad on the outside of the cylinder transfers the ink to the stencil which, in turn, transfers it to a sheet of paper. It may be used for the reproduction of typewritten matter, ruled forms, line drawings, charts, graphs, music, etc. Many grades of paper can be used on a stencil duplicator.

MIMEOGRAPH PAPER—A grade of writing paper made from chemical wood and/or mechanical pulps in basis weights of 16, 20, and 24 pounds (17 x 22 inches – 500). It is used for stencil duplicating and usually differs from mimeo bond (q.v.) in that printing and writing qualities are not as important as good mimeographing quality.

MIMEOTYPE STENCIL PAPER—See STENCIL TISSUE.

MINERAL FILLER—See FILLER.

MINING NOZZLE—A device for adding dilution water to the bottom of downflow towers. See DILUTION ZONE.

MISALIGNMENT—The deviation from the ideal geometry of the components of a process machine. Normally, all functional components are specified to be level and square to the machine centerline within some specified dimensional or angular tolerance. See ALIGNMENT.

MIST BOARD—A paperboard used largely for making suit boxes such as are supplied department stores for packaging wearing apparel. It is made of woodpulp and reclaimed paper stock on a cylinder machine and is 0.020 to 0.040 of an inch in thickness with one or both outer plies composed of a mixture of two different colored fibers. The board is strong, has good bending quality, and usually a finish suitable for ordinary printing. Mist gray is a board lined with a mixture of white and black fibers.

MIST GRAY BOARD—See MIST BOARD.

MITSCHERLICH PULP—A pulp made by a long, severe sulfite cook. The resulting pulp has high hemicellulose content. Its uses include glassine and greaseproof papers.

MITSUMATA—A low shrub (*Edgeworthia papyrifera*) which grows in temperate Asia and which is cultivated in Japan for its bark, the fibers of which are used in papermaking.

MIXED PAPER RECYCLING—Mixed paper is sorted into various quality grades for recycled products. The lowest quality is not limited as to fiber type and can contain up to 10% outthrows (q.v.) and 2% of prohibitive materials. Higher quality or "super mixed" paper is limited in the amount of groundwood content. A number of paper stock dealers have developed special grades of mixed papers collected from business offices. These are designated as mixed office papers or office pack. Their composition can vary widely depending upon the sourcing and subsequent sorting of the recovered papers.

MIXED PAPERS—A grade of wastepapers or paper stock not sorted or from which the better grades have been taken out. They are collected from department stores, offices, schools, etc.

MIXER—Equipment that disperses chemicals uniformly with the pulp. Mixing of pulp fibers with bleaching fluids is one of the most important operations in bleaching or brightening of pulp. Two types of mixers are presently used: dynamic and motionless (static) mixers. In the dynamic mixer, mixing results from the physical motion applied to the fluid mass by an agitator or mixer, mixing results from the physical motion applied to the fluid mass by an agitator or mixer which is generally a rotating impeller located on a shaft. Power is supplied by an external motor. High shear or medium consistency (MC) mixers are a special type of mechanical mixers. In the static mixer, mixing results from the stock passing over a stationary series of baffles. Power requirement is furnished by the feed pump.

MIX TANK—A tank usually used to mix black liquor with a variety of dry materials, such as makeup saltcake (q.v.), electrostatic precipitator dust, and ash hopper catch from the boiler bank and economizer (q.v.). Often referred to as a chemical ash tank.

MM—A rarely used system of expressing the basis weight of paper on a basis of 25 x 40 – 1000.

MO—An abbreviation for making order (q.v.).

MOBILITY—Is a measure of the surface potential (q.v.), or electrokinetic charge (q.v.), of suspended particles as derived by a microelectrophoretic measurement. It is defined as the ratio of the particle speed to the applied electric field potential gradient. Typical units for mobility are mm/second/volt/cm. See ELECTROKINETIC CHARGE; MICROELECTROPHORESIS.

MODE—The most frequent value in a data set.

MODEM—Short for MOdulator-DEModulator. A device that converts/reverts computer input and sends the information over telephone lines.

MODIFIED AERATION—A modification of the activated sludge process in which a shortened period of aeration is used with a reduced quantity of suspended solids in the mixed liquor.

MODIFIED CONTINUOUS COOKING—(1) An extended cooking technique in which cooking liquor is progressively added to the wood, permitting lower kappa numbers to be obtained. (2) See EXTENDED DELIGNIFICATION.

MODIFIED STARCHES—Starches which have been subjected to physical, biological, or chemical treatment primarily to effect a change in their viscosity or chemical characteristics, or both.

MOIRE—An interference pattern between two (printing) screens of different frequencies.

MOISTURE-BARRIER BUILDING PAPER—A heavy paper, usually kraft for strength, treated with asphalt or other moisture barrier material and used in wall construction to create a moistureproof membrane to prevent migration of moisture through the wall. See BUILDING PAPER.

MOISTURE CONTENT—The percentage by weight of water in sawdust, pulp, pulpboard, paper, or paperboard.

MOISTURE PROFILE—A graph indicating the variation of moisture content across the width of the web.

MOISTUREPROOFNESS—The property of a high degree of resistance to the passage of liquid water and water vapor. In paper and paperboard, it describes a sheet with unusually low water-vapor permeability (q.v.).

MOISTURE WELTS—Regularly spaced bands of soft circumferential buckles in the outer layers of a roll caused by paper taking on moisture from the air. The result looks like a corrugated culvert pipe.

MOLD—A general term to describe a frame over which is stretched a porous fabric, usually a wire screen, on which fibers are separated from a fluid suspension to form a sheet. There are several types (1) cylinder mold (q.v.); (2) for manufacture of molded pulp products (q.v.); (3) sheet mold, see HANDSHEET; (4) hand mold; see HANDMADE PAPER.

MOLDABILITY—The ability of a wrapping paper to retain a crease after being folded.

MOLDED PULP PRODUCTS—Contoured products such as egg packaging cartons, food trays, plates, and bottle protectors, made by depositing fibers from a pulp slurry onto a perforated mold, using either pressure applied to the slurry or a vacuum behind the mold, and then drying the preform with heat.

MOLD MACHINE—See CYLINDER MACHINE.

MOLD-MADE PAPER—A deckle edged paper resembling that made by hand but produced on a machine. It is made on a cylinder or cylindrical mold revolving in a vat of pulp, the various sizes being arrived at by dividing the surface with rubber bands to imitate the thinning of the deckle edge of handmade paper or by cutting the web by means of a jet of water. See MACHINE DECKLE EDGE PAPER.

MOLD-RESISTANT PAPERS AND BOARDS—Papers and boards that have been treated with mold-inhibiting chemicals.

MOLECULAR WEIGHT—The sum of the atomic weights of all the constituent atoms in the molecule or an element or a compound.

MONEY—See CURRENCY PAPER.

MONOCARBOXYLIC ACIDS—See ROSIN ACIDS.

MONOSULFITE PULP—See NEUTRAL SULFITE PROCESS.

MONOTYPE PAPER—A white book paper to be perforated with small holes and used on a monotype keyboard and casting machine. It is also called keyboard paper.

MORDANT—A substance which, when applied to fiber in conjunction with a dye, causes increased dye fixation.

MORSE PAPER—See TELEGRAPH PAPER.

MOTHER-OF-PEARL PAPER—A cover, book, or decorative paper having an iridescent effect. This was formerly produced by various chemical recipes, for example, surface treatment of the sheet with silver, lead, or other metallic salts followed by exposure to a gas such as hydrogen sulfide and fixing the iridescent effect with a lacquer. Today, those effects are achieved by special printing techniques including use of fluorescent coloring materials.

MOTTLE—An undesired pattern in printing caused by uneven absorption of printing ink.

MOTTLED—A variegated effect produced on the surface of paper by the introduction of a small amount of heavily dyed fibers (mottling fibers) into stock of another color. It is used for fancy and other effects. See GRANITE.

MOTTLED COLOR—Nonuniform coloring of a sheet of paper caused by irregularities in formation, calender pressure, dye application, drying, or plating.

MOTTLED FINISH—A finish which is characterized by high and low spots or by glossy and dull spots.

MOTTLED WHITE LINERBOARD—An unbleached kraft linerboard (q.v.) in which the top ply is made from bleached pulp or white grades of recycled fiber. The top layer is formed to give a mottled appearance. Also called oyster white linerboard (q.v.).

MOTTLING—(1) Uneven dyeing of pulp fibers, caused by adding the dyestuff in a hot concentrated solution to the beater in such a manner as to color a small number of fibers immediately, or by dyeing a mixed furnish in which certain fibers will fix the color more quickly than the others. (2) Uneven dyeing of paper produced by addition of coloring matter in drops to the paper on the paper machine wire or to the dry paper after it has left the paper machine. (3) Uneven formation (q.v.)

MOUNTING BOARD—(1) A paperboard upon which printed or lithographed sheets are pasted. It is made of reclaimed paper stock and is from 0.030 to 0.050 of an inch in thickness. It may be pasted to obtain required thickness. It has a high smooth finish and is stiff to resist warping. (2) See MAT BOARD.

MOURNING PAPER—A paper used for correspondence as an outward manifestation of grief by those suffering loss by death. It is a writing grade made of cotton fiber or chemical woodpulps or mixtures of these. It is usually printed with a black or gray border.

MOUSE—A small, hand-held device used in graphics computer programs to select on-screen features. Movement of the mouse rolls a small ball in the mouse which moves a pointer on the screen. Buttons on the mouse are used for item selection.

MOW—Mixed office waste.

MUD WASHER—A tank or pressure filter wherein the underflow lime mud/white liquor slurry from the white liquor clarifier (q.v.) is first mixed with fresh or recycled water and then the lime mud is separated from the resulting weak liquor by settling or pressure filtration.

MULCH PAPER—Plain, asphalt- or pitch-treated paper, usually kraft, which is spread on the ground to protect the roots of plants from heat, cold, or drought, to keep fruit clean, and to re-

tard the growth of weeds. It may be impregnated with disinfectants or insecticides to protect plants from molds, insects, and parasites, or with nutrient salts to give fertilizing value.

MULLEN—Bursting strength (q.v.). So called from the name of the instrument used in the test.

MULTICOLORED CREPE—Colored crepe paper that is colored two or more shades on the same piece. It is also made in various shades which are blended together. See also CREPED PAPER.

MULTICYLINDER MACHINE—A cylinder machine (q.v.) consisting of two or more vats arranged in tandem formation to facilitate the manufacture of thick papers, duplex papers, bristols, and filled boards.

MULTI-FORMER FABRICS—See FORMING FABRIC.

MULTIFUNCTIONAL CLEANER—See COMBINATION CLEANER.

MULTIGRAPH—A machine designed primarily for reproduction of typewritten matter, although other typefaces may be used. The principal part of the machine consists of a slotted revolving cylinder; the matter to be reproduced is set in type which is transferred to the slots of the cylinder. Impression may be made by a special typewriter ribber or by an inking device, the paper passing between a rubber-covered platen and the ribbon or the inked type. Linotype slugs may be used in place of movable type; curved electrotypes may be also used. It was used principally for business letters or for the preparation of reports, advertising matter, etc.

MULTIGRAPH PAPER—A bond or writing paper used for printing by the multigraph process.

MULTILITH—The multilith is a small (offset) press suitable for office work. The plate may be prepared by a typing directly on the plate, by drawing on it, or by photography. Multilith and rotoprint are names of presses, not the names of printing processes.

MULTILITH PAPER—A printing paper designed for use on multilith, rotoprint, and other small offset presses.

MULTIMEDIA—Using a combination of communication forms, such as print, audio tape, video tape, CD-ROM, film, or projection.

MULTIMETAL PLATE—A lithographic printing plate composed of two or more metals, the image metal being preferentially ink receptive and the nonimage metal being either preferentially water-receptive or easily desensitized to ink. Copper is the usual printing surface metal. The nonprinting metals are aluminum, chromium, or stainless steel. The base plate may be copper, but is usually iron or zinc with copper and chromium electroplated on it. Multimetal plates are used primarily for long runs or when printing with inks and papers that are unusually abrasive.

MULTIPLE-EFFECT EVAPORATOR—A set of two or more single-effect evaporators arranged so that the steam evaporated in one effect or body is used as the heating medium in another effect or body.

MULTIPLE FOURDRINIERS—A type of forming section which has two or more separate fourdriniers, each with its own headbox and white water system. The sheet is made as separate plies which are consolidated by transferring one ply onto another ply by bringing the plies and fabrics together and by making one ply transfer to another by means of a vacuum box (q.v.). The two most common configurations are two fourdriniers for linerboard production and three or more fourdriniers for folding boxboard. When two fourdriniers are used, the second one is generally smaller and is located over the first one. It is sometimes called a "mini fourdrinier" and the configuration is sometimes called "piggy back." The plies are consolidated at 6% to 8% solids and this is called "dry on dry". (For wet on dry consolidations see SECONDARY HEADBOX FORMER.) When three fourdriniers are used, the one which is producing the middle ply is called filler fourdrinier or "filler ply." Lower value fibers can be used in this ply because they are not seen or printed on in the final product. The most popular three fourdrinier configuration is called a "folded fourdrinier." In this configuration, two fourdriniers are on top and run toward each other. The top ply is consolidated with the filler ply first (dry on dry) then these two plies are consolidated with the bottom ply (again dry on dry).

MULTI-PLY—Made up of two or more plies. Depending on method of manufacture and use requirements, the plies may be firmly or partially bonded or completely unbonded.

MULTI-PLY FORMING—Refers to any type of forming in which the sheet can be manufactured in distinct layers and the layers combined during the manufacturing process. The most common types are: cylinder formers (wet on wet); Inverform formers (wet on dry); secondary headboxes on fourdriniers (wet on dry) and multiple fourdriniers (dry on dry) (q.v. all foregoing formers). The cylinder and Inverform formers are older; the Inverform was an early twin wire former and the multiple fourdrinier approach is more recent. The terms wet on dry are relative and refer to the consistency at the point where plies are "consolidated." Wet refers to a headbox consistency of about 1% solids. Dry refers to a consistency where the paper slurry begins to look like a sheet or web, which is in the range of 6% to 8% consistency.

MULTISTAGE BLEACHING—A bleaching operation carried out in two or more stages. Many combinations of stages can be used, depending upon the characteristics which are needed in the finished pulp. The most common bleaching steps and their symbols are: chlorination (C), extraction with sodium hydroxide (E), sodium or calcium hypochlorite (H), chlorine dioxide (D), hydrogen peroxide (P), oxygen (O), chelation (Q), ozone (Z), and peroxy acids (Pa).

MULTIWALL BAG KRAFT PAPER—See SHIPPING SACK KRAFT PAPER.

MULTIWALL CORRUGATED BOARD—A corrugated board containing a multiplicity of corrugated members made by adhering one or more plies of single-faced board to a double-faced board in such a manner that each corrugated members are combined into a single board of much greater strength than a double-faced board of the same material will exhibit. The type or number of walls is designated in terms of the number of corrugated members. See DOUBLE WALL CORRUGATED BOARD; TRIPLE WALL CORRUGATED BOARD.

MULTIWALL SHIPPING SACK—A shipping sack made of three to six walls of heavy-duty shipping sack kraft paper (q.v.). See DUPLEX; SHIPPING SACK.

MUNICIPAL WASTES—Usually refers to a solid waste generated by cities or towns, but could refer to any type of waste generated by these sources.

MUNITION BOARD—See CARTRIDGE PAPER; WAD STOCK.

MUNITION PAPER—See CARTRIDGE PAPER.

MUNSELL COLOR SYSTEM—System for color specification of surfaces illuminated by daylight and viewed by an observer adapted to daylight. The system yields approximations to scales of variables of hue, saturation, and lightness, with uniform perceptual spacing. In this system, the scales are designated hue, chroma, and value, respectively.

MUSIC COVER—A light or medium weight opaque cover paper used for the protection of sheet music. It is either coated or uncoated.

MUSIC LITHOGRAPH PAPER—See MUSIC PAPER.

MUSIC MANUSCRIPT PAPER—Paper printed with lines denoting the staff and used by musicians for writing musical scores. It is made from rag and bleached chemical woodpulp or mixtures of these. The paper is stiff enough to stand in a music rack and has a surface that will take pencil or ink. The common basis weights are from 60 to 80 pounds (25 x 38 inches – 500) or 24 to 30 pounds (17 x 22 inches – 500). Erasability and good opacity are important characteristics.

MUSIC PAPER—A paper used to print or lithograph sheet music. It is made largely of bleached chemical woodpulp. The paper is well formed, has good opacity, and is sized and processed to give a good pick strength and minimum curl. A flat English finish is desired and the paper must have stiffness to stand in a music rack. The basis weights in most common use range from 60 to 90 pounds (25 x 38 inches – 500).

MUSIC-ROLL PAPER—A paper used for the manufacture of music rolls for player pianos and organs. It is made from chemical woodpulp and may contain cotton or manila hemp fiber. The paper is resistant to change in dimension from the humidity of the air and gives clean perforations. Strength and resistance to abrasion are also important properties.

N

NACREOUS PAPER—See MOTHER-OF-PEARL PAPER.

NAIL STRENGTH—See EDGE NAIL STRENGTH.

NAPKIN PAPERS—Tissue papers used in the manufacture of paper napkins (q.v.).

NAPPING—The process of raising a fibrous surface on a fabric or nonwoven.

NATURAL COLORED—A term applied to papers whose colors result from the nature of the stock used, with little or no coloring matter added.

NATURAL FIBERS—Refers to fibers obtained in essentially usable form directly from plant, animal, or mineral sources. See SYNTHETIC FIBERS.

NCASI—The National Council of the Paper Industry for Air and Stream Improvement, a trade group concerned with environmental pollution abatement.

NCR PAPER—See CARBONLESS PAPER.

NEEDLED FELT—See NEEDLING.

NEEDLE PAPER—A black wrapping paper free from chemical matter which might cause rust or tarnish. It belongs to a class of papers variously termed, according to the use to which they are put, antitarnish, acid-free, blacks, cutlery, pin, etc.

NEEDLE PUNCH—The process of bonding or forming nonwovens using barbed or forked needles. Barbed needles are utilized to mechanically interlock and bond the fibers, while forked needles are utilized to produce velour and patterned nonwovens.

NEEDLING—A felt making procedure in which a "batt" of fibers is attached to a base cloth by repeatedly punching the batt into the base cloth with special shaped barbed needles. In the case of one form of nonwoven cloth, a batt is attached to another batt (or re-attached to itself) by the same punching process.

NEGATIVE HELIOGRAPHIC PAPER—See NEGATIVE PAPER.

NEGATIVE PAPER—A paper normally used instead of photographic films or plates for making negatives from which reproductions are made. It is a cotton fiber content paper manufactured from high-quality white cotton cuttings in basis weights of 24, 28, or 32 pounds (17 x 22 inches – 500); heavier weights may be made for original or permanent use in (brown) prints not for reproduction purposes. The paper has a very hard surface, a smooth finish, good formation, and relatively low opacity. It is glue sized to contribute to its wet tensile strength and to minimize penetration of the photographic emulsion with which it is coated.

NEGATIVE PHOTOGRAPHIC PAPER—See NEGATIVE PAPER.

NEPS—Fibrous balls formed while carding due to immature cotton, excessive through-put, fine fibers, straight fibers, or improper fiber crimp and poor carding.

NET REFINER LOAD—The total power load on a refiner less the circulating load—typical units are horsepower-hour or kilowatt-hour.

NEUTRAL GUMMED ELECTRICAL PAPERS—These papers are used as sticker tapes and coil wrappers. Sulfate paper or fish paper coated with 0.5 to 1 mil of high grade, water-soluble animal glue are the usual combinations. For the kraft base material, thicknesses of 0.0015 to 0.010 inch and for the fish paper base 0.005 to 0.010 inch may be used. The important properties are breaking strength, dielectric strength, freedom from metallic inclusions, uniform thickness, and a neutral pH value. They are used in the manufacture of small electric coils and other electrical devices.

NEUTRALIZATION—In pulp bleaching, a mild alkaline extraction stage, usually done at pH 7 to 8, used to achieve specific paper grades of pulp where it is necessary to retain as much of the hemicellulose as possible.

NEUTRAL KRAFT—A kraft paper with a pH of 7.0 and produced so as to be relatively acid and sulfur free. It is used in the textile industry where contact with wet materials precludes use of regular kraft which may give rise to staining and discoloration of textiles, in the stainless steel and aluminum industries (see INTERLEAVING PAPER), and for manufacture of certain multiwall shipping sacks.

NEUTRAL PAPERMAKING—See ALKALINE PAPERMAKING.

NEUTRAL SIZE—A form of rosin size which is neither acidic nor alkaline in nature.

NEUTRAL SIZING—Includes the use of synthetic (alkenyl succinic acid anhydride, alkyl ketene dimer, or polymeric sizing agents) as well as certain forms of rosin size (dispersion). If the chemicals are added in the right sequence and their action is enhanced by adding special cationic polymers, sizing of neutral or alkaline papers (containing $CaCO_3$) can be accomplished using rosin sizes near pH 7.0.

NEUTRAL SULFITE PROCESS—A modified sulfite process using sodium sulfite as the active cooking chemical with the sodium carbonate added so that the initial conditions are neutral (neither alkaline or acid). The yields are higher than with sulfite pulps, and refining is often used to defiber the pulps. When this is done, the process is called neutral sulfite semichemical (NSSC) pulping. Hardwoods respond better to the neutral sulfite process than conifers. Corrugating medium is a common NSSC product.

NEUTRAL SULFITE SEMICHEMICAL PULPING—See NEUTRAL SULFITE PROCESS.

NEWS—See NEWSPRINT.

NEWSBOARD—A paperboard used largely in the setup box trade. It is made on a cylinder machine from printed news, generally from 0.016 to 0.055 of an inch in thickness, and is stiff, with a surface of clean appearance such as is required for the better grades of boxes. (1) Combination newsboard is a board having a news base or center and lined on one or both sides with a higher grade of stock. (2) Solid newsboard is a board made entirely from printed news.

NEWS BOGUS PAPER—A bogus paper (q.v.) made entirely from old newspapers. See RECYCLED NEWSPRINT.

NEWS-LINED BOARD—See NEWS VAT-LINED CHIP.

NEWSPAPER RECYCLING GRADES—Newspapers to be recycled into new products are classified into various quality sorts. The lowest quality can contain up to 5% of other papers and a specified percentage of prohibitive materials and outthrows (q.v.). In the higher quality grades of special news, deink quality, and over issue news, no prohibitive materials are allowed and the percentage of outthrows

diminishes as the quality is increased. A normal percentage of rotogravure and colored sections is permitted in all grades.

NEWSPAPER WRAPS—A strong unbleached or semibleached kraft wrapping sheet approximately 35 to 40 pounds (24 x 36 inches – 500) used for wrapping newspapers and magazines in mailing. Desirable properties are strength, sizing, and scuff resistance.

NEWSPOSTER—See POSTER PAPER (2).

NEWSPRINT—A generic term used to describe paper of the type generally used in the publication of newspapers. The furnish is largely mechanical woodpulp, with some chemical woodpulp. The paper is machine finished and slack sized, and it has little or no mineral loading. It is made in basis weights varying from 25 to 32 pounds (24 x 36 inches – 500), the greatest preponderance being 30 pounds. The term includes standard newsprint and also paper generally similar to it and used for the same purpose, but which may exceed to slight degrees the limitations of weight, finish, sizing, and ash applicable to standard newsprint. It does not include printing papers of types generally used for purposes other than newspapers such as groundwood printing papers for catalogs, and directories.

NEWSPRINT SHEETS—Newsprint paper in sheet form as contrasted to roll news. The sheets are used in printing many weekly newspapers on flat-bed presses.

NEWS VAT-LINED CHIP—A paperboard used for setup boxes and for general purposes. It is a combination board of chip, news lined on one or both sides. The board is stiff and has the clean appearance characteristic of news.

NEWTON—See JOULE.

NINEPOINT SEMICHEMICAL BOARD—Term replaced by corrugating medium (q.v.).

NIP—(1) The contact, or near contact, of any two machine components that create a potential pinch point that is a safety liability for equipment or operating personnel. (2) The contact region or zone between two rolls or a roll and a shoe loaded against each other to form a region of pressure. (3) The "line" of contact between two rolls, such as press, calender, or supercalender rolls. Because of the compressibility of the fabric and/or the web of paper, the "line" of contact is actually a narrow zone. Wet nip refers to those at the presses; dry nip refers to those at the calenders. (4) The physical contact of two machine components such as a winding nip or intermeshing gears. (5) The unit loading or average linear distribution per unit width as a nip contact line, expressed in pli (English units) or N/m (ISO units). (6) In printing, it is the contact point between two cylinders, specifically between the print cylinder and its impression roller. See NIP PRESSURE; NIP WIDTH.

NIP IMPRESSION—The measurement of the nip contact length in the machine direction under loaded conditions and across the full width of the rolls. Usually accomplished by placing a prepared sandwich of white paper and carbon paper into the open, stationary nip followed by loading the rolls to the desired load. The imprint of the carbon paper shows the loaded zone or nip impression.

NIP PRESSURE—The measurable pressure in a nip expressed in pounds or grams per unit of length or per unit of area. A wet press could exert a nip pressure of 200–300 pounds per lineal inch (pli), whereas a supercalender could exert a nip pressure from 2000 to 3000 pli. See NIP; NIP WIDTH.

NIP VENTING—Press fabrics or felts often reach saturation under the heavy loading in the press nip, especially with large water removal from the sheet. It is beneficial to water removal from the sheet to provide a momentary storage space or vent on the side of the felt opposite the sheet. This is commonly accomplished using grooves or holes in the backing roll surface or by supporting the felt on a very stiff, open mesh fabric structure.

NIP WIDTH—The measurable width of a nip between two rolls under lineal pressure. The width can be measured statically by placing impression paper (NCR paper, carbon, or embossed aluminum foil) in the nip and subjecting it to a momentary nip pressure. See NIP; NIP PRESSURE.

NITRATING PAPER—A waterleaf sheet ranging from about 6 to 14 pounds (17 x 22 inches – 500) in weight, prepared for use as the raw material in the manufacture of the very highest grades of nitrocellulose (intended for conversion into films, lacquers, celluloid, etc.), where clarity, cleanliness, and freedom from coloring matter are of special importance. The paper is made from new white cotton cuttings, cotton linters, or from highly purified woodpulp (cellulose nitrate chemical conversion grade). It may be made on a fourdrinier or cylinder machine, and may or may not be calendered, according to the requirements of the converter.

NITROGEN DIOXIDE (NO_2)—A compound produced by the oxidation of nitric oxide in the atmosphere; a major contributor to photochemical smog.

NITS—Also called knits. (1) Small twists of fibers that resist dispersion into individual fibers, typically formed during processing at consistencies higher than 15%. Nits can also be formed during flash drying of pulp. (2) Small undefibered flakes of paper which are particularly resistant to defibering during repulping of dry paper, particularly papers that contain wet-strength additives.

NOISELESS PAPER—Paper used in any place where the rustle or rattle of paper is objectionable, such as for theater programs, radio manuscripts, and paper pillowcases. Such papers cannot be limited to any specific grades as any paper can be treated to eliminate noise caused by rattling.

NOISELESS PROGRAM PAPER—Paper used as the name indicates. This refers to a paper suitable for a program, so treated to eliminate noise caused by rattling.

NO-LOAD POWER—See CIRCULATING LOAD.

NOMINAL WEIGHT—The weight per ream or 1000 sheets at which paper is billed, whether the ream or 1000 sheets actually weigh, more or less than the weight stated, so long as it is within customary tolerances. See ACTUAL WEIGHT.

NONBENDER—A term applied to paperboard in which the top liner will rupture when folded on a score A grade frequently used in setup cartons that are cut scored.

NONCOMBUSTIBLE PAPER—See FIRE-PROOF PAPER. The term is also applied to cigarette paper (q.v.).

NON-CONDENSABLE GASES—Gases released during the processes of pulping, blowing, evaporation, and stripping which cannot be condensed to liquids at ordinary temperatures. Usually composed of organic compounds and sulfur compounds along with some nitrogen, oxygen, and steam.

NONCONFORMITY—The nonfullfillment of a specified requirement.

NON-CONTACTING COOLING WATER—Water which has not been treated chemically, and is used only for cooling purposes, before being returned to a receiving body. Since it has had no contact with the process and is unchanged other than in temperature, it is usually discharged with little or no treatment.

NONCORROSIVE GREASEPROOF WRAPPING—A barrier material which is grease resistant, acid free, and noncorrosive. This material may be laminated when necessary to meet requirements. It also may be coated to secure self-adhering and moldable characteristics. See BARRIER MATERIALS.

NONCURLING GUMMED PAPER—A gummed paper which has been treated to prevent curling. Gummed paper has a strong tendency to curl because of the unequal expansion and contraction of the gum and the paper with humidity changes. In noncurling paper, the gum has been broken into small particles after drying so as to present a discontinuous surface. See GUMMED FLAT PAPERS.

NONFADING POSTER—An offset-type printing paper designed for billboard use. It is usually made from chemical woodpulps and light-fast dyestuffs. See also POSTER PAPER.

NONIMPACT PRINTING—A type of printing where no pressure or impact is required to transfer ink to the paper, such as copy machines, laser printing, and ink-jet printing. Electrostatic charges in the paper cause ink images on the drum or plastic film to be transferred to the paper. The application of heat then bonds the ink to the paper. These inks are more difficult to remove from fiber because of their hot melt adhesive type of attachment to the paper.

NONPASTED BLANKS—A term applied to coated or uncoated blanks which are usually made on a cylinder machine and generally vat lined, such as patent-coated board. See BLANKS; CARDBOARD.

NON-SULFUR PULP—Pulp produced by digestion of wood with chemicals that do not contain sulfur. The most common application of this terminology is for a semichemical process (q.v.) in which the pulping chemicals are sodium hydroxide and sodium carbonate and the resulting pulp is used in the manufacture of corrugating medium (q.v.).

NONRATTLE PAPER—See NOISELESS PAPER.

NONRETURNABLE CORE—A core, usually of fiber, which is normally used only once. It cannot be returned to the mill for credit.

NONTEST CHIP—A paperboard used as a filler by makers of solid fiber container board and for general use where no minimum bursting strength is required. It is made of wastepapers and varies in caliper and finish according to requirements.

NONWOOD FIBERS—General term for natural fibers obtained from fibrous plants other than wood and includes bagasse (from sugar-cane), bamboo, straw, reeds, cotton, esparto, etc.

NONWOOD PULP—See WOODPULP.

NONWOVEN—A fabric-like structure produced using fibers that are bonded in some way or other.

NONWOVEN FABRIC—See NONWOVEN.

NOSE BAR—A bar shaped to fit the contour of the feed roll on a card or air-lay machine that assists in gripping the fiber.

NOTE PAPER—Writing or tablet paper usually folded. There is quite a wide range of quality and finishes used.

NOTEBOOK PAPER—Strictly speaking, any paper used for various types of notebooks whether bound, loose-leaf, spiral-wound or otherwise. Normally, however, this term is applied to a well-sized bond or tablet-type chemical woodpulp writing paper which is converted into various items such as punched, ruled, and edge-reinforced cut-size products for school and office use. There are many variations of these most common items.

NOTION-BAG PAPER—An MF or MG paper for notion bags (small, simply formed "flat" paper bags) usually made from bleached and unbleached sulfite and kraft pulp. It is normally made in a basis weight of 25 or 30 pounds (24 x 36 inches – 500). The paper is generally striped by watermarking or decorated by printing. The bags are used for packaging "notion" purchases.

NOVEL NEWS—See NOVEL PAPER.

NOVEL PAPER—A grade of paper once predominantly used in "pulp" magazines but now primarily used in pocket-sized paperback novels and in children's coloring books. It is made from approximately the same fiber furnish as newsprint in such a way as to produce a rough surface and maximum bulk. The usual basis weight is 32 pounds (24 x 36 inches – 500). Thicknesses range from a minimum of 0.004 to as high as 0.0055 inch compared with 0.003 inch for newsprint.

NOx—Collective term for the nitrogen oxides NO, NO_2, N_2O. NOx is often found as a pollutant species in flue gases. NOx can be generated during fuel combustion.

NP CHART—See NUMBER OF AFFECTED UNITS CHART.

NPEA—The National Printing Equipment Association, a trade association in the graphic arts field.

NPTA—The National Paper Trade Association, the U.S. trade association of paper merchants.

NSPS—New Source Performance Standards.

NSSC PULPING—Neutral sulfite semichemical pulping. See NEUTRAL SULFITE PROCESS.

NUMBER 1 MANILA—A kind of wrapping paper, pale straw in color, which is made wholly or principally of chemical woodpulps.

NUMBER 2 WHITE MANILA—See WHITE MANILA PAPER.

NUMBER OF AFFECTED UNITS CHART (NP CHART)—A control chart for evaluating the stability of a process in terms of the total number of units in a sample in which an event of a given classification occurs.

NUMBER ONE FABRIC—See TWIN WIRE FORMING FABRICS.

NUMBER TWO FABRIC—See TWIN WIRE FORMING FABRICS.

NUTRIENTS—Elements or compounds essential as raw materials for organism growth and development; for example, carbon, oxygen, nitrogen, and phosphorus.

O

OATMEAL PAPER—A term used to describe a grade of hanging paper in which fine sawdust is added to the furnish to give the sheet a coarse effect.

O BLEACHING STAGE—See OXYGEN DELIGNIFICATION (O).

OBSTETRICAL SHEET—A soft, waterproof, sanitary paper, used as the name implies to replace a rubber sheet. It may be plasticized, waxed or parchmentized, and possesses high wet tensile strength.

OC—Oxygen Consumed. See CHEMICAL OXYGEN DEMAND.

OCC—Old corrugated containers. See RECYCLED CORRUGATING MEDIUM.

OCR—Abbreviation for optical character recognition. A special alphanumeric character set, legible by both human and machine readers.

OCR PAPER—Generally, a writing paper suitable for optical character recognition printing.

OCTAVO—Printed paper folded three times, giving eight sections or sixteen pages.

ODD—In papermaking, not in accordance with regular or standard sizes, weights, finishes, colors, etc.

ODOR CONTROL—A process for removing objectionable smelling substances from industrial operations by either mechanical or chemical means.

ODORS—Air emissions that offend the sense of smell.

ODOR THRESHOLD—The point at which after successive dilutions with odorless water, the odor of a water sample can just be detected. The threshold odor is expressed quantitatively by the number of times the sample is diluted with odorless water.

OFF COLOR—Not matching the color of the sample or specification.

OFFCUT—In making paper, it is not always possible to make the size required without waste, so frequently a sheet of an alternative standard size is cut from the waste. These sheets are termed offcuts. The expression is also applied to remainders of reams which have been cut down to a smaller size.

OFF-LINE TESTING—Evaluations made of the physical properties of paper, paperboard, etc., made on samples removed from rolls, reels, or sheeters. The tests are usually performed in a conditioned test laboratory environment.

OFF-MACHINE BROKE PULPER—Broke pulpers located away from the paper machine, but which handle all or mostly broke. They may be operated in a batch or continuous mode, and they may handle broke from a variety of broke-generating locations, and in a variety of forms (split or culled rolls, trim from various locations, sheeted broke, etc.).

OFF-MACHINE COATING—See CONVERSION COATING.

OFFICE PAPER SORTED GRADE—Dry, baled papers, as typically generated by offices, containing primarily white and colored groundwood free paper, free of unbleached fiber. May include a small percentage of groundwood computer printout and facsimile paper. (Institute of Scrap Recycling Industries, Inc. scrap specifications for 1994.)

OFFSET BLOTTING—A duplex commercial blotting with surface intended for printing by the offset lithographic process. See DUPLEX OFFSET BLOTTING PAPER.

OFFSET BOOK PAPER—See OFFSET PAPER.

OFFSET BRISTOL—Any bristol that has been specially sized for offset lithography.

OFFSET CARTRIDGE—A hard, well-sized, strong printing paper, with a rough finish, suitable for the offset printing press.

OFFSET GRAVURE—See INTAGLIO.

OFFSET LITHOGRAPHY—An adaptation of the principles of stone (or direct) lithography, in which the design is drawn or photographically reproduced upon a thin, flexible metal plate which is curved to fit a revolving cylinder. The design from this plate is transferred to or offset onto a rubber blanket carried upon another cylinder which, in turn, transfers the design to paper, cloth, metal, etc. See PHOTO-LITHOGRAPHY.

OFFSET PAPER—An uncoated or coated paper designed for use in offset lithography. The kind, type, and combinations of pulps used in its manufacture depend upon the sheet qualities desired. Important properties are good internal bonding, high surface strength, dimensional stability, lack of curl, and freedom from fuzz and foreign surface material.

OFFSET POSTCARD—A bristol or cardboard, made of bleached chemical hardwood and softwood pulps, with a kind of vellum finish and used for postcards, or special jobs, where it is desired that work will stand out and still show a soft finish.

OFFSET PRINTING—(1) A technique in printing by which the ink images are transferred from the plate first to an intermediate rubber blanket and then to the material being printed. This technique, which reduces plate wear and permits printing on rougher material, is most commonly associated with lithographic printing. For this reason, the word "offset" alone is sometimes used to indicate offset lithography. (2a) Undesirable transfer of ink in any printing process from a printed surface to an adjacent surface, sometimes referred to as set-off to differentiate it from offset. It frequently occurs in the bottom of a pile of printed sheets. It is caused by lack of ink receptivity of the paper or slow drying of the ink. (2b) The result of undried or excess ink accumulating on some portion of the press after the paper leaves the impression cylinder. This ink is transferred to the paper at the second impression and, if the registration is not perfect, the offset will give a shaded edge effect to the first impression type. See DOUBLING; DIRECT LITHOGRAPHY.

OFFSET SHEET—(1) Offset paper (q.v.). (2) See SLIP-SHEET PAPER.

OFF SQUARE—Cut or trimmed so that two or more corners of the sheet deviate from a 90° angle.

OHM—A unit of electrical resistance equal to the resistance of a circuit in which a potential difference of one volt produces a current of one ampere.

OIL ABSORBENCY—See ABSORBENCY.

OILED—Treated with any kind of oil; the type of oil depends on the intended use.

OILED MANILA—See LOIN PAPER; OILED PAPER.

OILED MANILA TYMPAN—See TYMPAN PAPER.

OILED OFFSET PAPER—An oiled jute and kraft paper used in printing processes for the prevention of offset from freshly printed pages.

OILED PAPER—(1) A strong paper treated with boiled linseed oil, or a mixture of oil and turpentine. Its chief use is in copying letter books. (2) Thinner pages, generally manila wrapping saturated with a neutral mineral oil, used for packaging purposes. See also LOIN PAPER.

OILED STENCIL—See STENCIL BOARD.

OILED TRACING PAPER—A tracing paper in which the transparent properties are secured by impregnating it with oil.

OILED TYMPAN PAPER—See TYMPAN PAPER.

OIL PENETRATION—See HOLDOUT; PERMEABILITY.

OILPROOF PAPER—A specially treated paper which resists penetration by oil, such as vegetable parchment paper and greaseproof paper (q.v.).

OIL SPOTS—See GREASE SPOTS.

OIL WETTABILITY—See WETTABILITY.

OK SHEET—(1) A proof sheet against which press sheets are compared to determine when make ready is completed and the press run may begin. (2) An okayed press sheet.

OLD CORRUGATED CONTAINERS—This grade consists of baled corrugated containers having liners of either kraft, test liner or jute. The term is usually abbreviated to OCC. See RECYCLED CORRUGATING MEDIUM.

OLEORESIN—Exuded gum from living pine trees. See ROSIN; TURPENTINE.

OMG—Old magazines.

ON DEMAND PRINTING—The ability to generate only the number of copies required, which is made possible by digital data storage and duplication techniques.

ONE HUNDRED SHEET SEALED—A package of 100 sheets of cardboard or bristol.

ONE-SIDE FINISH—Paper finished or decorated on one side only.

ONE-TIME CARBON PAPER—A paper made of chemical woodpulp or a mixture of chemical and mechanical pulps with a basis weight of 10 pounds (20 x 30 inches – 500) and heavier, which is coated with carbon and used with manifolding forms. See CARBON PAPER.

ONIONSKIN PAPER—A lightweight writing paper used primarily for making carbon copies of typewritten matter. It is usually made from chemical wood and/or cotton pulps in basis weights of 7 to 10 pounds (17 x 22 inches – 500), and in smooth, glazed, plated, supercalendered, or cockle finishes.

ON-LINE TESTING—Evaluations made of the physical properties of paper, paperboard, etc.,

made on the paper machine as part of the papermaking process. Typical measurements are moisture content, basis weight, and caliper.

ON-MACHINE COATING—See MACHINE COATING.

ONP—Old newspapers.

OPACIFIED BOND PAPER—A grade of writing or printing paper originally used where strength, durability and permanence are requirements and which contains an opacifying agent.

OPACIFIED BOOK PAPER—An uncoated paper used in the manufacture of magazines, books, pamphlets, and brochures and which contains an opacifying agent.

OPACIFIER—A compound added to paper during its manufacture to reduce the translucency of the sheet.

OPACITY—The property of a sheet that obstructs the passage of light and prevents seeing through the sheet objects on the opposite side. This property is especially important for printing paper. See PRINTED OPACITY; PRINTING OPACITY; SHOWTHROUGH; TAPPI OPACITY.

OPACITY BOOK PAPER—Usually a lightweight book paper made to have maximum opacity for a given thickness or weight. See BIBLE PAPER.

OPACITY PAPER—See OPAQUE PAPER.

OPAQUE—The property of being impervious to light and nontransparent. See OPACITY.

OPAQUE CIRCULAR PAPER—A term sometimes applied to writing or book paper with more than the ordinary opacity features. The standard basis weights are 16, 20, 24, and 28 pounds (17 x 22 inches – 500).

OPAQUE PAPER—Paper that has been manufactured to have more than ordinary opacity for a given weight, thickness, etc. See BIBLE PAPER.

OPAQUE WRAPPING—See BREAD WRAPPERS.

OPENING—The process of separating, disentangling, fluffing and sometimes partially cleaning compressed fibers to enhance processing characteristics on subsequent processing equipment.

OPPONENT COLOR SCALES *(Lab)*—The symbols *L, a, b* are used to designate measured values of three attributes of surface-color appearance: *L* represents lightness, increasing from zero for black to 100 for perfect white. *a* represents redness when plus, greenness when minus, and zero for gray. *b* represents yellowness when plus, blueness when minus, and zero for gray.

OPTICAL BRIGHTENERS—See FLUORESCENCE.

OPTICAL SENSORS—In pulp bleaching, optical devices which, by measuring the color changes in pulp as delignification progresses, continuously monitor the degree of bleaching or brightness of a pulp slurry at different stages along the bleaching operation so that the flow of bleaching agent required in a bleaching stage can be controlled either manually or electronically. The optical information obtained can be used to control flows upstream from the measuring point, which is called a "feedback" system, or the information can be used to control flows downstream from the measuring point, which is called a "feed forward" system. With an optical control system, the flow rate of a bleaching reagent is automatically and simultaneously adjusted so that the optimum dosage for the demand is applied, thus ensuring quality pulp at minimum cost. Optical sensors are commonly used to control chemical addition in the chlorination, hypochlorite and chlorine dioxide stages.

OPTICAL WHITENERS—See FLUORESCENT DYES.

ORANGE PEEL—(1) A pebbled surface similar to the skin of an orange in texture. (2) A term commonly used in paper mills to describe film split pattern (q.v.). (3) See also INK MOTTLE.

ORDER-BLANK PAPER—A term used to describe the sheet of paper frequently enclosed with advertising pieces and catalogs. The sheet is sized for pen and ink writing, has a smooth hard surface for printing and writing, and is strong enough to withstand considerable usage. A variety of grades is commonly used including bond, tablet, manilas, and railroad manilas. It is usually supplied in various light colors. When used in multiple forms where carbon copies are desired, it is known as sales-book paper (q.v.).

ORDINANCE PAPER—See BARRIER MATERIAL.

ORGANIC CHLORINE COMPOUNDS—In the bleaching of chemical pulp (kraft, sulfite, or soda) the chlorine compounds, chlorine hypochlorite and chlorine dioxide, react with lignin and other compounds in the pulp to form organic-chlorine compounds. Depending on the molecular weight size and polarity of these compounds, some of them enter the water wash. These are known as AOX (adsorbable organic halogens). Because not all the organochlorines are soluble, some remain in the pulp. These are higher molecular weight chlorolignins and chlorinated resin compounds. They are called OX, and are the organic chlorine compounds left in the pulp.

ORGANIC FELT—A felt made from rags, wood fiber, or paper. It is distinguished from mineral felt such as asbestos. See ASPHALT FELT; SATURATED FELT.

ORGANOCELL PROCESS—See SOLVENT PULPING.

ORGANOCHLORINE—See ORGANIC CHLORINE COMPOUNDS.

ORPHAN—See WIDOW.

OSCILLATION—In finishing operations, the process of mechanically moving the unwind stand back and forth in the cross-machine di-

211

rection a controlled displacement and at a controlled speed while the web is unwinding. The purpose of this repeated displacement is to distribute the effects of non-uniform paper with the intent of reducing wound roll defects. In some winding operations, the windup is oscillated during the winding process, rather than the unwind.

OSMOSIS—The tendency of a fluid to pass through a semi-permeable membrane into a solution where the concentration is lower.

OUTDOOR SIGN PAPER—See SIGN PAPER; POSTER PAPER.

OUTER BACKING FABRIC—See TWIN WIRE FORMING FABRICS.

OUTER CONVEYING FABRIC—See TWIN WIRE FORMING FABRICS.

OUTFALLS—Liquid discharges from a treatment system into a stream or lake.

OUT-OF-CONTROL PROCESS—A process in which the statistical measure being evaluated is not in a state of statistical control (i.e., the variations among the observed sampling results cannot solely be attributed to a constant system of common causes and the existence of assignable causes is highly probable). See also IN-CONTROL PROCESS.

OUT-OF-ROUND ROLL—A roll that is non-concentric with the core, or flat on one side, or squeezed by a clamp tractor. This condition usually occurs after concentric winding and is due to on-side storage procedures or poor handling coupled with poor roll structure. It is also caused by impacts during shipping.

OUTTHROWS—The term is used in the recovered paper industry to designate all papers that are so manufactured or treated, or are in such a form as to be unsuitable for consumption in the grade specified.

OUTTURNS—Samples which represent the paper made on different runs of the paper ma-chine; they are kept by the mill and may be sent to the customer.

OUTTURN SAMPLES—See OUTTURNS.

OVENDRY—Containing practically no moisture. A paper or pulp is said to be ovendry when it has been dried in an oven at 105°C until its weight has become constant within 0.1%.

OVERCOOK—See SOFT COOK.

OVERDRIED—A term applied to paper which has been excessively dried with a resulting increase in brittleness and a loss of inherent strength.

OVERISSUE NEWS—Consists of unused, over-run regular newspapers printed on newsprint, baled or securely tied in bundles, containing not more than a normal percentage of rotogravure and colored sections.

OVERLAY—A piece of paper put on the tympan of a printing press to give more impression to a letter, line, or engraving.

OVERLAY PAPER—(1) A high purity paper for impregnation with a synthetic resin and molded as the to player of a decorative laminate. The treated overlay paper becomes substantially transparent during the molding procedure and gives added protection to the underlying layer. (2) See CHALK OVERLAY PAPER. Other materials than chalk may be used as a coating.

OVERLIMING—The practice of using more lime in the slaker than is chemically required to convert sodium carbonate to caustic (q.v.).

OVERRUN—A quantity of paper made in excess of the amount ordered. Trade custom allows a certain tolerance for overruns and underruns.

OVERSIZE—(1) Paper made to allow for trimming to the size ordered. (2) Paper larger than ordered.

OVERWEIGHT—Heavier than ordered or specified.

OXFORD BIBLE PAPER—See BIBLE PAPER.

OXFORD INDIA PAPER—See BIBLE PAPER.

OXIDATION POND—A basin used for retention of wastewater before final disposal, in which biological oxidation of organic material is effected by natural or artificially accelerated transfer of oxygen to the water from air.

OXIDATION TREATMENT—The process whereby, through the agency of living organisms in the presence of oxygen, the organic matter contained in wastewater is converted into a more stable or mineral form.

OXIDATIVE BLEACHING AGENTS—Bleaching chemicals which act by oxidizing the lignin. The relative rates of reaction with lignin and carbohydrates are different, and minimum damage to the pulp is achieved by maintaining optimum conditions. Oxidizing agents used in bleaching are chlorine dioxide, chlorite or chlorous acid, hypochlorite, peroxide, oxygen, and ozone.

OXIDATIVE EXTRACTION STAGE (Eo)—Eo. In pulp bleaching, a caustic extraction stage enhanced by injecting gaseous oxygen into the pulp at the beginning of the stage. The pulp is maintained at low pressure, typically 20 to 30 psig, for 3 to 10 minutes. It has been shown to improve the extent of delignification as measured by the kappa number of permanganate number (q.v.).

OXIDIZED WHITE LIQUOR—Kraft white liquor (q.v.) in which the sodium sulfide is oxidized to sodium thiosulfate or to sodium sulfate or to both using air or oxygen as the oxidizing agent. Oxidized white liquor is used as a substitute for purchased caustic in the oxygen delignification stage of pulp and, to a lesser extent, in the extraction stage where oxygen is used to enhance further delignification of a chlorinated pulp.

OXYGENATION—The act of putting oxygen into a waste or stream.

OXYGEN BALANCE—(1) The dissolved-oxygen level at any point in a stream, resulting from the opposing forces of deoxygenation and reaeration. (2) The relation between the biochemical oxygen demand of a wastewater or treatment plant effluent and the oxygen available in the diluting water. (3) The quantity of oxygen used in the biochemical oxidation of organic matter in a specified time, at a specified temperature, and under specified conditions. See BIOCHEMICAL OXYGEN DEMAND.

OXYGEN BLEACHING—A relatively modern bleaching process for wood and other pulps, involving the use of oxygen in an alkaline medium to reduce lignin and other dark-colored components. This process reduces the overall level of bleaching chemicals and lowers the biological oxygen demand of the effluent.

OXYGEN CONSUMED—See CHEMICAL OXYGEN DEMAND.

OXYGEN DELIGNIFICATION (O)—The treatment of wood or pulps with oxygen in an alkaline medium oxidizes, degrades, and solubilizes the lignin. Oxygen delignification is an established technology as a stage between the pulping digesters and the bleach plant to remove up to half of the residual lignin in the brownstock without unduly damaging the strength properties of the pulp. The remaining lignin is removed by the conventional chlorine bleaching methods, but with less chlorine-based chemicals, which reduces the pollution caused by the chlorinated organic compounds in the effluent. Magnesium may be added to protect the cellulose from degradation during oxygen delignification. The chemicals used in oxygen delignification are compatible with the kraft process, making it possible to recycle the oxygen stage filtrate to the recovery furnace.

OXYGEN DEMAND—The quality of oxygen utilized in the biochemical oxidation of organic matter in a specified time, at a specified temperature and under specified conditions. See BOD.

OYSTER WHITE LINERBOARD—See MOTTLED WHITE LINERBOARD.

OZONE (O$_3$)—An unstable, toxic, bluish gas with a very high electronegative oxidation potential used commercially in various chemical oxidation processes, such as bleaching of textiles, waxes and starch, and as a bactericide and algaecide. It is an effective delignifying and brightening agent for woodpulp but, because it is not specific to reactions with lignin, the process variables must be carefully controlled to minimize cellulose degradation.

OZONE PAPER—See TEST PAPERS.

OZONE STAGE (Z)—The ozone stage is used to delignify pulps to very low kappa numbers, normally in a total chlorine compound free bleaching sequence. Due to the very fast reaction rate and the large amount of oxygen gas present, mixing is very important. The ozone stage is closely coupled to the ozone generating plant and, in many cases, to downstream users of oxygen gas. Typical conditions in an ozone stage are temperatures of 50°C to 60°C and pH of 2 to 4. The reaction is so rapid that only several minutes of retention time is required.

P

P&J PLASTOMETER—See Pusey and Jones Plastometer.

Pa BLEACHING STAGE—See PERACETIC ACID.

PACKAGING—A process whereby paper products are packaged or wrapped for protection during the shipping or storage phase.

PACKAGING PAPERS—Paper used for bags, pouches, and other packaging. Often subdivided into "food" and "non-food" categories. Includes wax paper, polyethylene-coated papers, grease-resistant papers, and others.

PACKERS OILED MANILA—A manila paper used for wrapping certain types of meat. It is made from chemical and mechanical woodpulps and given a special oil treatment.

PACKING PAPER—A general term applied to any paper used for packing purposes.

PACKLESS MAT—See MATRIX BOARD (1).

PADS—Corrugated or solid fiberboard used as added protection or for separating tiers of articles, when packed for shipment. Pads usually are cut to meet the inside dimensions of the container. See INTERIOR PACKING.

PAGE COMPOSITION—On-screen assembly of the components of a page.

PALLET—A prefabricated platform or skid (q.v.) used to support rolls, cartons, or piles of paper during transport or storage of the product.

PALLETIZING—The process of placing paper rolls, cartons, or piles of paper onto a skid or pallet.

PAMPHLET COVER—Cover papers made of varying quantities of cotton fiber and chemical woodpulps in various thicknesses, colors, designs, and finishes. They are used to cover pamphlets, booklets, catalogs, and the like, which are saddle stitched but not sewn.

PAMPHLET PAPER—Any uncoated or coated printing paper used for pamphlets.

PANEL BOARD—A paperboard used in the automobile and building trade. It is a solid woodpulp board or a combination board, made on a wet machine and lined on one or both sides with woodpulp. It may be pasted to give the required thickness. It is rigid, resistant to abuse, and usually treated so that it is water- and moisture-proof. It may also be a tough leatherboard of 2 to 6 plies, of various compositions according to the use to which it is to be put; it is sometimes made of rags and wastepaper saturated with oil and baked. See AUTOMOBILE BOARD.

PAN LINER—(1) A sheet of paper made of bleached or unbleached chemical pulp, in basis weights 30, 35, and 40 pounds (24 x 36 inches – 480). (2) A release coated vegetable parchment paper, in basis weights 27 and 35 pounds (24 x 36 inches – 500), used in the bakery trade for pastry tray liners and bakery pan liners.

PANTONE MATCHING SYSTEM—Abbreviated PMS, a color matching system that uses a book of various color patches to describe color. A typical color designation, such as PMS 288, describes a dark blue of a specific hue and intensity.

PAPER—(1) (General term). The name for all kinds of matted or felted sheets of fiber (usually vegetable, but sometimes mineral, animal, or synthetic) formed on a fine screen from a water suspension. Paper derives its name from papyrus, a sheet made by pasting together thin sections of an Egyptian reed *(Cyperus papyrus)* and used in ancient times as a writing material. (2) (Specific term). One of the two broad subdivisions of paper (general term), the other being paperboard (q.v.). The distinction between paper and paperboard is not sharp but, generally speaking, paper is lighter in basis weight, thinner, and more flexible than paperboard. Its largest uses are for printing, writing, wrapping, and sanitary purposes, although it is also employed for a very wide variety of other uses.

PAPER ADDITIVES—Materials that are added to paper and paperboard to provide special end-use characteristics, such as wet-strength resins, fungicides, plasticizers, flame retardants, water repellents, antimycotic agents, and antioxidants.

PAPER-BAG LINERS—See BAG LINERS.

PAPER-BASE LAMINATE—The product obtained by impregnating or coating a paper with a thermosetting resin solution, and drying and pressing a number of layers of the treated paper until the resin is fully cured. Decorative laminates are used for counter and table tops, wallcoverings, etc. Industrial laminates are used for gears, refrigerator inner doors, etc. A given product may contain more than one kind of paper with different quantities or kinds or resins.

PAPER-BASE PLASTICS—See PAPER-BASE LAMINATE.

PAPERBOARD—One of the two broad subdivisions of paper (general term) the other being paper (specific term) (q.v.). The distinction between paperboard and paper is not sharp but, broadly speaking, paperboard is heavier in basis weight, thicker, and more rigid than paper. In general, all sheets 12 points (0.012 inch) or more in thickness are classified as paperboard. There are a number of exceptions based upon traditional nomenclature. For example, blotting paper, felts, and drawing paper in excess of 12 points are classified as paper, while corrugating medium, chipboard, and linerboard less than 12 points are classified as paperboard. Paperboard is made from a wide variety of furnishes on a number of types of machines, principally cylinder and fourdrinier. The broad classes are: (q.v.) (a) containerboard, which is used for corrugated boxes, (b) boxboard, which is principally used to make cartons, and (c) all other paperboard.

PAPER CALENDER—See CALENDER; SUPERCALENDER.

PAPER CLAY—A white or light-colored clay, very low in free silica, typically called kaolin. Most paper clays, as marketed, have been beneficiated to attain properties requisite for filling or coating use. See ATTAPULGITE CLAY; BENTONITE; CLAY; COATING CLAY; FILLER CLAY; KAOLIN.

PAPER CLOTH—(1) A kind of cloth faced with paper. (2) A fabric made by the Polynesians from the innerbark of the paper mulberry and other trees. (3) Twisted paper woven or knitted into fabrics. (4) Any of several types of papers specially processed for use as window draperies, shades, dusting cloths, polishing cloths, table coverings, bookbinding, and the like.

PAPER MACHINE CHEST—See MACHINE CHEST.

PAPER MACHINE COATED PAPER—See MACHINE COATED.

PAPER MACHINE FORMING FABRICS—See FORMING FABRIC.

PAPER MACHINES—See CYLINDER MACHINE; FOURDRINIER MACHINE; HARPER MACHINE; INVERFORM; STEVENS FORMER; TWIN WIRE FORM-ER; WET LAP MACHINE; YANKEE MACHINE; YANKEE MG MACHINE.

PAPER MACHINE WINDER—See REWINDER; WINDER.

PAPER-MACHE—(1) A molding material made from woodpulp or more typically from repulped paper stock to which glue or other adhesive has been added. It is used for a variety of articles of merchandise, models, relief maps, etc. (2) A lightweight product molded from the above material to which linseed oil, varnish, lacquer, or other protective or decorative finish has been applied after drying.

PAPER MACHINE SCREEN—A screen placed immediately before the paper machine headbox or former, to protect the former and forming fabric, deflocculate the stock, and remove tramp debris just before the sheet is formed. Paper machine screen systems may include primary, secondary, tertiary, etc., screens. See also SCREENING SYSTEM.

PAPERMAKERS ALUM—See ALUM.

PAPERMAKERS' FELT—See PRESS FELT.

PAPER MULBERRY—A small tree (*Broussonetia papyrifera*) of the Far East whose bast fiber from the innerbark is of considerable importance in the manufacture of special papers, especially hand-made papers. It is also known as kozo.

PAPER NAPKINS—Special tissues, white or colored, plain or printed, usually folded, and made in a variety of sizes, for use during meals or with beverages. Single ply napkins, usually embossed, are made in basis weights from 12 to 16 pounds (24 x 36 inches – 500). Cellulose or facial tissue napkins are made from special facial-type stock, usually possessing wet strength, in 2 or more plies, each approximately 10 to 10.5 pounds in basis weight (24 x 36 inches – 500). Plain tissue napkins are usually supplied flat, in size 12 x 12 inches and in a basis weight of about 10 pounds (24 x 36 inches – 480).

PAPER SHIPPING SACK—See SHIPPING SACK.

PAPER SPECKS—In papers made from recovered stock, the term refers to the undefibered pieces of the reclaimed paper. Also called flakes and, sometimes, nits or knots.

PAPER STOCK—Recovered papers, specifically such material sorted or segregated at the source into various recognized grades such as special news, new kraft corrugated cuttings, old corrugated containers, manila tabulating cards, and coated soft white shavings. Paper stock is used as a principal ingredient in the manufacture of certain types of paperboards, particularly boxboard made on cylinder machines where the lower grades may go into filler (q.v.) stock, and the higher grades into one or both liners (q.v.). Selected grades are also used in the manufacture of various papers.

PAPER TAPE—See PERFORATOR TAPE.

PAPER TEXTILES—A general term descriptive of various fabrics made of paper which has been twisted into yarn and then woven or knitted. See also NONWOVEN.

PAPER TEXTILE SPOOLS—Spools made from a combination of lightweight news-base stock, with surfaces of vatliners in various colors. They are used for winding many types of textile products, also for winding wire, solder, belting, etc.

PAPER TOWELS—Paper toweling in folded sheets, or in roll form, for use in drying, or cleaning, or where quick absorption is required.

It is sometimes embossed during the converting process. See TOWELING.

PAPER TWINE—Narrow strips of paper, twisted into form like twine. The paper is moistened just before twisting to preserve the twist; it is used for tying small packages. See TWISTING PAPER.

PAPER WASTE—In printing, that paper consumed that does not become part of a good product. Common subcategories include: (1) White waste—that part of consumed paper (not including wrappers) which has not been printed, such as that remaining on the core of a roll when it is removed from a reel (core waste); (2) Printed waste—all printed paper that does not leave the pressroom as work in process; (3) Handling and transit waste—in web printing (especially newspaper), all paper (not including wrappers) taken from a roll before the paper is started through the press (part of white waste); (4) Makeready waste—in web printing, that part of the paper printed after startup until the moment that OK signatures or sheets are achieved (part of printed waste); (5) Run or running waste—paper printed after OK signatures or sheets begin but that does not leave the pressroom as good work in process (part of printed waste).

PAPER WRAP—See WRAP.

PAPETERIE PAPERS—(1) Papeterie papers, which are generally uncoated, are widely used in the manufacture of greeting cards and invitations. They are designed to perform in a variety of printing, embossing, die cutting, stamping, and laminating operations. Papeteries must have superior appearance characteristics, excellent folding properties, sufficient stiffness, and contain sizing for writing inks. (2) A class of offset papers which are typically low strength, high bulk, and have good folding qualities. They are made in basis weights from 16 to 31 pounds (17 x 22 inches – 500).

PAPYRUS—A tall sedge *(Cyperus papyrus)* native to the Nile region, the pith of which was sliced and pressed into matted sheets and used for writing material by the ancient Egyptians, Greeks, and Romans. This is the forerunner of paper and the origin of the word.

PARAFFIN—A white waxy substance obtained from petroleum. It is commonly used as a waterproof coating or laminant in the paper industry. See MICROCRYSTALLINE WAX.

PARAFFINED BOARD—A paperboard made of long-fibered stocks or lined with such stocks, which has been subsequently treated with paraffin. Such boards have a smooth surface and are relatively waterproof. See WAXED PAPER.

PARAFFIN PAPER—Paper treated with paraffin wax to render it waterproof. See WAXED PAPER.

PARALLEL-LAID—A nonwoven produced using one or more cards in tandem where the fibers are oriented primarily in one direction.

PARALLEL LAMINATED—A laminate in which all the layers of material are oriented approximately parallel with respect to the grain or strongest direction in tension. See also CROSS LAMINATED.

PARCHMENT—(1) A sheet of writing material prepared from the skins of goats, sheep, and other animals. (2) See VEGETABLE PARCHMENT PAPER. (3) See ARTIFICIAL PARCHMENT; IMITATION PARCHMENT.

PARCHMENT BOND—(1) A parchment-like writing paper used as a substitute for animal or vegetable parchment for bonds, posters, deeds, etc. It is a high-quality sheet made from cotton and bleached chemical woodpulps which are usually well beaten to produce mechanical parchmentizing (hydration) of the fiber. The paper is tub-sized, and loft, air, or machine dried. The basis weights range from 24 to 40 pounds (17 x 22 inches – 500). Durability, toughness, and velvet surface are significant properties. (2) See ERASABLE PARCHMENT BOND.

PARCHMENT DEED—(1) A well-hydrated bond paper with the appearance of parchment. (2) See DOCUMENT PARCHMENT.

PARCHMENT FINISH—A finish resembling genuine parchment, produced in a plater by bunched plating. This finish is obtainable only with a very hard paper. The surface is smooth but uneven and has little glare.

PARCHMENTIZING—(1) Chemical process. The treatment of unsized cotton or purified chemical woodpulp paper by sulfuric acid or other chemicals under controlled conditions to produce vegetable parchment paper (q.v.). (2) Mechanical process. The beating of sulfite or sulfate pulp of the proper quality for a sufficient time under controlled conditions to produce a pulp suitable for greaseproof paper.

PARCHMENTIZING PAPER—A waterleaf sheet used in the production of vegetable parchment. See VEGETABLE PARCHMENT PAPER.

PARCHMENT PAPER—A grade of genuine vegetable or imitation parchment paper having a vellum finish for writing or printing end-uses.

PARCHMENT REPOUSSÉ—Typical parchment with a special finish or surface to imitate hammered metal.

PARCHMENT TRACING PAPER—Tracing paper in which the transparent properties are obtained by parchmentizing with chemicals, such as sulfuric acid, or by prolonged beating or refining.

PARCHMENT VELLUM—A parchment with a vellum finish.

PARCHMENT WRITING—A vegetable parchment paper (q.v.) used for documents such as deeds. It is made from cotton fiber and/or chemical woodpulps in basis weights of 28 to 36 pounds (17 x 22 inches – 500). Permanence, durability, and strength are significant properties. See also DOCUMENT PARCHMENT.

PARENCHYMA—Any tissue of thin-walled living cells. These cells are closely associated with the resin canals, the ray tissue in the softwoods, and with the vessels and the ray tissue of hardwoods. In the living bark tissue, they form an important part; but, in the dead bark in their collapsed state, they are located only with difficulty. These small cells add no strength to the paper and may contain extractives that appear as pulp dirt.

PARETO CHART—A graphical tool for ranking causes from the most significant to the least significant.

PARTICLE BOARD—A rigid board consisting of small discrete particles of wood or similar material combined with a synthetic resin adhesive and bonded or cured under controlled heat and pressure. Customarily made in thicknesses of 0.25 to 1 inch.

PARTICLE SIZE ANALYSIS—A test to determine the distribution of different size pulpwood chips by means of round-hole and slotted screens.

PARTICLE SIZE DISTRIBUTION—The percentages expressed in terms of weight of the various sizes of particles in a powder sample when classified in terms of size ranges and measured in terms of screen mesh or microns.

PARTICULATE EMISSIONS—The emission of solids or liquids in a gas stream. Current concern is over particles less than 10 microns in size because of their ability to penetrate the respiratory system.

PARTITION CHIPBOARD—A highly sized, low-density, stiff chipboard with a caliper of 50 to 80 points commonly used as partitions in beer and soft drink cases.

PARTITIONS—Corrugated or solid fiberboard pieces scored and/or slotted so as to form separate cells within a container. Also called dividers (q.v.).

PASSE PARTOUT—A strong embossed paper, gummed on one side, cut into narrow strips, and sold in rolls about two inches in diameter and one inch wide. It is made in many colors and is also coated with metallic bronzes; black is quite frequently used for the binding of lantern slides, though the other colors are frequently used in the mounting of pictures.

PASSIVE—In corrosion, the state of the metal surface characterized by low corrosion rates in a potential range that is strongly oxidizing to the metal. Passivity is typically observed in metals or alloys protected by a thin or invisible oxide film (e.g., aluminum, titanium, stainless steels).

PASTED—Formed of two or more layers (which may or may not be of the same stock) that have been pasted together as a separate operation from their manufacture on the paper machine.

PASTED BLOTTING—Any blotting paper which has been laminated to another paper or to another blotting.

PASTED BOARD—A paperboard used for many purposes where a stiff, thick board is required. It is made of two or more sheets of board or of board and paper pasted together in a subsequent operation.

PASTED BRISTOL—A term applied to bristols in which two or more thicknesses of paper, which may or may not be of the same stock, are pasted together as a secondary operation. This is generally done by machine, but in the case of some wedding bristols, hand pasting is used.

PASTED CHIPBOARD—A paperboard consisting of two or more layers of chipboard pasted together. See CHIPBOARD.

PASTED COVER PAPER—A term applied to a number of layers of cover paper prepared by pasting two or more sheets together. It is often called double thick, triple thick, or double-double thick, depending upon the number of thicknesses pasted together. In many cases, the individual plies have a basis weight of 65 pounds (20 x 36 inches – 500).

PASTED INDEX BRISTOL—Index bristol in which two or more thicknesses of dry paper have been pasted together as an operation after the paper has left the paper machine. The different sheets pasted together may or may not be of the same stock. Flour paste, starch, or

sodium silicate (water glass) may be used as the adhesive. The bristol may be roll or sheet pasted. See also PASTED WEDDING BRISTOL.

PASTE DOWNS—See END-LEAF PAPER.

PASTED WEDDING BRISTOL—A high grade heavyweight paper comprising two or more pasted plies of cotton content bristol paper. It is designed for wedding invitations, announcements, etc., and is characterized by superior aesthetic features such as brightness, cleanliness, and overall appearance.

PASTER—A machine for pasting two or more sheets together, either in a continuous roll or as separate sheets. See SPLICE.

PASTING—The process of uniting, by means of an adhesive, two or more sheets of paper or paperboard, paper to board, or coated or offset paper to blotting paper. Paper may be pasted off the reel or in the web (roll machine) or in sheets (sheet-pasting machine). See also LAMINATING.

PASTINGS—(1) Any thin paper used for pasting purposes, such as facings for pasteboards. The finish on one side is usually rough to enhance the adhesion of the paste. (2) Cuttings or clippings of paper from the pasting machine.

PATCH MARK—A watermark made with a wire mark patch sewed into the wire of a mold on a cylinder machine or into a dandy roll above the forming fabric.

PATCH STOCK—A heavy paper or paperboard, usually of a basis weight of 120 pounds (22 x 28 inches – 500) and gummed on one side, made of rope or chemical woodpulp or mixtures thereof. It is used for reinforcement of eyelets on large, open-end envelopes and tags.

PATENT COATED—A cylinder grade of boxboard which has been vat lined with white fibers. The board is not clay coated. The term is frequently used in conjunction with additional descriptive terms to specify a grade of board.

For example, patent coated news manila back has a white top liner, a news filler, and a manila-colored back liner.

PATTERN BOARD—A paperboard used as a pattern for cutting various shapes, such as parts of shoes, and clothing. It is made of woodpulp and/or reclaimed paper stock into a strong, dense sheet of 0.025 of an inch or more in thickness, which is heavily calendered. The board is hard and capable of withstanding considerable handling; it should give a clean edge when cut and resist warping or shrinking.

PATTERN-BONDED—A nonwoven bonded with binder or a heated calender roll where the bonding points are in some form of a pattern.

PATTERN FIBER—A grade of vulcanized fiber (q.v.) having high dimensional stability and resistance to warping. Its principal use is as patterns in cutting cloth, leather, and other materials. It is made in thicknesses from 0.030 to 0.060 inch, usually in black, gray, and red colors.

PATTERN PAPER—Paper used by designers and tailors for the making of patterns ranging in thickness from 0.007 to 0.034 inch. The surface is suitable for pencil or crayon markings. The furnish is usually kraft but may contain rope stock, depending upon the strength required. The paper, manufactured in a variety of colors, is sometimes referred to as x paper.

PATTERN TISSUE—A tissue paper used primarily for cutting patterns for household use. It is usually made of unbleached sulfite or sulfate pulps, most of it in basis weight 9 pounds (24 x 36 inches – 480). Pattern tissue is generally offered only in its natural color which is similar to manila. It is supplied in jumbo rolls and in large sheets from 50 x 80 to 50 x 120 inches in size.

PAY-OUT—The process of rotating the roll of paper in the unwind to add more slack to the draw, or to facilitate the web threading process.

P BLEACHING STAGE—See HYDROGEN PEROXIDE STAGE (P).

PC—Abbreviation for personal computer. In practice, refers to the DOS-based family of computers developed by IBM and others. Consists of a keyboard, a video display panel or screen, and processing unit.

PCB—An acronym covering a group of chemicals known as polychlorinated biphenyl compounds.

PCC—Precipitated calcium carbonate.

P CHART—See PERCENT CHART.

PC NUMBER—See BRIGHTNESS REVERSION; POST COLOR NUMBER.

PEARL FILLER—Anhydrous calcium sulfate used as a filler in papermaking. It occurs naturally as anhydrite or may be obtained by dehydrating gypsum (q.v.).

PEARL HARDENING—See CROWN FILLER.

PEARL STARCH—Unmodified cornstarch in the form of small granules, as contrasted with powdered starch. The term was originally used more broadly to describe starch in granular form.

PEARL WHITE—See BARIUM SULFATE.

PEAR WRAPS—See APPLE AND PEAR WRAPS; FRUIT WRAPS.

PECTINASES—Enzymes capable of cleaving linkages between the Beta-1, 4-linked glucuronic acid residues of pectin. Pectinases are hydrolytic or transeliminative.

PEELED WOOD—Pulpwood from which bark has been removed by peeling (q.v.). Obsolete practice.

PEELING—(1) Manual removal of bark from pulpwood, either with or without preliminary treatment with chemicals. Obsolete practice. (2) Scaling off of the surface of paper.

PELLETIZING—Conversion of materials into small, regularly shaped pellets.

PEN CARBON PAPER—See CARBON PAPER.

PENCIL CARBON PAPER—See CARBON PAPER.

PENCIL MASTER PAPER—Pencil master paper is designed for use with a hectograph pencil and for masters where a combination typewriter ribbon and pencil or carbon and pencil is used. It usually has an antique finish which draws the maximum amount of graphite from the hectograph pencil to produce long runs and bright copies. It is made in weights of 13, 16, and 20 pounds (17 x 22 inches – 500).

PENCIL PAPER—See LEAD-PENCIL PAPER; PENCIL-TABLET PAPER.

PENCIL POINT—See SLIPPED ROLL.

PENCIL TABLET PAPER—A low-grade paper made largely from mechanical pulp and designed primarily for pencil writing, although some variations thereof are suitable for pen and ink writing. It is usually made in white and canary shades, in a basis weight of 32 pounds (24 x 36 inches – 500), and is frequently used for scratch pads, tablets, and the like.

PENETRATION—See ABSORBENCY; IMPREGNATION; PERMEABILITY; STRIKE-IN.

PENETRATION OF INK VEHICLE—(1) The process by which vehicle is absorbed from a printed ink film into the printed material causing the ink film to set. (2) The ability of an ink vehicle to penetrate or be absorbed by a paper surface.

PERACETIC ACID—CH_3CO_3H. An oxidative bleaching agent which, at neutral pH, will brighten groundwood to high brightness by destroying the chromophores without dissolving the lignin. Bleaching of pulp by peracetic acid (Pa stage) is also achieved by adding hydrogen peroxide and acetic anhydride to pulp. Peracetic acid can be used also to delignify chemical pulps although strength losses are generally higher than with chlorine dioxide. It is not in commercial use.

PERCENT CHART (P CHART)—A control chart for evaluating the stability of a process in terms of the percent of the total number of units in a sample in which an event of a given classification occurs. Also referred to as a proportion chart.

PERCENT POINTS—An expression sometimes used for a strength factor calculated to a basis weight of 100 pounds. The factor is calculated by dividing the actual test result by the actual basis weight and multiplying by 100.

PERCENT WET TENSILE—The tensile strength of a paper when thoroughly saturated with water, expressed as a percentage of the air dry tensile strength of the same paper. The percent wet tensile of waterleaf paper is on the order of 5% which may be increased to 30% or higher by treatment with wet-strengthening agents. See WET-STRENGTH PAPER.

PERFECT—In papermaking, a term applied to paper free from defects.

PERFECT BINDING—A method of binding books, sometimes called adhesive binding, in which the signatures have been converted to individual pages by cutting, milling, or sanding off the folded edges while being held in a clamp. They are attached to a paperboard cover with an adhesive.

PERFECTING PRESS—A press that prints both sides of a sheet in one press pass.

PERFORATED ROLL—See DISTRIBUTOR ROLL.

PERFORATING—Punching lines of small holes or slits in a sheet so that it may afterwards be torn off with ease.

PERFORATING PAPER—A paper which will give clean-cut edges on punching.

PERFORATOR TAPE—A converted paper product in small roll or fan-folded form used for communications equipment, automatic machinery, business machines, typesetters, etc. Such tape is perforated (i.e., "punched") in use and these perforations transmit appropriate signals to actuating mechanisms, etc. Perforator base paper is usually made from chemical woodpulps to exacting specifications, the most important being uniform caliper, freedom from grit and mineral filler, high tensile strength, clean perforating ability, and good oil receptivity. The usual basis weight is 51 pounds (24 x 36 inches – 500) and the standard caliper is about 0.004 inch. Perforator tape may be oil-treated (usually from 12 to 22% mineral oil content by weight) or plain.

PERGAMYN—See GLASSINE PAPER.

PERMANENCE—(1) That property ascribed to a material which, under specified conditions, resists changes in any or all of its properties with the passage of time. (2) The permanence of paper refers to the retention of significant use properties, particularly folding endurance and color, over prolonged periods. The permanence is affected by temperature, humidity, light, and the presence of chemical agents. The probable permanence of paper is estimated by an accelerated oven-aging test or by tests under other specified conditions of temperature, light, and humidity. The evaluation of permanence is based on measurement of folding endurance, resistance to water penetration, color, solubility in aqueous alkaline solutions, and viscosity of a solution of the fibers in a cellulose solvent.

PERMANENT PAPER—Paper that is chemically stable, that is, resistant to deterioration caused either by internal chemical reactions or by environmental factors such as humidity and light, under normal conditions of storage and use. In practice, a minimum durability (resistance to wear and tear) is often assumed or specified. Perhaps the most important factors in permanence are an alkaline pH and an alkaline reserve, such as calcium carbonate; but paper made by very stable fibers such as cotton can last for centuries even at a slightly acidic pH.

PERMANENT WHITE—See BARIUM SULFATE.

PERMANGANATE NUMBER (K NUMBER)—A test value related to the lignin content of the pulp and therefore the bleaching chemical demand. The K number is the number of milliliters on $0.1N$ $KMnO_4$ (potassium permanganate) solution reduced by 1 gram of ovendried pulp. Two volumes of $KMnO_4$, 25 and 40 ml, are used to suit pulps of lower and higher lignin contents respectively. The reaction time is shorter (5 min.) than in the kappa number test, and no constant temperature is specified. The K number is not as precise as kappa number (q.v.) over a wide range of lignin content. The method continues to be used as a control tool because of its relative simplicity. For the lignin content of pulp entering most conventional bleach plants, the K number is approximately equivalent to kappa number/0.66.

PERMEABILITY—(1) The ease with which a fluid strikes into, permeates, or transudes paper. (2) A value indicating how freely a gas or liquid can pass through a material. (3) In paper machine clothing, permeability is usually measured in cubic feet of air per minute (CFM) that pass through a defined area of the felt or fabric when subjected to a differential pressure of 1/2 inches water.

PEROXIDE BLEACHING STAGE (P)—See HYDROGEN PEROXIDE BLEACHING STAGE (P).

PGW—Pressurized groundwood (q.v.).

pH—An expression of the hydrogen-ion concentration, and thus the acidity or alkalinity, of an aqueous solution. The pH value is the negative logarithm, to the base ten, of the hydrogen-ion concentration. A pH of 7 represents a neutral solution; decreasing pH values below 7 represent increasing acidity, and increasing values above 7 represent increasing alkalinity. The pH values of hot or cold aqueous extracts are empirically correlated with properties of paper such as its permanence, its reaction with the fountain etch in offset printing, and others.

PHARMACEUTICAL PAPER—See DRUG WRAPPING.

PHENOLIC SHEET—A laminated sheet produced by impregnating a base material, such as paper, canvas, or linen, with a phenolic resin and subjecting the sheet to heat and pressure, which fuses the material into a dense, insoluble, and homogeneous product. See also PAPER-BASE LAMINATE.

PHLOEM—The conducting tissue present in trees and other plants, which is chiefly concerned with the aqueous transport of food materials within such plants.

PHOSPHORESCENT PAPER—A paper which has been treated to glow after irradiation with light. A fluorescent paper sometimes exhibits weak phosphorescence. See also FLUORESCENT PAPER.

PHOTO ALBUM PAPER—See ALBUM PAPER.

PHOTO BLACK PAPER—See BLACK ALBUM PAPER.

PHOTOCOPYING PAPER—See COPY PAPER.

PHOTOELECTRIC PROCESS BASE STOCK—A coated base stock to be converted into paper for use in photoelectric reproduction equipment. The coated base stock should have the properties of good toluene holdout, a resistivity which is not higher than 108 ohms/square, and be little influenced by ambient conditions of relative humidity. It should also have good smoothness, wrinkle resistance, folding and creasing properties, and reasonable stiffness. Photoelectric base stock is converted into photoelectric process copy paper by further coating with a mixture of a photoconductive material, resin of high resistance, and material to control the spectral sensitivity of the finished paper.

PHOTOELECTRIC PROCESS COPY PAPER—See PHOTOELECTRIC PROCESS BASE STOCK.

PHOTOGELATIN BRISTOL—A high-grade heavyweight paper used for printing by the photogelatin process.

PHOTOGELATIN PAPER—A special paper used for printing by the photogelatin process. It is made from cotton and/or chemical woodpulps in basis weights ranging from 80 to 100 pounds (25 x 38 inches – 500) and is characterized by high brightness, cleanliness, smooth uniform finish, pick strength, dimensional stability, and freedom from curl.

PHOTOGELATIN PRINTING—See AQUATONE PRINTING; COLLOTYPE PRINTING.

PHOTOGRAPHIC BACKING PAPER—(1) A red and black duplex paper used for protecting photographic films. (2) A special paper for photo mounting.

PHOTOGRAPHIC BLOTTING PAPER—An absorbent, lint-free paper normally made from chemical woodpulps, in basis weights ranging from 100 to 120 pounds (19 x 24 inches – 500). It is used for drying photographic prints, and must therefore be free from water-soluble contaminants.

PHOTOGRAPHIC PAPER—A very high grade of light-sensitive paper used in the preparation of prints from exposed photographic films. The base stock is usually made from purified chemical wood ("alpha") pulps and/or cotton fiber and this base is normally pre-coated with baryta (barium sulfate) formulations, followed by sensitization with silver halide emulsions. The grade is characterized by chemical purity, wet-strength, and permanence, and is made in a wide range of basis weights, for contact printing, copying, enlarging, etc. Some grades are also embossed to provide textured surface.

PHOTOGRAVURE PAPER—See ROTOGRAVURE PAPER.

PHOTOGRAVURE PRINTING—Any intaglio (q.v.) printing in which the design is placed photographically on the printing plate or cylinder. See ROTOGRAVURE; SHEET-FED GRAVURE.

PHOTOLITH—See PHOTOLITHOGRAPHY.

PHOTOLITHOGRAPHY—Lithography, or offset printing, using plates created by photographic means.

PHOTOLITHO PAPER—See LITHOGRAPH PAPER.

PHOTOMOUNT BOARD—See PHOTO-MOUNT STOCK.

PHOTOMOUNT COVER—See PHOTO-MOUNT STOCK.

PHOTOMOUNT FOLDER—See PHOTO-MOUNT STOCK.

PHOTOMOUNT STOCK—Heavy paper or more commonly a paperboard used to mount photographs. It is made of various paperboard grades usually a newsboard lined with a plain, decorated, or embossed facing paper. It also may be a single sheet or two-ply cover usually in thicknesses of 15 to 35 points, with plain, antique, decorative, or embossed surface. Depending upon its thickness and use, it is variously termed photomount board, cover, folder, or paper.

PHOTO-OFFSET—See PHOTOLITHOGRAPHY.

PHOTOSENSITIVE COATING—A coating sensitive to light or similar radiation.

PHOTOSTAT PAPER—A silver-sensitized copying paper used for duplicating records, etc., by photographic methods. The term was originally a trademark. See PHOTOGRAPHIC PAPER; PHOTOCOPYING PAPER.

PIA—(1) The Printing Industries of America, a trade association. (2) Absolute pressure, pounds per square inch absolute (psia).

PICK—(1) See PICKING. (2) The adhering of pulp or fibers to the wet or drying sections of the paper machine.

PICKING—The lifting of coating, film, or fibers from the surface of the body stock during printing.

PICK STRENGTH—See BONDING STRENGTH.

PICKUP FELT—See PRESS FELT.

PICTURE-MAT BOARD—See PHOTOMOUNT STOCK.

PID CONTROLLER—A conventional controller capable of adjustment in a proportional gain (P), integral (I), and derivative (D) manner. The P-controlled output is capable of being in direct (or reverse) proportion to the input error signal. The integral (I) function is an integration of the error signal and drives any offset from setpoint back to alignment. This is sometimes referred to as the "reset" function. The derivative (D) function provides an anticipatory reaction to the error signal by sensing the rate of error change and overcorrecting the output signal until the rate of change decreases. This is sometimes referred to as the "rate" function. Controllers may consist of P alone, P plus I, P plus D, or of all three.

PIE FED PAPER—Paper, usually coated, the sheets of which have been inspected on an inspection machine consisting of a mechanical automatic feeder and a mechanical layboy. Obsolete procedure.

PIE-PLATE BOARD—Paperboard which can be pressed into the shape of a plate and to which a metal rim is frequently added. It has good forming qualities and stiffness, is well sized, and is resistant to discoloration by heat and greases. See also PLATEBOARD.

PIE TAPE—A pliable, creped sheet of vegetable parchment paper used around the edges of pies to retain the juices.

PIGGY BACK—See MULTIPLE FOUR-DRINIERS.

PIGMENT—A finely divided solid coloring material which is insoluble in the medium in which it is applied. Pigments are used in paper to alter physical properties such as bulk, porosity, and bonding, as well as to add color and improve brightness and opacity. Pigments are also used in printing inks.

PIGMENTED SURFACE SIZE—A term for a surface size that includes a mineral pigment, such as kaolin clay, titanium dioxide, or calcium carbonate along with adhesive materials such as starch or latex. Pigmented surface size is applied to improve the optical and printing qualities of paper. A sheet which has been pigmented surface sized is often referred to as film coated.

PIGMENT TRANSFER PAPER—See TRANSFER PAPER (2) c (2).

PILING—See COATING PILING.

PILING ON THE BLANKET—In offset printing, the accumulation of lint, dust, or picked particles on the offset blanket in sufficient quantity to cause defective impressions or to require stoppage of the printing process for cleaning the blanket.

PILLING—The formation of small balls of fibers on the surface of a material due to mechanical abrasion.

PIMA—The Paper Industry Management Association.

PIMARIC ACIDS—See ROSIN ACIDS.

PIN ADHESION—A measure of the force required to separate corrugated board at the bonds between the flute tips of the corrugating medium and its facings. This is accomplished by inserting pins in the flute openings to support the test piece and exert the force applied by a compression machine to perform the separation.

PINE OIL—A colorless to amber-colored volatile oil with characteristic pinaceous odor, consisting principally of isomeric tertiary and secondary cyclic terpene alcohols ($C_{10}H_{16}O$). The four principal kinds of pine oil are:

1. Steam distilled pine oil: obtained from pinewood by steam distillation or solvent extraction.

2. Destructively distilled pine oil: obtained from the lighter distillate from the destructive distillation of pinewood.

3. Synthetic pine oil: obtained by chemical hydration of terpene hydrocarbons, or dehydration of terpene hydrate.

4. Sulfate pine oil: a high boiling fraction obtained in the refining and fractional distillation of crude sulfate turpentine.

PINE TAR PAPER—A paper treated with pine tar and used for preventing attack by vermin and moths on furniture and other objects in storage.

PINHOLES—(1) Holes caused by the crushing and falling out of fine foreign particles when the paper is calendered, or produced by grit imbedded in the calender rolls. (2) Pores in thin papers, where the spaces between larger fibers are not filled by smaller ones. (3) Minute pits in the surface of coated papers.

PIN PAPER—Various qualities of antitarnish paper used for making up packets of pins. It may be coated and is usually made in a blue, black, yellow, or green shade. See also NEEDLE PAPER.

PIN SEAM—A particular way of "on machine" joining of machine clothing (q.v.) to facilitate installation and to convert the clothing into an "endless" belt. The general method is to construct the woven clothing such that both ends have loops that can be pushed together. A special yarn or cable is inserted through these loops, often with a special tool to join the ends. First used for dryer fabrics (q.v.), it is now used with some press felts. Other types of seams and "seam assists" (e.g., velcro), can be used for installing the pin.

PIRSSONITE—A double salt of sodium carbonate and calcium carbonate that precipitates from green liquor (q.v.) at concentrations of total titratable alkali (q.v.) above approximately 7.5 pounds per cubic foot. Frequently found as a deposit on green liquor lines.

PITCH—(1) In the paper industry, pitch is largely a mixture of fatty and resin acids and unsaponifiable organic substances that can be extracted from the wood, mechanical pulp, and chemical pulp by means of organic solvents, such as alcohol and ether. Pitch is associated mainly with ray cells of the wood and under certain conditions it accumulates on the paper machine fabric or on the press felt in the press section and causes trouble in the papermaking operation. (2) Hydrolysis products of synthetic sizes such as alkenyl succinic acid anhydride (ASA) and their calcium or magnesium salts in combination with (1). (3) The residue from the distillation of coal tar, which is used in the manufacture of roofing, sheathing, etc., as waterproofing material. The term is also used for the residues from the distillation of wood, petroleum oils, rosin, etc.

PITCH SPOTS—Dirt specks in paper, resulting from resins in the wood used to make the pulp.

PITH—In trees and other plants, a cylinder of parenchymous cells, lying centrally in an axis and surrounded by vascular tissue.

PITTING—Corrosion of a metal surface, confined to a point or small area, that takes the form of cavities. Typically, a pit is at least as deep as it is wide.

PIXEL—Short of picture element, the smallest unit of a digital picture. It is a spot with an associated gray level or color.

PLACEMAT PAPER—A grade of base paper made from bleached chemical woodpulps, for conversion into plain, embossed, or printed place mats, used by restaurants, etc. Such paper is generally fairly light in weight [40 to 50 pounds (25 x 38 inches – 500] and has offset paper characteristics.

PLAID FINISH—A finish obtained by pasting strips of paper between two sheets of cotton and using the sheet in a plater in the normal manner. Variations may be obtained by pasting thread between the cotton sheets, so as to form any desired pattern of straight lines or checkered design.

PLAIN—(1) Made throughout from one grade of stock. Plain chipboard and plain strawboard are examples. (2) Uncoated, as book paper for example. (3) Unprinted or embellished. See also PLAIN PAPER.

PLAIN PAPER—Copier (q.v.) and other use cut size paper, usually 8 1/2 inches x 11 inches, without any special print receptive coating.

PLAIN PAPER COPIER—A copier (q.v.) that does not require especially treated paper to produce copies.

PLAIN ROLL PRESS—A water removal device where the sheet and press fabric pass through a nip between two solid surface rolls. In very limited commercial use and generally replaced by more modern press types such as the grooved roll press (q.v.).

PLAIN SETTLING TANK—A tank or basin in which water, wastewater, or other liquid containing settleable solids is retained for a sufficient time, and in which the velocity of flow is sufficiently low, to remove by gravity a part of the suspended matter.

PLANCHETTE PAPER—Usually a thin (0.001 inch) lightweight [7 pounds (17 x 22 inches – 500)] paper in red or blue, from which small discs are punched in a perforating operation. These discs or planchettes are used to give distinctive features to currency paper to protect against counterfeiting. The term is also used for a cotton-content paper (25 to 100% cotton) in which these discs or planchettes are incorporated.

PLANETARY PRESS—See COMMON IMPRESSION PRESS.

PLAN PAPER—See MAP PAPER.

PLANT-CAP PAPER—Paper used to protect plants from exposure to the sun when first transplanted or from the frost in the spring or fall. The paper is crimped so that it may be set over the plants. It is usually waxed. The usual sizes are 16 x 18, 18 x 18, and 20 x 20 inches. It is also called capping paper.

PLANT DRIER—A gray absorbent paper used by plant collectors in their field work and in the laboratory to press and carry plant specimens. It is usually made from a mixture of cotton and chemical woodpulps in a basis weight of 100 pounds (18 x 24 inches – 500).

PLANT-PROTECTOR PAPER—See PLANT-CAP PAPER.

PLASTERBOARD—A thick building board designed to be fastened to studs or joists and to be used as a backing for the application of plaster. See GYPSUM BOARD.

PLASTER-SACK PAPER—See CEMENT-SACK PAPER; SHIPPING SACK KRAFT PAPER.

PLASTIC—Any of a large number of organic, synthetic, or natural polymeric materials, such as polyethylene, that may be molded, extruded, or cast into various shapes or films without rupture, or coated as film upon paper or board.

PLASTICIZER—(1) A relatively low molecular weight product used to soften or render flexible another material. (2) An agent added in the manufacture of certain papers, such as glassine, or employed in papermaking compositions or protective coatings, such as nitrocellulose lacquers, to impart softness and flexibility. Typical plasticizers of interest to the paper industry are glycerol, sorbitol, invert sugar, phthalic acid esters, various types of mineral oils, organic esters of phosphoric acid, and the like.

PLASTISOL—A suspension of a finely divided resin in a plasticizer (q.v.) that can be converted to a continuous film by the application of heat.

PLATE—A thin plastic, metal, or paper sheet that is the image carrier for many types of printing.

PLATEBOARD—Paperboard used in making plates for serving food. The board must be capable of being formed on a die press and be stiff. It is frequently printed and plastic coated. See also PIE-PLATE BOARD.

PLATED FINISH—See PLATE FINISH; PLATING.

PLATE FINISH—A smooth or polished surface on sheets of paper, especially writing papers, produced by placing the sheets between polished plates of zinc or copper and passing a pile of these (called a "book") under high pressure and slight friction between the rollers of the plating machine. The finish varies greatly in degree, depending upon the pressure used and the number of times the book is passed through the plater, sometimes being a very flat finish without shine and sometimes being a very glossy finish. Many smooth or glossy finishes now obtained by supercalendering are called plate finish and are indistinguishable from the finish obtained by the use of flat metal plates.

PLATE-FINISHED PAPER—Any paper finished on a sheet plater. The finish may be a smooth or a fancy finish such as linen, ripple, or coarse finish.

PLATE GLAZED—See PLATE FINISH.

PLATEN DRIER—A machine for drying paper or paperboard. The drying is effected by pressing stationary sheets between plane, heated surfaces or by passing the moving web of paper or board through the narrow space between a pair of parallel, heated surfaces (platens).

PLATE PAPER—A thick, soft, high-quality printing paper, lightly sized, having a smooth flat surface without a high gloss. The thicker kinds are made by pressing two or more webs together in the wet state. It is used for taking impressions from engraved copper and steel plates and also for woodcuts and lithographic printing.

PLATER—A machine for plate finishing which consists of two chilled-iron rolls, between which the form or book passes, forward and back, as the rolls reverse direction of turning; pressure is exerted only on the top roll. See PLATE FINISH.

PLATER BOOK—A pile of paper interleaved with fabric and zinc, etc., to be pressed between plater rolls. See also PLATE FINISH.

PLATER FINISH—See PLATE FINISH; PLATING.

PLATER VELLUM FINISH—A finish with the same surface characteristics as vellum except that the surface is smoother; it is produced by a plater press. It is distinguished from the calender vellum finish by the fact that the low spots in the finish are smoother. See CALENDER VELLUM FINISH; VELLUM FINISH.

PLATE WIPING PAPER—A machine-creped absorbent paper with low stretch and good strength. It is used primarily by engravers and printers for wiping plates. The basis weights are generally 40 to 60 pounds (24 x 36 inches – 500). See DIE-WIPING PAPER; LITHOGRAPHER'S PLATE WIPER.

PLATING—The process of producing special finishes on paper by subjecting the sheet to pressure between plating rollers while it is made up in books and is in contact with some material different from the sheet itself, thus producing on the paper an impression that is characteristic of the material used. See also BUNCH PLATER FINISH.

PLATINUM PAPER—A sensitized photographic paper (using salts of iron and platinum) used for making prints.

PLAYER-PIANO PAPER—A paper used as a "stencil" in mechanical or self-playing pianos. It is made from chemical wood or rope pulps when highest strength and durability are required, and it is usually unbleached. A tough, partly parchmentized paper is also used. Significant properties are high tensile strength, durability, and minimum elongation. See also MUSIC-ROLL PAPER.

PLAYING-CARD BOARD—See PLAYING-CARD STOCK.

PLAYING CARD STOCK—A heavy coated paper made for conversion into playing cards. It is usually a two-ply sheet made from chemical woodpulps, and colored paste is used to provide maximum opacity. Some grades are also varnished and/or plastic coated during manufacture or after printing in order to provide maximum durability. The usual thickness is from 10 to 12 mils.

PLEATING PAPER—A strong, well-sized, high-finish or water-finished paper used by tailors in pleating cloth. It is frequently made of jute and kraft, in thicknesses of 0.007 to 0.012 of an inch. It must have a smooth surface and good folding qualities and the durability to withstand repeated usage under heat and pressure.

PLEATING TISSUE—Usually a semicreped sheet, made in varying basis weights, for fabric pleating. The furnish may range from groundwood and kraft to the highest grades of fully bleached pulps, depending on the fabric material and weight and the nature and design of the pleat. The paper must be free of imperfections or chemicals that might spoil the pleat or discolor the fabric. The fabric goes through the pleating operation, including the setting of the pleats, between two thicknesses of pleating tissue.

PLUCKING—See PICKING.

PLUG FLOW—The movement of a fiber suspension, either in a pipe or tower, as a unit component.

PLUGS—See CORE PLUGS.

PLUME—The flow pattern made by a stack emission or liquid discharge as it disperses in the environment.

PLY—(1) One of the separate webs which make up the sheet formed on a multicylinder machine or multifourdrinier machine. Each cylinder or fourdrinier adds one web or ply, which is pressed to the other, the plies adhering firmly upon drying. (2) One of the sheets which are laminated to build up a pasted board of given thickness. (3) One of the separate layers which together make up a multilayer aggregate such as multi-ply tissues, multiwall shipping sacks, and carbon-interleaved business forms.

PLY ADHESION—(1) The bonding strength between the plies of a paper or paperboard. (2) The bonding of multi-ply tissues after embossing. See BONDING STRENGTH.

PLY SEPARATION—(1) The separation of paper or, more usually, paperboard produced on a multi-ply paper machine, into layers or plies. (2) The separation of pasted solid fiberboard into the component layers or plies. See INTERNAL BOND; MULTI-PLY FORMING.

PMS—See PANTONE MATCHING SYSTEM.

POCKET GRINDER—A machine for producing mechanical pulp, groundwood. It consists of a rotating pulpstone against which debarked logs are pressed and reduced to pulp. The two feet long logs are fed by hand from the sides into three or four pockets located around the upper portion of the stone and pressed against the rotating pulpstone by hydraulic cylinders.

POCKETS—In paper drying, the closed area between dryers in a multi-dryer-cylinder geometry caused by dryer fabrics.

POCKET VENTILATION—Air introduced into the pockets (q.v.) of a dryer section to provide a low humidity, uniform drying environment.

POINT—(1) One thousandth of an inch. It is used in expressing the thickness of paper or board. (2) A term used for expressing certain values of paper properties—e.g., points per pound, pounds per point. (3) The printer also uses the point as a unit of measurement; in the United States, it is 0.013835 of an inch. (4) Unit of size for type; 72 points equal one inch.

POINT BONDING—The bonding of a nonwoven fabric in the form of a pattern, usually by printing or heated engraved calender rolls.

POINTS PER POUND—A ratio obtained by dividing the results of a given strength test by the basis weight of the paper or paperboard in pounds. The term is most frequently applied to the bursting strength (q.v.).

POISSON'S RATIO—The ratio of the unit deformation in the transverse direction divided by the unit deformation in the axial direction when a body is subjected to a uniaxial stress. Poisson's Ratio describes the necking of a tensioned web or an increase in width of a wound roll resulting from the residual radial stress inside the wound roll.

POLARITY PAPER—A filter paper impregnated with an indicator and salt that is moistened and used to determine the positive or negative pole of a direct electric current. If the paper, saturated with sodium chloride and phenolphthalein solution, has two wires placed upon it a little distance apart, electrolysis takes place and the hydroxyl ion will color the phenolphthalein red; hence, the corresponding wire is the positive electrode.

POLAR LIQUIDS—Polar liquids contain covalent bonds whose electrons are not equally shared between the two involved atoms. This lends the molecules dipole character. This lends the molecules dipole character and promotes molecular association through hydrogen bonding and other mechanisms. Water is a good example of a polar liquid because the electrons in the O–H bond tend to reside closer to the oxygen than the hydrogen. Consequently, the bond is polarized with the negative end located at the oxygen and the positive end at the hydrogen. When two water molecules approach one another, the negative oxygen sites are attracted to the positive hydrogens, giving rise to association. Water will also associate strongly with other molecules that have O–H groups, such as cellulose.

POLE DRYING—See LOFT DRYING.

POLE MARK—See BACK MARK; STICK MARK.

POLICY PAPER—A high-quality bond or writing grade commonly used for insurance policies.

POLISH—A term referring to a high gloss.

POLISHED DRUM COATING—A process in which the coating is applied in any suitable manner and immediately placed with the coated side against the surface of a highly polished, heated drum which smooths and dries the sheet without the necessity of subsequent calendering or polishing. This is a type of cast coating (q.v.).

POLISHING POND—The pond in a treatment system used just prior to discharge to remove the last remaining small amount of pollution.

POLLUTANT—Any introduced gas, liquid, or solid that makes a resource unfit for a specific purpose.

POLLUTION—The presence of matter or energy whose nature, location, or quantity produces undesired environmental effects.

POLLUTIONAL INDEX—A criterion by which the degree of pollution of a stream or other body of water may be measured, such as bacterial density, plankton, benthos, biochemical oxygen demand, dissolved oxygen, or other index of water quality.

POLLUTION INDEX—A criterion by which the degree of pollution of a stream or other body of water may be measured, such as bacterial density, plankton, benthos, biochemical oxygen demand, dissolved oxygen, or other index of water quality.

POLLUTION INDEX (AIR)—An index of air pollution that is comprised of a weighted set of pollutants, usually based on national ambient air quality standards.

POLLUTION LOAD—A measure of the strength of a wastewater in terms of its solids or oxygen-demanding characteristics, or in terms of harm to receiving waters.

POLYALUMINUM CHLORIDE (PAC)—This chemical is a very useful form of cationic alumina which makes it unnecessary to operate in the acid pH range. PAC possesses a higher cationic charge than the various forms of alum and retains its cationic charge at neutral, or even slightly alkaline, conditions. PAC is formed by neutralizing $AlCl_3$ to about 60–80% forming a stable colloid, which acts as a cationic retention aid and is capable of reacting with rosin acid to form a sizing compound under neutral pH conditions. The preformed aluminas polymer is useful in water treatment and other applications, as well as papermaking.

POLYANIONS—An atom group that carries a number of negative charges.

POLYCATIONS—An atom group that carries a number of positive charges.

POLYCHROME PAPER—See CHROMO PAPER.

POLYCYCLIC HYDROCARBONS—See CRUDE TALL OIL.

POLYELECTROLYTES—Large molecules or ion clusters, or colloids that are charged (often containing ionic groups such as $-COO^-$ or $-NH\pm$ groups) and are capable of effecting absorption of other materials by strong ionic (electrostatic) interaction. Some polyelectrolytes may contain both anionic and cationic groups, with the type in excess controlling the net charge. Examples included cationic polyacrylamides, cationic starch, the polyacrylates, polyphosphates, and other ionic polymers used in the paper industry.

POLYETHYLENE—A plastic compound formed by the polymerization of liquid ethylene at high temperatures, and used as a paper or paperboard coating to provide liquid resistance and other qualities. Paper milk cartons are examples of products commonly coated with polyethylene.

POLYMERIZATION, DEGREE OF—See DEGREE OF POLYMERIZATION.

POLYOLEFIN—A synthetic resin polymerized from an olefin such as ethylene, propylene, etc. It is used as an extrusion coating, a film, or a molded object.

POLYPROPYLENE—A synthetic resin polymerized from propylene used as an extrusion coating, a film, or a molded object.

PONDS—Earthen impoundments used to contain wastewater. In modern practice, they are usually lined with clay or plastic film.

POND SIZE PRESS—A surface sizing device consisting of two rolls and a pair of headers to supply surface sizing chemicals to a pond formed by the nip between the rolls. The sheet passes through the nip ponds and is surface sized. A pond size press causes pressure induced migration of the sizing material into the sheet and effectively saturates the sheet. This is in contrast to the action of a metering size press, which transfers a film to the surface of the sheet, resulting in very limited penetration.

POOR CORE START—A loose or non-uniform start of the full width of a web during the beginning of a wind onto a core. This undesireable situation usually results in defective paper at the core and the potential for "telescoping."

POOR START—Slang for when the winding near the core is not tight enough.

POP STRENGTH—A synonym for bursting strength (q.v.).

POROSITY—The property of having connected pores or minute interstices through which fluids (liquids and gases) may pass. It is dependent on the number of pores and their distribution in size, shape, and orientation. The porosity of paper is commonly evaluated by measuring its air permeability. See AIR PERMEABILITY; PERMEABILITY.

P.O.S. COMPUTER PAPER—A term usually applied to small rolls of paper used on electronic cash registers, for "Point of Sale" recording of business transactions. Such rolls may be of single copy construction (bond or tablet base stock) or multi-copy (carbonized, carbon-interleaved, or carbonless). P.O.S. computer paper is also available in fanfolded makeup.

POST COLOR NUMBER—Post color (pc) number is a way of expressing the brightness loss or reversion independently of the original brightness.

$$pc \text{ number} = \frac{(100 - R_2)^2}{2R_2} - \frac{(100 - R_1)^2}{2R_1}$$

where

R_1 = brightness before aging, e.g., 89.7

R_2 = brightness after aging, e.g., 85.2

$$pc \text{ number} = \frac{(100 - 85.2)^2}{2 \times 85.2} - \frac{(100 - 89.7)^2}{2 \times 89.7} = 0.69$$

The lower the number, the more stable is the pulp brightness.

POST CONSUMER SOLID WASTE—Any product that has gone through its useful life, served the purpose for which it was intended, and been discarded by the user. Waste or scrap created in a manufacturing or converting operation is not considered post consumer waste.

POSTAGE STAMP PAPER—Paper used in the production of various types of postage stamps. It is usually made from a mixture of various bleached chemical woodpulps, and its special properties include offset or intaglio printing qualities, noncurling characteristics, and good perforating and gumming qualities. Some grades must also have wet strength. It is usually made in an off-white color in basis weights ranging from 16 to 24 pounds (17 x 22 inches – 500) or 40 to 60 pounds (25 x 38 inches – 500).

POSTAL MONEY-ORDER PAPER—A paper used by the U.S. Government Post Office for money orders. Bleached chemical woodpulps are used in making the base stock in a basis weight of 32 pounds (17 x 22 inches – 1000). The surface of the paper contains a dye or other substance sensitive to the action of chemical

231

reagents such as ink eradicators, to show clearly any attempt at mechanical erasure. The dyed or impregnated fibers form a pattern only on the surface of the paper. Sensitivity and bond paper characteristics are significant properties. See also SAFETY PAPER.

POSTCARD BRISTOL—See POSTCARD PAPER. It can be made in antique, vellum, and English finishes. Important properties are color, finish, rigidity, and sizing. See also COATED POSTCARD STOCK.

POSTCARD PAPER—Paper used for making postcards which may be at least three different types: writing, printing, and sensitized, depending on use. The first is for mailing brief, written, typed, or printed nonconfidential messages; the second, for picture cards which serve as souvenirs and message carriers; the third, for photographic purposes whereby a typical picture card is produced. Card stock, coated or uncoated, or pasted, and photographic paper, are used.

POSTCONSUMER PAPER—Paper and/or paperboard products that have gone through their intended final use and have been discarded. Includes used corrugated boxes, old newspapers, old magazines, mixed paper waste, and tabulating cards. Paper waste created in converting operations is generally but not always excluded from post consumer paper. See PAPER STOCK.

POSTER—Advertising matter printed on paper or paperboard for display.

POSTER BOARD—A stiff cardboard, usually 0.024 or 0.030 inch in thickness, lined on one or both sides with white or colored book paper (q.v.), in sizes 22 x 28 inches and 28 x 44 inches. It is used for indoor and outdoor advertising posters, games, cutouts, etc. See MILL-BLANKS.

POSTER PAPER—(1) A term applied to a paper especially suited for billboard poster work. It is generally made of chemical woodpulp in basis weights of 32 to 63 pounds (24 x 36 inches – 500). It is strong and well sized to enhance water resistance. It possesses wet strength and

should not curl after paste is applied. It should have good light fastness. The paper is usually rough on the underside in order that it may accept paste more readily. (2) A term applied to a special mechanical-pulp printing paper, the major use of which is in printed flyers, throwaways, and similar inexpensive advertising pieces and publications, although it is sometimes used in the converting trade in manufacturing sales books, order forms, etc. This grade commonly contains a substantial portion of refined mechanical pulp mixed with unbleached pulp. It is normally unsized for high-speed, flatbed presses, has a smooth printing surface, and is almost entirely used in a range of six standard colors. The usual basis weights range from 32 to 45 pounds (24 x 36 inches – 500). See also SIGN PAPER.

POSTER PARCHMENT—See PARCHMENT BOND.

POSTING BRISTOL—See INDEX BRISTOLS.

POSTING LEDGER PAPER—See BOOKKEEPING MACHINE PAPER.

POTENTIAL—See MOBILITY; ELECTROKINETIC CHARGE.

POTTERY TISSUE—Tissues specially prepared for printing transfers for pottery decoration. See DECALCOMANIA PAPER.

POUCH PAPER—Paper for food pouches used generally in combination with other packaging films; e.g., paper/poly/foil/poly.

POULTRY BAG PAPER—A vegetable parchment or an imitation parchment, usually in basis weights of 27 to 35 pounds (24 x 36 inches – 500), used as an inner liner for kraft or sulfite outer bag.

POULTRY BAND—A wet-strength, water-resistant, or a vegetable parchment paper in basis weights of 27 to 35 pounds (24 x 36 inches – 500), which is used to provide identification for poultry.

POULTRY BOX LINER—A vegetable parchment paper or wet-strength treated or waxed chemical woodpulp paper used for lining poultry boxes, usually in basis weights from 27 to 35 pounds (24 x 36 inches – 500).

POULTRY HEAD BANDS—Vegetable parchment, sulfite, or kraft paper, usually in basis weights of 27 to 35 pounds (24 x 36 inches – 500), used to conceal the head and prevent blood from staining other poultry.

POULTRY WRAPPER—A paper similar to poultry band used for completely wrapping poultry.

POUNCING PAPER—(1) A paper used by artists and printers in outlining letters and figures for large signs. The outlines are punched in a sheet of paper with a tracing wheel, thus forming a pattern which is used to outline the letters or figures on the baseboard by applying a pouncing powder (such as charcoal dust) to the pattern. Any grade of paper of basis weights 16 to 24 pounds (17 x 22 inches – 500) may be used, provided it gives clean perforations. (2) An abrasive paper (q.v.) which is coated with extremely sharp silicon carbide grains or very fine-grained sand and used in increasing the speed and improving the finish in hat-pouncing operations.

POUNDS PER POINT—The ratio of the basis weight in pounds divided by the thickness in mils. It is used to describe the density. See APPARENT DENSITY; BASIS WEIGHT; THICKNESS.

POWDER PAPER—(1) Manila paper, sometimes waxed or treated, for wrapping charges of powder used in mining. The basis weight is 50 pounds (24 x 36 inches – 500) and heavier. See also BLASTING PAPER. (2) A paper similar to cigarette paper but which is coated with a facial powder that is easily rubbed off.

POWER BOILER—A steam generator that produces high pressure steam for electric power generation using a steam turbine. Power boilers may be fitted with any fuel such as coal, fuel oil, natural gas, wood-waste or hog fuel, and peat. Recovery boilers also produce high-pressure steam for electric power generation using black liquor as a fuel but, usually, they are not referred to as power boilers.

POWER FACTOR—(1) The power factor of a dielectric material, such as paper, is a measure of the heat generated with the material when the latter is subjected to an alternating electric field of a given strength and frequency. It is the cosine of the phase angle between the applied a.c. voltage and the current resulting from the applied a.c. field, e.g., when the material is the dielectric of a condenser or cable. This property is especially important in condenser and cable papers. It is a measure of dielectric loss. (2) Power factor is the ratio of working power (watts) to apparent power (volt-amperes). Working power is the power to do work; apparent power includes the component of power to magnetize transformers, supply induction power losses, etc.

PRACTICAL MAXIMUM CAPACITY—Practical maximum capacity, as defined by the AF&PA is the tonnage of paper, paperboard or pulp of normal commercial quality that could be produced with full use of equipment and adequate supplies of raw materials and labor, assuming full demand. No allowance is made for losses due to unscheduled shutdowns, strikes, temporary lack of power, etc., which cause decreases in actual production, but not in production capacity.

PRAYER-BOOK PAPER—See BIBLE PAPER.

PREBLEACHING—The initial treatments in a pulp bleaching sequence (chlorination, extraction, oxygen delignification) which complete the process of delignification to facilitate the actual brightening of the pulp in later stages by oxidants such as chlorine dioxide, hypochlorite or peroxide. See ALKALINE EXTRACTION STAGE (E); CHLORINATION STAGE (C); OXYGEN DELIGNIFICATION (O); OZONE STAGE (Z).

PRECIPITATED CALCIUM CARBONATE (PCC)—See CALCIUM CARBONATE.

PRECIPITATOR CATCH—Fine particulate collected by electrostatic precipitators (q.v.), which are often used to collect dust from the flue gases of recovery boilers (q.v.) and power boilers.

PRECIPITATORS—In pollution control work, any number of air pollution control devices usually using mechanical/electrical means to collect particulates from an emission.

PRECISION—The extent to which a measurement repeats its results when making a series of repeat measurements on a single unit of material. See also ACCURACY.

PRECISION CUTTER—Designed to cut the web (or multiple webs) and to pile individual sheets accurately enough so they can be sold as finished product without subsequent guillotine trimming. Sheet to sheet accuracy can be within ±1/64 of an inch or as high as ±1/8 of an inch.

PRECISION RECORDING PAPER—Similar to chart paper but made so that it will expand and contract very little when exposed to changes in atmospheric conditions. It is used on precise scientific instruments where changes in dimensions of the paper would record inaccurate measurements or values.

PRECOAT—A preliminary coating that is impervious enough to give satisfactory coating hold-up on final coatings.

PRECOAT FILTER—A drum attached to a vacuum pump which rotates in a vat of slurry picking up a cake on its surface the outer layer of which is removed by a stationary or moving doctor blade (q.v.). Most often used as the final step in preparing the lime mud feed to the calciner. Also used to filter green liquor dregs.

PRECOATING—See BASE COATING.

PREHYDROLYSIS—An additional step in a kraft cooking cycle in which the chips in a digester are presteamed under pressure without pulping chemicals. The cellulose is degraded to some extent during this prehydrolysis treatment. Following prehydrolysis, a white liquor is added and the kraft cook completed. The resulting pulp is used to make dissolving pulp for acetate or viscose products. Pulp strength is low as a result of the cellulose degradation.

PRE-IMPREGNATION—See IMPREGNATION.

PREMETERING SIZE PRESS—See METERING SIZE PRESS.

PREPARED ROOFING—See ROOFING PAPER.

PREPREG—A nonwoven mat often produced by the wet laid (q.v.) process using glass, carbon or aramid (q.v.) fibers that is later used to produce resin inpregnated rigid composites.

PRE-PRINT—Linerboard that has been printed and rewound before combining into corrugated or solid fiberboard.

PRESERVATIVES—Chemical additives that prevent or inhibit the growth and development of spoilage organisms.

PRESS—In a paper machine a pair of rolls between which the paper web is passed for one of the following reasons: (a) Water removal at the wet press (q.v.). (b) Smoothing and leveling of the sheet surface at the smoothing press (q.v.). (c) Application of surface treatments to the sheet at the size press (q.v.).

PRESS ARRANGEMENT—The press type generally classified according to the number of rolls used, such as two or three roll (or more) arrangements, and often given specific names, both generic and trademark names.

PRESSBOARD—See IMITATION PRESSBOARD; INDEX PRESSBOARD; TRANSFORMER BOARD.

PRESS COPY PAPER—See COPYING PAPER.

PRESS DRYING—A drying process where the wet paper or paperboard is held under constant z-direction (perpendicular to the paper) pressure against the dryer cylinders until it reaches final dryness. See DRYER SECTION.

PRESS FABRIC—A woven fabric made chiefly from polyester and used in the press section to provide void volume in the nip. The press fabric is run inside, or under the felt (e.g., between the felt and the roll). Most of these press fabrics have been replaced with grooved or blind drilled (q.v.) press rolls.

PRESS FELT—A continuous belt generally made of synthetic filaments and/or fibers but frequently made of a combination of two or more of the following fibers: wool, polyester, and polyamide. Press felts perform the function of mechanical conveyors or transmission belts, provide a cushion between the press rolls and serve as a medium for the removal of water driven from the sheet by mechanical pressure. Felts may be divided according to use, as follows: Bottom felt: a felt which, at one point in its travel, contacts the bottom side of the sheet of board or paper. Bottom board felt: the bottom felt on a board machine performs the function of transmitting power, picks up and conveys the plies of the board, and acts as a cushion and medium through which the water can pass. Endless felt: a felt that is woven in the form of a continuous belt; it is sometimes called a jointless felt. Pickup felt: a dense, firm, closely woven felt which picks up a wet sheet from a forming fabric or from another felt. Top felt: a felt which, at least at one point of its travel, contacts the top side of a sheet of paper or board. Top board felt: the top felt on a board machine performs the function of supplying a cushion and assists the bottom felt to a marked degree in removing water from the sheet. Felts that are becoming increasingly obsolete are Harper pickup or Harper top felts used in conjunction with a Harper machine (q.v.), and cylinder tissue felts used on cylinder machines making tissue.

PRESS IMPULSE—The unified influence of press loading and dwell time of the sheet on water removal in the press nip. It is the product of (average nip pressure x the dwell time) and is expressed as psi.s or MPa s. It can also be calculated by dividing press loading by machine speed and expressed in appropriate units.

PRESSINGS—Thick, though, tinted covers made on a single-dryer cylinder machine, therefore with machine-glazed finish. Made originally for the silk and dyeing trades, they are now used for packing, covers of exercise books, box coverings, and similar uses.

PRESSMARK—A mark or design impressed into the wet web of paper, usually at the second or third press, by means of a rubber collar which carries the design and which is suitably mounted on a steel shaft. See also WATERMARK.

PRESS MARKS—Marks in paper caused by nonuniformities in press rolls or press felts (q.v.).

PRESSPAHN—See TRANSFORMER BOARD.

PRESS PIT BROKE—See WET BROKE.

PRESS PIT PULPER—An UTM broke pulper (q.v.) located in the pit under the press section of the paper or board machine. Multiple press sections on the same paper or board machine often are served by a single press pit pulper.

PRESS ROLL—(1) A general term for a roll used in the press section of a paper machine. Press rolls are usually covered with an elastomeric composition or a plastic composite. (2) A device to increase the consistency of the stock on a washer or thickener before being discharged to the shredder/conveyor or repulper. The pulp sheet is broken into small pieces, and the broken pulp is moved for further treatment or storage.

PRESS ROLL DOCTOR—A mechanical wiper applied with light loading to the surface of a press roll to deflect the sheet, remove crumbs picked from the sheet, and to deflect water from the surface of the roll. The positioning, material, configuration, and running conditions of the doctor depend upon specific functions and roll surface material.

PRESS SECTION—The press section is a dewatering unit used on a paper machine (between the sheet-forming (q.v.) equipment and the dryer section) to remove as much water as practical from the sheet ahead of the dryer section. Most commonly, the sheet is sent into the nip (q.v.) of loaded rolls (a press) with one or two fabrics or felts (q.v.). The wet sheet is transported through the nip of each press on a fabric or felt which is bulky and porous to absorb water from the sheet under pressure and allow this water to flow to venting or be removed by vacuum. Each nip is formed by a pair of heavy rolls rotating against each other with provision for controlling the pressure to provide a graduated increase in pressure for each successive nip. Especially in stand-alone presses, the first nip is called the first press, the second nip the second press, etc.

One roll of each pair is usually rubber covered and may be of the following type:

Plain Press Roll (Solid roll surface, q.v. plain press)

Suction Press Roll (Drilled surface with a vacuum chamber inside the roll, q.v. suction press)

Blind Drilled Press Roll (Drilled surface with partial depth holes for void volume, q.v. blind drilled roll)

Grooved Press Roll (Roll surface with grooved void volume, q.v. grooved roll press, grooved roll)

Some combination of these types.

Alternatively, if one of the rolls is replaced with a stationary shoe and elastomeric belt or shell, it is referred to as a shoe press (q.v.)

Press sections come in a variety of configurations to remove the maximum amount of water without adversely affecting sheet properties like bulk, two-sidedness, smoothness, etc. The press section normally consists of two or more pressure nips in various press arrangements (q.v.). The reader is referred to *The Paper Machine Wet Press Manual,* Third Edition, Revised, TAPPI Press, Chapter Four, for an extensive listing of press section arrangements; over twenty in common use are given and patented configurations number much higher.

STRAIGHT THROUGH, BOTTOM FELTED—A conventional press with a top roll, a bottom roll and one felt on the bottom. The sheet feeds "straight through."

STRAIGHT THROUGH, DOUBLE FELTED—A straight through, bottom felted press with a top felt added.

TANDEM—A configuration where one felt (usually a bottom felt) runs through two presses and carries the sheet from the first press to the second.

INCLINED PRESS—In this configuration the pickup felt forms top felt in first press between suction roll and a center roll with second press inclined on the other side of the center roll with a grooved roll and a second top felt.

INCLINED PRESS WITH STAND ALONE THIRD PRESS—An inclined press followed by a straight through press.

TWIN NIP—Similar to an inclined press with a double felted first configuration press made by adding a bottom grooved or blind drilled roll and a bottom felt, and removing the grooved roll and the second top felt. The pickup felt then serves as the first top felt in the double felted press and the second top felt in the single felted second press.

TWIN NIP CONFIGURATION WITH STAND ALONE THIRD PRESS—A twin nip press followed by a straight through press.

TRIPLE NIP—Similar to the twin nip but with a grooved roll and a second roll and a second top felt added to the center roll to form the third nip. See TRIPLE NIP PRESS.

TRIPLE NIP CONFIGURATION WITH STAND ALONE FOURTH PRESS—A triple nip press followed by a straight through press.

TRIPLE NIP WITH GROOVED ROLLS—Similar to the triple nip press but with the center roll moved away from the suction roll, and a top press roll is added inside of the pickup felt and against the center roll. The result is that there are still three nips with three felts but there is a fifth press roll. This allows a higher second press loading with reduced danger of shell marking.

TRIPLE NIP WITH GROOVED ROLLS AND STAND ALONE FOURTH PRESS—A triple nip with grooved rolls press followed by a straight through press.

SMOOTHING PRESS—The last press in the press section, it has two solid or plain rolls and no felt. It is used to improve surface properties of the sheet.

SHOE PRESS—A press in which one of the rolls has been replaced with a shaped shoe and a shoe press blanket or sleeve added to the clothing run. The advantages are a much longer nip, higher loading, and controlled pressure profile.

LONG NIP PRESS—A roll press with one or both of the rolls having a much larger roll diameter (jumbo roll) than usually employed. A soft rubber cover is often used to create a longer press nip.

PRESS TRIMMING—The trimming of paper to accurate size and shape on a guillotine. Obsolete. See GUILLOTINE TRIMMER.

PRESSURE COUCH ROLL—See COUCH ROLL.

PRESSURE FILTER—A filter used to separate solids from a slurry. It consists of a vessel enclosing a set of fabric socks supported by wire frames, which acts as the filter medium. Often used to clarify white liquor (q.v.) or wash lime mud.

PRESSURE ROLL—A press roll (q.v.) running against a Yankee dryer (q.v.) in a tissue machine. It presses the felt and paper web against the dryer surface. Depending on the application, the particular roll may be a suction pressure roll (q.v.) or a blind drilled pressure roll

(q.v.). Pressure rolls are covered with a medium hardness elastomeric composition.

PRESSURE SCREEN—A screen that is fed under pressure and which removes oversize debris by size separation, while passing fibers. Pressure screens are used in many positions and locations. See also BROKE SCREEN; PAPER MACHINE SCREEN; PRIMARY SCREEN; SCREENING SYSTEM; SECONDARY SCREEN; THICK STOCK SCREEN.

PRESSURE-SENSITIVE PAPER—A paper impregnated or coated on one or both surfaces with a pressure-sensitive adhesive. It is used in the manufacture of pressure-sensitive tapes and also labels as, for example, price or identifying labels on various articles of merchandise. See also MASKING TAPE.

PRESSURIZED GROUNDWOOD (PGW)—A mechanical pulp produced by pressing debarked logs against a rotating pulpstone under elevated temperatures and pressures. The resultant pulp will have more long fibers than in groundwood produced at atmospheric conditions.

PRETREATMENT—In wastewater treatment, any process used to reduce pollution load before the wastewater is introduced into a main sewer system or delivered to a treatment plant for substantial reduction of the pollution load.

PRIMARY ARM—The first moveable support for an empty reel-spool on a conventional continuous reel drum which is located on the dry end of a paper machine, a coater, or a re-reeler, or any similar reel which winds a web onto a reel spool.

PRIMARY CLEANER—The first stage in a multiple stage cleaner system. The debris-rich rejects of the primary cleaner stage are further cleaned in the secondary cleaner. The primary cleaner stage accepts are sent downstream from the system. Normally, a large number of primary cleaners operate in parallel, for capacity reasons. See also CENTRIFUGAL CLEANER SYSTEM.

PRIMARY SCREEN—The first stage of screens in a multistage screening system. Multiple screening units are often in the primary stage position, for capacity reasons. The primary screen accepts are fed downstream. See also SCREENING SYSTEM.

PRIMARY SLUDGE—Sludge obtained from a primary settling tank.

PRIMARY TREATMENT—(1) The first major treatment in a wastewater treatment works, usually sedimentation. (2) The removal of a substantial amount of suspended matter but little or no colloidal and dissolved matter.

PRIMARY WALL—The thin, usually lignified, outer layer of the cell wall composed of cellulose fibrils oriented a various angles to the principal axis of the fiber.

PRIME COATING—See BASE COATING.

PRIMER—The first layer of coating applied to a surface upon which a second layer of the same or a different coating will be applied.

PRIMER PAPER—A strong, single or double deckle-edged sheet similar to cartridge paper (q.v.); it is usually red and is made in basis weights of 80 to 100 pounds (24 x 36 inches – 500). It is used for sealing the open end of the primer.

PRINTABILITY—That property of a paper which yields printed matter of good quality. This complex property is not accurately defined; it is judged by uniformity of color of the printed areas, uniformity of ink transfer, contrast between the printed and unprinted areas, legibility of the printed matter, and rate of ink setting and drying. Ink receptivity, compressibility, smoothness, opacity, color, and resistance to picking are among the simple properties that jointly determine printability. Printability should be distinguished from runnability, which refers to the efficiency with which the paper may be printed and handled in the press.

PRINT BONDING—The process of bonding where the binder is printed onto the nonwoven.

PRINT-ON COATING—A process in which the coating is applied to paper or paperboard by means of applicator rolls which print an accurately metered evenly distributed film of high density coating mixture directly on the paper. Print-on coating has been generally replaced by blade coating (q.v.).

PRINTED BOX COVER—A lightweight cover paper, plain or coated and embellished by a printed design in one or more colors, which is commonly used for box covering and other purposes; the basis weight is approximately 25 to 40 pounds (20 x 26 inches – 500).

PRINTED GUMMED TAPE—Gummed tape that has been printed for advertising or instruction use. See GUMMED SEALING TAPE.

PRINTED OPACITY—The ratio of the diffuse reflectance of the uninked side if a sheet printed in a solid color to the diffuse reflectance of an opaque pad of the unprinted paper. It is a measure of the showthrough exhibited by a printed sheet. This quality depends upon the color and density of the ink, the strike-in of the ink and ink vehicle or both, and upon the opacity of the paper.

PRINTED WASTE—See PAPER WASTE.

PRINTER ROLLS—See TELETYPE PAPER.

PRINTING BRISTOL—A term applied to mill bristol which is manufactured primarily for printing purposes.

PRINTING MANILA—A term generally applied to papers of manila color, containing up to 50% of mechanical woodpulp, with the balance of chemical woodpulps, and usually supplied in MF finish. It is used for pamphlets and various printed forms. The basis weights range from 32 to 90 pounds (24 x 36 inches – 500). See ENVELOPE MANILA.

PRINTING OPACITY—The ratio of the diffuse reflectance of the sheet when backed by a black body to that when backed by an opaque pad of the paper itself. It is important in book papers where the sheet is viewed when backed by printed pages; it is a measure of the visibility of printed matter on a sheet lying below the viewed sheet.

PRINTING PAPER—Any paper suitable for printing, such as book paper (general definition), bristols, newsprint, writing paper, etc.

PRINTING PROCESS—In general, there are four fundamental printing processes: letterpress or relief, intaglio, planography, and silk screen. (Other methods or reproduction, used as a substitute for printing when a relatively small number of copies are required, include the Mimeograph stencil, Multigraph, and spirit duplicator.) Letterpress or relief printing includes printing from raised type, halftone, or woodcuts by a platen, cylinder, or rotary press. Intaglio or gravure printing includes printing from engraved plates, etching, photogravure, and rotogravure. Planography or flat-surface printing includes lithography, offset, aquatone, collotype, etc. Silk-screen printing uses a stencil (silk or other material) through which ink, paint, etc., is forced by a rubber squeegee. See AQUATINT; AQUATONE PRINTING; COLLOTYPE PRINTING; DEEP-ETCH OFFSET; DESIGN PRINTING; DIE EMBOSSING; DIE STAMPING; DIRECT LITHOGRAPHY; DRIOGRAPHY; ELECTROSTATIC PRINTING; ETCHING; FLEXOGRAPHIC PRINTING; HOT SMASHING; HOT STAMPING; INK-JET PRINTING; INTAGLIO PRINTING; LETTERPRESS PRINTING; LITHOGRAPHY; MEZZOTINT; MIMEOGRAPH; MULTIGRAPH; MULTILITH; OFFSET LITHOGRAPHY; PHOTOGRAVURE PRINTING; PHOTOLITHOGRAPHY; PLANOGRAPHIC PRINTING ROTOGRAVURE PRINTING; SCREEN PRINTING; THERMOGRAPHY.

PRINTING QUALITY—See PRINTABILITY.

PRINTINGS—A general term used by some printers to indicate various grades of book paper.

PRINTING SMOOTHNESS—See SMOOTHNESS.

PROCESS ACTIVATED SLUDGE—A biological sewage treatment process in which a mixture of sewage and activated sludge is agitated and aerated. The activated sludge is subsequently separated from the treated sewage (mixed liquor) by sedimentation, and wasted or returned to the process as needed. The treated sewage overflows the weir of the settling tank in which separation from the sludge takes place.

PROCESS CAPABILITY—A statistical measure of the inherent process variability for a given characteristic. The most widely accepted for process capability is six sigma.

PROCESS COATED PAPER—See MACHINE COATED.

PROCESS COLOR—The process of producing printed materials with infinitely variable gradations of color (hue) and tone by the combination of varied tints of three colors (cyan, magenta, and yellow) and sometimes black. This is a subtractive system, using transparent inks to filter out some of the wavelengths from the white light reflected from the paper.

PROCESS CONTROL—A system where a measuring component communicates the measurement to a controlling force so that this force, in turn, will produce the required changes in the process.

PROFESSIONAL TOWELS—Towels especially designed for use by physicians, dentists, barbers, and beauticians. Some are made from two-, three-, or four-ply wet-strength facial-type tissue in white and green.

PROFILE (CROSS DIRECTION)—A longitudinal section (cross direction).

239

PROFILE PAPER—(1) A black cover paper to be used for making silhouettes, basis weight 60 pounds (20 x 26 inches – 500). (2) A drawing paper specially ruled for engineers and surveyors, in which case it might be considered as a cross-section paper. The usual pattern is 0.25-inch squares subdivided 5 times in each direction; if metric, centimeter squares subdivided 10 times in each direction.

PROGRAM PAPER—(1) Any paper used for programs. A very wide range of qualities, finishes, etc., is used. No one grade or specification dominates this field. (2) Soft printing paper so manufactured that it can be handled without rustle or rattle. It may be unsized, drying paper being sometimes used for this purpose. See NOISELESS PROGRAM PAPER.

PROGRESSIVE PROOF SHEETS—Proving paper, one sheet of which is used for each of the colors used in multicolor printing.

PROHIBITIVE MATERIALS—Any material which by its presence in a packaging of paper stock, in excess of the amount allowed, will make the packaging unusable as the grade specified. Also any material that may be damaging to equipment. (Institute of Scrap Recycling Industries, Inc. scrap specifications for 1994.)

PROMOTERS—Strongly cationic polymers, of moderate molecular weights, which are used in papermaking to offset the excess negative charge of a system, and thus enhance the performance of other chemical additives, such as sizing emulsions, wet strength resins, dyes, and dry strength agents. See ANIONIC CONTAMINANTS; CATIONIC DEMAND.

PROOF CORRECTION MARKS—A standard set of marks used by editors, proofreaders, etc., to mark-up draft printed copy, prior to final printing.

PROOF OR PROOFING PAPER—(1) A high-grade, smooth-finish cardboard or heavyweight coated paper used for making engravers' proofs. (2) A cheap book paper or newsprint, used for making galley proofs. See DRY-PROOFING

PAPER; GALLEY-PROOF PAPER. (3) Chromo as used by block-makers.

PROPRIETARY MILL BRAND—Paper sold under the name of the owner of a mill.

PROTECTIVE PAPER—(1) A writing paper which has been treated or modified to prevent fraudulent duplication or alteration of the matter printed or written thereon. See SAFETY PAPER. (2) Any grade of paper which is used in a protective sense, such as wrapping paper.

PROTOLIGNIN—Wood lignin in an unchanged form.

PSID—Differential pressure, pounds per square inch differential (psid).

PSIG—Gauge pressure, pounds per square inch gauge (psig), referenced to local atmospheric pressure. Equal to (absolute pressure – atmospheric pressure).

PUBLICATION PAPERS—Papers used in publications, periodicals, magazines, inserts, direct mail, etc. Publication papers can be either coated or uncoated; they are made from chemical woodpulps, mechanical pulps, recovered fiber, or a combination thereof.

PUCKER—A cockle-like surface effect on paper which has contracted unevenly during drying.

PULP—A general term used to describe the fibers after they are liberated by pulping from a fibrous raw material source such as wood chips, straw, cotton, or grasses.

PULP AIR DRYER—A form of a dryer section in which the sheet is fed into a large enclosed chamber where it is supported by rolls or air flotation devices and is dried by the counter circulation of hot, low humidity air (air dryer).

PULP CONTENT—The fiber content by analysis of any paper or paperboard exclusive of clay or any other filler or coating materials.

PULP BLEACHING—The purpose of bleaching pulp is to whiten it with minimal degradation

240

of the pulp fibers. Pulps are produced either by chemical or mechanical means, or combinations of the two. The bleaching procedure depends on the pulping process, the wood species and the end use of the bleached product. The bleaching of chemical pulps to high brightness is accomplished in several stages. A bleaching stage consists of equipment to bring the stock to the desired consistency, a mixer that will blend the chemicals, fibers and steam, a retention vessel wherein the reaction may proceed, and a washer to separate the treated fibers from the waste products. A series of such stages is called a bleaching sequence. The bleaching of chemical pulps is accomplished in several main steps:

1. The preparation for dissolution of the ligneous compounds not removed during the digestion of wood.

2. The dissolution of the lignin fragments made alkali-soluble and, where desired, the non-cellulosic carbohydrates.

3. Oxidation of the remaining color groups to develop a white pulp.

Mechanical pulps are brightened by chemically transforming the color-producing compounds to a less colored state without their removal, thus retaining the high yield advantage of this pulping process.

PULP COLOR—Color lakes; sometimes also pigments sold in paste form, such as the chrome yellow.

PULPBOARD—See PAPERBOARD.

PULPER STOCK—Pulp from such sources as pulp sheets, dry broke, and reclaimed paper, which has been mechanically disintegrated by agitation in water using a device known as a pulper (q.v.).

PULPER—(1) A machine designed to break up, defiber, and disperse dry pulps, mill process broke, commercial recovered papers, or other fibrous materials into slush form preparatory to further processing and conversion into paper or paperboard. It normally consists of a tank or chest with suitable agitation to accomplish

the dispersion with a minimum consumption of power. It may also be used for blending various materials with pulp. (2) In recycled fiber pulping, the pulper converts recovered paper to a pumpable slurry by the addition of water and chemicals, in some applications. The pulping action defibers paper, loosens inks, and some gross contaminants are removed by means of a junk trap, tower or auxiliary equipment. Most pulpers are circular tubs equipped with a bottom agitator and with baffles to redirect the stock flow. A rotating drum type pulper is used in some applications, notably for pulping old newspapers and magazines. See PULPER STOCK; SLUSH; SLUSH PULPER.

PULP FELT—See PRESS FELT.

PULP GRINDER—See GRINDER.

PULPING—The process of separating a cellulosic raw material, such as pulpwood, chips, rags, straw, and recycled paper, into a fibrous pulp suitable for manufacture of paper or paperboard or for chemical conversion into materials such as rayon and cellophane. Pulping may vary from simple mechanical action to complex digesting sequences. Pulps made by treating cellulosic raw material with chemicals, under pressure, and at high temperature to dissolve a substantial amount of lignin are called chemical pulps. The pulping process may be done in batch or continuous equipment in production sizes up to several thousand tons per day.

PULPING INDEX—An arbitrarily defined and now obsolete term used to express the degree to which paper or broke has been defibered into individual fibers. The currently preferred term is flake content (q.v.).

PULP SHEET—(1) See HANDSHEET. (2) Pulp, usually market pulp, in sheets from a pulp dryer prepared in this form for convenience in shipment and handling.

PULPSTONE—A natural or artificial stone with high grit content. It is grooved and used for the manufacture of groundwood mechanical pulp in a grinder (q.v.).

241

PULP SUBSTITUTES—A general classification of recovered paper grades that can be used directly in paper manufacture without cleaning, bleaching, or deinking. Darker colored pulp substitutes could be used in dark toweling wipes. Many tissue mills call pulp substitutes direct entry grades.

PULPWOOD—Those woods which are suitable for the manufacture of woodpulp. The wood may be in the form of logs as they come from the forest or cut into shorter lengths suitable for the chipper or the grinder. The term may also be applied generically to chips produced from roundwood or from whole trees remote from the pulp mill.

PULSATION ATTENUATOR—A device installed in the paper machine approach flow circuit (q.v.), usually just before the headbox or former, to reduce the magnitude of pressure pulsations, or to shift their frequency to a range less likely to cause problems such as barring (q.v.).

PULSATIONS—Variations in pressure within the paper machine approach flow circuit (q.v.), which can cause barring in the sheet. Pulsations typically originate in rotating equipment, such as the fan pump, or the paper machine screen, or from vibrations in the structure.

PULVERIZED COAL BOILER—Finely pulverized coal is burned while in suspension in the combustion air. Similar boilers have been fired with finely pulverized wood or other solid fuel. This type of firing may be combined with other types, such as grate firing.

PUMICE STONE PAPER—An abrasive paper (q.v.) which is coated with pumice stone or varying degrees of fineness.

PUNCHBOARD—A paperboard used for advertising purposes in selling articles upon chance. It is made of pasted chipboard or builders' board. Holes are punched out of the board and into each is inserted a small ticket. A printed form indicating where the board is to be punched is then pasted over the top. The board is stiff and gives smooth edges where punched.

PUNCHBOARD PAPER—A thin, hard paper, in basis weights of 12 to 16 pounds (17 x 22 inches – 500) used for filling the holes in punchboards. It has a rather high finish and good folding qualities.

PUNCHED TAPE—See PERFORATOR TAPE.

PUNCTURE—The resistance of paper or paperboard to perforation, as measured with a puncture tester having a pyramidal steel head, and expressed as the energy in puncture units (one puncture unit is 0.265 inch-pound).

PURCHASED LIME—Lime derived from calcining limestone by driving off carbon dioxide from calcium carbonate at high temperature. Also referred to as fresh lime or makeup lime.

PURITAN FILLER—See CALCIUM SULFATE.

PURITY—See COLORIMETRIC PURITY; EXCITATION PURITY.

PUSEY AND JONES PLASTOMETER—A portable device used to measure the hardness of paper machine roll covers. This instrument measures penetration into the cover surface of a 3.175 mm (0.175 inches) spherical point under a load of one kilogram (2.2 pounds). A dial indicates penetration in hundredths of a millimeter. The dimensions are not recorded and the reading is noted as P&J units. The hardness of cover compositions vary from 0–1 P&J (rock hard) through 15–40 P&J (medium hard) to 150–250 P&J (extremely soft) depending on the function of the roll and the desired nip action. See ROLL COVER HARDNESS; SHORE DUROMETER.

PVA—A common abbreviation for polyvinyl alcohol, an additive which is used in certain papers to improve sizing, etc.

PYROLYSIS—The process of chemically decomposing an organic substance by heating it in an oxygen-deficient atmosphere.

PYROXYLIN-COATED PAPER—A paper which is coated with a pyroxylin lacquer (cellulose tetranitrate dissolved in a solvent) used for box coverings, greeting cards, book covers, labels, menus, food wrappers, tobacco wrappers, artificial leather, etc. The lacquer may be clear or colored by dyes or pigments. Such papers are also manufactured with gold, silver, and copper metallic finishes.

Q

Q BLEACHING STAGE—See CHELATION STAGE (Q).

QUALITY ASSURANCE—All the planned and systematic activities implemented within the quality system that can be demonstrated to provide confidence that a product will fulfill requirements for quality.

QUALITY CONTROL—The operational technique and activities used to fulfill requirements for quality. (Note that the terms "quality assurance" and "quality control" are often used interchangeably to refer to actions performed to ensure the quality of a product, service, or process.)

QUALITY ENGINEERING—The analysis of a manufacturing system at all stages to maximize the quality of the process itself or the product it produces.

QUARTO—Paper that is folded twice, producing four sections, or eight pages.

QUICK-SET INKS—Inks set by a short period of heating, inks applied in melted liquid phase which set on cooling, inks set by steaming (water vapor precipitates ink resins).

QUIRE—One twentieth of a ream. Twenty-five sheets in the case of a 500-sheet ream of fine papers, and 24 sheets in the case of a 480-sheet ream of coarse papers.

QUIRED—In a ream packing, having sheets folded in half (in sections of twelve), instead of being put up flat.

R

RACKING STRENGTH—The resistance of a panel, having a prescribed wood frame and sheathed with structural insulating board, to a racking load such as would be imposed by winds blowing on a wall oriented at 90° to the panel.

RADIATION CURE INKS—Inks which require radiation of a specific wavelength or type to dry or cure. Common types of radiation used for these purposes include ultraviolet and electron beam.

RAG BOOK—A paper used for high-grade books when a quality or a longevity feature is important. The term is applied to a wide range of book papers with cotton fiber content of 25, 50, 75 or 100%. See also BIBLE PAPER; COTTON FIBER CONTENT PAPER.

RAG CONTENT—A term used to indicate that a paper contains a percentage of cotton fiber pulp from virgin or waste textile materials. The cotton fiber content normally used may vary from 25 to 100%.

RAG-CONTENT PAPER—Papers containing a minimum of 25% rag or cotton fiber. These papers generally are made in the following grades: 25, 50, 75, 100% and extra No. 1 (100%). They are used for bonds, currency, writing, ledgers, manifold, and onionskin; papeteries; and wedding, index, carbonizing, blueprint, and other reproduction papers, maps and charts, and other industrial specialties. See COTTON FIBER CONTENT PAPER.

RAGGER—A long cable (rope) of primarily baling wires that dangles down into the fiber slurry in a pulper. The ragger slowly removes, by winding around the cable, large contaminants consisting of bale wires, strings, plastics, etc. A hoist gradually pulls the long cable out of the pulper, usually on a discontinuous basis.

RAG INDEX BRISTOL—An index bristol containing 25% or more of cotton fiber. Rag index bristols are usually made with 25, 50, 75, and 100% cotton fiber content.

RAG PAPER—See RAG-CONTENT PAPER.

RAG PERFORATING PAPER—A paper used as a base for a pattern formed by pricking the paper with a punching wheel. Originally it was made from 100% cotton fiber, but at present the furnish may include chemical woodpulps. The basis weight is from 16 to 24 pounds (17 x 22 inches – 500). The stock is well beaten and the sheet has uniform formation. Clean perforations are the most significant property. See also POUNCING PAPER.

RAG PLATE PAPER—Rag book paper with a supercalendered finish.

RAG PRINTING PAPER—See RAG BOOK.

RAG PULPS—Papermaking fibers made from new or old cotton textile cuttings. The term may also apply to cotton linters, i.e., the short fibers which adhere to the cotton seed after the ginning process. Rag pulps are used in papers where permanence and durability are needed, e.g., ledger, blueprint, map, currency papers, etc.

RAIL FREIGHT CLASSIFICATION—See RULE 41.

RAILROAD BOARD—Clay-coated, white-lined paperboard, the coating raw stock being a cylinder-machine product made of over-issue news. The clay coating is usually applied to both sides of the raw stock and is tinted to give a wide range of colors. The board is generally waterproof to withstand handling by wet hands. The basis weight ranges from 130 to 300 pounds (22 x 28 inches – 500). It is used in general for mailing cards, showcards, tickets, checks, tags, and car signs. It is also known as car-sign board, colored railroad ticket, colored ticket stock, colored railroad stock, and ticket stock.

RAILROAD BOARD (UNCOATED)—Railroad board (uncoated) is manufactured on a cylinder machine in white and colors in 4-ply and 6-ply (0.018 and 0.024 inch) thicknesses, usu-

ally in sizes 22 x 28 and 28 x 44 inches. This grade is made with a news center and the two outside liners are either white or colored. Uses are for games, signs, hat checks, display advertising, car signs, tickets, and general show card work.

RAILROAD COPYING TISSUE—Originally made from rag pulp, today it is a tissue made from chemical woodpulps. It is made on a fourdrinier machine, although in the past a cylinder machine was used. After the paper is made, it is treated with oil to make it transparent. It is used in pads with carbon paper between the sheets. The writing is done with a metal stylus. The carbon impression is on the back side of the paper and read through the sheet. The paper must be strong so that the stylus will not tear it. It is made in several weights, 10 to 15 pounds, and oiled. It is also called train-order tissue.

RAILROAD MANILA—A writing paper containing a substantial quantity of mechanical pulp (book grade) mixed with long-fibered chemical pulp. It is sized for pen and ink writing and is usually supplied in a canary color. It is commonly used in school tablets, for printing business forms, for advertising throw-aways and flyers, for second sheets in typing, and in manufacturing sales books, order blanks, etc. The usual weights are 32 to 45 pounds (24 x 36 inches – 500).

RAILROAD WRITING—See RAILROAD MANILA.

RAINBOW PRINTING—Printing using split-fountain ink application, which provides a continuous shift from one color to the other.

RAM—Abbreviation for random access memory. The part of computer memory that provides quick access, and user-modifiable memory.

RAMIE—A plant of the nettle family native to tropical Asia, but cultivated in other sufficiently warm regions. The botanical name is *Boehmeria nivea* (especially important is variety *tenacissima*). The bast fiber from the decorticated material is commercially known as China grass and is used as a textile fiber. It is a potential source of papermaking fibers.

RANDOM CAUSES—See COMMON CAUSES.

RANDOM SAMPLING—A sampling technique in which sample units are selected in such a manner that all combinations of units under consideration have an equal chance of being selected.

RANGE—The difference between the largest and smallest value in a data set.

RANGE CHART (R CHART)—A control chart in which the sub-group range, R, is used to evaluate the stability of the variability within a process.

RATIO CONTROL—A special case of feed-forward control. Typical application would be where one measurement is used to control blending streams (two or more) to a final point, one stream being controlled and the others being wild. Applications would include stock flows and additive flows.

RATTLE—Noise produced when a sheet is shaken. While it is sometimes considered indicative of the quality of bond paper, its absence is important in some papers used before an audience, especially when the speaker uses a microphone.

RAW COOK—See HARD COOK.

RAW GREEN LIQUOR—Green liquor (q.v.) prior to clarification to remove dregs.

RAW STOCK—Stock that has not been blended or refined. See COATING RAW STOCK; HANGING PAPER.

RAW WEIGHT—See UNCOATED WEIGHT.

RAY CELLS—Short cells, chiefly parenchyma, which make up the wood ray. The wood ray is a ribbon-like strand of tissue extending in a radial direction across the annual rings of the wood.

RAYON REJECTS—Cellophane and rayon pulps that fail to measure up to specific requirements for use in chemical conversion plants. Their physical characteristics make them usable in papers where strength is a minor consideration and softness and absorbency are desired.

RAZOR-BLADE SLIT—The process of slitting a web or film with a stationary or oscillating razor blade or with a single driven circular blade. The slit quality and dimensional stability may be enhanced by having the blade penetrate a narrow groove with no contact with the groove bottom or sides.

R CHART—See RANGE CHART.

REACTION PAPER—See TEST PAPERS.

REACTION WOOD—Wood with abnormal characteristics formed in leaning or crooked stems and in branches of trees. It consists of tension wood in hardwoods and compression wood in softwoods. See TENSION WOOD; COMPRESSION WOOD.

REACTIVE SIZES—See CELLULOSE REACTIVE SIZES.

REACTOR—In electrical systems, a reactor is an electromagnetic protective device that stores energy and, based on the charging rate, limits the instantaneous flow of current through the downstream circuit.

REAM—(1) A number of sheets of paper, either 480 or 500 according to grade. (The Federal government specifies 1000-sheet reams.) (2) A sealed or banded package containing the required number of sheets.

REAM LABELS—Labels pasted on the ends of sealed packages of paper (each containing a ream) to describe the contents.

REAM MARKERS—Slips of paper placed in a pile of paper to mark its division into reams.

REAM SEALED—See REAM WRAPPED.

REAM WEIGHT—The weight of one ream of paper, either actual or nominal. See ACTUAL WEIGHT; NOMINAL WEIGHT.

REAM WRAPPED—Term applied to sheet paper. Packaged in bundles of 480 or 500 sheets.

REAM WRAPPER—A machine which wraps reams of paper into shipping packages.

REAM WRAPPERS—A paper used in wrapping reams of paper. It is made from sulfate, sulfite, jute or rope pulp, in natural and many colors, and often printed. It is frequently asphalt or wax laminated and may be polyethylene coated. In addition to holding the contents intact and protecting against contamination, the wrapper frequently must provide protection against excessive changes in moisture; therefore, low moisture vapor transmission properties maybe required. See SEALING PAPERS.

REBURNED LIME—Lime that is derived by calcining lime mud (q.v.) in special rotary kilns or other reactors by driving off carbon dioxide from calcium carbonate at high temperature.

RECEIVING WATER—Water into which an effluent is discharged.

RECEPTIVITY—Acceptance of oil, water, or other liquid by the surface of paper; it depends on the wetting of the surface by the liquid and the initial rate of penetration of the liquid into the surface. Ordinarily this term implies penetration under capillary forces alone, but in some printing processes, penetration is aided by pressure. See ABSORBENCY; INK RECEPTIVITY; HOLDOUT.

RECESSES—See VOID FRACTION.

RECORD PAPER—See LEDGER PAPER.

RECORDER PAPER—See METER PAPER; STRIP CHART PAPER.

RECORDING INSTRUMENT PAPER—See METER PAPER; STRIP CHART PAPER.

RECORDING PAPER—See CHART PAPER; STRIP CHART PAPER.

RECOVERED FIBER—Fiber that has been processed into a usable form that had been either source separated or reclaimed from the solid waste stream. The recovered fiber can be either from recovered paper or other papermaking fibers, e.g., cotton rags, jute.

RECOVERED PAPER—Paper, paperboard and packaging grades in their final consumer form that are collected by various means in recycling programs which serve as the raw material for paper recycling facilities.

RECOVERY—Deriving a substance in usable form from refuse material, such as recovered paper.

RECOVERY BOILER—A boiler used to recover pulping chemicals by burning off the organic material in kraft black liquor. The inorganic pulping chemicals are tapped from the lower portion of the recovery boiler furnace as a molten salt mixture of sodium sulfate and sodium carbonate while the heating value of the black liquor organics is recovered as high-pressure steam through heat exchange between the flue gases and the boiler tubes.

RECOVERY FURNACE—An older term used to indicate a recovery boiler. Now specifically used to indicate only the furnace portion of the boiler wherein black liquor combustion is carried out.

RECTIFIER ROLL—See DISTRIBUTOR ROLL.

RECTIFY—The adding of sheets to a broken ream to bring it to the required standard of a full ream.

RECYCLED CORRUGATING MEDIUM—A corrugating medium (q.v.) made entirely, or predominantly, from reclaimed paper fiber, primarily from old corrugated containers (OCC).

RECYCLED FIBER—Cellulose fiber reclaimed from recovered material and reused.

RECYCLED FIBER LINERBOARD—Paperboard used chiefly for the facings of corrugated board (q.v.) and solid fiberboard made totally from recycled fiber. See LINERBOARD.

RECYCLED NEWSPRINT—A newsprint grade made predominantly from deinked old newspapers. See NEWSPRINT RECYCLING.

RECYCLED PAPER—Recycled paper and paperboard produced with a percentage of recycled fiber, with the percentage so specified. Examples include recycled corrugated medium and liner, recycled paperboard, recycled newsprint.

RECYCLED PAPERBOARD—Paperboard manufactured from a combination of fibers from various grades of recovered paper, with the predominant portion of the total furnish being recycled fibers.

RECYCLED PAPER PULP—A papermaking fiber made from recovered papers, such as old newspapers, magazines, corrugated containers, office waste, and printing plant wastes.

RECYCLING—The return of once-manufactured-and-used material for reprocessing into new products. In the paper industry, recycling refers to the process involved in making new paper out of previously used paper including in-plant, commercial, and postconsumer waste.

RED PATCH PAPER OR BOARD—A red paper or paperboard, made from virgin chemical pulp and/or selected reclaimed paperstock in a basis weight of 110 pounds (22.5 x 28.5 – 500), although it may be lighter or heavier than this. It is used for patches, reinforcement for the eyelets of tags, and buttons on envelopes.

RED ROPE PAPER—See RED WALLET.

RED ROSIN SHEATHING PAPER—A sheathing paper colored red. It is made of wastepapers on a cylinder machine, well sized and with a hard finish; it is commonly supplied in rolls 36 inches wide, varying in weight from 20 to 60 pounds to the roll of 500 square feet. The basis weight varies from 125 to 300 pounds (24 x 36

inches – 500). The paper protects against wind and dust and is water and moisture repellent. It is used under sidings, between floors, for lining walls to keep out dust, dirt, and drafts, as lining for box cars, and as protective coverings over new floors before completion where a waterproof paper is not required. See SHEATHING PAPER.

RED ROSIN-SIZED SHEATHING—See RED ROSIN SHEATHING PAPER.

RED WALLET—A flexible paperboard used principally in making expanding envelopes or wallets and also for file folders. It is made from rope, jute, sulfite, or sulfate pulps, sometimes with an admixture of mechanical pulp. It may be made on a cylinder or fourdrinier machine. It varies in thickness from 7 to 20 pounds.

REDUCIBLE SULFUR—Any form of sulfur or sulfur compounds in paper that can be converted to hydrogen sulfide on treatment with a metal such as aluminum and an acid. It is a measure of the quantity of sulfur compounds in paper that may react with metals to cause tarnishing.

REDUCTANTS—Lignin-brightening and lignin-preserving bleaching agents such as sodium hydrosulfite (dithionite), borohydrides, and sulfurous acid salts, used almost exclusively on lignin-rich pulps. Reductants convert colored compounds to less colored ones by adding hydrogen to unsaturated chromophoric groups or by reducing carbonyl groups to hydroxyl groups. These pulps are sensitive to oxidation, which causes brightness reversion on aging. See BOROHYDRIDE; HYDROSULFITE.

REDUCTION EFFICIENCY—The ratio of the sodium sulfide to the total quantity of sodium/sulfur compounds in a material. Often applied to smelt from a recovery boiler as a measure of its performance in producing the kraft cooking chemical sodium sulfide.

REEL—(1) The roll winding device that converts a continuously running web of paper, of full paper machine width, onto a reel spool or core shaft. This "reel" connotation usually applies to the final roll winding operation on the end of a papermaking machine, but identical components can be found at the end of an off-machine coater operation, or at a supercalender, or on a re-reeler, which is simply a reel designed for intermittent operation to remove defects. Several different reel designs have evolved over the years ranging from the pope reel, the level-rail reel, the two-drum reel, the center-wind reel, the combination surface-centerwind reel, to the surface-centerwind reel without primary and secondary arms. In some designs, the full deckle web can be slit before the reel and wound onto cores split into two or more rolls on the same reel spool for direct to converting operations. (2) The untrimmed roll of paper of full paper machine width wound on a large shaft at the dry end of the paper machine.

REELING—The operation of winding paper into a reel.

REEL SAMPLES—Samples taken from the reel as the paper is being manufactured; usually one sample is taken from each reel; such samples are used for physical tests of the paper or for reference purposes.

REEL SPOOL—The full width core shaft which is required to span the width of the web on a reel and provide support for the winding web from a start at the core to the finished diameter, which may exceed 110 inches. Reel spools are sometimes covered with a hard elastomeric composition.

REFINED STOCK—Stock that has been refined with the main refiners, but not with the tickler refiners (q.v.). Refining may take place before or after stock blending, or in both places.

REFINER—A machine used to rub, macerate, bruise, and cut fibrous material, usually cellulose, in water suspension to convert the raw fiber into a form suitable for formation into a web of desired characteristics on a paper machine. The many types of refiners differ in size and design features, but most can be classified as either conical refiners or disc refiners. Beaters are not usually referred to as refiners although in a broad sense they serve a similar function. Refiners may be used in various combinations of types and numbers of units depending on the type of stock to be treated and the capacity required. See also DEFLAKER; DISC REFINER; JORDAN.

REFINER BAR CROSSINGS—The length of bar crossings per unit time in a refiner, determined primarily by the refiner tackle geometry, and rotor speed. Typical units are inch-contacts/minute, or meter-contacts/second.

REFINER BLEACHING—A bleaching sequence which uses the intense mixing action of a refiner for bleach mixing. Often, additional reaction time is provided after the pulp exits the refiner.

REFINER CHEST—The chest that feeds the refiners in a stock preparation system.

REFINER (DISC REFINER)—See DISC REFINER.

REFINER GROUNDWOOD—A variety of papers with substantial proportions of mechanical woodpulp processed in a refiner.

REFINER GROUNDWOOD (RGWD)—See REFINER MECHANICAL PULP (RMP).

REFINER INTENSITY—The relative amount of energy applied per bar crossing length during refining – how severely refining energy is applied. Typical units are hp/inch contacts per minute, or Watts sec/meter. Higher refining intensity results in more energy applied to a single fiber per impact. Also called specific edge load. See also REFINER BAR CROSSINGS.

REFINER MECHANICAL PULP (RMP)—A mechanical pulp produced from wood chips in an atmospheric refiner with an open inlet and discharge and without any pretreatment.

REFINER PLATES—See REFINER TACKLE.

REFINER TACKLE—Bars or other working surfaces which operate in close proximity to each other during refining. Because of high wear rates, refining tackle is usually easily replaceable. Refining tackle for disc refiners is often called refiner plates, or refiner plates, or refiner plate segments, for larger refiners, because the size and weight of the entire circle of tackle would be awkward to work with.

REFINING—In mechanical pulping, a mechanical treatment of papermaking pulp to condition the fibers, or to separate fibers from wood chips. Any of several operations, all of which involve the mechanical treatment of pulp in a water suspension to develop the papermaking properties of hydration (q.v.) and fibrillation (q.v.), and to cut the fibers to the desired length distribution. See REFINER.

REFINING TACKLE—See REFINER TACKLE.

REFLECTANCE—The ratio of the intensity of the light reflected by the specimen to the intensity of the light similarly reflected by a standard reflector. The instrument and the conditions of measurement must be carefully specified. The spectral reflectance curve gives the reflectance of the specimen as it varies with wavelength throughout the whole visible spectrum. Reflectance is of importance in the physical measurement of color and opacity.

REFLECTIVITY—The reflectance of a sample, e.g., a pile of sheets, thick enough so that no change in reflectance is observed when the thickness is doubled.

REFRIGERATOR PAPER—Paper used in the walls of a refrigerator as an insulating material. It is made of two or more sheets of kraft paper which have been treated to render them impervious to moisture and then creped, the sheets being combined by means of asphalt. The paper may or may not be reinforced. Such a product is durable and does not allow the passage of vapors.

REGENERATED CELLULOSE—See CELLULOSE II.

REGENERATIVE DRIVE—An electric motor drive, often found on unwinds, that is able to return roll or roller braking energy back to the power lines.

REGENERATIVE UNWIND—A type of tension producing mechanism for an unwind stand in a winding process. This type of braking has the ability to convert much of the breaking effort back to electrical power, hence, the name regenerative unwind. Normally this type of unwind has the capacity to accelerate the unwind. This acceleration capacity is particularly beneficial to operations winding high density grades at low web tension.

REGISTER—(1) Exact correspondence in the position of pages or other printed matter on two sides of a sheet or in its relation to other matter already ruled or printed on the sheet. (2) To print a succeeding form or color so that it is in correct position with reference to matter already printed on the sheet. (3) In paper ruling, a sheet is said to register when ruled on both sides so that when the sheet is held up to the light the lines exactly coincide.

REGISTER BOND—A common type of lightweight writing paper designed for single and multi-copy business forms and variations thereof such as computer output forms, snap-aparts, unit sets, carbon interleaved order books, invoice sets, and the like. The grade is usually made from chemical woodpulps in basis weights of 9, 10, 11, 12, 13, 14, 15 pounds (17 x 22 inches x 500) although companion items for multi-copy forms sets are also made in 16-, 20- and 24-lb weights. Important product qualities include printability, good tensile and tearing strength, perforating, folding, and manifolding qualities. See also FORM BOND.

REGISTER MARKS—Lines, crosses, or other patterns used to show register between print colors or to aid further processing.

REGISTER PAPER—See REGISTER BOND.

REGISTER TEST SHEET—Paper run through the press to test its register.

REGRESSION ANALYSIS—A statistical technique for determining the best mathematical expression describing the functional relationship between one response and one or more variables.

REGULAR SIZE—A regular standard size of any kind of paper, as distinguished from irregular sizes. In roll paper, a regular size is one in which the roll width corresponds to either of the two dimensions of a standard sheet size.

REGULAR SLOTTED CONTAINER—The type of paperboard box most generally in use as the outer container in the shipment of a wide variety of articles. It is made from a single sheet of corrugated or solid fiberboard, slotted and scored. The two side edges are taped or stitched together leaving the end flaps to be folded inward when the box is to be closed. Rule 41 of the Consolidated Freight Classification specifies the strength and qualification of the box for freight shipments and the requirements for express and parcel post are separately set forth. The container is rigid, usually water resisting, and built to withstand the hard usage incident to transportation.

REGULAR NUMBER—The number of sheets of boxboard, 25 x 40 inches, required to make a bundle of 50 pounds. For book cover board, sheets 25 x 30 inches are used.

REGULAR WEIGHT—Any weight that is standard for the grade as established by trade custom.

REINFORCED BUILDING PAPER—Two plies of strong kraft paper laminated together with asphalt and reinforced with fiberglass, jute, or other fibrous material. It is strong and reasonably moistureproof. It is generally sold in rolls of selected footage. A roll of 500 square feet weighs approximately 33 pounds. See also REINFORCED PAPER.

REINFORCED FILLER PAPER—A notebook filler paper with some type of reinforcement designed to prevent tearing at the holes. Common reinforcing materials are metals, cloth, paper, or polyester film applied locally as a patch, or as a strip along the margin of the sheet where the holes are located.

REINFORCED PAPER—(1) A multi-ply paper, of varying weights depending upon its use, which is united by means of asphalt or adhesive, embedded in which is the necessary reinforcing material, such as string, yarn, sisal, synthetic fibers, or glass fibers. (2) A paper, such as filler paper which has been reinforced by cloth, copper, or other material to prevent tearing at the punched areas. (3) Waterproof sheathing paper to which is bonded a continuous layer of metal; the paper is a highly effective moisture and vapor barrier and can be used as a termite shield. (4) A single-ply paper in which has been incorporated reinforcing material such as scrim.

REJECT RATE—The reject quantity divided by the feed quantity, usually in reference to screens or cleaners. Quantity may be defined as suspended solids or fiber or ash, expressed as mass or mass flow rate. Reject rate then can be further defined as reject rate based on solids, reject rate based on fiber, reject rate based on ash, etc. Quantity may also be defined as liquid flow or liquid flow rate, and reject rate then can be further defined as reject rate hydraulic, reject rate based on flow, hydraulic split, water split, etc. All of the foregoing may also be expressed as percent by multiplying by 100%. Some of the above terms are used in acronym form, e.g., RRBOS for reject rate based on solids.

REJECT REFINING—Mechanical pulps contain undeveloped fibers and fiber bundles which need to be removed by screening. The rejects from the screens can be converted into useful pulp by further refining in atmospheric or pressurized refiners at high consistencies (25%+).

REJECTS—Material rejected by a separator. Typically, rejects consist of debris (q.v.) plus some otherwise usable fiber.

REJECTS SULFONATION—A means for treating screen rejects with sodium sulfite. Sulfonation will soften the fibers and, by making them flexible, increases their bonding strength.

RELATIVE HUMIDITY—The ratio of the amount of moisture in the air at any temperature to the amount required at that temperature to saturate the air. See also ABSOLUTE HU-MIDITY; HUMIDITY.

RELEASE PAPERS—A group of papers specifically designed for easy stripping from or non-adherence to tacky surfaces. The types vary widely from glassine or vegetable parchment to kraft papers and paperboards, all of which are usually treated with a release agent such as silicone, fatty acid metal complex, or acrylic polymer. Release papers are used for packaging wax, asphalt, rubber, certain foods, as backing for pressure-sensitive tapes and labels, and as casting surfaces for plastic films, etc.

RELIEF PAPER—Paper with a suitable printing surface to permit the reproduction of an object to appear in relief. It is usually a highly finished coated paper.

RELIEF PRINTING—Printing from raised surfaces, such as type, woodcuts, zinc and half-tone plates, as contrasted with intaglio work, such as copper and steel plates, and lithography. See LETTERPRESS PRINTING.

RELIEF PRINTING PAPER—See RELIEF PAPER.

REP FINISH—A ribbed or corded finish or surface, somewhat like coarse linen or the weave of felt, produced by passing paper through grooved steel rollers or by felts with heavy warp threads, or in platers in which the books are made up with rep cloth, which is similar to corduroy.

REPRINT PAPER—A printing paper used for reprints of books or other publications. It may be the same as the original paper used in publication, or of a higher or lower quality.

REPRODUCTION PAPER—Base paper used in various reproduction processes or systems. Such paper may be sensitized (as in the cases of whiteprint, negative, blueprint, photographic, or zinc oxide coated) or plain (as in the case of electrostatic reproduction paper). It is generally made from bleached chemical woodpulps and/or cotton fiber pulps in basis weights ranging from 11 to 32 pounds (17 x 22 inches – 500). It is usually very well sized and characterized by chemical purity, good wet or dry strength, and a high degree of permanence. See also DUPLICATING PAPER; MIMEO-GRAPH PAPER.

REPRODUCTION PROCESSES—See HECTO-GRAPH; MIMEOGRAPH; MULTIGRAPH; PRINTING PROCESSES; SPIRIT DUPLICATION.

REPROGRAPHIC PAPER—Generally a writing paper used for office copying by xerography, spirit, stencil, or offset duplicating.

REPROGRAPHY—(1) A general term for office copying, etc., including the reproduction of printed, typed, or handwritten material by processes other than printing or photography. This includes stencil printing, direct photocopying, xerography, heat copying, and the like. (2) The applied physics and chemistry related to such processes of reproduction.

REPTISSUE—An embossed or ribbed tissue.

REPULPING—The operation of rewetting and fiberizing (q.v.) pulp or paper for subsequent sheet formation. See also PULPER.

REREELER—An additional unwind and reel which accepts a full reel of paper and rewinds it to improve roll structure and to remove defects before it progresses to the next function, which may be an off-machine coating operation.

RESIDUAL ALKALI—Residual caustic remaining at the end of chemical digestion of wood.

RESIDUAL CARBONATE—A measure of the calcium carbonate in reburned lime. Used as a measure of the conversion efficiency of the calcium carbonate in lime mud (q.v.) to calcium oxide in reburned lime.

RESILIENCY—That property of a sheet which allows it to recover from a distortion of its shape. See also ELASTICITY.

RESIN—A general term applied to various amorphous solid or semisolid organic extractives found in wood. They are insoluble in water but soluble in organic solvents. Gum resins that contain carbohydrate gums and oleoresins are mixtures of resins and volatile oils. The non-volatile residue of the conifer resins is called rosin (q.v.), which is the most important natural resin used in the paper industry. See also SYNTHETIC RESIN.

RESIN ACIDS—See ROSIN ACIDS.

RESIN-IMPREGNATING BLOTTING PAPER—A paper into which resinous compounds are introduced. Such paper has the ability to absorb a considerable volume of these compounds and still retain a moderate wet strength. Frequently, this paper is required to be free from metallic conducting substances. It is made from either rags or chemical woodpulp or combinations of these, in a variety of thicknesses from 0.003 up to about 0.025 of an inch. See also PAPER-BASE LAMINATE; PHENOLIC SHEET.

RESISTANCE TEMPERATURE DETECTOR (RTD)—A sensor that varies resistance value nearly linearly with temperature.

RESISTANCE TO PENETRATION BY A LIQUID—See HOLDOUT. See also PERMEABILITY; RECEPTIVITY.

RESISTANCE TO WEAR—That property of a sheet which withstands abrasion or, more generally, changes in physical properties during use. Tests for evaluating resistance to wear are usually intended to simulate use conditions. See also ABRASION RESISTANCE.

RESISTIVITY, ELECTRICAL—The resistance to direct current between opposite parallel faces of a centimeter cube of the material. The electrical resistance of paper is often measured by passing a current through a square specimen between electrodes attached to opposite edges of the square. This leads to an expression referred to as "ohms per square." This property is strongly influenced by the presence of foreign conducting particles and moisture, and is important in all electrical papers. See ELECTRICAL CONDUCTIVITY.

RESOURCE RECOVERY—The extraction and utilization of materials and values from municipal or industrial solid waste or from both. In the paper industry, the term generally refers to the recovery of cellulose fiber from such waste, or the recovery of energy from such waste (or components thereof) when used as a fuel source.

RETENTION—(1) The amount of filler or other material which remains in the finished paper expressed as a percentage of that added to the furnish before sheet formation. (2) The process of retaining soluble and insoluble stock components in the forming paper web on the paper machine. See FIRST PASS RETENTION.

REVERSE CLEANER—A type of hydrocyclone that removes floating type contaminants (specific gravity less than one) from a slurry of fibers in water. In this cleaner, the contaminants collect in the center of the cone and are drawn off the top center of the cleaner. The cleaned fiber moves down to discharge at the apex of the cone. The bottom discharge orifice is much smaller than for the forward cleaner (q.v.). Because the accept and reject streams depart the unit in the opposite manner, as in the more traditional and previously established centrifugal cleaner (q.v.), now called a forward cleaner, the low-specific-gravity debris removal centrifugal cleaner application became known as reverse cleaning, beginning around 1970. See also THROUGH-FLOW CLEANER.

REVERSE ROLL COATING—A method of applying and distributing or spreading a coating by the use of revolving rolls. The machine generally utilizes three rolls, all running in the same angular direction (reverse surface directions). Two of the rolls are metal rolls referred to as a metering roll and a casting roll. The third roll is a resilient or rubber-covered roll referred to as the backing roll. The metering roll determines the amount of coating carried by the casting roll and transferred to the web in the opposite direction to the travel of the web as it passes around the backing roll.

REVERSING NIP—The term applied to an arrangement of rolls in a supercalender whereby the web contact to the steel and filled rolls is reversed by arranging two filled rolls together instead of the usual alternating chilled-iron and filled rolls from top to bottom. The intent is to obtain uniform gloss on both sides, i.e., a "one-sided" sheet.

REWET—The process of water returning to the sheet of paper after being removed by the compression zone of the press nip. This can occur in the expanding or exiting side of the press nip or after the press nip if the sheet contacts the press fabric or other source of water.

REWINDER—The term applied to a winder that rewinds fractional deckle defective rolls or rolls which must be "cut down" from wider rolls. The full deckle winder that accepts the reel spool from the paper machine reel is normally called the paper machine winder; however, it has also been referred to as a rewinder, as it rewinds the parent roll from the reel.

REWINDER TRIM—Trim produced at the rewinder.

REWINDING—The process of rewinding a roll of paper to produce a proper size for the customer and/or to remove defects and to splice ends together.

RGWD—Refiner groundwood (q.v.).

RHEOLOGY—The science of deformation and flow of matter under the influence of various forces. The rheological properties of pigments and binders under the high shear forces of commercial paper and paperboard coating processes are extremely important to successful and economic operation.

RHOMETER—A hardness measuring instrument used to measure the hardness profile of a roll of paper, which correlates well with the cross-deckle caliper profile. The objective is to produce a uniform cross deckle web thickness and basis weight profile, which translates into fewer winding defects (q.v.) and improved runnability in subsequent converting operations.

RIBBING—See REP FINISH.

RIBBON PAPER—A paper which is interwound with textile ribbons such as silk or satin. The paper protects the ribbon and aids in keeping it in a roll that can be easily handled. It is made from chemical woodpulps with a machine finish, and it possesses good tensile strength. Color and formation are not highly important factors. The common basis weights are 40 to 50 pounds (25 x 38 inches – 500).

RICE PAPER—A misnomer for the sheet material cut from the pith of a small tree *(Aralia papyrifera)* which grows in the swampy forests of Taiwan. The cylindrical case of pith is rolled on a hard flat surface against a knife, by which it is cut into thin sheets of a fine ivorylike texture. Dyed in various colors, it is used extensively for the preparation of artificial flowers, the white sheets are employed by native artists for watercolor drawings.

RIDER ROLL—A metal roll (usually associated with a two-drum winder) designed to "ride" above the centerline and parallel to the winding roll, and to exert a progressively declining pressure so the supporting winder drums see a more uniform nip pressure from start to finish of a wound roll to improve roll structure. Some designs may be segmented in order to provide more uniform pressure for all slit rolls in a set.

RIGHT SIDE OF PAPER—The side of a paper from which the watermark is read correctly. It is the wire side in handmade papers and the top or felt side in machine-made papers.

RIGID BOXES—See SETUP BOXES.

RIGIDITY—See STIFFNESS.

RING CRUSH TEST—The resistance to edgewise compression of a short cylinder of paper or paperboard. It is also called ring compression resistance, and is usually measured as the force in pounds required to collapse the cylinder. See EDGEWISE COMPRESSION STRENGTH.

RINGELMANN CHART—A series of illustrations ranging from light grey to black used to measure the opacity of smoke emitted from stacks and other sources. The shades of gray simulate various smoke densities and are assigned numbers ranging from one to five. Ringelmann No. 1 is equivalent to 20 percent dense; No. 5 is 100 percent dense. Ringelmann charts are used in the setting and enforcement of emission standards.

RINGERS—"Rings" or circumferential grooves in a chilled-iron roll in a supercalender. These "ringers" are caused by indentations caused by metal particles imbedded in the adjacent rotating filled rolls.

RING GRINDER (ROBERTS GRINDER)—A machine for producing mechanical pulp, groundwood. It consists of a rotating pulpstone against which debarked logs are pressed and reduced to pulp. The debarked logs are fed from the side and are pressed against the rotating stone by an outer ring, which is set eccentric to the grinder stone.

RING MARKS—Rings, clouds, or blotches in the color in pigment-colored papers, caused by the formation and breaking of bubbles on the surface of the stock on the wire.

RING POROUS HARDWOOD—Hardwood in which the pores (vessels) at the beginning of the growing season (in the springwood) may be more or less contiguous, forming a ring and in which the pores tend to be smaller and less conspicuous in summerwood, such as oak, elm, and ash.

RING STIFFNESS—See RING CRUSH TEST.

RIPPLE FINISH—Originally a plater finish made by plating paper between sheets of sulfite pulp with a crushed formation. The result was an undulated glossy finish with the resulting indentations darker in color than the higher spots. More recently, this finish has been obtained by passing a continuous web of paper (1) through the nip of two embossed steel rolls, one male and one female, or (2) through the nip of one steel embossed roll and one rubber, plastic, or paper backing roll.

RISERS—Dunnage (q.v.) used to raise rolls off the floor during shipment.

RMP—Refiner mechanical pulp (q.v.).

ROBERTS GRINDER—See RING GRINDER.

ROBUSTNESS—The relative ability of a control loop to deliver acceptable performance after process dynamics have changed from those conditions in place when the loop was tuned.

ROD COATER—A rod is used instead of a blade to meter the coating. Rods can be either wire-wound or grooved.

ROD METERING SIZE PRESS—A metering size press which uses a rod as the metering element to control the amount of wet film transferred to the sheet. The rods used may be either threaded rods, which meter volumetrically, or smooth rods, which allow control of wet film by varying rod loading pressure. The rods are rotated continuously to minimize wear. Applicators used with a rod metering size press include the enclosed pond applicator and the short dwell coater.

ROD MILL—A machine, used first in metallurgy, consisting of a heavily lined horizontal cylinder at least twice its diameter in length and half-filled with iron rods ranging from 2 to 3 inches in diameter. Stock is fed to one end of the revolving cylinder and leaves at the other end, usually larger than the inlet in diameter and partly closed by a heavy door to prevent the rods from leaving the mill. The rod mill was used for fiberizing pulps in lap or sheet form and for the conversion of screenings into acceptable stock; today, except for a few instances of the latter use, it has been replaced by other fiberizing or refining devices.

ROE-GENBERG CHLORINE NUMBER—See CHLORINE NUMBER.

ROLL CALIPER GAUGE—A device utilized to measure the roll crown or contour of a roll before and/or after a roll grinding process.

ROLL COATING—A process in which coating color is applied to either one or both sides of a paper web by transfer from a rubber applicator roll onto which the coating color has been metered. This process may be carried out on or off the paper machine. See also MACHINE COATED; REVERSE ROLL COATING.

ROLL COVER—The elastomeric, rubber, fiberglass, or other type of composite cover on a steel roll which will provide a soft nip or a noncorrosive (protective) or elastic surface. See also ROLL COVER HARDNESS.

ROLL COVER HARDNESS—The hardness of a roll covering is the only easily obtained measurement to indicate how much the covering will deform as the roll operates. This is an important property because it controls the nip width or contact area between a pair of rolls and the pressure distribution in the nip, two important roll-operating parameters. Hardness is measured with the Pusey & Jones Plastometer (q.v.) or the Shore Durometer (q.v.).

ROLL CROWN—The difference in diameter between the middle and ends of a roll, expressed in thousands of an inch, is called the roll crown. The increase in diameter of the middle of the roll compared to the ends allows for deflection under load so that the nip pressure will be uniform over the full width of the roll. The exact crown is more critical on hard covers than on softer covers. The amount of crown for a particular roll depends on the operating load and the natural deflection of the roll. The crown must be changed when the load is increased or decreased beyond a small range. See VARIABLE CROWN ROLL.

ROLL DENSITY—The measured or calculated mass per unit of volume of a roll of paper. The mode of winding or tightness of wind can increase or decrease the roll density or hardness which is a measure of roll structure or quality. See ROLL HARDNESS.

ROLLED—Glazed by passage between rolls. See GLAZED.

ROLLED EDGES—Edges of the sheet that curl (q.v.).

ROLL EDGE DAMAGE—Cuts, punctures, or indentations on the end of a roll, which cause nicks or tears on the web edge. See also DEFECTIVE ROLL ENDS OR EDGES.

ROLL EJECTOR—A roller or beam used to push a wound roll set out of a winder onto a cradle, table or floor.

ROLL END—See ROLL HEADER.

ROLL END PROTECTOR—See ROLL HEADER.

ROLL–ENGRAVED—See ENGRAVED ROLL.

ROLLER—A rotating cylinder used for web transport. The terms drum and roll are also used to denote a roller.

ROLLER TOP CARD—A carding machine having workers and strippers on top of the main cylinder.

ROLL GRINDING—A process used to grind a desired contour or diameter profile on a roll surface to compensate for normal deflection due to roll weight or applied nip pressure or operating temperatures.

ROLL HANDLING—The method by which rolls of paper (wrapped or unwrapped) are transported from one location to another. Clamp trucks, conveyors, rails, and overhead yokes and hoists are typical roll handling devices.

ROLL HARDNESS—The measured density or hardness of wind of a roll of paper. Roll hardness is typically measured by pounding the surface with a wooden baton or, more accurately, measured with a Rhometer (q.v.) or Schmidt hardness tester (q.v.). See also ROLL DENSITY.

ROLL HEAD—See ROLL HEADER.

ROLL HEADER—A protective disc of paperboard of approximately the same diameter as the shipping roll to be protected. A wrapped roll typically has an inside and an outside header on both ends of a wrapped roll of paper installed for end protection during the roll handling and shipping process.

ROLL NIP—The contact area between two rolls.

ROLL PAPER—Paper in rolls of any required width and diameter.

ROLL PROTECTOR—See ROLL HEADER.

ROLLS—Paper or paperboard made in a continuous sheet and wound to a specified diameter. See REEL; ROLL PAPER.

ROLL SET—See CURL.

ROLL–SWIMMING—See VARIABLE CROWN ROLL.

ROLL TICKET—An identification ticket placed in a roll of paper.

ROLL TISSUE—See TOILET PAPER.

ROM—Abbreviation for read only memory. Part of the computer program that provides information, but cannot be modified by the user.

ROOFING FELT—A very porous, soft paper, made largely from the lowest grades of old cotton and woolen rags and some old paper stock, the quality depending on the nature and quantity of rags used. The rags are beaten as quickly as possible, inasmuch as the freest possible sheet is essential. It is generally made on a single-cylinder machine, although it may be made on a fourdrinier machine. No sizing or loading is used. The basis weights run from 15 to 110 pounds (12 x 12 inches – 500). It is used as a base for saturating in the manufacture of roofing papers. See DRY FELT.

ROOFING PAPER—A general term used to designate any material used in waterproofing upper decks of buildings. There are several varieties. (1) Prepared roofing: felts saturated and coated with asphalt, plain or crushed slate or other grit, embedded in an asphalt-coated surface. (2) Built-up roofing: felts saturated but not coated with asphalt; used in plies or layers and coated or built up at the time of application. (3) Roofing shingles: prepared roofing cut into various sizes and styles of shingles.

ROOFING RAGS—Used rags consisting of cotton and wool garments, etc., employed in the manufacture of dry felt (q.v.). Several grades are recognized.

ROOFING SHINGLES—Prepared roofing cut into various shapes generally composed of roofing felt saturated and coated on both sides with compounded asphalt coating and surfaced on the weatherside with mineral granules.

ROOFING WRAPPERS—A heavyweight sheet, usually kraft, used to package roofing materials. It is often printed and furnished in sheets and in rolls for use on automatic wrapping machines.

ROOF-INSULATING BOARD—(1) A type of slater's felt or asphalt-saturated paper used under roofing. It is 0.016 of an inch or more in thickness and is made in small rolls 36 to 48 inches wide and containing 500 square feet. See ASPHALT SHEATHING PAPER; SLATER PAPER. (2) A type of structural fiber insulating board, 23 or 24 inches wide, 47 or 48 inches long, and 0.5 to 2 inches thick, with a natural finish.

ROPE ARMATURE PAPER—A clear, strong rope sheet of high di-electric strength used as insulation in the manufacture of electric motors.

ROPE BAG PAPERS—See ROPE MANILA PAPER.

ROPE BRISTOL—A bristol in which the furnish contains at least 50% of manila fibers. It is a sheet of unusual strength, having high tearing resistance, used for various purposes where durability is essential. See BRISTOLS; TAG BOARD.

ROPE MANILA PAPER—A strong, durable paper used for envelopes, tags, etc. It is made of manila hemp with a manila color and has good folding qualities. The basis weights range from 60 to 140 pounds (24 x 36 inches – 500).

ROPE MANILA SANDPAPER—A paper made from manila rope and chemical woodpulp on a cylinder or fourdrinier machine, in basis weights of 60 pounds and heavier (24 x 36 inches – 480) and used as a basis for abrasive papers. It is usually highly sized, and of high tensile strength, smooth surface, and resistance to abrasion are important properties. The heavier weights are used in manufacturing abrasive papers for machine use.

ROPE MARKS—Equally spaced angular wrinkles usually oriented at a 45° angle to the axis and located between hard and soft spots around the periphery of a roll of paper. The defect derives its name from its appearance; i.e., the angular welts look like a rope. The angular wrinkles or "rope marks" form because the larger diameter area moves ahead of the smaller diameter area during the dynamic winding process.

ROPE PAPER—Any paper made from manila hemp (commonly called rope). It may be composed entirely of rope fibers or it may contain some chemical pulp. Such papers are made on both cylinder and fourdrinier machines in practically all weights and thicknesses. They are used as cable papers, shipping tags, saturating papers; and for other purposes where strength is an important property.

ROPE SACK PAPER—A paper used for flour, feed, pigments, and rock products. It is made of rope pulp, or in some instances, mixed with a percentage of chemical woodpulp. It may be enamel coated and may have a blue lining. The basis weight is from 50 to 100 pounds (24 x 36 inches – 500). See FLOUR-SACK PAPER; SHIPPING SACK KRAFT PAPER.

ROPE TAG—A tag board which contains manila rope fiber, the amount depending on the purpose for which the tag is to be used. See TAG BOARD.

ROPE WRAPPING—A term for a strong paper in brown color made from rope or a combination of rope and chemical woodpulp. It is used in wrapping, especially in the hardware trade.

ROPING—The longitudinal wrinkling of a web of paper caused by tension as the web is drawn over the drying drums and also by the tendency of the web to creep transversely upon the drum surface.

ROSIN—A specific kind of natural resin obtained as a vitreous water insoluble material from pine oleoresin (exuded gum) by removal of volatile oils, or from tall oil by the removal of the fatty acid components thereof. It consists primarily of tricycle monocarboxylic acids having the general empirical formula $(C_{20}H_{30}O2)$. The three general classifications of rosin in commerce are:

1. Gum rosin: obtained from the oleoresin collected from living trees.

2. Wood rosin: obtained from the oleoresin collected from dead wood such as stumps and knots.

3. Tall oil rosin: obtained from tall oil.

ROSIN ACIDS—Principally monocarboxylic acids with the empirical formula $C_{19}H_{29}COOH$. They are classified into two groups: the abietic type and the pimaric type. Both types and their derivatives are found in wood, gum, and tall oil rosins.

ROSIN SIZE—A soap type solution or dispersion of rosin used to give water resistance to paper and paperboard.

ROSIN SIZED—Treated with rosin size, and alum.

ROSIN-SIZED SHEATHING PAPER—Red or blue paper for sheathing houses, protecting new woodwork, etc. It is made in four weights: 4, 5, 6, and 8 pounds per 100 square feet. See RED ROSIN SHEATHING PAPER.

ROSIN SPECKS—Translucent, amber-colored specks of rosin in paper caused by the incomplete emulsification of the rosin size or by precipitation of the size before it is uniformly dispersed in the stock.

ROTARY JOINT—See STEAMFIT.

ROTARY NEWSPRINT—Obsolete term. Newsprint in roll form for use on rotary printing presses.

ROTARY PHOTOGRAVURE PRINTING—See ROTOGRAVURE PRINTING.

ROTARY PRINTING PAPER—A paper used for rotary printing presses. The term indicates that the paper is delivered in rolls; any grade of paper meeting specifications required for the subject matter to be printed may be used. Tensile strength and uniformity of quality are important. See WEB OFFSET PAPER.

ROTARY SLITTER—See SLITTER.

ROTOFORMER—A paper-forming device used to manufacture single or multi-ply grades of paper and paperboard. It consists essentially of a wire-covered drilled shell on which the stock is formed, suction boxes inside the shell to remove water, and a restricted forming area in the upper quadrant of the upturning side of the cylinder.

ROTOGRAVURE PAPER—A general term used to describe paper manufactured for rotogravure printing. The outstanding characteristic of the paper is the highest possible printing smoothness for both coated and uncoated grades. Consequently, a supercalendered sheet is frequently used although English finish and machine finish papers are also used. The term rotogravure paper in general may be applied to a variety of coated or uncoated grades ranging from paper containing no mechanical pulp to one containing a substantial proportion of mechanical pulp. The weights of the supercalendered sheets vary from 26 to 90 pounds (25 x 38 inches – 500), and of the machine finish and English finish sheets from 22 to 70 pounds (25 x 38 inches – 500).

ROTOGRAVURE PRINTING—An intaglio printing (q.v.) process for rotary web presses, used by newspapers and magazines for printing catalogs, and also for much specialty printing and paper converting. A photographic positive is prepared on a film and is printed upon sensitized gelatin paper (carbon tissue). A reversed line screen is then printed over the picture and the so-called carbon resist is ready to be transferred to the cylinder. The tissue is dampened and rolled onto a copper cylinder or plate to which it adheres. After drying, the tissue receives a flow of warm water, which loosens the paper backing, leaving the gelatin film on the copper surface. In the shadows, or dark tones, this film is very thin, the thickness increasing as the tone lightens, so that the etching solution penetrates the film more rapidly where the "cups" are to be etched deeper as in the shadows or tones. The etched copper cylinder is finally inserted in position on the press. In recent years, instead of using solid copper cylinders, which can be reground and re-etched, the design has been etched on flat sheets of copper which are later curved and fitted to a permanent plate cylinder.

ROTONEWS—See SUPER NEWS.

ROTOPRINT—See MULTILITH.

ROUNDWOOD—Whole tree trunk sections which are harvested for commercial uses such as lumber or pulpwood.

RPTA—The Recycled Paperboard Technical Association, formerly known as the Boxboard Research and Development Association.

RRBOS—Reject rate based on solids.

RSC—Regular slotted container (q.v.).

RUBBER MARK—See WATERMARK.

RUBBER SPOTS—Dirt specks in paper, often sticky, composed principally of rubber. Rubber spots originate (a) in rag-content papers from elastic yarns in the rags or cuttings used;

(b) in papers or paperboards containing recovered paper stock; (c) in latex type adhesives used on mailing labels, self-seal envelopes, etc., present in the recovered paper furnish. See also STICKIES.

RUBBER-STAMP MARK—A mark simulating a watermark (q.v.) but impressed in the wet web of paper by a rubber band on the press rolls or baby dryer, usually on high-speed machines.

RUB RESISTANCE—The ability of printing to resist the transfer of ink when scuffed or rubbed.

RULE 41—The portion of the Uniform Freight Classification of the national Freight Committee of the Western Railroad Association which specifies the construction and application of corrugated and solid fiber shipping containers (q.v.) for shipping products by rail. An almost identical specification is published in the National Motor Freight Classification as Item 222.

RULING AND WRITING QUALITIES—Ability to accept lines ruled with ink or characters written with a pen (not ball-point) without feathering or other distortion of the marks.

RUNNABILITY—The capability of operating trouble-free, reliably, and with minimal operator interaction. The general term runnability could be expressed in terms of percent of time available for service, the absence of production or service interruptions, such as sheet breaks, purges, or backflush cycles.

RUN (RUNNING) WASTE—See PAPER WASTE.

RUST-FREE PAPER—See ANTITARNISH PAPER.

RUST-PREVENTIVE PAPER—See ANTITARNISH PAPER.

S

SACK—A term used for a flexible container, usually larger than a bag (q.v.). See SHIPPING SACK.

SACK PAPER—Any paper used in making sacks, usually stronger and heavier than bag paper. It is made from rope or kraft pulps or mixtures of these in basis weights of 40 pounds (24 x 36 inches – 500) and heavier. See also SHIPPING SACK KRAFT PAPER.

SAD COLORS—Colors which are toned down, flattened, or dulled by the application of certain colors or chemicals.

SADDLE STITCHING—Staples through the single fold of a booklet or magazine. See BINDING.

SAFETY PAPER—A special grade of paper having a surface design or hidden warning indicator both of such chemical composition as to make obvious any attempt at fraudulent alteration of writing thereon by ink eradicators, mechanical erasure, etc. It is used for bank checks, coupons, money orders, lottery tickets, trading stamps, transportation tickets, and other paper items having a negotiable value. Safety paper is made by either a wet or dry process. In the wet process, the surface design is applied to the paper by immersing it in a dye bath and then removing the excess dye solution usually by passing the paper through a press comprising one steel engraved pattern roll and a rubber composition backing roll. Wet process safety papers have a positive surface pattern on one side and a reverse pattern on the other. Such papers must be dried after processing. In the dry process the surface design and warning indicia, if any, are applied by a flexographic press employing spirit-soluble inks or dyes. Dry process safety papers can be produced with positive-reading patterns on both sides of the sheet, with positive and negative patterns on the top and bottom of the sheet, or with multi-color or multi-pattern combinations of the two. Dry process safety papers require little or no drying during conversion because of the fact that alcohol instead of water is the principal ink (i.e., dye) solvent. The English equivalent of safety paper is cheque paper. See DESIGN PRINTING; PROTECTIVE PAPER.

SAFETY BRISTOL—A bristol which has safety features. See SAFETY PAPER.

SAFETY CHECK PAPER—See SAFETY PAPER.

SAFETY COUPON PAPER—See SAFETY PAPER.

SAFETY-PAPER BASE STOCK—A typical bond or writing paper which may or may not be treated with chemicals sensitive to chemical ink eradicators. It is usually made for surface treatment with sensitive inks by a printing process or with sensitive dyestuffs. It is made in basis weights of 20 and 24 pounds (17 x 22 inches – 500). The paper has typical bond properties with good strength, especially fold and tear, uniform formation, and smooth surface.

SAFETY-TICKET PAPER—A grade of paper or paperboard having special surface markings or containing uniquely reacting material to permit authentication and thus minimize forgery. Multiply board is often used with differently colored or reacting plies.

SALESBOOK MANILA—A grade of paper used in sales books and order books. It is usually made from a furnish containing considerable mechanical pulp, in a manila color, though frequently other colors, such as canary, pink, green, blue, and salmon, are used. The usual basis weights are 35 and 37 pounds (24 x 36 inches – 500).

SALESBOOK PAPER—A term applied to various grades of paper used in the manufacture of sales books and other books which are used in wholesale and retail stores. The various grades are (q.v.) carbon paper, poster, railroad manila, register bond, salesbook manila, and salesbook tissue.

SALTCAKE—Sodium sulfate (Na_2SO_4).

SAMPLE—A limited number of measurements taken from a large source to make a prediction about the defined population from which the sample was taken. See also RANDOM SAMPLE; STRATIFIED SAMPLE.

SAMPLE CARDS—Cards made from bristols or cardboard, used for display purposes by pasting thereon samples of commodities offered for sale.

SAMPLING—See RANDOM SAMPLING.

SAMPLING PAPER—Paper used for wrapping samples of cotton, wool, etc. It is made from rope, sulfate, or sulfite pulp and furnished in natural color and in duplex; the duplex sheet is usually blue and white. The basis weights are 70 pounds (24 x 36 inches – 500) and heavier.

S AND SC—Sized and supercalendered.

SAND FILTER—Device used in the treatment of water or wastewater to remove suspended solids by filtration through a bed of sand.

SANDPAPER—See ABRASIVE PAPERS; FLINT PAPER.

SANDWICH PAPER—Usually a waxed, bleached chemical woodpulp sheet, white in color, in a basis weight of 30 to 35 pounds (24 x 36 inches – 500), although a waxed or unwaxed glassine or a vegetable parchment paper may be used. As the name implies, it is used in wrapping sandwiches and for making sandwich bags.

SANITARY LANDFILL—A site for solid waste disposal using sanitary landfilling techniques.

SANITARY TISSUE—Any of a group of papers used for sanitary disposable purposes. Generally, these papers are absorbent, bulky, and have a rough finish. Included in these papers are facial tissue, paper napkins, toilet tissue, towels, and the like.

SANITARY TUBE STOCK—Paper manufactured for the construction of tampon tubes. Paper must be caliper controlled and free from contaminants.

SANITARY WALLPAPER—A variety of wallpaper printed from engraved or etched copper rolls with oil inks. It may or may not be washable.

SAPI—An acronym for the Sales Association of the Paper Industry and the South African Paper Industry.

SAPWOOD—The outer living portion of a tree stem, root or branch, through which water is conducted; it is usually distinguished from heartwood by its lighter color. The parenchyma, or ray cells, remain alive in the sapwood.

SATIN FINISH—A smooth finish of paper or bristol, suggestive of satin.

SATIN FOLDING BRISTOL—A bristol resembling translucents (q.v.) but with a manila or dark base. It is particularly adapted for folding purposes.

SATIN PAPER—A term used for mica paper (q.v.).

SATIN WHITE—A coating pigment produced by the interaction of aluminum sulfate and slaked lime. It is used as a pigment in coating mixtures, particularly in coated paper of high white color requiring an enamel finish.

SATURATED FELT—A porous, bulky felt sheet made from wood fibers, paper and low-grade rags and saturated with tar or asphalt. See ASPHALT FELT; TARRED FELT.

SATURATED STEAM—Water vapor at a temperature where it can condense to liquid water without changing temperature. The "saturation" temperature increases with increasing pressure.

SATURATING FELTS—A term applied to those dry felts which are used as vehicles to carry and hold various tars, asphalts, or other water-proofing compounds. They are sometimes designated as waterproofing felts and form the base for roofing papers, etc. See DRY FELT.

SATURATING PAPERS—Open, porous papers that are to be saturated or impregnated with solutions or compounds of various types.

SATURATING PROPERTIES—The properties of an impregnating paper that determine the quantity of impregnating materials that the paper will take up, and the rate of impregnation. For roofing felts, saturating capacity and saturating rate are used to evaluate these properties.

SATURATION—(1) The process of impregnating a sheet, usually waterleaf, with a liquid, sometimes to the maximum possible extent but usually to a controlled degree. (2) An attribute of color. See COLORIMETRIC PURITY.

SATURATION BONDING—A method of bonding whereby the entire web being bonded is immersed into the binder.

SAVEALL—A device that recovers and thickens fibers and other suspended solids from low-consistency broke, trim, or white water (q.v.), and thickens it to the point that it can be conveniently reintroduced into the stock preparation system. Many types of savealls are used commercially, including disc filters, vacuum drum filters, deckers, dissolved air flotation clarifiers, and sidehill screens.

SAWTIMBER—Commercial tree species having dimensions large enough to cut for lumber. Trees as small as 6 inches in diameter, breast-height are sawn in small-log mills.

SBS—See SOLID BLEACHED SULFATE.

SCALE PAPER—See CHART PAPER; PROFILE PAPER.

SCANNER—An optical electronic device to digitize printed or pictorial work, for subsequent on-screen processing.

SCATTER DIAGRAM—A graphical technique to analyze the relationship between two variables. Two sets of data are plotted on a graph, with the y-axis being used for the variable to make the prediction.

SCD—Streaming current detector. See STREAMING CURRENT.

S CHART—See STANDARD DEVIATION CHART.

SC PAPERS—A type of uncoated printing paper, typically used for catalogs and magazines, which has been subjected to a supercalendering operation to improve its gloss and printability.

S-LAYERS—A reference to the three layers which make up the secondary cell wall of fibers, usually designated as S1, S2, and S3 layers.

SCHMIDT HARDNESS TESTER—One of several wound roll hardness testers. This tester operates on a principle of measuring the rebound height of a mass impinging on the roll. See HARDNESS TESTER (PAPER).

SCHOOL CONSTRUCTION PAPER—See CONSTRUCTION PAPER.

SCHOOL DRAWING PAPER—A grade of drawing paper used in school work. It is manufactured from a mixture of mechanical and chemical pulps. The usual colors are white, gray, and manila, and the basis weights range from 40 to 80 pounds (24 x 36 inches – 500).

SCHOOL FLATS—Table paper manufactured with a high finish for pen and ink writing. It is usually ruled.

SCHOOL PAPERS—A term covering the various types of papers used in schools, colleges, and universities. School papers include such items as typewriting, mimeographing, and duplicating papers; ruled writings, such as tablets, copybooks, composition books, and looseleaf notebooks; drawing papers of all kinds; construction paper; cross-section, profile, logarithmic, and other papers of this type;

index bristols in the form of cards; stenographer notebooks, etc. These are supplied in various grades and weights, depending upon their use.

SCHOOL POSTER PAPER—See CONSTRUCTION PAPER.

SCHOPPER-RIEGLER FREENESS—In Europe, the rate at which water drains from a stock suspension is measured on the Schopper-Riegler (SR) tester. The SR test is reported as SR degrees. See FREENESS. 140

SCHWEIZER'S REAGENT—See CUPRAMMONIUM HYDROXIDE.

SCLEREIDS—Clusters of layers of variably shaped, often rounded, bark cells with thickened, lignified and often mineralized cell walls. Sometimes referred to as "stone cells" which may give rise to dirt or specks in calendered papers.

SCORE—See SCORING.

SCORE CRACK—A crack originating at or near a score or crease in paperboard or heavyweight paper.

SCORE CUTTER—Common terminology is score slitter (q.v.).

SCORE SLITTER—A web separation device composed of a thin steel disc with a rounded edge which is pressed against a driven, hardened steel mandrel which runs at web speed and conveys the paper web. A number of adjustable score slitter blades can engage the same mandrel and thus separate the running web into numerous strips slit in the machine direction. The pressure at the contact point glassines and separates the web and has been used at speeds up to 6000 feet/minute (1800 meters/minute)

SCORING—The production of a score or crease in a sheet of (a) heavy-weight paper or paperboard by pressing it between two metal surfaces, one of which has a recessed groove and the other a tongue; (b) or in corrugated board by compressing it with a rule or wheel against the groove, grooved wheel, or softer surface. The score may be produced by scoring plates or by scoring rollers and is made along the line on which the sheet is to be folded. It alters the sheet structure in such a way that it will fold more readily with less tendency to crack or break.

SCRAP IN ROLL—Trim or scrap paper wound into a roll.

SCRATTED PAPER—(1) An early form of wallpaper. (2) A cheap imitation of marbled paper, prepared by "spiriting" or spotting various colors on paper by means of a brush. It is used for lining boxes, for end leaves, etc.

SCREEN—A device which separates debris that is larger than fibers, from fibers, by a size separation, in which the fibers pass through a small hole or slot in the screening media, while the debris particles do not. See also BROKE SCREEN; PAPER MACHINE SCREEN; PRESSURE SCREEN; PRIMARY SCREEN; SCREENING SYSTEM; SECONDARY SCREEN; TAILING SCREEN; THICK STOCK SCREEN; VIBRATING SCREEN.

SCREEN (HALFTONE)—See HALFTONE SCREEN; SCREEN PRINTING.

SCREENING—(1) The operation of passing pulp or paper stock through a screen to reject coarse fibers, slivers, shives, knots, adhesives, plastics, and other unwanted materials. (2) In mechanical pulping, it is used to separate fiber bundles and stiff fibers from the pulp slurry. Also see SCREENING (CHIPS); SCREENING SYSTEM.

SCREENING (CHIPS)—To improve pulping uniformity, chips are screened to remove overlength, overthick, and fines fractions. A gyratory or vibratory screen system fitted with round, square, or oval shaped holes is used to remove overlength and fines chips. Disc, roll, or bar screens are used to remove overthick chips, which are reprocessed by slicing or crushing.

SCREENINGS—(1) Coarse materials that do not pass through pulp or stock screens. The term refers chiefly to woodpulp, where screenings consist mainly of undefibered fragments of the wood. (2) A paper sheet made predominantly from refined chemical pulp screenings, usually containing a small proportion of screened fiber broke, or waste pulp for strength, usually in basis weights of 40 to 125 pounds (24 inches x 36 inches – 500) depending on its use. It is intended for such coarse-paper uses as roll heads, car liners, and inner-bundle wrappings.

SCREENINGS BOARD—A paperboard made of woodpulp screenings and used as a mill wrapper (q.v.). It is strong and sufficiently pliable to be formed about the edges of a bundle of paper.

SCREENING SYSTEM—A system consisting of multiple stages of screens; the first, or primary stage, passes fibers downstream, and rejects a debris-rich fiber slurry to a secondary screen. The secondary screen passes good fibers to the primary screen feed, or downstream, and a debris-rich fiber stream to the tertiary screen. After multiple stages of screening, the highly concentrated debris-rich reject stream is further concentrated in a tailing screen, the final stage, prior to disposal or refining.

SCREENING WRAPPER—A wrapping paper made form screenings. See also MILL WRAPPER.

SCREEN PRINTING—A stencil printing process in which ink or paint is forced through the fine meshes of a special silk or other material by means of a squeegee onto the surface to be printed directly beneath the screen. Most of the screens are hand prepared; in the hand-blocked screen, the design is traced on the silk and all parts not to be reproduced are painted out with a filler; in the hand-cut type the outline of the color to be painted can be cut out of transparent paper or cellulose acetate film and applied by heat on the underside of the screen. Fine-line cuts and halftone subjects can be photographically rendered as stencils by first applying a sensitized coating to the screen. The process may be used for advertising displays of all kinds and for the printing of dress goods, draperies, wallpaper, and the application of trademarks and designs on all kinds of manufactured articles. It is usually used for short runs (store cards, etc.), but there is an automatic press for printing screen work at speeds up to 1800 impressions per hour.

SCREEN RESIDUE—(1) Dirt, fine materials, uncooked chips, and knots retained on the screen through which chemical pulp is passed as it comes from the storage tank. (2) Residue on a laboratory screening device, the amount of which is a general indicator of the debris level in pulp.

SCREW PRESS—A device for dewatering sludge where the dewatering action is accomplished by a variable pitch screw contained within a cylinder.

SCRIBBLING PAPER—A low grade of tablet paper used for student practice writing, longhand calculations, scratch pads, etc.

SCRIM PAPER—A paper laminate using a base paper (sometimes creped) and a fiberglass or polyethylene scrim (for strength).

SCRIPT—A term sometimes applied to writing paper.

SCRUBBER—An air pollution control device that uses a liquid spray to remove pollutants from a gas stream by absorption or chemical reaction. Scrubbers also reduce the temperature of the emission.

SCUFFING—The raising of the fibers on the surface of a paper or paperboard when one piece is rubbed against another or comes in contact with a rough surface. Paper and paperboard are more susceptible to scuffing when wet. See WET RUB.

SCUFF RESISTANCE—The resistance to scuffing of paper or paperboard, usually measured in terms of the number of cycles required to produce a designated degree of scuffing on a designated area with a designated abrasive ob-

ject of designated size and weight rotating or reciprocating at designated speeds. See SCUFFING.

SEALABILITY—That property of a paper or paperboard which renders it capable of being sealed by adhesives, heat, pressure, or other means.

SEAL BOX OR SEAL TANK—A tank under a pulp washer used for separating the air and water that are drawn down the dropleg. It also provides a reservoir for white water (q.v.) pumps to distribute the white water selectively to a desired location.

SEALING PAPERS—(1) Thin sulfite or kraft machine-glazed wrapping papers used as parcel wrappers, manufactured in various substances and colors. (2) Heavy mill wrappers of chemical woodpulp, with a high finish, sometimes duplex in color, used for sealing ream packages of book or writing paper. The basis weight is 60 pounds (24 x 36 inches – 500) or heavier.

SEALING TAPE—A strong gummed paper used for sealing purposes. See GUMMED CLOTH TAPES; GUMMED SEALING TAPE; GUMMED STAY.

SEALING WRAPPERS—See SEALING PAPERS.

SEAM—The overlapped and pasted portion running the length of a tube (q.v.) to be formed into a bag or sack.

SEAMING CORD—A twisted kraft paper twine made in a variety of sizes and used for welting and seaming upholstery work.

SEAMLESS DISPLAY PAPER—A special grade of cover stock made from chemical and mechanical woodpulps or both in a large variety of pastel and deep shades, in substance 50 to 55 pounds (20 x 26 inches – 500). This grade is usually converted into rolls 107 inches wide and 12 feet in length, and is used as a backdrop in display windows of department stores. Other uses include backgrounds in television scenes, photographic studios, etc.

SEASONING—Exposure of paper or board to relatively uniform conditions of atmospheric temperature and humidity, to allow its moisture content to reach equilibrium with the atmosphere and become uniformly distributed throughout the sheet. See CONDITIONING; MATURE.

SECONDARY ARM—The mechanical device that accepts the reel spool from the primary arm on a continuous winding reel immediately following the transfer of the web onto the new reel spool.

SECONDARY CLEANER—The second stage in a multiple stage centrifugal cleaning system. The secondary cleaner stage is fed debris-rich rejects from the primary cleaner stage, and sometimes also the accepts from the tertiary cleaner stage, if present. The debris-rich rejects from the secondary cleaner stage is often sent to a tertiary cleaner stage for further concentration, and fiber recovery. The cleaned secondary cleaner stage accepts are sent to the primary cleaner stage feed, or downstream. Often, a large number of secondary cleaners operate in parallel, for capacity reasons.

SECONDARY CREPING—See CREPING TISSUE.

SECONDARY HEADBOX FORMER—A former for making a two-ply sheet common in the production of linerboard. The former consists of a flat fourdrinier and a second headbox located down the fourdrinier in the vicinity of the flat box area. In this area, the primary sheet has a consistency of 6% to 10% solids. The secondary headbox discharges onto the primary sheet, usually over the first or second flat box. The secondary sheet is drained through the primary sheet over flat boxes and the couch. Since the secondary headbox usually has a consistency of about 1% solids, this is called "wet on dry" forming.

SECONDARY SCREEN—The second stage screen in a multistage screening system. The second stage screens are fed with primary screen rejects, sometimes mixed with tertiary screen accepts. The second stage accepts are sent to the primary stage feed, or sometimes downstream, with the primary stage accepts. See also SCREENING SYSTEM.

SECONDARY STOCK—Fibers which have been previously used in the papermaking process. The term includes paper stock reclaimed from recovered papers collected from all generators of post consumer papers.

SECONDARY TREATMENT—Wastewater treatment, beyond the primary stage, in which bacteria consume the organic parts of the wastes. This biochemical action is accomplished by use of trickling filters or the activated sludge process. Effective secondary treatment removes virtually all floating and settleable solids and approximately 90% of both BOD and suspended solids. Customarily, disinfection by chlorination is the final stage of the secondary treatment process.

SECONDARY WALL—That portion of the cell wall that makes up most of the fiber structure, usually formed in three layers differentiated by their fibril alignment to the main axis of the fiber. See S-LAYERS.

SECONDARY WASTEWATER TREATMENT—The treatment of wastewater by biological methods after primary treatment by sedimentation.

SECOND ORDER SYSTEM—One whose output can be modeled by a second order differential equation.

SECOND PRESS—See PRESS SECTION.

SECONDS—Paper which is inferior to the established standard quality but which is merchantable at some lesser value.

SECOND SHEETS—(1) A paper used where one or more carbon copies of the same letter or writing are desired. It is made in various colors and frequently in lighter weight than the ribbon sheet. (2) A paper of the same character as the letterhead used but without printing or simply carrying the name of the firm. It is used for the continuation of a letter requiring more than one page.

SECTIONAL LINEN FINISH—A linen finish (q.v.) obtained by passing two or more sheets of paper which are contained between two linen sheets and in turn between zinc plates or pressboards through a plater. See PLATE FINISH. If only one sheet is used, a lawn finish (q.v.) is obtained.

SECTIONAL PAPER—See CHART PAPER; PROFILE PAPER.

SECURITY PAPER—(1) A paper with bond characteristics similar to currency paper. It is used for stocks, bonds, and other securities, and it is usually made for engraving by the wet intaglio process. It may contain distinctive features and safety features to protect against counterfeiting. (2) See SAFETY PAPER.

SEDIMENTATION—The process of subsidence and deposition of suspended matter carried by water, wastewater, or other liquids, by gravity. It is usually accomplished by reducing the velocity of the liquid below the point at which it can transport the suspended material. Also called settling.

SEDIMENTATION BASIN—A basin or tank in which water or wastewater containing settleable solids is retained to remove by gravity a part of the suspended matter. Also called sedimentation tank, settling basin, settling tank.

SEDIMENTATION TANKS—In wastewater treatment, tanks where the solids are allowed to settle or to float as scum. Scum is skimmed off; settled solids are pumped to incinerators, digesters, filters or other means of disposal.

SEDIMENT VALUE—A measure of the rate of settling, varying with the different fibers or pulps and the fiber length.

SEED-BAG PAPER—A strong, supercalendered sulfite or sulfate paper, usually 50 to 60 pounds (25 x 38 inches – 500), used to make seed bags.

SEED FIBERS—Fibers which grow attached to the seed coat, such as cotton.

SEED-GERMINATING PAPER—A blue absorbent paper which does not bleed or run when wet. It is used by seedmen as a medium upon which to grow seeds for the purpose of determining the percentage which will germinate. This paper is usually made of a mixture of cotton and chemical woodpulp, and it is free of chemicals injurious to seeds. The usual basis weight is 120 pounds (19 x 24 inches – 500).

SEIDLITZ PAPER—See DRUG WRAPPING.

SEL—Specific edge load.

SELF-ADHESIVE PAPER—Paper either plain or coated to which a pressure-sensitive adhesive has been applied. See PRESSURE-SENSITIVE PAPER.

SELF BONDING—See INHERENTLY BONDED.

SELF-COLORED—See NATURAL COLORED.

SELF-SEAL WRAPPER—Wrapping paper coated on both surfaces so that, when two parts of a sheet are pressed together and heated or pressed, the overlapping parts become sealed together. Wax, rubber, or thermoplastic materials may be used in the coating operation.

SEMIBENDING CHIP—A paperboard used for the lower grades of folding boxes. It is made of chip and does not possess bending qualities equal to bending chip.

SEMIBLEACHED PULP—Any papermaking pulp which has been partially bleached and therefore has a brightness in the range from GE 45 through 75. See BRIGHTNESS.

SEMICHEMICAL BOARD—A paperboard, usually corrugating medium, made from pulp produced by a semichemical process. See also CHESTNUT BOARD; SEMICHEMICAL CORRUGATING MEDIUM.

SEMICHEMICAL CORRUGATING ME-DIUM—A corrugating medium (q.v.) made from a furnish (q.v.) which is principally woodpulp produced by the semichemical process. See SEMICHEMICAL PULP.

SEMICHEMICAL PULP—A pulp produced by combining mild chemical treatment and mechanical defibering. See NEUTRAL SULFITE PROCESS.

SEMICREPED—See CREPING TISSUE.

SEMICREPE TISSUE—A tissue or lightweight paper that resembles crepe but which lacks the characteristic stretch and strength. It is used for napkins, paper towels, tablecloths, toilet paper, etc. See CREPING.

SEMIDULL FINISH—See DULL FINISH.

SENSITIZATION—The process, in stainless steels, in which the alloy becomes susceptible to intergranular corrosion because of prolonged heating in the 425–815°C (800–1500°F) temperature range, because of chromium depletion at the grain boundaries. The effect may be general, as by stress-relieving heat treatment, or localized in the heat-affected zone alongside welds.

SENSITIZED—Treated with chemicals that change color on exposure to light, heat, or chemicals. See SENSITIZED PAPERS.

SENSITIZED PAPERS—A general term for papers which have been coated, impregnated or otherwise treated with chemicals to render them responsive to light, heat, moisture, erasure, reagents, etc. Examples are safety paper, photographic paper, blueprint paper, diazotype paper, etc.

SENSORS—A device environmentally designed to be in close contact with the process to sense (measure) variables such as pressure, temperature, flow, pH, humidity, and consistency. The measurement is then transduced into a useable signal for transmission of the process variable values into process control systems. The trans-ducer function is sometimes in a separate housing from the sensor.

SEPARATING TISSUE—A creped tissue, white or colored, used on the cutting table in the garment industry to keep the cuts from different bolts of cloth separated from each other. This prevents pieces from different bolts, which vary constantly in shade, from being used in the same garment. One class of interleaving tissue (q.v.).

SEQUESTERING AGENT—In pulp bleaching, transition metals (manganese, iron, copper, etc.) and other salts of heavy metal ions restrict brightness gain and adversely affect pulp brightness stability. Chelating agents (q.v.) and sequestering agents, such as $Na_5P_3O_{10}$, EDTA (ethylenediaminetetraacetic acid), and DTPA (diethylenetriaminepentaacetic acid), are used to immobilize the metal ions.

SERIFS—Small finishing strokes on printed letters.

SERPENTINE DRYER SECTION—See DRYER SECTION.

SERPENTINE PAPER—A fairly strong, medium weight paper, generally made from chemical woodpulp or with mechanical woodpulp. It is made in a large variety of colors and put up in small rolls about 3/8 of an inch wide. It is used on festive occasions for throwing and decorating purposes and is often made flame resistant.

SETOFF—See OFFSET PRINTING (2a).

SET POINT—An independent process variable used to advise the control system where it should be operating. It may be a fixed manual setting, or driven by remote signals determined by measurement of other process needs. The setpoint is compared with the value of the dependent variable to obtain an error signal that determines the controller output signal to the final control element.

SETTLING BASIN—Containment area which is used to settle suspended matter from an effluent stream.

SETTLING TANK—Tank used to settle suspended matter from an effluent stream.

SETUP BOXBOARD—Paperboard used in making boxes in rigid form as contrasted with a folding or collapsible box. It may be a solid or combination board depending on the style of box; it ranges in thickness from 0.016 to 0.065 of an inch and weighs 60 to 206 pounds per 1000 square feet. Stiffness, rigidity, and resistance to abuse are essential qualities. The class includes plain chipboard, filled newsboard, single news vat-lined chipboard, and single white vat-lined chipboard.

SETUP BOXES—Boxes which are manufactured in the form and shape in which they are to be used, as distinguished from folding cartons which are manufactured in a collapsed form and not set up until used.

SETUP DRINKING-CUP STOCK—See CUP PAPER.

SEWAGE—Waste material discharged from an industrial process or a municipality.

SEWAGE DISPOSAL—The process of disposing of material from an industrial process or a municipality.

SEWAGE TREATMENT—The process of treating waste material from an industrial process or a municipality.

SGW—Stone groundwood (q.v.).

SHADE-CRAFT WATERMARK—See WATERMARK.

SHADOW MARKS—An undesirable series of marks in the paper web duplicating the drilling pattern or groove pattern in a suction roll (q.v.) or grooved roll (q.v.). This condition can be eliminated by carefully controlling such factors as hole or groove size, hole spacing, and drilling pattern.

SHAFTLESS UNWIND—An unwind stand that supports an unwinding roll by directly engaging a cantilevered core chuck into each end of the roll and lifts the roll off the floor. Typically, the brake or motor is engaged permanently to the chuck spindles.

SHAFTLESS WINDING—A process of winding on a two-drum winder where, instead of winding all rolls on a common shaft, each roll is permitted to wind to its own preferred diameter. This process was introduced in the late 1950s and early 1960s.

SHAKE—A device that causes oscillation of the fourdrinier fabric in the plane of the fabric but at right angles to the machine direction. Its purpose is to assist in securing the desired formation of the sheet. The oscillations may be varied in frequency and length of stroke to obtain the desired result. Shake can be characterized by shake number, which is $(frequency)^2 \times$ amplitude/machine speed.

SHAKE NUMBER—See SHAKE.

SHAVINGS—(1) A class of reclaimed paper stock, consisting of unused, unprinted trimmings from converting operations. Sized white writing, bond and ledger paper shavings free from mechanical pulp are called hard white shavings. Book or coated paper shavings are called soft white shavings. (2) Mixed groundwood shavings are the trim from magazines, catalogs, etc; containing groundwood fiber, some printing, and color. (3) Fly leaf shavings are similar to (2), except that the fiber is predominantly of the chemical type. (4) The narrow strip of paper with the deckle edge trimmed off the edge of the web by the slitting knives of the roll winder or sheet cutter.

SHEAR SLITTER—A web separation device that uses a top and bottom rotary knife to slit the sheet by a method similar to scissor action. The top thin knife, which can be 4 to 10 inches in diameter, contacts the edge of a hardened, driven bottom slitter which provides torque and a cutting position for the web. Both top and bottom slitters can be moved by hand or by remote controls to change slit positions and widths as required.

SHEARING—The process of cutting long fibers or pile to form a uniform pile height and sometimes pattern effects.

SHEARING STRENGTH—The maximum shear force required to produce failure in a paper or paperboard member. The shear force is the internal force acting along a plane between two adjacent parts of a body when two equal forces, parallel to the plane considered, act on each part in opposite directions.

SHEATHING—A type of fiber board manufactured in various sizes and thicknesses and used in building construction as a structural and/or insulating material.

SHEATHING BOARDS—See BUILDING BOARD; INSULATING BOARD.

SHEATHING PAPER—A paper used between rough boards and finish in outside walls of a frame building. The paper is closely felted and relatively compact to provide protection against wind and dust. See ASPHALT SHEATHING PAPER; HOUSE SHEATHING PAPER; K-B SHEATHING; RED ROSIN SHEATHING PAPER.

SHEET CALENDERED—The result of a process of applying a finish or glaze to sheets of paper or paperboard by passing them through a calender stack (but not in a continuous web) with the aid of a sheet feed system. The stack consists of three to five rolls; chilled iron and cotton rolls are alternated in the stack.

SHEET CRUSHING—A rearrangement or displacement of fibers in the wet sheet of paper caused by excessive wet press hydraulic pressure gradients in the plane of the paper. Sheet crushing can reduce sheet strength or in severe cases result in wrinkles or holes in the sheet. Compare BLACKENING.

SHEET FLUTTER—The tendency for part or all of a moving sheet to take an oscillating path as opposed to a straight path. The most common form of sheet flutter is edge flutter in the open draw (sheet unsupported) between dryer cylin-

ders. Excessive sheet flutter can cause folds or sheet breaks or both.

SHEET—A term used extensively in the paper industry meaning: (1) A single piece of pulp, paper or board. (2) The continuous web of paper as it is being manufactured. (3) A general term for a paper or board in any form and in any quantity which, when used with appropriate modifying words, indicates with varying degrees of specificity, attributes of the product such as quality, class, use, grade, or physical properties. Examples: a bright sheet, a kraft sheet, a folding boxboard sheet. (4) To cut paper or board into sheets of desired size from roll or web.

SHEETER—See CUT-SIZE CUTTER; DUPLEX CUTTER; FOLIO CUTTER; PRECISION CUTTER.

SHEET-FED GRAVURE—See INTAGLIO PRINTING; ROTOGRAVURE PRINTING.

SHEET-FED PRINTING—The process of printing on material that is fed into the press in the form of individual sheets instead of rolls. See WEB PRINTING.

SHEET FLATNESS—A web normally comes to the reel of the paper machine in a relatively flat state; however, after a winding process, a stretched-out, full-width web can display baggy areas or wrinkles which are due to the web being wound tightly over variations in diameter across the face of the winding roll. This web distortion, caused by overly stretched areas, is known as sheet flatness and is due to nonuniform caliper of the web across the deckle.

SHEET FORMING—See FORMING SECTION.

SHEETING—The process of unwinding a roll or a number of rolls into a slitting and cutting station and overlapping and piling the resultant sheets into an appropriate pile.

SHEET LINED—Any paperboard to which a liner has been pasted after being sheeted.

SHEET LINING—The process of lining sheets of paperboard with sheets of lining paper.

SHEET MOLD—See HANDSHEET; MOLD.

SHEET SEALING—A drainage condition usually found under undesirable operating conditions on a fourdrinier or twin wire former. Sheet sealing is believed to be caused by the rapid draining of fines-containing furnishes, especially groundwood. The rapid drainage is believed to cause the fines to plug the drainage pores of the sheet. The result is a slurry that is exceedingly difficult to drain.

SHEET TRANSFER—The passage of the sheet between machine sections such as from the press section to the dryer section or between elements within a machine section. An open draw is a transfer without sheet support while a closed draw is a transfer with full sheet support. See DRAW.

SHELF LINING PAPER—See SHELF PAPER.

SHELF PAPER—Any paper used principally for shelf coverings. It may be made from mechanical and/or chemical woodpulps and may be specially treated and decorated, oil- and water-resistant, coated or uncoated. It is usually hard sized. Much of this paper is bought in rolls by printers and converters for processing to make fancy shelf paper. It may be plain, printed, creped, or embossed. The basis weight ranges from 30 to 70 pounds (25 x 38 inches – 500).

SHELL-PACKING PAPER—See CARTRIDGE PAPER.

SHELL PAPER—See CARTRIDGE PAPER.

SHERATON ROLL—A sawtooth metal roll used to pulse the forming fabric and introduce turbulence into the draining stock for better formation (q.v.).

SHINER—(1) A glossy fiber bundle occurring in mechanical pulp or in an undercooked chemical woodpulp. (2) Particles of fillers which are compressed to a translucent spot upon passage through the calenders.

SHIPPING CONTAINER—A box made of corrugated board or solid fiberboard used as an outer container in the shipment of commodities. The most common style of container is the regular slotted container (RSC) (q.v.). Other common styles include half, overlap and center special slotted, telescope, folders, self-locking, self-erecting or automatic bottom, double and triple slide and bliss (q.v.) and design-style boxes.

SHIPPING SACK—A flexible container made by forming paper or other flexible material into a tube, closing one or both ends but leaving an opening for the introduction or more of the material to be packaged. The term generally refers to a package of a size to contain 30 or more pounds of material, usually up to 100 pounds used for the packaged shipment of a large variety of agricultural and food products, chemicals, building materials, minerals, pigments and the like. Paper shipping sacks are made from one to six plies (or walls) of shipping sack kraft paper (q.v.). Multi-wall (q.v.) refers to using several walls of relatively light weight, rather than fewer walls of heavier paper. Sacks may be flat or gusseted. If one end is open (open mouth) the other end may be closed by sewing and/or taping or it may be pasted. The latter with pasted square bottom is called an automatic (SOS) sack. If both ends are sealed, the opening for filling is called a valve and the sack is designated a pasted valve style. Sacks may be lined (See LINER) and/or the paper may also be coated or treated to meet specific end-use requirements.

SHIPPING SACK KRAFT PAPER—Paper manufactured of kraft pulp against specifications which take into consideration the high strength requirements of the completed multiwall shipping sack. This paper must not only protect the contents of the filled multiwall bag during transit, storage, and subsequent handling, but must also withstand high stress during commercial bag filling operations. Domestic production is manufactured in basis weights of 40, 50, 60, and 70 pounds (24 x 36 inches – 500) but different weights are commonly used in countries other than the United States.

SHIRT BOARD—A board sufficiently rigid for use as a stiffener in packaging laundered shirts. It is generally made in sizes 8 x 16 and 9 x 18 inches and may be of any grade of boxboard (0.020 of an inch or heavier) suitable for the purpose.

SHIVES—(1) Small splinters of undercooked wood found in mechanical, semichemical and chemical pulps. They form when large, thick chips are not completely penetrated by cooking chemicals or when refiners do not sufficiently defiber the wood particles. Shives should be screened from the pulp and reprocessed or discarded to prevent dirt specks in the product. (2) The nonbast fiber portion of the flax plant.

SHOE BOARD—A fiberboard formed from a single web on a wet machine from woodpulp, reclaimed paper stock, leather waste, or other waste materials, or a combination of such materials, with or without the addition of chemicals. To meet the different requirements needed for different kinds of shoes, shoe board is produced in a range of characteristics achieved primarily by changing the material blend in the board. Depending upon their end use, shoe boards may be termed counterboard, heeling board, innersole board, leather fiber, midsole board, reinforcement board, shank board, or tuck board. Each type of board is divided into one or more of three classifications which reflect the three major material blends. (1) Cellulose fiber shoe board—shoe board made principally from cellulose fibers containing no additives (other than sizing agents and coloring) in such quantities as to alter the basic solid fiber characteristics of the materials. (2) Specialty shoe board—shoe board made from cellulose fiber, leather fiber, or a combination thereof, with chemical additives that modify the physical properties of such fibers. (3) Leather fiber shoe board—shoe board made from leather, with or without cellulose fiber, and containing no additives (other than coloring and sizing) in such quantities as to alter the basic solid fiber characteristics of the material.

SHOE COUNTER STOCK—See SHOE BOARD.

SHOE NIP BLANKETS OR BLANKETS—Machine clothing in a shoe nip press (q.v.) which, typically, consists of a synthetic base fabric coated with a form of plastic, and subjected to a surface treatment such that a smooth, endless, impervious belt results. This belt can be long and have an open run like a press felt, or it can be short, having the circumference of a roll so that it can be sealed on each end by mechanical means to contain the lubricating oil.

SHOE NIP PRESS—See SHOE PRESS.

SHOE PRESS—A press type, first introduced in 1980, that consists of a press roll and a hydraulically actuated stationary shoe separated from the felt and web by a rotating impervious belt or sleeve. Nip lengths of 10 inches (254 mm) and nip loads of 8000 pli (1400 kN/m) have been achieved. See also PRESS SECTION.

SHOPPING BAG—Heavy, single ply bag, constructed with top handle to facilitate carrying. This bag is typically distributed or sold at department stores and similar retail establishments to enable customers to carry out their purchases.

SHOPPING-BAG PAPER—Kraft paper, bleached or natural, in heavy basis weights, i.e., 70 pounds and higher (24 x 36 inches – 500), used for the manufacture of shopping bags.

SHORE DUROMETER—A series of spring loaded, hand held devices used throughout the rubber industry to measure the hardness of all types of elastomeric and plastic products. Because they are pocket size, easy to use, and quick reading, durometers are sometimes used in place of the P&J Plastometer (q.v.) for measuring hardness of covered paper mill rolls. However, the two instruments operate so differently that there is no absolute conversion from one to the other. All conversions should be considered as rough guides only. Still, two durometers are useful for measuring paper mill rolls. The Shore Durometer A will measure soft and medium hard covers. The Shore Durometer D will measure very hard elastomeric compositions and plastic composites, actually beyond the limits of 0–1 P&J. However, the penetrating pin of the Shore Durometer D is a sharp cone and may leave a slight mark on the cover surface. The scale on each instrument reads from 0 at the soft end to 100 at the hard end.

SHORT CIRCULATION SYSTEM—The white water (q.v.) recirculation loop immediately adjacent to the forming section.

SHORT COLUMN TEST—See COLUMN STRENGTH; EDGEWISE CRUSH RESISTANCE.

SHORT DWELL COATING—Coating is applied through an orificed chamber and metered by the blade immediately, with very little dwell time between coating application and metering.

SHORT SHEETS—Paper under the size ordered. They sometimes are inadvertently included in a ream with full-sized sheets.

SHORT SPAN COMPRESSION TEST—The maximum compressive force in the plane of the paper which a 15-mm strip of paper can withstand without failure when it is clamped between two jaws with a 0.7-mm span.

SHORT STOCK—(1) Beaten or refined pulp in which the length of the individual fibers has been greatly reduced by the mechanical treatment. (2) Pulp from naturally short-fibered sources such as hardwoods.

SHORTWOOD—Log segments cut prior to delivery or in the woodyard or woodroom to relatively short lengths of 4 to 8 feet long. This length allows wood to flow around corners in conveyors and to tumble in debarking drums. Shortwood is gradually being phased out of many operations in favor of tree length processing systems. However, for smaller scale private forest owners, shortwood is more easily handled.

SHOT-SHELL TOP BOARD—A paperboard used as a round plug at the top of a shotgun shell. It is made of reclaimed paper stock and/or chemical woodpulp in thicknesses of 0.030 of an inch and up. The fibers must be sufficiently short so that the small circles may be readily punched out of a larger sheet. The product must be water resistant.

SHOWERS—Various forms of sprays or showers used to provide mechanical decompaction, lubrication, flushing, and chemical cleaning. Four commonly used cleaning showers are high pressure, flooding, suction box lubricating, and chemical application showers.

SHOWTHROUGH—A condition where the printing on one side of the sheet can be seen from the other side when the latter is viewed by reflected light. See OPACITY; PRINTED OPACITY.

SHREDDED TISSUE—See EXCELSIOR TISSUE.

SHRINKAGE—(1) The change in the width of the paper sheet as it passes from the wet end of the paper machine to the reel. The magnitude of shrinkage will vary, depending upon the weight of the paper, the degree of refining, and the type of fibrous raw material used, as well as the tension of the wet draws. (2) Any decrease in the dimensions of paper. (3) The loss in weight incurred between the dry solids content of the paper machine furnish and the paper or paperboard produced. (4) The loss in weight resulting from the removal of pulp fiber constituents, mostly lignin, as a result of chemical treatment during a bleaching stage. Shrinkage is usually expressed as a percent of lost production or spoilage. See also YIELD.

SHRINK SLEEVE PRESS—A simplified version of the fabric press (q.v.) where a fabric jacket or sleeve is shrunk over the rubber-covered press roll. In limited commercial use.

SHRINK WRAP—A biaxially oriented film which shrinks with applied heat after being wrapped around a paper product (roll or skid or carton). This type of wrapper provides a conforming tight package with good moisture proof properties.

SHUTE STRAND OR WEFT STRAND—See FORMING FABRIC.

SHUTE WIRE—See FORMING FABRIC.

SIDE CUT—(1) See SIDERUN. (2) Cuts on the edges of the web caused by markings on the ends of a roll.

SIDE ENTRY AGITATOR—An agitator, usually a propeller agitator, which enters the mixing vessel through the side of the vessel.

SIDEHILL SCREEN—A washing device to remove ink and inorganic fillers from the fiber/water slurry. A fine mesh screen of fabric or wire is mounted at a 38° angle to the horizontal. The fiber/water slurry is introduced to the screen by a headbox at the top. The pulp tumbles and slides down the screen to a discharge box, and the dirty filtrate drains through the screen and is collected in a water compartment. See also WASHING.

SIDELAY—To move an unwind sideways for purposes of aligning a web with downstream components. See UNWIND.

SIDERUN—A roll differing in width from that of the rest of the rolls being made at the time on the slitter. It is made for the purpose of utilizing the full width of the machine when the multiple of roll widths desired does not equal the normal operating width of the paper machine.

SIDERUN NEWS—Sideruns of standard newsprint paper.

SIDES—Right and wrong sides of the sheet. The term may also refer to the top and undersides or the felt and wire sides. See RIGHT SIDE OF THE PAPER.

SIDE STITCHING—A binding method where the folded, collated sheets are stapled along the (left) edge, from the front to the back of the signature. See BINDING.

SIEVE ANALYSIS—See PARTICLE SIZE ANALYSIS.

SIGMA—The lower case Greek letter used to describe the variation of a process. It is estimated from the standard deviation. See STANDARD DEVIATION.

SIGNATURE—A section of a book or magazine as folded ready for binding with other sections. It often has 16 pages but may have from 4 to 64 pages in multiples of 4. Thicker stock is usually run in signatures with a fewer number of pages, and thin stock with greater number. But, especially in magazine printing, other factors may affect the number of pages in a signature.

SIGN BOARD—A paperboard upon which signs or advertising matter is printed. It is made of woodpulp and reclaim paper stock, usually 0.020 to 0.040 of an inch in thickness. It may be a white patent-coated or clay-coated board, it is a rigid board not susceptible to warping and has a surface adapted to receive fine printing. It is frequently treated to render it water-resistant. See CAR-SIGN BOARD; POSTER BOARD.

SIGN PAPER—A paper used for outdoor and indoor sign and poster work, especially for advertising purposes. It is generally made of bleached chemical woodpulp and is surface sized or treated to enhance the properties required for its use. The normal basis weights are 90, 100, or 110 pounds (24 x 36 inches – 500). Significant properties include rigidity, color, finish, formation, fastness-to-light, water resistance, and fair strength.

SILENT PAPER—See NOISELESS PAPER.

SILENT PROGRAM PAPER—See NOISELESS PAPER.

SILICATE—See SODIUM SILICATE.

SILICATED PAPER—A paper which has been coated with sodium silicate to give hardness and finish.

SILK-SCREEN PRINTING—See SCREEN PRINTING.

SILK PROTECTION PAPER—(1) A writing paper having silk threads incorporated to give protection against duplication or counterfeiting. See SAFETY PAPER. (2) A sulfite or sulfate paper used as an inner wrapper for bolts of silk.

SILK WRAPPER—A paper made of sulfite, sulfate, or mixed furnishes on a multiple cylinder or combination cylinder-fourdrinier machine. It is duplex in color, the most common combinations being buff and white, blue and white, and green and white. It is sold by caliper, the range being from 0.006 to 0.015 of an inch. It is sufficiently strong to wrap and protect heavy silk bolts and resists the passage of moisture.

SILVER EXPRESS—A pearl or gray water-finished chemical woodpulp wrapping paper, used for the same purposes as express wrapping.

SILVER LABEL PAPER—A label paper (q.v.) which has been aluminum coated. It is used for box and gift wrap purposes as well as general label work.

SILVER NITRATE PAPER—See TEST PAPERS.

SILVER TISSUE—A paper used for wrapping metal objects that are subject to tarnishing. It is chemical woodpulp and/or cotton fiber paper having a basis weight of 8 pounds (20 x 30 inches – 480) or 12 pounds (24 x 36 inches – 480). It is free from chemical impurities which would cause tarnishing. It is generally made on a fourdrinier machine. See ANTITARNISH PAPER.

SILVER WRAPPING PAPER—See ANTITARNISH PAPER.

SIMULATED FELT MARK—A pattern or texture produced in certain grades of paper such as cover, text, offset, and papeterie, by patterned rubber composition rolls usually located in the press section of the paper machine. Such rolls are generally used on fairly high-speed machines where the use of marking felts would not be feasible. See WATERMARK.

SIMULTANEOUS TWO-SIDED COATING—Coating both sides of the sheet at the same time. Many types of equipment are available for this purpose.

SINGLE COATED—A term used to indicate that a paper or paperboard has been coated once, either on one or both sides. The term is sometimes used (incorrectly) to designate a paper or board that is coated on one side only; such a product should be called coated one side.

SINGLE-CYLINDER MACHINE—(1) See CYLINDER MACHINE. (2) An incorrect designation sometimes used for the Yankee machine (q.v.), in reference to the single drying cylinder.

SINGLE DISC REFINER—A refiner with single rotating disc and a single stationary disc, each having one surface with refining tackle.

SINGLE FACED BOARD—A fluted medium faced on one side only, usually with linerboard. See also CORRUGATED BOARD.

SINGLE FACED CORRUGATED BOARD—See CORRUGATED BOARD.

SINGLE FACED ROLL—Single-faced corrugated board furnished in rolls for use as a packing material. See CORRUGATED BOARD.

SINGLE-LINED BOARD—Paperboard vat lined on one side with a stock different from the remainder of the board. Single manila-lined chip and single manila-lined newsboard are examples in the folding grades. Single news vat-lined chipboard and single white vat-lined chipboard are examples of setup box grades.

SINGLE PLY—Having only one ply or layer. See PLY.

SINGLE-SHEET CUT—Obsolete term. See PRECISION CUTTER.

SINGLE-THICK COVER—A cover paper of a single thickness of paper, not pasted.

SINGLE-TIER DRYER SECTION—See DRYER SECTION.

SINGLE WALL BOARD—Corrugated board in which a single faced (q.v.) structure is adhered to one facing, usually linerboard. See DOUBLE FACED CORRUGATED BOARD.

SINGLE WALL CORRUGATED BOARD—See CORRUGATED BOARD.

SINGLE WHITE VAT-LINED CHIPBOARD—A paperboard used for cartons, etc. It is made on a cylinder machine. The top liner is made of virgin woodpulp or paper stock or a combination of both (usually a newsprint color); the back is made of wastepaper. The liner is usually fairly well sized and has a smooth finish. Calipers range from 0.016 of an inch and up.

SINTERED METAL SPARGER—A device used to inject gas into the turbulent zone after a medium consistency pump. The sintered metal is produced in such a way that the pore size is only several millimeters.

SISAL—A plant *(Agave sisalana)*, and the fiber obtained from its leaves and used for hard fiber cordage. Native to Central America, it is grown extensively in the West Indies and Africa. Some is used in rope papers and is obtained from cordage waste. The fiber has also been called sisal hemp.

SI UNITS—See INTERNATIONAL SYSTEM OF UNITS.

SIZE—Any material used in the internal sizing or surface sizing of paper and paperboard. Typical sizes are rosin, alkyl ketene dimer (AKD), alkenyl succinic acid anhydride (ASA), styrene maleic anhydride (SMA), glue, gelatin, starch, modified celluloses, synthetic resins, latexes, and waxes.

SIZED AND SUPERCALENDERED—A term denoting a supercalendered book paper with ordinary sizing. Also called SSC.

SIZE PRESS—A surface sizing device usually installed in a paper machine. Most size presses consist of two rolls with the sheet passing through the nip between the rolls. Types of size presses include pond size presses and metering size presses. Size press configurations are classified by the relative positions of the two size press rolls. A vertical size press has one roll on top of the other. A horizontal size press has the two rolls side by side. An inclined size press has one roll higher that the other with the sheet passing through on a downward incline, usually at an angle of about 30° from the vertical.

SIZE PRESS PICKUP—A term used to describe the amount of surface sizing chemicals added to the sheet at the size press. It is normally expressed in terms of dry chemical per unit of paper or paperboard and may be expressed in pounds per ton, pounds per ream, or percent; in these cases, it is assumed that all of the water in the surface sizing solution or suspension has been evaporated.

SIZE PRESS ROLL—Used in size press (q.v.) units to form a pressure nip for the application of sizing chemicals. Usually one roll of the combination is metal or is covered with a hard composition and the mating roll is covered with a softer elastomeric composition. The cover specification for the softer roll is very critical for the proper operation of the size unit. The elastomeric covering must be designed to resist the sizing chemicals and the hardness chosen to produce the nip conditions required for the sizing application.

SIZES OF PAPER—The following are some of the common sheet sizes in the United States:

Bible paper: 25 x 38, 28 x 42, 28 x 44, 32 x 44, 35 x 45, 38 x 50 inches.

Blanks (plain): 22 x 28, 22 x 24, 22 x 28, 23 x 43, 28 x 44, 34 x 43 inches.

Blanks (coated): 22 x 28, 22 x 42, 28 x 44 inches.

Blotting paper: 19 x 24, 24 x 38 inches.

Bogus bristol: 22 x 28 inches.

Bond paper (wood pulp and cotton content): 8 x 11, 8 x 13, 8 x 14, 10 x 14, 11 x 17, 17 x 22, 17 x 28, 17 x 22, 19 x 24, 20 x 28, 22 x 25, 22 x 34, 22 x 35, 23 x 29, 24 x 38, 28 x 34, 34 x 44, 35 x 45 inches.

Book paper (uncoated): 17 x 22, 19 x 25, 22 x 35, 23 x 29, 23 x 35, 25 x 38, 28 x 42, 28 x 44, 32 x 44, 35 x 45, 38 x 50 inches.

Book paper (coated two sides): 22 x 35, 24 x 36, 25 x 38, 26 x 40, 28 x 42, 28 x 44, 32 x 44, 35 x 45, 36 x 48, 38 x 50 inches.

Boxboard: 25, 40 inches.

Butcher paper: 12 x 18, 18 x 24, 20 x 30, 34 x 36, 30 x 40 inches.

Cover paper (wood pulp and cotton content): 20 x 26, 23 x 29, 23 x 35, 26 x 40, 35 x 46 inches.

Document manila: 22 x 28, 34 x 36 inches.

Glassine: 24 x 36, 25 x 40, 30 x 40 inches.

Gummed papers: 17 x 22, 20 x 25, 25 x 38 inches.

Index bristol (wood pulp and cotton content): 20 x 24 3/4, 22 x 28, 22 x 35, 25 x 30, 35 x 45 inches.

Kraft wrapping: 18 x 24 , 20 x 30, 24 x 36, 30 x 40, 40 x 48, 48 x 60 inches.

Label paper (coated one-side book): 20 x 26, 25 x 38, 26 x 40, 28 x 42, 28 x 44, 32 x 44, 35 x 45, 36 x 48, 38 x 50, 41 x 54 inches.

Ledger paper (wood pulp and cotton content): 17 x 22, 17 x 28, 19 x 24, 22 x 34, 22 x 22, 22 x 34, 24 x 38, 24. x 24., 28 x 34 inches.

Litho label paper (coated one side): 25 x 38, 28 x 42, 28 x 44, 32 x 44, 35 x 45, 36 x 48, 38 x 50, 41 x 54 inches.

Manifold paper: 17 x 22, 17 x 26, 17 x 28, 19 x 24, 21 x 32, 22 x 34, 24 x 38, 26 x 34, 28 x 34 inches.

Manuscript cover: 18 x 31 inches.

Mill blanks: 22 x 28, 28 x 44 inches.

Mill bristols: 22 x 28 inches (2, 3, and 4 ply, 125, 150, and 175 pounds).

Newsprint: 21 x 32, 22 x 24, 24 x 36, 25 x 38, 28 x 34, 28 x 42, 34 x 44, 36 x 48, 38 x 50 inches.

Offset book paper (uncoated): 17 x 22, 22 x 35, 25 x 38, 28 x 42, 32 x 44, 35 x 45, 36 x 48, 38 x 50, 38 x 52, 41 x 54 inches.

Offset book paper (coated): 22 x 29, 22 x 35, 25 x 38, 28 x 42, 28 x 44, 32 x 44, 35 x 45, 38 x 50 inches.

Opaque circular: 17 x 22, 17 x 22, 22 x 34, 23 x 29, 23 x 35, 25 x 38, 28 x 34, 35 x 45, 38 x 50 inches.

Photomount board: 23 x 29 inches.

Postcard bristol: 22 x 28 inches.

Postcard (coated): 22 x 28 inches or double.

Railroad manila: 17 x 22, 17 x 28, 19 x 24, 22 x 34, 24 x 38, 28 x 34, 34 x 44 inches.

Rotogravure paper: 25 x 38, 28 x 42, 28 x 44, 32 x 44, 35 x 45, 38 x 50 inches.

Safety paper: 17 x 22, 17 x 28, 19 x 24, 19 x 26, 19 x 28, 22 x 34, 24 x 38, 28 x 34, 28 x 38 inches.

Tag board: 22. x 28., 24 x 36, 30 x 40 inches or double.

Text paper: 23 x 29, 23 x 35, 25 x 38, 26 x 40, 35 x 45, 38 x 50 inches.

Thick china: 22 x 28 inches or double.

Tough check: 22 x 28 inches or double.

Translucents: 22 x 28 inches or double.

Waxed paper: 9 x 12, 12 x 12, 12 x 13, 12 x 18, 18 x 24, 24 x 36 inches.

Wedding papers: 17 x 22, 22 x 34, 35 x 45 inches.

Wrapping tissues: 10 x 15, 10 x 30, 12 x 18, 12 x 24, 15 x 20, 18 x 24, 20 x 30, 24 x 36 inches.

Writing paper: 17 x 22, 17 x 28, 19 x 24, 22 x 34, 24 x 38, 28 x 34 inches.

SIZE SPECKS—Specks appearing in the sheet as transparent or glazed spots normally of different color from the rest of the sheet, caused by undispersed particles or agglomerations of sizing materials carrying through into the sheet. See also ROSIN SPECKS.

SIZE TUB—See TUB-SIZE PRESS.

SIZING—(1) A property of paper resulting from an alteration of fiber surface characteristics. In internal sizing, the entire furnish is uniformly treated before the paper is formed to increase its resistance to the penetration of polar liquids. The surface sizing relates to treating the finished sheet, on one or both sides, to increase such properties as water resistance, abrasion resistance, abrasiveness, creasability, finish, smoothness, surface bonding strength, and printability, and the decrease of porosity and surface fuzz. (2) The addition of materials to a papermaking furnish or the application of materials to the surface of paper and board to provide resistance to liquid penetration and, in the case of surface sizing, to effect one or more of the properties listed in (1).

SIZING POLYMERS—Water dispersable polymers that contain hydrophobic groups (which resist the penetration of water) and can act as effective sizing agents when adsorbed on fibers.

SKATING—Flow-streaking of the stock on the fourdrinier wire, whereby small irregularities at the slice discharge grow into steaks that migrate across the wire, diagonally to the machine direction. Such flows result in streaks of greater mass in the finished sheet.

SKETCHING PAPER—A paper used for rough drawings and plans. Almost any type of writing or low-grade drawing paper is suitable, but it should have a surface suitable for pencil marks.

SKID—A fabricated platform of wood designed to support paper products for storage or for transport by a fork tractor. A skid can also refer to a pile of sheeted paper on the prefabricated wooden platform and wrapped and labeled for shipping. A standard sheeted skid can weigh in excess of 3000 pounds. See also PALLET.

SKIM—Term applied to a thin top or back liner (q.v.).

SKIN—(1) See SKIM. (2) A lightweight wallpaper, usually printed without design. (3) A high, hard finish on a paper or paperboard.

SKIN COAT—A very thin coating on a sheet of paper or paperboard. Generally applied as a precoat to prepare the web for further coating. Skin coat has been replaced by the terms film coat and prime coat.

SKIN PARCHMENT—Animal parchment. See PARCHMENT (1).

SKIPPED COATING—Discontinuities in the coating.

SLAB—Dry broke which has been culled or split from a roll.

SLABBING—The act of slicing off several outer layers of a roll of paper to examine the inner layers.

SLACK EDGES—(1) One or both edges of a roll that are soft or slack, usually with thin paper at the edges. (2) In the process of coating paper, a condition in which the middle of the web carries the tension, the edges being slack.

SLACK SIZED—Lightly sized and somewhat water absorbent. Also, having a degree of water resistance below standard.

SLAKER—An agitated tank with an attached classifier which receives clear green liquor (q.v.) and lime. The initial reaction of slaking the lime with water and a portion of the causticizing occurs in this unit as does the separation of grit (q.v.) from lime/liquor slurry.

SLAT DRYING—A method of air-drying paper, accomplished by running the web over a series of iron spiders carrying wooden slats on their circumferences. Obsolete as a commercial practice.

SLATERS PAPERS—A tarred or asphalt felt or sized sheathing used under slates in the roofs of buildings. See SHEATHING PAPER.

SLICE—That part of a fourdrinier machine or twin wire former which regulates the flow of stock from the headbox (q.v.) onto the fabric as a sheet of liquid of even thickness or volume traveling at or very close to the speed of the fabric. The slice extends across the fabric and forms that side of the headbox adjacent to the fabric. Variations in the thickness of the sheet of liquid passing the slice are obtained by adjustment of its top edge (or lip) by means of closely spaced slice screws (q.v.).

SLICE MARKS—Uneven surface and look-though, resulting from maladjustment of the slice (q.v.).

SLICE SCREW—Used to adjust the slice (q.v.) flow, slice screws are manually or computer-controlled screws that adjust the position of the top lip of the slice relative to the bottom lip (or apron) to control the volume of headbox discharge locally in order to regulate basis weight variations at local zones of the cross machine direction of the sheet. On most machines, the slice screws are located every 15 cm (6 in.) or so across the slice. On newer machines, the trend is to locate them at about half that spacing.

SLICK FINISH—A smooth finish.

SLIDE BOX—The single lined slide box (also known as a two-piece Lambert) is made of two pieces of double faced corrugated board, one forming the top and bottom, the other the sides. The double lined slide box is made of two shells of double faced corrugated board which slide one within the other. It has two thicknesses of board on two sides only. The triple lined slide box is similar to the double slide box but is supplied with a liner which gives all surfaces protection of two thicknesses. The characteristics of double faced corrugated board are all found in the slide box. Slide boxes are used largely for parcel post and express shipments of small articles.

SLIME—An aggregation of heterogeneous material, sometimes having a slippery feeling, found at various points within a pulp or paper-making system. It may be caused by microbial growths or deposits of nonbiological materials.

SLIME HOLE—A hole in a paper web caused by slime which was incorporated inadvertently during the formation of the web, breaking out of the dried web when it is run through the calender stack. A slime hole is often identified by the occurrence of translucent fragments around the edge of the hole.

SLIME SPOTS—Spots or smears in paper caused by fungi or bacterial growths in the pulp stock.

SLIMICIDE—A toxic material to control microbiological growths in pulp and paper mill systems.

SLIP—A fluid or semisolid mixture of a pigment, such as clay, and water.

SLIPPED ROLL—A roll of paper which has slipped off its core or has been pushed out, giving the roll a cone shape or "pencil point."

SLIP SHEET—A large flat sheet of corrugated or solid fiberboard used as a base upon which goods and materials may be assembled, stored, and transported usually as a replacement for a skid or pallet.

SLIP-SHEET BOARD—A paperboard which is used in connection with a mimeograph to prevent setoff of the mimeographed sheets, especially when a bond paper is used. The term slip-sheet board is now obsolete.

SLIP-SHEET PAPER—A paper used by a printer to protect a wet printed surface from setoff by the next printed sheet that is piled up on it, or in the case of rolls, which is wound on it. It must have a high finish and be free from lint.

SLITTER—Any mechanism used to separate a web in the machine direction. In addition to the familiar score and shear slitters, there are razor blade slitters, water jet slitters, and laser slitters in use for specialty paper products. See SCORE SLITTER; SHEAR SLITTER.

SLITTER DUST—Small particles of fibers or coating or both which are chipped off during the slitting operation which may adhere to the edge of the sheet and later work their way into subsequent converting operations. This dust is very similar to cutter dust (q.v.).

278

SLITTER EDGE—That edge of the paper web which is made by the slitter. See KNIFE EDGE.

SLITTING—The operation of separating a single web (or a multiplicity of layers of webs) in the machine direction by means of a slitter (q.v.).

SLIVER—A small splinter of wood found in mechanical and chemical pulps as a result of incomplete fiber from fiber separation during the pulping process. Slivers need to be removed from the pulp by screening (q.v.).

SLOT PAPER—A cotton fiber paper with a glazed finish and high density, used for various types of formed electrical insulation such as slot tubes. High tearing resistance, high dielectric strength, and uniform thickness are very important.

SLOTTED CONTAINER—See REGULAR SLOTTED CONTAINER.

SLOWNESS—See FREENESS.

SLOW SHEET—A paper made from "slow stock," or slow draining stock, usually a low freeness stock.

SLOW STOCK—A pulp suspension from which the water drains slowly. It usually results from refining.

SLUDGE—The solids removed from the effluent stream of a pulp or paper mill.

SLUDGE DISPOSAL—The process of disposing of sludge through depositing it in its final location.

SLURRY—A suspension, usually aqueous, of pigments or other insoluble materials used in coating or papermaking.

SLUSH—(1) A suspension of paper pulp of such consistency that it will flow or can be pumped, usually containing from 1 to 6% of dry stock. (2) The action of forming a slurry from wet or drylap pulp. See also SLUSHER.

SLUSHER—A furnish pulper (q.v.) designed to reduce to a slurry only wet lap or dry lap pulp. Wet and dry lap pulps generally require much less energy and dwell time to reduce to a slurry than dry broke, or recovered paper, because the fibers in pulp laps are generally weakly bonded, compared to fibers in dried broke or recovered paper.

SMASHED BULK—The bulk of a given number of sheets of paper under such a pressure as will eliminate the air between the sheets. This bulk is usually specified by book publishers.

SMELT—High temperature, liquid inorganic salts derived from pulping chemicals and separated from the black liquor organic solids during combustion in the recovery boiler furnace. Molten salts that flow out of the bottom of the recovery furnace through specially designed, water-cooled spouts.

SMITH NEEDLE—A handheld, interlayer, pressure measuring instrument that measures the pressure between layers of a roll.

SMOG—An air emission which is a combination of particulate matter, oxides of sulfur, and photochemical oxidants which occurs in cities and produces unhealthy effects, such as respiratory problems.

SMOKE—Opaque air emissions which occur as a result of combustion.

SMOOTHING PRESS—A type of press sometimes used in the press section (q.v.) of a paper machine usually located next to the dryer section. This press is equipped with smooth, low-porosity, hard-surfaced rolls having no felt running through the nip (q.v.). This press, usually a third or fourth press, is used to remove or reduce fabric and felt marks from the sheet and to increase its smoothness and density before drying. On some paper machines, the smoothing press is located partway through the dryer section.

SMOOTHNESS—The property of a surface determined by the degree to which it is free of irregularities. In printing, the smoothness of the paper in the printing nip is important and is referred to as printing smoothness. Smoothness improves as the paper is compressed and locally deformed under mechanical pressure.

SMOTHERED WATERMARKS—Watermarks which are so close together as to cover the entire surface of the sheet.

SMUT SHEET—See SLIP-SHEET PAPER.

SNAILING—Streaks or marks resulting from air bubbles or an excess of water in front of the dandy roll (q.v.), or from froth bubbles at the slice (q.v.).

SOAP—See SOAP SKIMMINGS.

SOAP SKIMMINGS (TALL OIL)—The curd, not acidified or otherwise processed, skimmed from the black liquor of the kraft pulping process. The soap separates readily in liquor tanks with a dissolved solids content greater than 12%. It reaches a minimum solubility at 25–30% dissolved solids and is usually skimmed from the liquor after three effects in the multiple effect evaporators (q.v.).

SOAP WRAPPER—Paper used as an inner or outer wrapper for cakes of soap. Depending on use requirements, it may be made from a wide variety of furnishes and may be printed, creped, waxed, laminated, or otherwise treated. Important characteristics include freedom from discoloration on contact with dilute alkali and resistance to growth of bacteria and fungi.

SOCKET PAPER—A strong, heat-resisting paper having a high dielectric strength and used in the manufacture of electric light sockets. See also ELECTRICAL INSULATION FIBER; HARD PAPER.

SODA—NA_2CO_3. A term for sodium carbonate. See SODIUM CARBONATE.

SODA ASH—A commercial anhydrous (contains little or no water) form of sodium carbonate. It is used as a component of the kraft white liquor.

SODA-CHLORINE PROCESS—A multistage chemical process for pulping straw, which includes an alkaline pretreatment, a chlorination step, an alkaline wash, and a final hypochlorite bleach.

SODA PULP—A chemical pulp produced by the high temperature digestion of wood with sodium hydroxide or caustic soda solutions.

SODA PULPING—See ALKALINE PULPING PROCESS.

SODA STRAW PAPER—See DRINKING STRAW PAPER.

SODIUM ALUMINATE—$NaAlO_2$. A strongly alkaline salt used in paper sizing.

SODIUM BISULFITE—$NaHSO_3$. An inorganic salt that is acidic in aqueous solutions. It is a component of semichemical cooking liquors, and is also used to remove excess chlorine from bleached pulp. It has a mild brightening effect on pulp.

SODIUM BOROHYDRIDE—$NaBH_4$. A lignin-preserving, powerful bleaching agent which brightens pulp by reducing carbonyl groups, thereby increasing the brightness stability. Its high cost makes commercial application prohibitive. However, it is widely used for on-site manufacture of sodium hydrosulfite.

SODIUM CARBONATE—Na_2CO_3. An inorganic salt that is strongly alkaline in aqueous solution. It is used in the preparation of soda base sulfite cooking liquor and is a component of the kraft and soda pulping liquors. See SODA; SODA ASH.

SODIUM CHLORATE—$NaClO_3$. One of the principle compounds used to manufacture chlorine dioxide. It is also one of the oxygen-containing chlorine compounds formed during chlorine dioxide and hypochlorite bleaching. Its presence is particularly significant in chlorine dioxide bleaching because its formations, which is pH dependent, constitutes a loss of oxidative power and becomes unavailable for bleaching since the chlorate alone does not react with the lignin in the pulp. See CHLORINE DIOXIDE MANUFACTURE.

SODIUM CHLORIDE—NaCl. Common table salt used in papermaking to control the electrical properties of xerographic papers and for the mordanting of dyes.

SODIUM CHLORITE—The sodium salt of chlorous acid, sodium chlorite forms chlorine dioxide when acidified with sulfuric acid or chlorine. Sodium chlorite, buffered at pH 4 with sodium acetate at room temperature, releases chlorine dioxide. This mixture can be used to selectively delignify unbleached pulp for viscosity determination or for other purposes. See CHLORITE.

SODIUM DITHIONITE—$Na_2S_2O_4$. A lignin-preserving reductive brightening agent commonly used to whiten groundwood (q.v.) and high-yield pulps. A brightness increase of about 10–12 points on some groundwood pulps and about 2–8 points on chemical pulps may be achieved, depending on pulp quality and species.

SODIUM HYDROSULFIDE—NaSH. Principal form of the chemical sodium sulfide (q.v.) in an aqueous solution. Often used as a purchased chemical for sodium and sulfur makeup.

SODIUM HYDROSULFITE—A reducing compound ($Na_2S_2O_4$) which is used for bleaching (especially mechanical pulp). It is also termed sodium dithionite.

SODIUM HYDROXIDE—(NaOH) Often referred to as caustic or caustic soda (q.v.).

SODIUM HYPOCHLORITE—See HYPOCHLORITES.

SODIUM PEROXIDE—Na_2O_2. An oxidizing agent sometimes used in bleaching mechanical pulp and in multistage bleaching of chemical pulps. It is a hazardous substance and ignites on contact with water.

SODIUM SILICATE—A compound added to a peroxide bleaching stage as a buffer and as a stabilizer of hydrogen peroxide. More commonly used in mechanical pulp brightening. Commonly called "water glass."

SODIUM SULFATE—Na_2SO_4. Primarily used in the preparation of kraft pulping white liquor. It is added to the concentrated black liquor before burning and is converted to sodium sulfide, one of the two active components of kraft white liquor. The other component is caustic soda or sodium hydroxide.

SODIUM SULFIDE—Na_2S. An inorganic salt that is one of the two active components of kraft cooking liquor. The other component is caustic soda or sodium hydroxide.

SODIUM SULFITE—Na_2SO_3. A mild bleaching agent sometimes used as a steep bleaching agent (q.v.). It is also being added to shower water of grinders in the mechanical pulping process to improve brightness and reduce power consumption.

SODIUM THIOSULFATE—$Na_2S_2O_3$. An inorganic salt used to neutralize excess chlorine in pulp bleaching. In this application, it is also referred to as "antichlor" or "hypo."

SOFT—A term applied to paper which has a soft surface and body and little or no sizing, requiring relatively little pressure for printing. News and common book paper are examples.

SOFT CALENDER ROLL—See SOFT-NIP CALENDER.

SOFT COOK—A batch cook or a period of pulping time in a continuous digester when the target pulping conditions are not maintained and the result is overcooking. High temperature, too long cooking time, and excess chemical will result in a soft cook. The pulp produced is called soft pulp. The kappa number of a soft cook is lower than the target. A common cause of soft cooks is a process upset upstream from the digesters that forces the holding of the pulp in the digesters longer than normal. These are called "held cooks."

281

SOFTENING (WATER)—In water treatment, high hardness (calcium and magnesium salts), alkalinity (carbonate, bicarbonate, etc.), and silica in feedwater can cause many scaling and corrosion problems in equipment. Softening is the process of reducing the concentration of these compounds in boiler feedwater by use of inorganic or organic compounds. Lime and zeolite are examples of softening agents.

SOFT FOLD—A method of folding a lift of large sheets of paper so there is no resulting crease.

SOFT HARDWOODS—A terminology used to differentiate the softer hardwoods of the hardwood group from the harder hardwoods. It includes such pulpwood species as yellow poplar, cottonwood, the gums (black, tupelo, and sweetgum), the aspens, the magnolias, and the soft maples (red and silver). See HARD HARDWOODS.

SOFTNESS—The property of a paper, usually tissue or toweling, that relates to the pleasing, soothing sensation perceived by tactility or handling. Subjectively, softness combines psychological and physical attributes, including limpness, smoothness, thickness, compressibility, and possibly others, in a combination not yet accurately defined. Correlation of test methods with subjective perception has not been very well established.

SOFT-NIP CALENDER—A soft-nip calender is usually composed of a single-nip, vertically oriented calender design with one of the mating rolls being covered with a durable elastomeric composite surface. The other roll is usually a variable crown metal roll. The conventional design incorporates a tandem arrangement so a one-sided sheet is produced. The soft-nip arrangement delivers a more uniform compaction to a non-uniform web. It can be installed "on-machine," and the results approach an off-line supercalendering operation. The system can also be used to micro-emboss a matte finish to the substrate.

SOFT PULP—See SOFT COOK.

SOFT ROLL—A roll which is soft because of loose winding and/or caliper variations across its face.

SOFT SIZED—See SLACK SIZED.

SOFT WHITE SHAVINGS—See SHAVINGS.

SOFTWOOD—Wood from coniferous trees whose leaves are needlelike such as pine, spruce, or hemlock or scale-like such as cedar. See CONIFER.

SOFTWOOD PULP—A pulp made from softwood or coniferous wood species.

SOLARBROMIDE PHOTOGRAPHIC PAPER—See BROMIDE PHOTOGRAPHIC PAPER.

SOLARCHLORIDE PHOTOGRAPHIC PAPER—See CHLORIDE PHOTOGRAPHIC PAPER.

SOLID BLEACHED SULFATE (SBS)—A board which is made totally from hardwood and/or softwood fibers pulped by the sulfate, or kraft, process. The board may be layered or homogenous as long as all of the fibers are made by the sulfate process. This grade, typically, is used in food packaging.

SOLID BOARD—A board made of the same material throughout as contrasted with a combination board where two or more stocks are used. A pasted board is not a solid board even though the same stock is used.

SOLID BOARD FRAME—A method of packing paper in which a solid wooden top and bottom are placed on the flat sides of a bundle and the unit is then tied in both directions.

SOLID BRISTOL—A term applied to index or mill bristols which are made of homogeneous stock in one operation.

SOLID FELT—See DEADENING FELT.

SOLID FIBERBOARD—A pasted board (q.v.) used for fabricating shipping containers in which several layers of paperboard or containerboard are pasted together to make a thicker structure.

SOLID FIBER SHIPPING CONTAINER BOARD—See SOLID FIBERBOARD.

SOLID FRACTION—The fiber, filler, sizing material, etc., that constitute what is generally thought of as paper. It is the ratio of the volume of solid material to the total volume of the measured sample of paper. See also VOID FRACTION.

SOLID INDEX BRISTOL—See SOLID BRISTOL.

SOLID WASTE—Discarded solid material from an industrial process or a municipality.

SOLID WASTE MANAGEMENT—The purposeful, systematic control of the generation, storage, collection, transport, separation, processing, recycling, recovery, and disposal of solid wastes.

SOLID WASTE STREAM—The continuous generation or output of refuse or solid waste, as for example the creation of such wastes by households.

SOLUBLE GLASS—See SODIUM SILICATE.

SOLVENT—The liquid in an adhesive, coating, ink, paint, or like product, which dissolves or disperses the film-forming ingredients and controls the viscosity and solids characteristics. It generally undergoes evaporation during final use of the product.

SOLVENT PULPING—Involves the use of organic solvents such as methanol or ethanol to chemically pulp wood chips. The Organocell process uses methanol with sodium hydroxide and anthraquinone; the Alcell process uses only ethanol and the Asam process uses methanol, sodium hydroxide, alkaline sulfite, anthraquinone and methanol; the Acetocell process uses only acetic acid, and the Milox process uses formic acid and hydrogen peroxide. The use of other organic solvents such as phenol,

ethyl acetate, and glycerol have not advanced beyond the laboratory or pilot plant level.

SOLVENT SIZING—Sizing (q.v.) by use of a solution of a suitable sizing agent. The solution being applied to the unsized paper and the solvent being removed by evaporation and recovered.

SOOT—Particulate matter which has collected on the inside wall of a stack or stovepipe.

SOOTBLOWER—A retractable lance used in the convective tube banks of boilers. Sootblowers are fitted at the end with two opposed nozzle openings through which high-pressure steam flows forming jets of steam that mechanically remove deposits from the outside surface of the boiler bank, superheater, or economizer tubes.

SOOTBLOWING STEAM—High pressure steam that is used in special lances used in the boiler convective tube banks to remove deposits that accumulate on the outside of the tubes due to contact with particle-laden flue gas.

SORTING—(1) In finishing, a procedure whereby sheets are manually inspected for defects by a statistical process or 100% inspected by an electronic sensor system that may be an integral part of a precision sheeter and which can automatically reject defective sheets, thereby eliminating the costly manual statistical inspection system. (2) Manipulation of data in a spreadsheet or database to provide compilation of information as desired to assess, compare, and place in preferential order.

SOUND ABSORPTION COEFFICIENT—The coefficient of sound absorption, or absorptivity of any given material, is the fraction of the intensity of the incident sound absorbed by the material. It is evaluated for a definite pitch or frequency.

SOUND TRANSMISSIVITY—The sound transmissivity (or transmission) or a partition is the fraction of the intensity of sound incident upon the partition which is transmitted through the partition and radiated into the space on the opposite side of the partition. It is evaluated for sound of a definite pitch or frequency.

SOUR—To clean a fourdrinier wire, dandy roll, etc., with an acid solution. This is an obsolete practice.

SOURCE SEPARATION—The segregation and collection of individual recyclable components at the point of generation before they become mixed into the solid waste stream (e.g., bottles, cans, newspapers, corrugated containers, or office papers).

SOUR COATED PAPER—Coated paper which has an offensive odor resulting from the use of decomposed casein in the coating mixture. Casein is now rarely used as a coating binder.

SOYA FLOUR—The flour or meal resulting from grinding soybeans and extracting the oil. It consists of about 45% soybean protein (q.v.), the remainder being hemicellulosic in nature. In paper coating, soya flour has been replaced by synthetic latexes, modified starches, and improved chemically modified soybean protein (q.v.).

SOYBEAN PROTEIN—Usually the alpha-protein fraction of the soybean which is used as an adhesive and as a sizing and coating material for paper.

SPANISH GRASS—Esparto (q.v.). See also ESPARTO PAPER.

SPC—See STATISTICAL PROCESS CONTROL.

SPECIAL FOOD BOARD—A variety of paperboard grades used for packaging foods. In this group are fourdrinier grades of bleached chemical pulps and cylinder grades of solid bleached pulps, single- and double-white-lined manilas. The boards are hard sized for water resistance and are frequently coated for certain applications and for high-quality printing. After printing, the cartons are frequently waxed, coated, or otherwise treated. Typical examples are butter, ice cream, and milk cartons.

SPECIAL LAWBOOK COVER—See LAWBOOK COVER.

SPECIAL MARKING ORDER—An order made to customer specifications as opposed to stock order (q.v.).

SPECIALTIES—(1) Grades of paper and/or paperboard made with specific characteristics and properties to adapt them to particular uses. (2) Grades of papers and/or paperboards made in a given mill which are not the primary products of that mill.

SPECIALTY COVER PAPER—A coated, waterproof, or semiwaterproof paper which may have a smooth surface or be embossed. It is made in a medium weight (90 pounds, 20 x 26 inches – 500) or double thick (180 pounds, 20 x 26 inches – 500). These papers are particularly well suited for printed matter that requires protection and unusual finished appearance. Specialty covers are also made in a lightweight, which is coated and waterproofed two sides; the basis weight is approximately 50 pounds (20 x 26 inches – 500).

SPECIFIC EDGE LOAD—See REFINER INTENSITY.

SPECIFIC ENERGY—(1) In mechanical pulping, a unit of electrical energy applied during refining per ton of pulp produced. Usually expressed as hpd/ton (horsepower days/ton) or kW h/ton (kilowatt hour/ton). (2) Energy consumed per unit of production; a measure of the energy consumption of a particular unit operation or process. Typical units are as above. Often used in connection with refining, screening, and centrifugal cleaning steps, or with a process as a whole.

SPECIFIC GRAVITY—The ratio of the weight of the specimen to the weight of an equal volume of water. In pulping, the specific gravity of wood is an important factor in determining how much wood can be put into a digester and its pulping production capacity. The specific gravity of chemical solutions varies with the concentration of dissolved and suspended components, and the specific gravity data are used in the design of storage and metering systems.

SPECIFIC INDUCTIVE CAPACITY—The ratio of the capacitance of a two-plate electrical condenser when the space between the plates is filled with the test sample to the capacitance of the same condenser when the space between the plates is filled with air (or, more strictly, when the space is evacuated). It is frequently called the dielectric constant. This property is of importance in condenser paper and in electrical insulating paper.

SPECK—A particle of contrasting appearance in pulp or paper. See DIRT.

SPECTRAL REFLECTANCE—The relative amount of incident light energy reflected from a surface at various wavelengths. The spectral reflectance curve gives a basic description of color which, for nonfluorescent materials, is independent of the illuminant. See REFLECTIVITY.

SPECTRAL REFLECTIVITY—The relationship between reflectivity and wavelength. The spectral reflectivity curve gives a basic description of color which, for nonfluorescent materials, is independent of the illuminant. See REFLECTIVITY.

SPECULAR GLOSS—The ratio of the intensity of light reflected from the specimen to that similarly reflected form an arbitrary standard, for specified and equal angles of incidence and reflection. It is an important measure of gloss, glare, and glossiness of paper; it is usually evaluated for incident and reflected rays of light making a small angle with the surface of the paper.

SPECULAR REFLECTION—That reflection which causes a surface to appear somewhat like a mirror. Specularly reflected light is that reflected from a surface at an angle equal to that at which the incident light strikes the surface.

SPENT LIQUOR—Pulping liquor at the end of a cook has only a small amount of the active

cooking chemicals left. It is called spent liquor. It is important not to completely exhaust the cooking chemicals, but to have a small amount of residual left so that lignin will not precipitate back onto the wood and fibers.

SPGW—Superpressure groundwood.

SPILLS—Accidental discharges of waste materials.

SPINNING PAPER—See TWISTING PAPER.

SPINNING PARCHMENT—A specially treated, lightweight vegetable parchment paper, usually red or pink. It is used to make small rolls about one inch in width for use in connection with the French system of mule spinning of worsted goods.

SPIRAL LAID DANDY ROLL—A term applied to a special type of dandy roll, where the laid wires run around the circumference of the roll producing lines parallel with the grain of the paper. This laid mark is characterized by the absence of chain lines (q.v.).

SPIRAL WOUND DANDY ROLL—See SPIRAL LAID DANDY ROLL.

SPIRAL WOUND TUBES—Paperboard tubes made by winding strips of paperboard on a mandrel to make tubes.

SPIRIT DUPLICATION—An obsolete reproduction process in which a master copy is prepared on a sheet of paper by means of a special carbon paper. The master sheet is clamped to a cylinder and the paper, slightly moistened with a special duplicating fluid, is fed to the cylinder. The moistened sheet dissolves a small amount of the ink from the master copy, thus giving an impression on the paper. See also DUPLICATING PAPER.

SPLICE—The joining of the ends of two webs of paper to make a continuous roll. Materials used are a variety of adhesives, gummed tapes, or splicing tissues (q.v.). Splices are marked to alert the converter of these defects.

SPLICE TAG—A marker used in roll paper to indicate the location of a splice. See also FLAG.

SPLICING TISSUE—Usually a thin (1–3 mils) tape composed of a material such as paper, cloth, or plastic film coated with an appropriate adhesive. It is used for splicing paper webs. The adhesive may be pressure sensitive, water- or solvent-soluble, thermoplastic, or thermosetting.

SPLINT STOCK—See BOOK-MATCH BOARD; MATCH-STEM STOCK.

SPLIT-COLORED PAPER—A paper one side of which is uncolored and the other colored.

SPLIT-RANGE CONTROL—A condition where one controlled output is split into one or more parts to control more than one final control element. A typical example is using tank level to control both the pump out and recirculation streams on a liquid tank.

SPLITTING—The separating of plies of paper or paperboard.

SPOILAGE—See SHRINKAGE.

SPONGY—A term used to describe (1) paper that is bulky and compressible, or (2) paper that is unsized so that it is absorbent.

SPOT COATING—A coating process in which the application of the coating material is confined to certain portions of the sheet, rather than covering the entire surface.

SPOT CROWN PAPER—An express paper or a water-finished sulfate paper of high density and varnish coated, the paper after coating being approximately 0.0055 of an inch in thickness. It is used in bottle caps for sealing beverages.

SPOT ORDER—A direct sale to one customer for shipment within thirty days from date of order or for shipment to begin within thirty days from date of order and to be completed within sixty days from date of order.

SPRAY BONDING—A method of bonding in which atomized binder is deposited on a web to produce high loft types of nonwovens.

SPRAY COATING—A process in which the coating is applied by batteries of paint guns, arranged to spray one coat on top of another in rapid succession until a sufficient weight of coating has been built up.

SPRAY DYEING—(1) A process of spraying a dyestuff solution onto a sheet of paper by means of spray nozzles, either before or after the web passes over the first suction box. (2) A process of spattering a dyestuff solution onto the paper web by means of rotating brushes.

SPREADER—(1) A device intended to affect the width of a tensioned web in a web process. (2) A device intended to reduce the development of pucker wrinkles after a long web draw. (3) A device intended to separate individual slit webs such that through the remainder of the web process, the individual webs do not overlap and do not become entangled.

SPREADER ROLL—See BOWED ROLL.

SPREADING—A procedure whereby a web under tension is spread in the cross-machine direction to minimize machine direction wrinkles. Special rolls or bent bars are used to spread a traveling web, which could be a felt, a fabric or a paper web in almost any finishing process.

SPRINGBACK—(1) The increase in thickness (after a certain interval of time) of pulp mat after it has been subjected to a definite pressure for a specified time. It is a special meaning of the more general term "resiliency." (2) The degree to which a sheet can return toward its original flat condition after being folded under specified conditions and then released.

SPRING ROLL—A paper-web-carrying roll mounted on spring suspension bearings, and located in a position to cushion sudden deviations in web tension. Spring rolls are also used to adjust tension in forming fabrics.

SPRINGWOOD—See EARLYWOOD.

SPUNBONDED—A nonwoven consisting of continuous filaments laid down in a random order on a moving belt or screen that are bonded and rolled up as a finished product without any interruptions in the manufacturing process.

SPUNLACED—A nonwoven that is bonded by a multitude of water jets impinging on the web that is supported by a moving porous belt having an appropriate pattern and resulting in fiber entanglement to achieve mechanical bonding within the web. See HYDROENTANGLING.

SQC—See STATISTICAL QUALITY CONTROL.

SQUARED—(1) Cut or trimmed on two or more sides to ensure exactness of angle. (2) Sectional or scale paper. See PROFILE PAPER. (3) A gusseted bag which, when formed for filling, assumes a generally rectangular shaped opening.

SQUARED PAPER—(1) Drawing or tracing paper having ruled or printed squares of various sizes. (2) A paper which has been guillotine trimmed square on four sides or on one side and one end.

SQUARE SHEET—A term used to describe a paper or paperboard which has equal tensile strength and tearing resistance in machine and cross-machine directions.

SQUIRTS—Also called squirt jets, squirt trimmers, squirt trims, trim jets, trim squirts, etc. A small-diameter, low-volume, high-pressure water jet located on each side of the sheet, usually near the end of the forming section, which is used to trim the sheet to the desired width. The trim is diverted to the couch pit.

SQUIRT TRIM—The trim produced by the squirts (q.v.).

SR FREENESS TEST—See SCHOPPER-RIEGLER FREENESS.

SSC—Sized and supercalendered (q.v.).

STABILITY—The ability of paper or paperboard to resist change in any of its properties on exposure to various conditions. See BRIGHTNESS REVERSION; COLOR FASTNESS; DIMENSIONAL STABILITY; DURABILITY; PERMANENCE; YELLOWING.

STABILIZED—A paper whose moisture content is in equilibrium with the moisture of the surrounding air. See also CONDITIONING.

STACKED DRYER SECTION—See DRYER SECTION.

STAINED PAPER—(1) A paper which has been surface stained with color as, for example, on the calender stack of the paper machine. (2) A semi-absorbent sheet of white or tinted paper run through a bath of color and sizing material to produce unusual colors and/or a depth of color not normally obtained in the beater. See SURFACE COLORING and CREPE PAPER (2). (3) An old term for a printed wallpaper.

STAMP PAPER—See GUMMED PAPER; LABEL PAPER; POSTAGE STAMP PAPER; TRADING STAMP PAPER.

STAND ROLLS—Rolls of wrapping paper 15 or 18 inches in diameter and 48, 60, or 72 inches in width wound on wooden plugs for use in a vertical holder. Smaller rolls or narrower widths and smaller diameters for use in horizontal holders are called counter rolls (q.v.).

STANDARD BROWN KRAFT WRAPPING PAPER—An uncolored kraft wrapping paper made from unbleached sulfate pulp.

STANDARD COVER PAPER—A cover paper used for tablet covers, envelopes, mailing folder, menus, etc. It is made from unbleached chemical and mechanical woodpulps in basis weights of 25 to 50 pounds (20 x 26 inches – 500). It usually has an antique finish and is made in a wide range of colors.

STANDARD DEVIATION—A measure of dispersion of sample results from their mean. It is

287

defined as the square root of the average of the squared deviations from the mean. See also SIGMA.

STANDARD DEVIATION CHART—A control chart in which the subgroup standard deviations, S, is used to evaluate the stability of the variation within a process.

STANDARD MOISTURE—See AIR DRY.

STANDARD NEWSPRINT—See NEWS-PRINT.

STANDARD OBSERVER—A hypothetical observer based on color mixture date obtained for 2 degree field of view for 17 real observers—adopted by CIE in 1964.

STARCH—A white, odorless carbohydrate found in various plants. When extracted and purified, primarily from tapioca, corn, potatoes, and wheat, it is used in paper as an adhesive.

STARCH PAPER—See TEST PAPERS.

STARRED ROLL—A roll that exhibits a "star" or symmetrical wavy appearance on the roll ends. This pattern is caused by the rim effect of a tightly wound exterior on interior layers which have been compressed from their original diameters and are then under negative tension.

STATEMENT LEDGER—A special grade of ledger paper used for bank statements, loose-leaf ledger sheets for use in bookkeeping machines and similar purposes, where the writing is done entirely by typewriters and bookkeeping machines. Since it may be in and out of the machine many times, it requires good wearing quality. Its primary use requirement, in addition to the normal requirements of ledger paper, is stiffness in order that it may stand upright in a posting file and feed to an automatic machine without bending or slipping out of position.

STATIC ELECTRICITY—The electrical charge that, sometimes, collects on paper and other electrical insulating materials owing to contact with other substances. It is occasionally troublesome wherever dry paper is handled, e.g., in the last dryer section of the paper machine, in calender stacks, and in printing. It is most evident at low relative humidities when natural dissipation of the charge by leakage to grounded objects is slow. Electrostatic charging of paper not only causes trouble in handling the paper, because of the tendency of charged sheets to stick together, but it is a hazard since a shock from a moving web or roll may cause a worker to move involuntarily into dangerous contact with nearby machinery.

STATIC FRICTION—The starting friction between the surfaces of papers or paperboards when like surfaces are placed in contact. See KINETIC FRICTION.

STATIC MIXER—See MIXER.

STATIONARY TOP CARD—A carding machine having non-moving working surfaces of emery or metallic card clothing positioned on top of the main cylinder.

STATIONERY—An inclusive term which, as related to paper products, includes papeteries, typewriter paper, packaged papers, billheads, and other papers sold by stationers.

STATISTICAL PROCESS CONTROL—The application of statistical methods to control a process.

STATISTICAL QUALITY CONTROL—The application of statistical methods to control quality. (Note that the terms statistical process control and statistical quality control are often used interchangeably, although statistical quality control includes acceptance sampling as well asstatistical process control.)

STATISTICAL SORTING—A method whereby a calculated number of sheets are inspected for defects. This proportion can be shown to represent a statistical accuracy for the entire lot. See SORTING.

STAY TAPE—See BOX STAY TAPE; GUMMED STAY.

STEAK INTERLEAVING PAPER—A paper that butchers place between cuts of meat. It should not darken the meat and should reduce shrinkage by loss of meat juices.

STEAM—Saturated: steam at a temperature equal to the saturation temperature for its pressure. Superheated: steam at a temperature higher than the saturation temperature for its pressure.

STEAM BLISTER—See BLISTER (1).

STEAM DISTILLED PINE OIL—See PINE OIL.

STEAM-DISTILLED WOOD TURPEN-TINE—See TURPENTINE.

STEAMED MECHANICAL PULP—An unbleached groundwood pulp produced by steaming the wood before grinding. Obsolete practice.

STEAM-EXPLODED WOOD—See EXPLODED FIBERS.

STEAM DRUM—That section of the boiler where the feedwater volatilizes into steam. It contains a large water-steam interface to prevent carryover of water droplets with the steam.

STEAM FINISH—See STEAM SHOWER.

STEAMFIT—Duplex: a double passage that allows steam to pass from stationary piping to a rotating dryer, and condensate/steam mixture to pass from the dryer to the piping. Also called a rotary joint. Simplex: a single passage device used to allow either steam to pass from the stationary steam piping to the rotating dryer, or the condensate/steam mixture to pass from the dryer to the piping. Also called a rotary joint.

STEAMING TUBE—A pressure vessel where wood chips are heated to an elevated temperature by steam prior to entering the primary pressurized refiner. It could be horizontal, but vertical tubes are more common.

STEAM SEASONED—Hung in a steam room prior to the plating operation.

STEAM SHOWER—A device that emits a high percentage of saturated steam onto a traveling web in order to increase the moisture content in the area treated. Steam showers are used on supercalenders to improve gloss levels. They are also used on wet ends of paper machines toimprove the efficiency of water removal systems that use vacuum boxes.

STEEL ENGRAVERS PAPER—A paper for printing from a hand-engraved intaglio plate, usually a high-grade cotton fiber content or chemical pulp bond used for letterheads, formal calling cards, wedding invitations, etc.

STEEL ENGRAVING—See DIE STAMPING.

STEEL INTERLEAVING PAPER—A kraft paper used in steel mills to separate sheets of steel. Basis weight generally ranges from 18 to 30 pounds (24 x 36 inches – 500). It is essential that the paper be chemically neutral and free of shives or other materials which might mar the steel surfaces.

STEEL-PLATE PAPER—A paper used for steel-plate engravings. It may be a ledger, chart, or bristol type, tub-sized to give uniform surface which is the most significant property.

STEEP BLEACHING—Bleaching done usually with peroxides at high density, ambient temperature, low alkalinity, and long retention times. The bleach liquor is applied on the surface of a feltless type web machine or mixed according to normal high-density procedures and stored. Bleaching takes place over a period of days. Because no neutralization step is required, the alkalinity in the bleach liquor must be carefully controlled so that the final pulp is neutral or slightly acidic when the peroxide has been fully consumed.

STEM FIBERS—Fibers from the main stem or trunk of the plant, such as wood fibers, straw, bamboo, bagasse.

STENCIL—(1) Material perforated with lettering or a design, through which ink is forced onto a surface to be printed. (2) The assemblage of a stencil sheet, a cushion sheet, and a backing sheet, before the stencil sheet is perforated with typewriter or handheld stylus.

STENCIL BACKING SHEET—A good grade of hard paper (essentially a tympan paper), used as the backing for a stencil sheet. It is usually oiled and is about 0.005 of an inch in thickness.

STENCIL BOARD—A paperboard used in making stencils for marking shipping cases and the like. It is made on a cylinder or fourdrinier machine from long-fibered chemical woodpulps, but may contain rope stock and is so formed that it will take oil uniformly. The basis weight is normally 150 to 250 pounds (24 x 36 inches – 500); it varies in thickness from 0.012 to 0.015 of an inch (12 to 15 pounds). It is ordinarily impregnated with linseed or other oil. The sheet is strong, hard surfaced, and long wearing. It gives a clean cut on the stencil machine or stamp and must resist penetration of ink and fuzzing action when the inkbrush is passed over the surface. A lighter grade of stencil board is used as a stencil for decorative work. Significant properties include high tearing resistance, high finish, and stiffness.

STENCIL-CUSHION SHEET—An unimpregnated tissue paper of the same grade as stencil paper which is placed between the stencil sheet and the backing sheet during the cutting of the stencil to act as a cushion.

STENCIL DUPLICATORS—See MIMEOGRAPH.

STENCIL PAPER—See STENCIL TISSUE; STENCIL BOARD.

STENCIL TISSUE—The base of a stencil sheet to be used on mimeograph, and other types of duplicating machines. Important properties are high oil permeability combined with high tensile strength.

STENOGRAPHERS' NOTEBOOK PAPER—As the name implies, a writing paper designed for use in stenographers' notebooks. It is normally made from chemical woodpulps in a basis weight of 16 pounds (17 x 22 inches – 500), and is characterized by a smooth writing surface, water and ink resistance, opacity, cleanliness, and bright color.

STENOTYPE PAPER—A low grade of bond paper supplied in the form of small rolls or folded packs for use on stenotype machines.

STEP-AND-REPEAT—Process of making a multiple image plate by imaging a single image and then iteratively moving it a predetermined amount.

STEREO BACKING—Obsolete term. See MATRIX BOARD (1); STEREOTYPE BACKING.

STEREOTYPE BACKING—A gummed felt sheet made in a range of thicknesses and cut into strips of various widths, used to back out or pack the spaces in the molded stereotype dry mat supplied in the 0.024 to 0.036 inch thickness range. Its purpose is to prevent the spaces from being pushed back by the hot metal when a stereotype plate is being cast. Stereotype backing is not required with packless mats. Obsolete term. See also MATRIX BOARD (1).

STERILIZABLE PAPER—Papers made for: (a) heat sterilization, e.g., in an autoclave; (b) gas sterilization with ethylene oxide; or (c) radiation sterilization.

STERILIZATION—The process of making nonwovens sterile by utilizing dry heat, steam, ethylene dioxide gas, or radiation.

STEROLS—See CRUDE TALL OIL.

STEVENS FORMER—An older paper-forming device used to manufacture a wide variety of papers in weights ranging from tissues to heavy paperboard grades. It consists of a wire-covered suction roll, or a vacuum cylinder upon which the stock is formed, and a unique approach flow (q.v.) section which brings the stock to a restricted forming area under pressure. The forming area is located in the upper quadrant of the upturning side of the cylinder.

STICKIES—Suspended particulate contaminants, usually of low specific gravity, in the finished pulp. These contaminants originate with the adhesive and coating residues, films, tapes, rubber-like particles, ink, and hydrolysis products of synthetic sizing materials. If not removed during processing, the stickies can cause sheets of paper to stick together. See also RUBBER SPOTS.

STICK MARK—(1) See BACK MARK. (2) Marks in coated paper caused by the rods or poles used in festoon drying. The term stick mark is not in current usage.

STICK PAPER—Paper which is rolled into a tight tube, generally with a starch "glue." Used for lollipops and cotton swabs for ears.

STICTION—In process control, an unpredictable binding characteristic of a final control element that imposes resistance to the control signal's ability to achieve repeatable response.

STIFF BLADE COATING—See BEVELED BLADE COATING.

STIFFENER BRISTOL—A piece of bristol board placed in envelopes to protect the enclosure from creasing or crushing.

STIFFNESS—The ability to resist deformation under stress. Resistance to a force causing the specimen to bend is termed bending or flexural stiffness. See also RING STIFFNESS.

STIPPLING—A type of embossing of paper to reduce the high gloss of a sheet by running the sheet between rollers with counter-grained surfaces. See EMBOSSED.

STITCH-THROUGH—The process of mechanically bonding nonwovens using a form of knitting.

STOCHASTIC SCREENING—A digital screening process using very small dots of equal size and variable spacing.

STOCK—(1) Pulp that has been beaten and refined, treated with sizing, color, filler, etc., and which after dilution is ready to be formed into a sheet of paper. (2) Wet pulp of any type at any stage in the manufacturing process. (3) Paper on inventory or in storage. (4) Paper or other material to be printed, especially the paper for a particular piece of work. (5) A paper suitable for the indicated use, such as coating raw stock, milk-carton stock, tag stock, towel stock, etc.

STOCK ORDER—An order to be filled directly from warehouse inventory of a standard grade, size, weight, and color, as opposed to a special making order (q.v.).

STOCK PREPARATION SYSTEM—The section of the papermaking process where the fibrous components are blended with the non-fibrous components of the furnish, and physically prepared for papermaking, typically by mechanical, chemical, or thermal treatment. The stock preparation system, traditionally, is considered to start at the high density storage chest discharge, in an integrated mill, or at the furnish pulper, in a non-integrated mill, and extend to the paper machine headbox. Broke reprocessing, especially the under-the-paper-machine broke pulpers, the white water system, the fines recovery system, and process control for all these areas are also considered to be within the stock preparation system.

STOCK SIZES—Common sizes of papers and boards which are usually stocked by producers, distributors, or consumers. They are sizes which are standard and which are reordered from time to time.

STOCK WEIGHTS—Common weights of papers and boards which are usually stocked by producers, distributors, or consumers.

STOICHIOMETRY—The branch of chemistry and chemical engineering that deals with the quantities of substances that enter into and are produced by chemical reactions.

STONE CELLS—See SCLEREIDS.

STONE GROUNDWOOD (CONVENTIONAL)—A method of preparing mechanical pulp in which the volume of cooling water used is so regulated that the temperature of the pulp in the grinder pits is from 130°F to 190°F. The process is in common use in North America.

STORAGE TOWER—See HIGH DENSITY STORAGE.

STRAIN—The deformation per unit length resulting from the application of a force. In the case of the tensile testing of paper, it is customary to express the deformation at rupture as the percent stretch.

STRATIFIED HEADBOX—A headbox comprised of two or more headboxes on top of one another in a single structure. The flows from the multiple headboxes merge just ahead of the overall slice feeding the forming section. Each headbox has its own stock delivery system so that a single sheet can be formed with a different composition in the different plies of the sheet.

STRATIFIED SAMPLING—A sampling technique in which sample units are selected in the same proportion of an attribute that can affect the result as contained in the lot (i.e., a scheme that accounts for all relevant layers or groups in the lot).

STRAWBOARD—A paperboard made largely from straw for specialty purposes; formerly used extensively for common items such as setup boxes, egg case partitions, globe-type maps, and to some extent as a corrugating medium.

STRAW PAPER—(1) See STRAW BOARD. (2) See DRINKING STRAW PAPER.

STRAW PULP—A papermaking pulp made from the straw of such plants as wheat, oats, rye, barley, etc.

STREAMING CURRENT—The streaming current detector (SCD) measures the relative magnitude and sign of particle charge by measuring the current that is generated when counter ions are sheared off an adjoining surface by an os-

cillating piston. The charge values derived from this technique correlate with other electrokinetic measurements, such as microelectrophoresis, at the point of zero charge only. Streaming current is often used as an endpoint detection method in colloid titration determinations. See also ELECTROKINETIC CHARGE.

STREAMING POTENTIAL—A streaming potential arrises when a liquid is passed through a preformed pad which is compressed between two electrodes under precisely controlled conditions (temperature, pressure drop across the pad, conductivity, dielectric constant, pad structure). It is an electrokinetic effect. See also ELECTROKINETIC CHARGE.

STRESS—See LOAD.

STRESS-CORROSION CRACKING—An anodic cracking process that requires the simultaneous action of a corrosive species and sustained tensile stress.

STRETCH—The elongation corresponding to the point of rupture in a tensile strength measurement. It is usually expressed as a percentage of the original length.

STRETCHABLE PAPER—See EXTENSIBLE PAPER.

STRETCH ROLL—One or more movable rolls used to create and adjust tension in machine clothing. Press felt and dryer fabric stretch rolls generally have 180° wrap for maximum "take-up."

STRIKE-IN—The penetration of liquid into a sheet of paper. The term is commonly used in printing where it refers to absorption of ink vehicle by the paper. Inks used in printing newspapers "dry" by strike-in or absorption. In other cases strike-in may be objectionable, as the absorbed vehicle increases showthrough.

STRIKETHROUGH—A case of extreme strike-in (q.v.) in which the ink vehicle penetrates the sheet and is visible as a stain on the opposite side. Pigment may also be evident.

STRING INSERTED—See REINFORCED PAPER.

STRIP CHART PAPER—A form of recording instrument chart paper made in light weights [less than 16 pounds (17 x 22 inches – 500)] and in narrow rolls for continuous recording of data by process control instruments. It is generally made of rag and/or chemical woodpulps, and must have uniform thickness and weight, good dimensional stability, and hard sizing. Translucency may be required where sections of the charts are to be reproduced by blue-printing or direct printing.

STRIPE-COATED CARBON—Carbon papers in which areas are left uncoated or clean of carbon so that a clean edge is available for gluing sets or to leave out certain information from the carbon copies.

STRIP FILM PAPER—A strong, well-formed chemical woodpulp sheet, with a high density, which is coated with a sensitized cellulose acetate film.

STRIPPED CONDENSATE—Contaminated condensate that has passed through a stripper which separates the organic contaminants from the condensate by heating.

STRIPPER GAS—Organic and sulfur-containing gas generated during stripping of contaminated condensate.

STRIPPING—(1) The process of one carding surface removing fibers from another carding surface. (2) In offset printing, the process of positioning negatives to make a plate layout.

STRIPPING QUALITY—(1) The ability of a paper to be removed from meat while still in a more or less frozen condition without sticking to the meat. (2) See RELEASE PAPERS.

STRONG BLACK LIQUOR—See CONCENTRATED BLACK LIQUOR.

STRUCTURAL FIBER INSULATION BOARD—A board manufactured principally from wood, bagasse, or other vegetable fiber by a felting or molding process to which is added a suitable sizing material to render it water resistant and a small amount of resin to improve fiber bonding. The board possesses structural, thermal-insulating, and sound-deadening qualities.

STUB ROLL—An incomplete roll of paper of small diameter which (1) remains from the use of the greater portion of the roll, or (2) is made and used for testing purposes.

STUCK WEB—Adherence of layers of roll to each other, whether caused by water, adhesive, or other tacky material.

STUFF—See HALFSTUFF; STOCK.

STUFF BOX—A gravity fed, gravity discharged flow distribution box used to provide stock at constant head and flow rate to a fan pump (q.v.) or to distribute flow among multiples of equipment operating in parallel, such as refiners or screens.

SUBJECTIVE GLOSS—See GLOSS.

SUBSTANCE—See BASIS WEIGHT.

SUBSTANCE NUMBER—Of cultural papers, the weight in pounds of a ream (17 x 22 inches – 500). Federal government agencies may use 17 x 22 inches – 1000.

SUBSTANTIVE DYE—See DIRECT DYES.

SUBSTRATE—The base material onto which something is applied, such as ink, coating, or adhesive. In the paper and converting industries, the substrate is usually a flexible web. The paper (base stock) or other material onto which printed matter is impressed.

SUCTION-BLANKET MARK—An undesirable mark produced by the endless perforated blanket which runs over the suction box on a conversion coating machine.

SUCTION BLIND DRILLED ROLL—A suction roll containing a combination of blind-drilled and through-drilled holes to achieve

293

more symmetrical sheet dewatering in the press nip. Open area can be increased as high as 32% by this technique.

SUCTION BOX—A device used to remove water from the sheet being formed on the fourdrinier fabric or twin wire former or from the wet felt of a cylinder machine prior to pressing. It is a box with a perforated top surface over which the fabric of felt passes. Water is removed from the stock or web by induced suction within the box.

SUCTION-BOX MARKS—Streaks in paper produced by uneven suction.

SUCTION BREAST ROLL FORMER—This is a tissue former in which the headbox is shaped to curve around a large suction breast roll. Most drainage takes place over this roll and the single fabric conveys the sheet to the pickup felt which, in turn, conveys the sheet to the Yankee dryer (q.v.).

SUCTION COUCH ROLL—A suction roll supporting the forming fabric and paper web at the end of the table immediately prior to transfer of the web to a felt for passage into the press section. The suction box inside the shell applies a high vacuum for removal of water from the sheet. Holes in the suction couch roll are usually countersunk on the surface to provide a greater open area for better water removal. On older machines without a wire turning roll (q.v.) the suction couch roll drives the table (q.v.). On newer machines, the driving power is usually shared between the two rolls. The suction couch roll is usually covered with an elastomeric composition to increase its coefficient of friction.

SUCTION DECKLE EDGE—A deckle edge produced by sucking stock off the fourdrinier fabric.

SUCTION DRUM RIDER ROLL—In a cylinder machine, this roll runs on top of the suction drum to form a nip for water extraction prior to the primary press section. Suction drum rider rolls are usually covered with a 32 mm (1.25 inches) thickness of a 70 P&J to 125 P&J (q.v.) elastomeric composition.

SUCTION DRUM ROLL—In a cylinder machine, this is a suction roll located just prior to the primary presses. In some cases it may contain an elastomeric cover.

SUCTION DUSTING—A method of eliminating dust from the ends of a roll or the sides of a pile of paper by means of a vacuum cleaner.

SUCTION FEED—A suction gripper that feeds sheets or blanks of paper or board into a printing press or converting or processing equipment.

SUCTION GROOVED ROLL—A suction roll containing a combination of surface grooves and through-drilled holes to increase the open area of the roll.

SUCTION HINGE—A thin section in a sheet of paper, which serves as a hinge (see HINGED LEDGER), produced by removing a part of the stock from the fourdrinier fabric by means of suction nozzles.

SUCTION PICKUP FELT—See PRESS FELT.

SUCTION PICKUP ROLL—A suction roll located inside the pickup felt. It presses the felt against the forming fabric to transfer the web to the pickup felt. The web is then carried underneath the felt to the first press nip.

SUCTION PRESS ROLL—A roll consisting of a corrosion resistant cylinder covered with an elastomeric composition and drilled with holes passing through the cover and the metal. It is used as one of a pair of rolls, the second being a solid roll. Liquid pressed out of the paper sheet carried through the nip of these rolls (on an endless press fabric of a transverse press) passes through the drilled holes and is collected in a stationary suction (vacuum) box inside the roll. The drilled holes are typically 3.2 mm to 4.8 mm (0.125 inches to 0.187 inches) in diameter with an open area of 11% to 22%. The roll body is a centrifugally cast shell of bronze or stainless steel and must be capable of withstanding both the corrosive environment and the operating stress in highly loaded press sections. The drilling limits the operating loading on suction rolls. See NIP VENTING.

SUCTION ROLL—See SUCTION BLIND DRILLED ROLL; SUCTION BREAST ROLL FORMER; SUCTION COUCH ROLL; SUCTION DRUM RIDER ROLL; SUCTION DRUM ROLL; SUCTION GROOVED ROLL; SUCTION PICKUP ROLL; SUCTION PRESS ROLL.

SUCTION ROLL MARK—Mark in paper, sometimes in a pattern, produced by excessive suction in the suction couch (q.v.) under the forming fabric or suction press roll (q.v.). See SHADOW MARKS.

SUCTION ROLL PRESS—The first improvement made to the plain roll press (q.v.) where one of the two solid surfaced rolls is replaced by a suction press roll (q.v.), which aids in water removal by venting the back side of the press fabric or felt. See NIP VENTING.

SUEDE PAPER—See VELOUR PAPER.

SUGAR-BAG PAPER—Paper made from bleached, semibleached, or unbleached sulfate pulp for manufacture into bags for sugar. See also SHIPPING SACK KRAFT PAPER.

SUGAR-WRAP PAPER—A bleached chemical woodpulp sheet manufactured for wrapping lump sugar. The paper should lie flat and be suitable for die cutting.

SUIT BOARD—A paperboard made for folding boxes such as are used for the packaging of wearing apparel. It is commonly made of mist board in various colors, is a good bender, and has a surface suitable for printing. See MIST BOARD.

SUITCASE BOARD—See FIBERBOARD.

SULFAMIC ACID—NH_2HSO_3. An inorganic acid used in flame retardant formulations and as an inhibitor in pulp bleaching in the bleaching of chemical pulp.

SULFATE BOARD—See KRAFT BOARD.

SULFATE LINER—See KRAFT LINERBOARD.

SULFATE PAPER—See KRAFT PAPER.

SULFATE PINE OIL—See PINE OIL.

SULFATE PROCESS—See KRAFT PULPING.

SULFATE PULP—See KRAFT PULPING.

SULFATE WOOD TURPENTINE—See TURPENTINE.

SULFIDITY—Ratio of the sodium sulfide concentration to the active alkali (q.v.) concentration in white liquor (q.v.). Occasionally, the ratio of the sodium sulfide to the total titratable alkali (q.v.) in green liquor (q.v.).

SULFITE—See SODIUM SULFITE.

SULFITE BAG PAPER—Bleached or unbleached sulfite papers used in the manufacture of paper bags. The paper may have an MG finish, may be colored or printed, and may be especially sized or treated depending on the use.

SULFITE BOND—A chemical woodpulp bond paper (q.v.). Originally a bond paper made entirely from sulfite pulp. As kraft bleaching technology advanced, the fully bleached sulfate pulps, both softwood and hardwood, which become readily available were used in many grades of paper including bond paper. Today the term sulfite bond is usually interpreted to apply to any bleached chemical woodpulp bond paper which meets the particular physical specifications, whether manufactured from bleached sulfite, sulfate, soda, or any other chemical woodpulp in any combination.

SULFITE MANILA TAG—See TAG STOCK.

SULFITE PULP—See ACID SULFITE PULP.

SULFONATION—A process for modifying water repellent (hydrophobic) lignin and rendering it water absorbent (hydrophilic), by attaching a sulfur molecule to lignin. This allows swelling of the fiber, making it more flexible for papermaking.

SULFUR—Sulfur is a yellow, naturally occurring element, also called brimstone. In the preparation of acid sulfite cooking liquors, it is burned to produce sulfur dioxide, the free component dissolved in the soluble pulping liquor base.

SULFUR DIOXIDE—SO_2. A colorless gas produced by burning molten sulfur. (1) It dissolves in water to produce sulfurous (H_2SO_3) acid, and when reacted with soluble bases, produces sulfite cooking liquors. (2) It is used to acidify pulps (a) to remove metal ions, (b) to reduce residual bleaching agent after a bleaching stage (chlorine, chlorine dioxide or peroxide) to prevent brightness reversion, (c) to create a pH favorable to brightness stability (pulps yellow on exposure to light and air in alkaline conditions), (d) to prevent further bleaching by dyestuff in paper mills when the pH is lowered by alum. See ACID TREATMENT.

SULFURIC ACID—H_2SO_4. Also called oil of vitriol, sulfuric acid is a commercially important mineral acid with widespread industrial uses. In the paper industry, it is used to parchmentize paper, to prepare chlorine dioxide bleach from sodium chlorite or chlorate to "sour" bronze paper machine wires, to dissolve certain wet strength resins, to assist in repulping wet strength broke, etc.

SULFUR IMPREGNATED BOARD—A board which has been impregnated with molten elemental sulfur to improve stiffness, performance in high humidity environments and impart acid resistance, high dielectric strength, and low heat conductivity.

SULFUR TRIOXIDE—SO_3. A gaseous compound formed when sulfur is burned in the presence of catalysts. It reacts with water to produce sulfuric acid and is also used in various chemical processes.

SUMMERWOOD—See LATEWOOD.

SUPERCALENDER—A specialized, off-line, calender stack (q.v.) used to increase the gloss, smoothness, and density of papers. The supercalendering process also improves the cross-deckle caliper profile and the ink holdout in subsequent printing operations. The machine is normally constructed of vertically oriented, chilled-iron calender rolls separated by alternatively arranged filled rolls except when the arrangement includes a reversing nip. Synthetic, soft-nip rolls are replacing the customary filled rolls of compressed paper in some supercalenders to improve the durability of the soft-nip rolls. Supercalenders can be designed to handle the full width web of a paper machine or a fractional width and can exert nip pressures up to 3000 pounds per lineal inch.

SUPERCALENDERED COVER PAPER—Uncoated cover paper that has a high or glazed surface, obtained by passage through a supercalender stack.

SUPERCALENDERED FINISH—A finish obtained by passing paper between the rolls of a supercalender under pressure. Supercalenders are usually composed of alternate chilled cast-iron and fibrous or polymeric rolls. The resulting finish will vary, depending upon the raw material used in the paper and the pressure exerted upon it, from that of the highest English finish to a highly glazed surface. Papers supercalendered to a very high gloss are sometimes referred to as "plate finished." Supercalendered finish is also known as supercalendering.

SUPERCALENDERED PAPER—See SC PAPERS.

SUPERCALENDERING—See SUPERCALENDER FINISH.

SUPER FINISH—Incorrect term for supercalender finish (supercalendering). Super finish can be confused with "superfinishing," a polishing process for steel or chilled-iron rolls.

SUPER NEWS—Newsprint having a smooth finish, making it suitable for gravure printing. Also called rotonews.

SUPER PATENT-COATED BOARD—A paperboard used for highgrade folding boxes. It is made in a similar manner to a patent-coated board but includes a larger proportion of bleached stock in the top liner to produce a whiter appearance. See PATENT COATED.

SUPERFINISHING—A polishing system which utilizes a vibrating stone of about 600 grit to produce a mirror-like finish on metallic products such as slitter blades, knives, or chrome-plated rolls.

SUPERGLAZED FINISH—An unusually highly glazed finish.

SUPERHEATED STEAM—Steam at a temperature above the saturation temperature for the given pressure. See also SATURATION STEAM.

SUPERHEATER—The section of the boiler tube banks used to raise the temperature of the steam generated in the boiler to a level above its saturation temperature.

SUPERPRESSURE GROUNDWOOD (SPGW)—A mechanical pulp produced by pressing debarked logs against a rotating pulpstone at temperatures above the boiling point of water (110°C) and high pressures over 4 Bar. The resultant pulp will have more long fibers than in groundwood produced at atmospheric or pressurized conditions, but will be 2 to 3 points lower in brightness and 1 to 2 points lower in light scattering coefficient.

SUPERSTANDARD KRAFT WRAPPING PAPER—A term more or less obsolete referring to a kraft paper containing color and sometimes brightened through partial bleaching.

SUPERSTANDARD NEWS—A grade of newsprint paper conforming to the United States Government Customers definition of standard newsprint paper (q.v.) but made with an MF rotogravure finish. Its primary use is in magazine supplements of newspapers. It is also used in some magazines.

SUPERVISORY CONTROL—A controller whose output provides setpoints to another related process controller(s).

SUPERVISORY MEASUREMENT—The measurement of process variables. The output is used to provide supervisory set points to control loops not under direct control by a computer. This form of measurement is routinely made for steam generation control systems and for paper machine control systems. See SUPERVISORY PAPER MACHINE MEASUREMENT.

SUPERVISORY PAPER MACHINE MEASUREMENT—This type of measurement refers to measurement sensors that measure machine direction (MD) and cross machine direction (CD) characteristics. Most measurements are made with non-contact (preferable) sensors usually mounted on a scanning frame to obtain specific localized information and resulting machine profiles. (Some sensors are still of the contact type.) The signals are used as input to a dedicated machine computer that provides operator information and can also provide remote setpoint (q.v.) adjustment to single station controllers or cascaded control loops. Typical variables measured for this purpose are basis weight, moisture, caliper, gloss, formation, color, temperature, ash content, smoothness (critical for printing papers), and other specialized parameters. As with all process control variables, once a measurement is successfully made, control algorithms can be developed, along with final control elements, to perform corrective action.

SUPPLEMENTARY OBSERVER—A hypothetical observer based on color mixture data obtained for 100° field of view for 76 real observers—adopted by CIE in 1964.

SUPPLEMENT PAPER—Paper used in newspaper supplements, comic, or photogravure sections.

SURFACE-ACTIVE AGENT—A substance which tends to reduce surface tension. The common term is surfactant (q.v.).

SURFACE BONDING STRENGTH—See BONDING STRENGTH; PICKING.

SURFACE COATED—A term applied to any paper or paperboard which has one or both sides coated with a pigment or other suitable material. See COATING.

SURFACE COLORING—The application of a color to paper after the paper web has been formed on the paper machine. The dyeing may be a part of the papermaking operation of the paper machine or it may comprise a separate operation. See CALENDER DYED.

SURFACE CONTOUR—The topography of the surface. See GLOSS; SMOOTHNESS.

SURFACE DYED—See SURFACE COLORING.

SURFACE HEAT RELEASE RATE—Ratio of the total heating value input to a recovery boiler furnace (q.v.) to the projected surface area of the furnace walls. Often used as design guideline and a measure of the boiler loading.

SURFACE LIFTING—See PICKING.

SURFACE PEELING—See PEELING (2).

SURFACE POTENTIAL—See ELECTROKINETIC CHARGE; MOBILITY.

SURFACE SIZE—A term used to describe surface treatment chemicals for application by size press or calender sizing. The most common surface size is starch solution, which forms a film on the surface of the sheet to resist liquid penetration and also acts as an adhesive to increase strength. Other surface sizing materials include synthetic surface sizing chemicals which provide resistance to liquid penetration by increasing contact angle to decrease wettability. See also PIGMENTED SURFACE SIZE.

SURFACE SIZED—An adjective describing paper or paperboard which has been treated by application of a sizing material to the surface of the dry or partially dried sheet; surface sizing is usually done on the paper machine but may also be done as a separate off machine operation. See also CALENDER SIZING; FILM TRANSFER SIZE PRESS; GATE ROLL SIZE PRESS; METERING SIZE PRESS; POND SIZE PRESS; SIZE PRESS; SURFACE SIZE; TUB-SIZED; TUB-SIZE PRESS.

SURFACE-SIZE PRESS—A unit of the paper machine, designed for relatively light applications of surface-sizing agents or other materials to paper or paperboard, usually located between two dryer sections, comprising a vertically or horizontally oriented set of press rolls and equipment for spraying or otherwise applying the sizing material to one or both sides of the sheet and removing the excess therefrom. Obsolete as a commercial practice. See also TUB-SIZE PRESS.

SURFACE STAINED—See SURFACE COLORING.

SURFACE STRENGTH—Resistance of the surface layer of a sheet or coating to the breakaway of surface fragments, when the sheet is separated from the inked plate or blanket in the printing process.

SURFACE TENSION—A force existing at various solid, liquid, and gas interfaces which tends to bring the contained volume into a form having the least superficial area.

SURFACE WIND—A driven reel drum or driven winder drum provides the torque to the surface of a winding roll of paper, and this action is called surface winding.

SURFACTANT—A material which when used in small amounts modifies the surface properties of liquids or solids. Detergents, wetting agents, emulsifying agents, dispersing agents, and foam inhibitors are all surfactants.

SURGE CHEST—A chest that receives a single flow stream at varying flow rates, continuously or irregularly, and discharges stock continuously, at a more steady rate. Surge chests are used to level out variations between supply and demand, and to level out residual process variations from upstream. A surge chest can be thought of as a chest which blends stock from a single source, over time. Also called a leveling chest.

SURGICAL DRESSING PAPER—A plasticized glassine sheet used to wrap surgical dressings and pressure-sensitive adhesive bandages. It must be able to withstand sterilization heat treatment without discoloration or embrittlement.

SUSPENDED SOLIDS—Fibers or colloidal particles suspended by beating (mechanical working) or refining.

SWEAT DRYER—A paper machine dryer cylinder that operates at a temperature sufficiently low to cause condensation of moisture or sweating on its surface. It serves to cool the sheet and slow the drying process. The sweat dryer is usually located at the end of a dryer section, and may be chrome coated or otherwise protected against corrosion.

SWEETENER STOCK—Stock with a high percentage of long fiber, or generally fast draining, which is used as a filtering aid in a saveall (q.v.).

SWIMMING ROLL—See VARIABLE CROWN ROLL.

SYNTHETIC FIBER—Filaments extruded or spun from man-made resins.

SYNTHETIC PAPER—A general term for noncellulosic sheet material resembling paper and used in a similar fashion. Most synthetic papers are made from thermoplastic materials such as polyolefins, nylon, and polystyrene, by direct film or foil extrusion methods or by bonding filaments thereof.

SYNTHETIC PINE OIL—See PINE OIL.

SYNTHETIC RESIN—A complex, substantially amorphous organic semi-solid or solid material prepared by chemical reaction or polymerization of comparatively simple compounds. The synthetic resins approximate natural resins in various physical properties. Types include

formaldehyde condensation products of phenol, urea, and melamine; reaction products of polyhydric alcohols and polybasic acids (alkyd and polyester resins); polymerization products of acrylic acid and its derivatives (acrylic resins) or styrene (polystyrene); polymers of butadiene and its derivatives or copolymers with other materials (synthetic elastomers), etc. These resins find use in the paper industry as adhesives in coating and laminating, as barrier materials, and as agents to impart special properties such as improved wet and dry strength.

SYPHON—Rotating: device for scavenging condensate off the inner surface of a steam-heated dryer cylinder. A rotating syphon is in physical contact with the dryer inner diameter and rotates with the dryer. Stationary: device for scavenging condensate off the inner surface of a steam-heated dryer cylinder. A stationary syphon remains in a fixed position; it does not rotate with the dryer.

SYSTEMS ENGINEERING—Constructing a complex of apparatus for a specific function.

T

T4S—A common abbreviation for "trimmed 4 sides."

TABLE—The flat top part of the fourdrinier section of the paper machine where the forming fabric carrying the sheet travels between the breast roll (q.v.) and the couch roll (q.v.). The forming fabric is supported by a series of hydrofoils (q.v.), table rolls (q.v.), and suction boxes (q.v.), which remove water from the sheet.

TABLE ROLLS—These small diameter, stiff rolls support the forming fabric on the fourdrinier table. They are covered with a hard composition and may be grooved to improve sheet formation. On wide, faster machines, table rolls have been reduced in number or replaced entirely by foils for better sheet formation.

TABLET-BACK BOARD—A paperboard used as stiffener for pads of paper. It is generally made of chipboard or newsboard of 0.020 of an inch and upward in thickness and its chief requirement is stiffness.

TABLET BLOTTING—A lightweight blotting paper used as the first sheet in a tablet or inserted in a box of papeterie. The basis weight ranges from 35 to 80 pounds (19 x 24 inches – 500).

TABLET BRISTOL—A lightweight bristol, either white or colored, which is used for the cover of school tablets, etc.

TABLET NEWS—See PENCIL-TABLET PAPER.

TABLE-TOP PAPER—Any paper, usually kraft, which is suitable for coating, printing, or lacquering and used in the manufacture of permanent table covers.

TABLET PAPER—A general term descriptive of any of the grades of paper used in the manufacture of tablets, but chiefly applied to book and writing grades. It has a fairly good writing surface. Uniformity of caliper is essential to facilitate the conversion process. Resistance to abrasion in erasing is also necessary. For ink, the basis weight is between 16 and 20 pounds (17 x 22 inches – 500); for pencil, it is usually 32 pounds (24 x 36 inches – 500).

TABLET WRITING PAPER—Hard-sized tablet paper for pen and ink writing.

TABULATING BOARD—A tag board used in automatic tabulating machines. There are two principal types—mechanical and electrical. The board is made from bleached, partly bleached, or unbleached chemical woodpulps, has a caliper of 0.0067 to 0.007 inch, and an even machine finish free from twist or curl. The two types of tabulating board have the same specifications except that the electrical-type tabulating board is free of electrically conducting material. Physical tests are usually specified for other properties, such as mullen, tear, fold, stiff-ness, sizing, dirt, dimensional stability and, in the case of the electrical type, electrical conductivity. See also COMPUTING MACHINE PAPER.

TABULATING CARD STOCK—See TABULATING BOARD.

TACK—Stickiness. For example, adhesives, some printing inks, varnishes, and freshly painted or coated surfaces. In printing, some high tack inks can pull particles from the sheet.

TACKLE—The replaceable assembly of bars, knives, or the like which provide the working surfaces in a refiner or beater to fibrillate and cut fibrous raw material preparatory to paper-making. The action of a refiner unit may be changed drastically by installing an appropriately different tackle design.

TAD—Through-air-dryer (q.v.).

TAD FABRICS—See THROUGH-AIR-DRYER FABRICS.

TAG AND FOLDER STOCK—See FOLDER STOCK; TAG STOCK.

TAG BLANKS—See TAG STOCK.

TAG BOARD—A paperboard used for printed forms, envelopes, shipping tags, file folders, and many other purposes. It is manufactured of rope, jute, sulfite, sulfate or mechanical woodpulp or various combinations of these, principally on a cylinder machine but sometimes on a fourdrinier machine. It is usually of a manila color and has a smooth finish. The basis weights range from 80 to 300 pounds (22 1/2 x 28 1/2 inches – 500), the most common being 80, 100, 125, 150, 175, 200, and 270 pounds. Tag board possesses good bending or folding qualities, high bursting and tensile strength, high tearing resistance, and a high water finish, suitable for writing or printing.

TAG STOCK—A cylinder or fourdrinier sheet ranging in weight from 100 to 270 pounds (24 x 36 inches – 500) suitable for the manufacture of tags. It may be made from a wide variety of furnishes, such as rope fiber, sulfite, sulfate, or mechanical woodpulp, and various types of wastepapers, such as manila clippings, bottle-cap waste cuttings, and reclaimed shipping sack kraft. The board is sometimes tinted or colored on one or both sides. It may be vatlined and is often coated. The more durable tag stocks, such as those used in foundries, machine shops, laundries, and nurseries, are made of rope and jute. Tag stock, depending upon its intended use, has the following properties to a greater or lesser extent: good bending or folding qualities, suitable bursting and tensile strength, good tearing and water resistance, and a surface adaptable to printing, stamping, or writing with ink.

TAG-WASHER MANILA PAPER—A tough manila paper used to border the eyelets of shipping tags in order to prevent the tags from being torn.

TAIL—The term applied to a machine direction strip of paper slit from the full-width web which is used as the conduit to route the entire web. This tail, or strip of paper, can be eventually widened to conduct the full web width through a web processing process.

TAILING SCREEN—The final stage screen in a multistage screening system. Tailing screens may be pressurized, or non-pressurized, and are fed debris-rich rejects from the previous screening stage. See also SCREENING SYSTEM.

TAIL PASSING—The process of routing the "tail" (q.v.) of a web through the machinery so the web follows the proper line of travel. This process is also known as threading.

TAKE-OFF ROLL—A small stainless steel or fluted roll which carries the pulp sheet from a washer drum over the lip of the vat and causes it to fall into the repulper. Also called doctor roll.

TALC—A white or greenish-white mineral substance which is primarily hydrous magnesium silicate. It is used as a filler in some grades of paper.

TALL OIL—See SOAP SKIMMINGS (TALL OIL).

TALL OIL, CRUDE—See CRUDE TALL OIL.

TALL OIL ROSIN—See ROSIN.

TAMALE WRAPPER—A greaseproof or vegetable parchment paper used in cooking tamales.

TANDEM FINISHER—A type of soft-nip calender composed of two similar, in-line, single-nip, calenders arranged to treat both sides of the sheet with both the hard roll and the soft-nip. This processing system is usually used to produce matte finished, coated papers and can be done in-line on the paper machine or off-line. See MATTE FINISH; SOFT-NIP CALENDER.

TANDEM PRESS—See PRESS SECTION.

TANDEM REFINER—Same basic design as the "twin refiner," but loading of the refiner is accomplished by positioning the non-rotating discs on both sides at different pressures. The refiner is equipped with large thrust bearings to take the load. Since the thrust load does not need to be balanced, the refiner can be operated in two stages (in tandem) in a single unit.

TANNING PAPER—An abrasive paper (q.v.) which is coated with silicon carbide and is used in the leather industry to condition tanned hides and skins for the final buffing operation.

TAPE—A narrow strip of paper or cloth, sometimes reinforced, coated on one side with an adhesive, used (1) to seal the manufacturer's joint (q.v.) or flaps or to reinforce a fiberboard container, or (2) to reinforce the sewn closure of a multiwall shipping sack. See GUMMED TAPES.

TAPE PAPER—See GUMMED SEALING TAPE; TAPE.

TAPER TENSION—The lowering of the web tension in a process, such as a winder, as a function of increasing wound roll diameter.

TAPPI—The Technical Association of the Pulp and Paper Industry (TAPPI) is an international professional association serving manufacturers of pulp, paper, paperboard, packaging and converted products, and producers of chemicals, equipment, components, and other materials used to manufacture these products. The membership comprises individual members and corporate memberships. TAPPI's primary goals are: (a) to advance technology in the paper and related industries, (b) to serve as a worldwide forum for the collection, dissemination, and interchange of technical concepts and information in the fields of interest to association members, (c) to improve the performance of these members, and (d) to provide high-quality, timely, and innovative products and services relating to the above purposes.

TAPPI OPACITY—Opacity (q.v.) measured by the TAPPI method.

TAR BOARD—See K-B BOARD.

TARGET PATCHES—See TARGET PAPER.

TARGET PAPER—A paper used for printing targets. It is similar in character to manila drawing paper and made in a natural oyster white or a light-cream color. It has a short fiber (usually mechanical pulp), which is punctured cleanly by the passage of the bullet. The basis weight is about 60 to 75 pounds (24 x 36 inches – 500). Stiffness is important.

TARGET POSTERS—See TARGET PAPER.

TARIFF PAPER—An MF book paper used for tariff rates for railroads and transportation companies. It is usually made of chemical and mechanical woodpulps in a basis weight of 40 to 45 pounds (25 x 35 inches – 500). Opacity is fairly important, and the paper should have a good printing surface and be able to stand considerable handling.

TARNISHPROOF BOARD—See ANTITARNISH BOARD.

TARNISHPROOF PAPER—See ANTITARNISH PAPER.

TAR PAPER—Any paper impregnated or coated with coal tar or asphalt.

TARRED FELT—An organic felt (q.v.) saturated with coal tar. The usual weight is of 14 pounds per 100 square feet. It is used in the same manner as asphalt felt (q.v.) except that, in the built-up roofs wherein it is used, coal tar pitch is used instead of asphalt.

TARRED SHEATHING—Paper saturated with coal tar. See TARRED SLATERS FELT.

TARRED SLATERS FELT—A tarred sheathing felt used as a liner under roofs and on walls. A roll of 500 square feet weighs about 30 pounds.

TARRED THREAD FELT—A tarred sheathing felt with threads running lengthwise of the sheets to increase resistance to tearing. Important properties are resistance to water, wind, and decay.

TCDD—2, 3, 7, 8–tetrachlorodibenzo-p-dioxin. See DIOXIN.

TCDF—2, 3, 7, 8–tetracholordibenzofinan. See DIOXIN.

TCF—See TOTAL CHLORINE COMPOUND FREE.

TDS—See TOTAL DISSOLVED SOLIDS.

TEA—Tensile energy absorption (q.v.).

TEA BAG PAPER—A lightweight absorbent paper usually made from bleached manila hemp and long-fibered chemical woodpulp and specially treated to give a high wet strength. The basis weight ranges from 8 to 12 pounds (24 x 36 inches – 480). Its chief characteristics are cleanliness, good absorbency, high wet strength, a sheet structure to permit rapid diffusion of the tea extract, and ability to perform satisfactorily in high-speed packaging equipment with which the tea bags are fabricated and filled.

TEA CARTRIDGE—See TEA PAPER.

TEA PAPER—A class of tough, nonporous papers, specially specified by the tea trade. They are also termed tea cartridge.

TEAR—See TEARING RESISTANCE.

TEAR FACTOR—The tearing resistance (q.v.) in grams (per sheet) multiplied by 100 and divided by the basis weight in grams per square meter.

TEAR INDEX—The numerical value found by dividing the value of the tearing resistance in milliNewtons (mN) by the basis weight in grams per square meter.

TEARING RESISTANCE—The force required to tear a specimen under standardized conditions. Two methods of measurement are in common use: (1) Internal tearing resistance, wherein the edge of the specimen is cut before the actual test; and (2) Edge tearing resistance. (3) Tearing resistance in milliNewtons (mN) is obtained by multiplying the force to tear a single sheet in grams by 9.807.

TEAR-OUTS—Portions torn from the paper web for inspection or test.

TEAR RATIO—The relationship between machine direction and cross-machine direction tearing resistance.

TECHNICAL ASSOCIATION OF THE PULP AND PAPER INDUSTRY—See TAPPI.

TELAUTOGRAPH PAPER—A grade of paper used in the manufacture of small rolls for telautograph recording machines. The base paper is usually an English finish book grade having a basis weight of 60 pounds (25 x 38 inches – 500). It is moderately sized to provide rapid recording ink absorption. Telautograph recorder rolls are generally 5 inches wide and 3, 4, or 5 inches in diameter.

TELEGRAPH BLANKS—See TELEGRAPH PAPER.

TELEGRAPH MANILA—See TELEGRAPH PAPER.

TELEGRAPH PAPER—Paper made for printing and converting into pads for telegraph messages. It is sized for pen and ink writing and offset printing, and must meet strict requirements of satisfactory mechanical handling in the central telegraph offices. Papers used range from 100% chemical woodpulp sheets, sometimes in the form of carbonless copy paper, to sheets containing approximately equal proportions of chemical and mechanical woodpulp. The latter, usually yellow in color, is predominately used.

TELEGRAPH TAPE PAPER—Paper somewhat similar in appearance to telegraph paper but satisfactory for gumming. After gumming it is cut into narrow-width rolls about 3/8 of an inch wide. The tape is threaded through the necessary device, which prints on the paper. It is then separated into individual messages and pasted on the telegraph blanks. Uniformity of caliper, good printing surface, and low tearing resistance (to enable the operator to tear off strips for pasting to the blank) are essential qualities.

TELEGRAPH WRITING—See TELEGRAPH PAPER.

TELEPHONE-DIRECTORY PAPER—See DIRECTORY PAPER.

TELEPHONE MEMO PAPER—Tablet or writing paper used in tablet or pad form at or near the telephone to make memoranda of conversations, etc.

TELEPRINTER ROLLS—See TELETYPE PAPER.

TELESCOPED ROLL—A defect found in an unwinding roll where the roll is axially displaced relative to the weight supporting core as a result of soft winding at or near the core. See also SLIPPED ROLL.

TELETYPE PAPER—Paper, usually in the form of small rolls or fanfolded packs, used on teletype (teleprinter) communications equipment. Such papers are usually made from mechanical and/or chemical woodpulps and have the usual characteristics of printing papers. The

303

usual basis weights vary from 10 to 20 pounds (17 x 22 inches – 500). In converted form, teletype papers are generally prepared in 8 7/16-inch wide rolls, 4 or 5 inches in diameter on 1-inch I.D. cores, or in folded packs of varying lengths. Such rolls or packs may be single ply or multicopy, i.e., carbon interleaved or carbonless.

TELETYPE PERFORATOR TAPE—See PERFORATOR TAPE.

TELETYPE TAPE—See TELEGRAPH TAPE PAPER; TELETYPE PAPER.

TELLER ROLLS—Small rolls of plain bond, tablet or other paper used on teller machines in banks, savings and loan companies, etc. These rolls are sometimes made in 2-ply carbonized or carbonless format and their use on teller machines is often tied into central computers.

TEMPLET BOARD—A stencil board which is treated with wax and resins or other materials. Dimensional stability is important.

TEMPLATE PAPER—See TEMPLET BOARD.

TENSILE ENERGY ABSORPTION—The energy absorbed when a paper specimen is stressed to rupture under tension. It is expressed in energy units per unit area, e.g., kg-cm/cm2. It is useful in evaluating packaging materials subject to rough handling.

TENSILE INDEX—The numerical value found by dividing the tensile strength by the basis weight in grams per square meter.

TENSILE STIFFNESS OF PAPER—Tensile stiffness is defined as the slope of the primary elastic region part of the load/elongation curve of paper.

TENSILE STRENGTH—(1) The maximum tensile stress developed in a specimen before rupture under prescribed conditions. It is usually expressed as force per unit width of the specimen. (2) The numerical value equal to 0.6538 times the tensile breaking load in kilograms of a 15-mm wide tensile strip. The units being kiloNewtons per meter, kN/m.

TENSION WOOD—A wood with abnormal structure, occurring as a rule on the upper, or compression, side of branches or leaning tree trunks of hardwood species. Tension wood has more cellulose, less lignin and hemicellulose than normal wood. An additional gelatinous layer, or G-layer, or almost pure cellulose may often be found at the fiber lumen. See also REACTION WOOD; COMPRESSION WOOD.

TERTIARY CLEANER—The third stage in a multiple stage centrifugal cleaning system, if present. The tertiary cleaner stage is fed debris-rich refects from the secondary cleaner stage, to be further concentrated. If the debris is sufficiently concentrated in the tertiary stage cleaner rejects, it may be disposed of. The tertiary cleaner stage accepted stock is sent to the secondary cleaner stage feed, normally. Often, multiple tertiary cleaners are operated in parallel, for capacity reasons.

TERTIARY SCREEN—The third stage screen in a multistage screening system. The tertiary stage screen is fed with rejects of the secondary stage screen, sometimes mixed with accepts from the tailing stage screen, if present. The tertiary stage screen rejects are sent to a tailing screen, if present; not all screening systems have three or more stages. The tertiary stage accepts are generally sent to the feed of the secondary stage screen. See also SCREENING SYSTEM.

TERTIARY TREATMENT—Wastewater treatment beyond the secondary, or biological stage that includes removal of nutrients such as phosphorus and nitrogen, and a high percentage of suspended solids. Tertiary treatment, also known as advanced waste treatment, produces a high-quality effluent.

TERTIARY WALL—The thin inner layer of the fiber cell wall surrounding the central cavity or lumen and sometimes referred to as the S3 layer. See S-LAYERS.

TEST BOARD—A container board (q.v.) which, when combined with a fluted corrugating medium, will allow that corrugated structure to meet the requirements of the applicable freight regulations.

TEST JUTE BOARD—In the United States, the term has been replaced by recycled fiber linerboard (q.v.).

TEST JUTE LINER—In the United States, the term has been replaced by recycled fiber linerboard (q.v.).

TEST LINER—A term used mainly in Europe for linerboard manufactured from recycled or secondary fiber meeting the test requirements of the applicable freight regulations.

TEST LINERBOARD—In the United States, the term has been replaced by recycled fiber linerboard (q.v.).

TEST PAPERS—Papers, usually unsized, that are impregnated with any of a variety of chemical reagents, and used for detecting the presence of certain substances in solutions or in gases and vapors by the appearance of color changes.

TEST SHEET—See HANDSHEET.

TEXT FINISH—A finish intermediate between antique and machine finish. It is closely akin to vellum finish.

TEXT PAPER—A paper of fine quality and texture for printing. Text papers are manufactured in white and colors, from bleached chemical woodpulp or cotton fiber content furnishes with a deckled or plain edge, and are sometimes watermarked. They are made in a wide variety of finishes, including antique, vellum, smooth, felt-marked, and patterned surfaces, some with laid formation. Common basis weights are 60, 70, and 80 pounds or heavier (25 x 38 inches – 500 sheets). Many of these papers are manufactured in matching cover weights. Designed for advertising printing, the principal use of text papers is for booklets, brochures, fine books, announcements, annual reports, menus, folders, and the like.

TEXTILE PAPER—(1) A general term for strong wrapping papers of various weights, colors, and furnishes, used for wrapping bulk textiles such as bolts of cloth. (2) A paper made of chemical or mechanical woodpulp or a mixture of these and used in the textile industry. The basis weights range from 50 to 70 pounds (24 x 36 inches – 500). It may be water or steam finished or machine finished on both sides; usually one side is rough. It is duplex in color, usually being colored on one side only.

TEXTILE WRAPPERS—A general term for papers of various weights, colors, and finishes used for wrapping bolts of cloth.

TEXTURE—Identifying characteristics of a sheet that pertains to its feel and appearance. See FINISH; FORMATION; GRAIN; LOOK-THROUGH; SMOOTHNESS; WILD.

THEME PAPER—A school paper designed for themes, reports, etc. It is normally a bond or writing grade made from chemical woodpulps in basis weights of 16 to 20 pounds (17 x 22 inches – 500).

THERMAL CONDUCTIVITY—The rate of heat flow under steady conditions, through unit area, per unit temperature gradient in the direction perpendicular to the area. This property is important for structural insulating board.

THERMALLY BONDED—A method of bonding using elevated temperature to fuse bond the nonwoven web.

THERMAL PAPER—Generally, any paper with a heat sensitive coating. Typical of thermal papers are those used in hot stylus recording instruments, etc.

THERMAL POLLUTION—Degradation of water quality by the introduction of a heated effluent. Primarily a result of the discharge of cooling waters from industrial processes, particularly from electrical power generation. Even small deviations from normal water temperatures can affect aquatic life. Thermal pollution usually can be controlled by cooling towers.

THERMAL PRINTING—Printing that uses a transfer sheet containing the inks and a heated print head, which causes the inks to transfer to the paper.

THERMAL SPRAY COATING—Metal or alloy coatings applied in the form of molten droplets propelled to the metal substrate by a process involving the generation of intense heat in an arc or plasma.

THERMOCOUPLE—A sensor that generates a voltage at the juncture of two dissimilar metals which varies nearly linearly with temperature.

THERMOGRAPHY—Offset or letterpress printing using a special ink which, while still wet, is dusted with a resinous powder. The sheets are baked, causing fusion of the powder and ink, and yielding a raised image, which simulates an engraved image. This is primarily used for letterhead, business cards, and announcements.

THERMOMECHANICAL PULP (TMP)—A mechanical pulp produced from wood chips, where the wood particles are softened by pre-heating in a pressurized vessel at temperatures not exceeding the glass transition point of the lignin, before a pressurized primary refining stage.

THERMOMECHANICAL TURPENTINE—See TURPENTINE.

THERMOPLASTIC—Polymeric materials that can be softened by heat and are therefore moldable and extrudable into sheet, pipe tanks, etc.

THERMOSETTING RESIN—Thermosetting resins are mixed with catalysts to cure into rigid structures. Because these resins are rather brittle, they are usually reinforced with glass or other fibers. Polyesters epoxies and phenolics are the most common resins used to manufacture process equipment.

THICKNESS—The space between opposite surfaces; the smallest of the three dimensions of a single sheet of paper or paperboard under specific conditions of measurement. Also called caliper.

THICK STOCK CIRCUIT—The portion of the stock preparation system from the high-density storage chest discharge, in an integrated mill, or the furnish pulper in a non-integrated mill, to the fan pump suction or primary cleaner feed pump suction. Typical operations in the thick stock circuit include consistency regulation, stock blending, refining, addition of wet end chemicals and non-fibrous furnish components and, sometimes, screening. The stock is typically 3–6% consistency in the thick stock circuit.

THICK STOCK PUMP—A large gear pump used to transport or pressurize pulp at consistencies over about 5%. In a typical bleach plant, one thick stock pump is needed to feed stock to each upflow tower (except the chlorination tower). It is commonly used to transport pulp to high-density storage tanks. This pump is being replaced by the medium-consistency (MC) pump.

THICK STOCK SCREEN—A screen placed in the thick stock circuit, prior to dilution at the fan pump to headbox consistency. A thick stock screening system may contain primary, secondary, tertiary, etc., screens. See SCREENING SYSTEM.

THINNINGS—Trees removed from a stand in its early years of growth to improve the growth rate of remaining trees. Trees removed in thinning are usually the smallest and lower quality trees. As a result, thinnings are often better suited for pulp and paper than for solid wood products. Thinnings contain a higher percentage of juvenile wood than at final harvest, and they have different pulping properties from those of mature wood.

THIN PAPER—Any lightweight paper. The term usually is applied to such papers as Bible, carbonizing, cigarette, condenser, manifold, and like papers, but not to facial or toilet tissue.

THIN STOCK CIRCUIT—See APPROACH FLOW SYSTEM.

THIRD PRESS—See PRESS SECTION.

THIXOTROPY—The phenomenon observed when the viscosity of a material decreases with time at a constant rate of shear. This reduction in viscosity is due to a temporary breaking down of an internal structure of the system under shear. The viscosity of thixotropic materials depends upon the shear history or amount of previous work to which the material has been subjected. The property is important in coating colors as it enables the working of the color formulation to a viscosity that permits the color to be applied to the sheet and allows surface leveling due to afterflow of the color on the sheet.

THRASHING—As once practiced, this operation involved passing rags as taken from a bale through a revolving cylindrical drum. It opened up the rags and removed loose dirt and dust that was present. This operation was also called dusting.

THREADED FELT—See TARRED THREAD FELT.

THREADING—See TAIL PASSING.

THREAD SPEED—A mode of winder operation where the speed is maintained low for the purpose of facilitating the threading of the tail through the winder web run.

THREE-PIECE LAMBERT—See SLIDE BOX.

THROUGH-AIR-DRYER (TAD)—A large diameter (3–6 meter) dryer cylinder with a perforated shell. A low level vacuum inside the dryer is used to pull air, typically from an air cap, through the web. This method is used on grades where high bulk is desirable and the paper has a high permeability. See also THROUGH-AIR-DRYER SECTION.

THROUGH-AIR-DRYER FABRICS—An open mesh dryer fabric (q.v.) which holds the sheet in contact with a through-air-dryer. In some configurations the TAD fabrics also transfer the sheet to the Yankee dryer (q.v.). These fabrics can also be called transfer fabrics (q.v.)

THROUGH-AIR-DRYER SECTION—A part of the Yankee machine located after the forming section and before the Yankee dryer. Hot air is passed through the sheet in order to dry it to 60–70% solids without compacting the sheet.

THROUGH-FLOW CLEANER—A reverse type centrifugal cleaner that removes contaminants of low specific gravity (less than 0.98) by centrifugal force. With a through-flow cleaner, both the rejects (contaminants) and the cleaned pulp are removed at the bottom of the cone; whereas, with a reverse cleaner, the rejects exit at the top. Because through-flow cleaners are functionally competitive with reverse cleaners (q.v.), some experts classify through-flowcleaners as a variety of reverse cleaner. The term "through-flow cleaner" was first applied in the paper industry about 1984 to distinguish this type of cleaner from the then well-established reverse cleaner.

THRU-AIR-DRYER FABRICS—See THROUGH-AIR-DRYER FABRICS.

THRUST LOAD—The force created in a refiner by the torque applied on the pulp, (pushing the disc surfaces together) and by the steam generated between them. This load is carried by the thrust bearing on the rotating element in a refiner.

TICKER PAPER—A communication paper made primarily from mechanical pulp in basis weight of 37 pounds (24 x 36 inches – 500) and converted into small, wide rolls for use on stock quotations, tickers, and the like.

TICKER TAPE—See TICKER PAPER.

TICKET BOARD—Various grades of paper and paperboard used in the manufacture of tickets. Much of this board now has safety or anti-falsification features.

TICKET BRISTOL—Any bristol used for ticket purposes. It is usually a bogus or mill bristol, depending on the quality of ticket contemplated.

TICKLER REFINERS—Refiners located after the machine chest and used for a minor additional mechanical treatment to the stock. Tickler refiners are usually located between the machine chest and the stuff box, and often are controlled by couch roll vacuum.

TILE HANGING—See TILE STOCK.

TILING—See TILE STOCK.

TILE LINER—A strong kraft paper used as a backing for tile in mosaic work.

TILE MOUNTING PAPER—Paper used to keep the pattern of small tile intact during the process of laying. It is commonly made from unbleached chemical woodpulps in a natural color and basis weight of approximately 50 pounds (24 x 36 inches – 500). It is usually calender sized and has a rather open formation. The paper washes clean from the laid tile. Significant properties include low sizing and porosity, and high strength and absorbency.

TILE STOCK—A paper made from bleached chemical woodpulp which is used for the manufacture of high-grade embossed wall-papers and paper moldings. It is strong to withstand the processing operations.

TIME-CARD BRISTOL—See TAG BOARD; TAG STOCK.

TIME-TABLE PAPER—Paper in widely varying grades used in arrival and departure schedules for various means of transportation. Quality varies according to subject matter used in the folder. It is usually made from chemical woodpulp, with a machine finish and of fair opacity and folding strength. This paper should be hard enough to give it some rigidity and to fold smoothly with and against the grain. The common basis weights are 50 to 60 pounds (25 x 38 inches – 500).

TINTED—A term applied to lightly dyed paper having relatively high visual brightness and little color depth, as distinguished from a more heavily dyed sheet having considerable color depth. See also TINTED WHITE.

TINTED WHITE—A term applied to "white" sheets containing small amounts of reddish blue, or blue and red dyes or pigments. The result is an almost imperceptible neutral gray which is more pleasing to the human eye than the yellowish cast of the corresponding untinted sheet.

TINTING—(1) In lithography, the production of an all-over light tint on the unprinted areas of paper, etc. It can be caused by ink emulsifying in the dampening water, or by soluble materials (usually proteins) in paper coating that sensitize the nonimage areas of the plate to ink. (2) In papermaking, the operation of adding dyes or pigments to the sheet furnish or to a coating composition to produce tinted white (q.v.) or pastel shades.

TIPPING FEE—The charge for depositing municipal solid waste at disposal sites or facilities.

TIRE WRAPPER—See AUTOMOBILE-TIRE WRAP.

TISSUE FORMING—See CRESCENT FORMER; SUCTION BREAST ROLL FORMER; YANKEE TISSUE MACHINE.

TISSUE PAPER—A general term indicating a class of papers of characteristic gauzy texture, in some cases fairly transparent, made in weights lighter than 18 pounds (24 x 36 inches – 500). The class includes sanitary tissues, wrapping tissue, waxing tissue stock, twisting tissue stock, fruit and vegetable wrapping tissue stock, pattern tissue stock, sales-book tissue stock, and creped wadding. Tissue papers are made on any type of paper machine, from any type of pulp including reclaimed paper stock. They may be glazed, unglazed, or creped, and used for a wide variety of purposes.

TISSUE WINDER—See COMBINING CALENDER.

TITANIUM DIOXIDE—The white oxide of titanium, TiO_2. There are two crystalline forms useful to the paper industry: the anatase form employed primarily as a filler pigment, and the rutile form used primarily in pigmented coating. Both types are particularly useful because of their white color, high brightness, and high refractive index (2.52–2.76), which makes them highly effective for improving both brightness and opacity. Commercial grades are usually specially treated to facilitate use in the many papermaking and coating applications and to provide that particle size necessary for optimum optical behavior.

TMP—Thermomechanical pulp (q.v.).

TNT—Abbreviation for tension, torque and nip. The TNTs are the three primary winder controls that determine the wound-in-tension of a roll.

TOBACCO PAPERS—Papers used for packing small quantities of tobacco, generally a low grade of enameled or highly glazed paper, fairly bulky but pliable and soft.

TOC—See TOTAL ORGANIC CARBON.

TOCl—See TOTAL ORGANIC CHLORINE.

TOILET PAPER—A sanitary tissue paper manufactured from a variety of furnishes primarily in bleached and semibleached grades. The most popular types are dry creped with some semicreped [basis weight 10 to 14 pounds (24 x 36 inches – 500)] and facial tissue type in two or more plies of dry creped [basis weight per ply about 10 pounds (24 x 36 inches – 500)]. Besides white, pastel shades and printed designs are in demand. Some qualities are developed through embossing during the converting process. The principal characteristics are softness, absorbency, cleanliness, and adequate strength (considering easy disposability). It is marketed in rolls of varying sizes, or in interleaved packages.

TOILET TISSUE—See TOILET PAPER.

TONER INK—A dry ink powder typically composed of small plastic beads containing carbon black and magnetite. It is applied to a substrate initially by electrostatic attraction, then fixed by heat, which fuses the plastic particles to each other and to the substrate. Toner inks are used for imaging in photocopy machines and in "laser" printers.

TONS/DAY/INCH—A number calculated by dividing daily tonage by paper machine width (in inches). Used to normalize production so that the performance of one machine can be compared to another machine.

TOOTH—A characteristic of the grain in the surface of various papers, especially drawing papers, handmade papers, and other papers of low finish. The term is used to describe their ability to take pencil or crayon marks. The roughness or surface contour of the paper is one factor in its tooth, and probably the fuzz and the stiffness of fibers projecting from the surface is another. Also referred to as bite.

TOP—(1) The correct term for the so-called felt side of machine made paper. (2) In paperboards composed of different stocks, the better quality side is usually referred to as the "top," and the rest of the board, composed of another stock, as the "back."

TOP BOARD—See PRESS FELT.

TOP ENTRY AGITATOR—An agitator, usually a propeller agitator, which enters the mixing vessel from the top.

TOP FABRIC—See TWIN WIRE FORMER FABRICS.

TOP FELT—See PRESS FELT.

TOP LINER—The liner on the top side of a multilayer board. It is usually of a cleaner and brighter grade than the filler or back liner (q.v.).

TOP PRESS ROLL—A roll which forms a pressure nip with a bottom roll, usually a suction roll, as part of a press section. The surface of a top press roll is exposed directly to the paper web and must possess fiber release properties so as not to disturb the sheet and cause sheet breaks. Special elastomeric and plastic compositions and ceramic coatings have been developed for this application.

TOP SIZING—Surface or tub sizing of paper which has already been internally sized. Obsolete as a commercial practice.

TOP WIRE FORMER—A class of formers characterized by having a second fabric which contacts the sheet during the forming process. Top wire formers are characterized by having the first part of the drainage occurring just as it would on a flat fourdrinier. This is called "open wire" forming because the sheet is not sandwiched between two forming fabrics. After the top fabric contacts the sheet, drainage can be induced through the top fabric by the action of roll, blades, pressure and vacuum.

TORN DECKLE—An imitation deckle edge made by tearing on a watermarked line either before or after the paper has been dried.

TORN SHEETS—Sheets of paper in a ream which are not full size because of tear-outs or broken ends fed to the rotary cutter and not removed during the sorting operation.

TORSION TEARING RESISTANCE—A test for corrugated board in which the test piece, clamped in jaws, is subjected to a twisting action. See also TEARING RESISTANCE.

TOTAL CHLORINE COMPOUND FREE—A term used to describe a pulp bleached without any chlorine compound, hence the name totally chlorine free.

TOTAL DISSOLVED SOLIDS (TDS)—The total solids which remain in a water sample after passage through a 0.5-micron glass fiber filter pad. It reflects the total solids which are dissolved in the water.

TOTAL ORGANIC CARBON (TOC)—An instrumental method for determining the organic content of a wastewater. It involves incinerating the sample in a catalyzed combustion chamber and detecting the amount of carbon dioxide that is produced.

TOTAL ORGANIC CHLORINE (TOCl)—An older test for the total amount of halogen present in a liquid sample. It consists of absorbing the halogen-containing compounds onto an exchange resin and measuring the amount absorbed.

TOTAL REDUCED SULFUR (TRS)—An air pollution parameter that monitors the total amount of reduced sulfur compounds present in a gas stream. The compounds in the gas stream may include a combination of hydrogen sulfide (H_2S), methyl mercaptan (CH_3HS), dimethyl sulfide (CH_3SCH_3), and dimethyl disulfide (CH_3SSCH_3). Often found in the flue gases of combustion devices during incomplete combustion of sulfur-containing fuels. Also generated during acidification of materials containing sodium sulfide.

TOTAL SUSPENDED SOLIDS—Solids in a water sample that are retained on a 0.5-micron filter pad.

TOTAL TITRATABLE ALKALI—A measure of the total sodium chemical in green and white liquor. Often expressed as equivalent concentration of sodium oxide, Na_2O.

TOTAL TRANSMITTANCE—A particular transmittance measured in a standardized instrument specifically designed for the purpose. It is of importance in the determination of the transparency of transparent papers. See TRANSMITTANCE.

TOUGH CHECK—A very strong paperboard made on a cylinder machine from extra-strong, usually unbleached, chemical woodpulps which may be blended with rope stock. It may be coated on one or both sides and comes in a variety of colors. The usual thicknesses are 3-, 4-, 6-, and 8-ply, corresponding to 12, 18, 24, and 30 points, respectively. It is used principally for tickets, shipping tags, and wherever toughness is essential.

TOUCH PAPER—A paper saturated with nitrates or other salts so as to control burning properties.

TOWELETTES—Generally refers to small, pre-moistened tissues or towels contained in a sealed pouch.

TOWELING—A creped, absorbent paper made from either bleached or unbleached chemical woodpulp, with or without the addition of mechanical pulp. It should have sufficient strength to withstand use without disintegration or tear. Fast absorbency and water holding capacity are prime requisites. Other important characteristics are softness and freedom from lint and unpleasant odors. Basis weights range from 15 to 35 pounds (24 x 36 inches – 500).

TOWEL PAPER—See TOWELING.

TOXICITY—The quality of being poisonous, especially the degree of virulence of a toxic microbe or of a poison.

TRACHEIDS—The long, tubular cells in wood that function to conduct moisture and support the stem. These cells have tapered, closed ends and are 1 to 5 mm long. When liberated by pulping, they are commonly called fibers.

TRACING PAPER—A paper used for tracing original drawings, figures, patterns, curves, etc. It is a translucent greaseproof sheet, or a bond or manifold sheet chemically treated or oiled to increase transparency. It is made from cotton fiber and/or chemical woodpulps in basis weights from 7 to 16 pounds (17 x 22 inches – 500). Significant properties are proper receptivity to drawing ink and transparency so that prints can be made successfully from the tracings.

TRACING TISSUE—See TRACING PAPER.

TRACTION–WEB—See WEB TRACTION.

TRADING STAMP PAPER—A lightweight printing paper designed for the production of trading stamps. It is usually made from bleached chemical woodpulps in basis weights of 16 or 18 pounds (17 x 22 inches – 500). The most important technical characteristics are caliper, finish, formation, uniformity, and adequate but not excessive tearing resistance.

Important functional qualities include good gumming and perforating characteristics, freedom from curl, ability to take flexographic or offset printing, and freedom from common web defects such as slime spots, holes, and foreign particles. Some trading stamp papers are actually variations of safety paper with surface designs, hidden warning indicia, and/or other properties designed to prevent fraudulent alteration, imitation, etc. Trading stamp papers are usually converted into books or coils of finished trading stamps for distribution to customers of supermarkets and other commercial establishments.

TRAILING BLADE COATING—See BLADE COATING.

TRANSDUCER—A device used to take a natural process variable measurement in its raw state and provide a normalized signal output (in the desired format, i.e., milliamps, volts, psig) which may then be used as a signal input to a control system.

TRANSFER BOX—A specialized type of blow-box (q.v.) used in press-to-dryer sheet transfers to prevent sheet blowing.

TRANSFER FABRICS—See TRANSFER FELTS OR TRANSFER FABRICS.

TRANSFER FELTS OR TRANSFER FABRICS—Press felts that simply convey the sheet without performing any other work like pressing or marking. These are called transfer felts. In through-air-drying, usually one fabric holds the sheet to the dryer, and a second fabric transfers the sheet to the Yankee dryer (q.v.). These are called transfer fabrics and generally have a "knuckle" and weave pattern which impart a pattern to the sheet.

TRANSFER FUNCTION—The mathematical relationship between the output and input of a process or control element.

TRANSFER MARBLE PAPER—See MARBLE PAPER.

TRANSFER PAPER—(1) A grade of mechanical pulp paper similar in its general characteristics to groundwood poster paper, with the exception that the range of colors is greater. This paper is used for streetcar or bus transfers. Uniformity of caliper is important to the converter so that he may produce pads of transfers of uniform thickness. The predominant weight is 35 pounds (24 x 36 inches – 500). (2) Broadly, this term covers: (a) paper coated with a pressure transferable mass, such as carbon paper; (b) paper coated with moisture transferable film, such as the slip film of decalcomania; (c) paper coated with a heat and pressure transferable film or mass which is furnished for hot die stamping of paper, cloth, leather, plastics, etc. The last group (c) constitutes the product generally known as transfer paper. It falls in two general classifications: (1) metal transfer paper which usually consists of a transferable film of bronze powder and (2) pigment transfer paper which, as the name implies, consists of a transferable pigment film. The paper backing material is normally glassine of about 25 pounds (24 x 36 inches – 500). The backing material may be other than paper; regenerated cellulose is sometimes employed.

TRANSFORMER—An electromagnetic device that transfers electrical energy at one voltage level to another voltage level, either higher (step-up transformer) or lower (step-down transformer). Primary and secondary coils are wound around a common core, and the ratio of the number of primary to secondary coil turns establishes the voltage that is developed.

TRANSFORMER BOARD—A dense paperboard used as layer insulation in power and distribution transformers and in many types of electrical apparatus. It is manufactured on a wet machine from sulfate pulp and cotton rags or a combination of these stocks. This material has a calender finish. The usual thicknesses range from 0.031 to 0.250 inch as made directly on the wet machine. Other thicknesses may be made by laminating. Particularly important are uniformity of thickness and density, freedom from conducting particles, and excellent ply adhesion which influences the forming and molding characteristics. The moisture content must be maintained at a low value compatible with good formation in order that shrinkage and warpage characteristics may be held at a minimum. General chemical purity and neutral reaction characteristics of all electrical papers are important. The European equivalent of this material is Presspahn.

TRANSFORMER COIL WINDING PAPER—See CABLE PAPER; LAYER INSULATION PAPER.

TRANSLUCENCY—That property of a material which permits it to transmit light with strong scattering of the light so that transparency is not obtained. This property must be carefully distinguished from transparency: a sheet of bond paper is translucent, whereas a sheet of high-grade glassine paper is fairly transparent.

TRANSLUCENT BRISTOL—See TRANSLUCENTS.

TRANSLUCENT MASTER PAPER—An unfilled bond-type paper made from fairly well-hydrated bleached chemical wood or cotton pulps or both for use as a master paper in the diazo (white-print) reproduction process. Lightweight grades are usually 11, 13, 15, or 16 pounds (17 x 22 inches – 500). Translucent bond grades are usually 20 pounds (17 x 22 inches – 500).

TRANSLUCENTS—A soft cardboard, pasted or nonpasted, made of chemical pulps with both sides coated with a pigment. It is used for booklet covers, window hanger, cards, etc. Standard sizes and weights are 320 pounds (26 x 40 inches – 1000), 350 pounds (23 x 35 inches – 1000), and 420 pounds (28 x 45 inches – 1000). Standard thicknesses are 2-, 3-, 4-, and 5-ply, corresponding to 8, 10, 12, and 15 points, respectively. These boards are usually white or India in color.

TRANSMITTANCE—The fraction of incident light which passes through a specimen. This quantity has definite meaning only when the nature of the incident light and the design of the measuring instrument are specified. The variation of this property with wavelength of the incident light is of great importance for several applications of paper, e.g., some wrapping and reproduction papers. See TOTAL TRANSMITTANCE.

TRANSPARENCY—The property of a material that transmits light rays so that objects can be distinctly seen through the specimen. The transparency ratio is a measure of transparency when a space separates the specimen and the object viewed through it. It is useful for glassine and other papers intended to be seen through, rather than to conceal.

TRANSPARENCY RATIO—A measure of transparency. It is the ratio of parallel transmittance to total transmittance, usually expressed in percent.

TRANSPARENT—Transmitting light rays so that objects can be distinctly seen through the material; said of certain thin papers.

TRANSPARENT CELLULOSE—See CELLULOSE FILMS. The term is also applied to glassine paper.

TRANSPARENT MANIFOLD PAPER—(1) A very thin semitransparent manifold paper for use with duplex carbon, so that the carbon impression on the back can be read through the paper. (2) A lightweight greaseproof paper with high transparency.

TRANSPORTATION LAG—See DEAD TIME.

TRANSVERSE POROSITY—See LATERAL POROSITY.

TRANSVERSE PRESS—Wet presses designed to provide the shortest path for water escaping from the roll nip. The main water flow is perpendicular to the felt, and lateral flow is minimized. The suction roll press (q.v.) was the first design of this type, followed by the fabric press (q.v.), the grooved roll press (q.v.) and the blind drilled roll press (q.v.).

TRAPPING—(1) The ability to print one image over another. Wet trapping is printing over an ink before it is dry. (2) The extent to which different print images overlap.

TRAY BOARD—See DISHBOARD.

TREATED BROKE—Broke that has been treated such that it can be reintroduced into the blend chest. Process steps which may be used to treat broke include, but are not limited to, deflaking, refining, color stripping, chemical treatments, cleaning, screening, washing, etc. See also BLENDED STOCK.

TREATED BUTCHERS PAPER—A strong semibleached or bleached wrapping paper [approximately 40 to 50 pounds (24 x 36 inches – 500)] used as a retail market paper. During manufacture it is usually treated in a separate operation with sizing materials. Desirable properties are wet and dry strength, bloodproofness, grease resistance, and freedom from sticking to the product.

TREATMENT—The remediation of a waste stream or emission.

TREE FARM—A area of private land managed primarily for forest products.

TREE-WRAP PAPER—A crinkled duplex kraft paper, a tar paper, or an asphalt-coated paper which is used to protect trees from sunscald, frost, borers, and from rabbits and other rodents which feed upon young, tender bark. It may remain on the tree for two years or more before it disintegrates.

TRICKLING FILTER—A fixed film wastewater treatment technique which involves dribbling the wastewater over a film of organisms attached to a medium, such as crushed rock (older process) or special high-surface-area plastic materials.

TRIM—(1) The widest sheet of paper, trimmed to remove deckle edges, that can be made on a given machine. (2) To cut true to exact size, by cutting away the edges of paper in the web or sheet. (3) The paper which is trimmed off the edges of a continuous web of paper or from the edges of sheeted paper. Trim can be produced at many locations, including the forming section (squirt trim), the winder or the rewinder, or during various coating or converting steps. See SQUIRT TRIM; WINDER TRIM.

TRIM CHUTE—A device, adjacent to the outside edge slitters on a winder, used to convey the trim away from the adjacent winding roll. The trim can be accumulated at the winder but is normally conveyed by high-velocity air in a tube to a recycling process.

TRIMETAL PLATE—See MULTIMETAL PLATE.

TRIMMED—(Of a sheet.) Cut on two or more sides to ensure exactness of angle at the corner.

TRIMMED SPLICE—A pasted joint in a paper web, the loose ends of which have been trimmed.

TRIMMER—See GUILLOTINE TRIMMER.

TRIMMING—(1) The process of cutting the edge of a pile of paper in a guillotine trimmer (q.v.). (2) The process of slitting off the excess width on both ends of a large reel of paper so the outside shipping roll ends are straight and clean on the paper machine winder. This trim is normally recycled in the papermaking process. A rewinder can also cut down or reduce the width of a shipping roll by trimming one or both outside edges.

TRIM PULPER—A pulper that handles only trim (q.v.), often from multiple locations. Few mills have pulpers exclusively dedicated to trim, but many mills use their UTM pulpers (q.v.) to pulp trim continuously when broke is not being generated at that location. When operating on trim only, it is common practice to operate the pulper rotor at a reduced speed, or to operate only one out of two or three rotors, to reduce the energy consumption.

TRIM REMOVAL—The process of conveying trim from the trim chute to a subsequent recycling process.

TRI NIP PRESS—See TRIPLE NIP PRESS.

TRIPLE NIP—See PRESS SECTION.

TRIPLE NIP WITH GROOVED ROLLS—See PRESS SECTION.

TRIPLE NIP PRESS—A "no draw" press configuration for lightweight sheets. A pickup felt carries the sheet from the former (q.v.) and becomes the top felt of the first press, which has a bottom felt dedicated to that function. That is, the bottom felt goes around the bottom first press roll, which is usually a suction roll run under high vacuum. There is a top first press roll with low vacuum to hold the sheet to the pickup felt. A hard solid roll rests on the top suction roll at about the one o'clock position. The pickup felt carries the sheet around the top suction roll and into the nip between this roll and the solid roll, called the "center roll." This nip is the second press, and the pickup felt serves as the "top" felt in this nip. After the second press, the sheet transfers to the smoother center roll. A third press roll sits on top of the center roll at about the two o'clock position. It has a separate top felt. The sheet can then be directed into a fourth press, a smoothing press (q.v.) or the dryer section (q.v.). The sheet is supported continuously until it reaches the open draw after the third press nip.

TRIPLE PAPER—Three separate sheets used together.

TRIPLE WALL CORRUGATED BOARD—A corrugated board made by combining three single face (q.v.) webs with an additional facing. The four facings and three corrugated members thus make a single board which possesses greater strength than single or double wall boards made from the same materials. See also DOUBLE WALL CORRUGATED BOARD.

TRIPLEX—Having three layers bound together by three couchings or by pasting.

TRISTIMULUS VALUES—See CIE COLOR SYSTEM.

TRS—Total reduced sulfur (q.v.).

TRUCK AND CASE FIBER—A grade of vulcanized fiber (q.v.) (sometimes known as truck fiberboard) having good bending and forming qualities and a smooth, clean surface. Its principal uses are in the manufacture of trunks, cases, suitcases, receptacles, material handling boxes, athletic goods, welders' helmets, etc. It is made in thicknesses from 0.020 to 0.125 inch in a variety of colors and color combinations.

TRUNK WRAPPER—A general term to designate wrapping paper used to wrap traveling bags and trunks during shipment and storage. It is generally a heavy sulfite, kraft, or rope paper.

TSS—See TOTAL SUSPENDED SOLIDS.

TUB COLORING—A method of dyeing paper usually in a converting operation in which the paper is passed by submergence through a suitable dye bath or into a flooded nip, the excess dye being removed by squeeze rolls and the paper then dried by suitable means such as steam-heated driers. This produces a deep brilliantly dyed sheet which in contrast to stock dyed paper is usually of poor fastness to crocking and bleeding. This operation is also called dipping and is a form of surface coloring (q.v.).

TUBE—(1) A sheet of corrugated board, scored and folded to a multi-sided form with open ends usually used as a form of inner packing to add strength to a container (shell). See LINER. (2) A length of paper formed into the shape of the finished shipping sack (q.v.) and cut to length but without the ends being formed and closed.

TUBE BOARD—Paperboard made of various raw materials in thicknesses from 0.009 of an inch up, generally unsized and fairly smoothly finished. It is suitable for slitting into narrow rolls for winding and pasting into spiral or convolute mailing tubes, cores, etc. See CAN BOARD.

TUBED—Formed into a tube.

TUBE PAPER—See TUBE BOARD.

TUB LINERS—See BUTTER-BOX LINER; LARD PAPER.

TUB-SIZED—A term applied to paper or paperboard which has been surface-treated and/or impregnated with natural or synthetic sizing materials in a tub-size press. The term is often incorrectly applied to surface-sized papers where a tub-size press, as such, is not used. See also SURFACE-SIZED; SURFACE SIZE PRESS; TUB-SIZE PRESS.

TUB-SIZE PRESS—Generally a unit of a paper machine, designed for relatively heavy applications of sizing agents to paper or paperboard, usually located between two dryer sections, comprising a tub or vat for holding the liquid sizing material, and a set of vertically oriented press rolls the bottom unit of which is usually partially submerged in the sizing material. In the customary operation of a tub-size press, the moving web enters the tub under a dip roll and is totally submerged in the liquid sizing bottom press roll into the nip of the press in reverse fashion. As it leaves the nip, it follows the contour of the top press roll, and then continues its forward travel into a second dryer section. The tub or vat is generally constructed of wood or metal, while the size press unit is usually, but not always, made up of a pair of rolls of differing hardness and composition. Tub-size press units also include such auxiliary elements as pumps, piping, doctor blades, liquid level devices, thermostats, viscosity controllers, spreading rolls, and the like. In addition to tub-size presses on paper machines, such units may also be used with converting or processing machines such as air-dryers, impregnators, etc. See also SURFACE SIZE PRESS.

TUNGSTEN CARBIDE—A thin flame deposition of a very hard material on a drum or roller. It can be used to enhance wear resistance, or in varying the surface roughness, to increase web/roller or roll/roller traction.

TUNNEL DRYER—A well insulated sheet-metal tunnel or large box through which paper or paperboard is passed for the purpose of drying. For boards, the dryers may be built in 8 or 10

315

tiers, with a wet saw (a saw with a knife edge used to slit or cut the paper while it is wet) and an automatic feeder. The former cuts the board into 12- or 16-foot lengths (3658 to 4877 mm) and the latter feeds one sheet into each tier of the dryer in turn. Thus, the board in the tunnel dryer advances at a small fraction of the speed of the wet end (q.v.). The dryer is usually in three sections: in the first, the board is brought to a high temperature by circulating hot air; in the second the temperature is maintained; in the third section the temperature is reduced to avoid burning the board as it approaches dryness. Obsolete as a commercial practice.

TURNED EDGE—An edge on a sheet of paper which has been doubled back, creased, and held in that position by subsequent layers.

TURN OVER—An edge of the web that folds over at an occurrence of an edge crack and is wound into the roll in the folded condition.

TURPENTINE—A volatile oil which consists primarily of a number of terpene hydrocarbons of the general formula ($C_{10}H_{16}$). Five kinds of turpentine are now recognized:

1. Gum turpentine or gum spirits: obtained by distilling the crude exuded gum or oleoresin collected from living pine trees.

2. Steam-distilled wood turpentine: obtained from the oleoresin exuded gum within pinewood stumps or cuttings, either by direct steam distillation or solvent extraction of mechanically disintegrated wood.

3. Sulfate wood turpentine: recovered during the conversion of wood to paper pulp by the kraft pulping process.

4. Destructively distilled wood turpentine: obtained by the fractionation of certain oils recovered by condensing the vapors formed during the destructive distillation of pinewood.

5. Thermomechanical turpentine: recovered during the conversion of wood to paper pulp by the thermomechanical pulping process.

TURPENTINE RECOVERY—A process whereby turpentine is recovered from digester relief gases by condensation and decantation.

TURPENTINE TEST—The procedure used to measure the time required for turpentine penetration, as an indication of the grease resistance of paper.

TWILL WEAVE—One of many weaves used in machine clothing (q.v.). The twill weave has distinctive patterns that are at an angle to the machine direction.

TWIN NIP CONFIGURATION—See PRESS SECTION.

TWIN REFINER—The rotating element of this refiner consists of a disc, with refiner plates installed on both sides. It operates between two non-rotating discs. Loading of the refiner is accomplished by positioning the non-rotating discs on both sides. To ensure a balanced thrust load, both sides are loaded from a common hydraulic system, at equal pressure. The refiner has limited thrust capability.

TWIN WIRE FORMERS—A class of formers (q.v.) having the common feature that both sides of the sheet come into contact with the forming fabrics. There are two general types: gap formers (roll and blade) and hybrid formers (roll and blade) or preformers. In gap formers, the headbox jet is injected directly between two converging fabrics. In hybrid formers, the jet is landed first in a fourdrinier (q.v.) configuration, and only subsequently does the partly formed sheet come into contact with a second fabric. There are six major dewatering configurations:

(1) Gap roll former: With this former, the stock (headbox jet) is injected between two fabrics which immediately wrap a solid or suction forming roll. This design was favored for newsprint production and tended to give a micro formation with a grainy appearance. When this configuration is used in the production of tissue, it is known as a C-wrap, with transfer boxes to help the sheet stay

on the inner fabric (the fabric closest to the roll). A variation of this configuration is an S-wrap in which an upper turning roll and transfer boxes help the sheet stay on the outer fabric.

(2) Gap blade former: With this former, the stock is injected between two fabrics which then run over alternating drainage shoes or alternating drainage boxes.

(3) Gap roll blade former: With this former, the fabrics immediately wrap a solid or suction roll (q.v.) at a low angle and then continue over bladed forming shoes and boxes.

(4) Hybrid roll former: This type of former starts like a fourdrinier until the sheet comes into contact with a top fabric. The fabric-sheet-fabric "sandwich" then wraps a series of suction forming rolls and solid rolls. A low wrap version is used for slow-speed twin wire forming.

(5) Hybrid blade former: This type of former starts like a fourdrinier, but then the sheet comes into contact with a top fabric and the fabric-sheet-fabric "sandwich" wraps or contacts bladed boxes in which blades can be on one or both sides, alternately or sequentially.

(6) Hybrid roll/blade variation: The former starts out like a fourdrinier then the sheet comes into contact with a second fabric. Further dewatering is done by rolls and bladed boxes.

Advantages of twin wire formers are their ability to reduce two-sidedness (q.v.), improve formation (q.v.), improve speed potential, and occupy less space than long fourdriniers. Gap formers are usually chosen for stratified forming where a stratified headbox (q.v.) has two or more separate stock feed systems that form the sheet in layers (plies), which then emerges as a sheet layered or stratified by fiber type.

Twin wire formers can have a variety of press and dryer sections designed to give the desired drainage and sheet characteristics. As with fourdrinier machines, they can have a size press (q.v.), on-machine coating, on-machine finishing or a combination of all these units.

TWIN WIRE FORMING FABRICS—These are forming fabrics (q.v.) with designs and names appropriate to their use. The wide variety of twin wire formers (q.v.) has given rise to a wide variety of fabric names, including: "number one fabric," "number two fabric," "bottom fabric," and "top fabric." One convention that can be applied to all twin wire formers is to use appropriately two of the following four terms: inner, outer, conveying, and backing. Inner and outer apply to the position of the fabrics in the initial dewatering zone be it a roll or bladed device. The inner fabric is the one closest to the initial roll or bladed dewatering device. The other is the outer fabric. After the forming zone, the sheet will go with one of the two fabrics. The name of this fabric is conveying fabric. The other fabric's name is backing fabric. Thus, all twin wire formers have two of the following four fabrics: inner backing; inner conveying; outer backing; outer conveying.

TWIN WIRE MACHINE—See TOP WIRE FORMER; TWIN WIRE FORMER.

TWIN WIRE PAPER—(1) Duplex paper made as two separate sheets on two different fabrics of the papermaking machine and later combined, thus giving twin sides or two top sides to the paper at the end of the machine. (2) Paper made between two forming fabrics, configured so that the paper is formed between such fabrics and therefore has two fabric sides.

TWISTING PAPER—Paper generally made from a sulfate pulp, bleached or unbleached, on either a cylinder or fourdrinier machine. The paper may be slack or hard sized, depending on the converter's requirements, and may contain wet-strength agents. Colors used should be fast to light and bleeding. High tensile strength is required in the machine direction and some converters prefer, especially in the heavier weights, a paper that is soft and pliable. The basis weights vary from 9 pounds to 60 pounds, depending largely on the end use of the paper. The paper is slit to various narrow widths depending on requirements, and is then twisted or spun into yarn or twine which may be woven or knitted into fabric. The paper may be treated to wear and moisture. In commercial usage, the paper is sometimes described by the end use as: carpet tissue, fiber rug paper, etc. These papers are distinct from candy twisting or kiss papers.

TWISTING TISSUE—See TWISTING PAPER.

TWO-DRUM WINDER—The typical paper machine winder is known as a two-drum winder because the front and back drums support the winding roll and provide the torque for the winding process. Some specialty paper machines also provide centerwind torque to the core, but the greater proportion of two-drum winders provide only surface wind torque for the winding process. A typical fine paper winder produces rolls to 40 inches in diameter (100 centimeters) and some machines can produce rolls to 50 inches in diameter (127 centimeters). See COMBINING CALENDER; SURFACE WIND; WINDER.

TWO-PIECE LAMBERT—See SLIDE BOX.

TWO-SHEET DETECTOR—In printing, a device that detects when more than one sheet is being fed into the press nip.

TWO-SIDEDNESS—The property of having appreciable difference in color or texture between the fabric and felt side. The term is commonly applied to dyed papers, where the felt side is usually darker. It may occur in paper prepared from a mixed furnish of long-and short-fibered stock, the latter being more evident on the felt side, or in filled sheets, where more pigment is retained on the felt side.

TWO-TIER DRYER SECTION—See DRYER SECTION.

TYLOSES—Growths in vessel elements that fill and plug the cell cavity in the heartwood of certain hardwoods.

TYMPAN PAPER—An oiled or unoiled paper used on printing presses for packing between the platen and the printed sheet. A hard even paper is desired which will serve for a maximum number of impressions. It is furnished in

rolls and also in cut and scored sheets to fit standard press sizes. It is generally a manila-colored paper or sulfite, kraft, or kraft and jute. The basis weight is approximately 100 pounds (24 x 36 inches – 500); the thickness is normally 0.007 inch, although it may range from 0.003 to 0.012 inch. Uniform caliper, permanent high finish, and freedom from dirt are important.

TYPE I ERROR—An incorrect decision to reject something when, in fact, it is acceptable. Also referred to as alpha risk.

TYPE II ERROR—An incorrect decision to accept something when, in fact, it is unacceptable. Also referred to as beta risk.

TYPEWRITER PAPER—A cut-sized bond paper designed for typewriting.

TYPEWRITER TISSUE—A lightweight manifold designed to be used when many copies are desired.

TYPOGRAPHY—(1) The art or process of printing from movable type. (2) The art and technique of setting and arranging written material in printable form.

U

U CHART—See COUNT-PER-UNIT CHART.

ULTIMATE BIOCHEMICAL OXYGEN DEMAND—The BOD test taken to its longest time limit, usually 30 days. Represents the maximum degradation which can be achieved by a mixed culture of bacteria in infinite time. See also FIRST-STAGE BIOCHEMICAL OXYGEN DEMAND.

ULTRAFILTRATION—A membrane process for purification of liquid waste streams, in which waste is filtered through a membrane at high pressures.

ULTRASONIC ENERGY—A form of energy that incorporates a rapidly vibrating horn often used for mixing, cleaning and welding or bonding.

ULTRAVIOLET SPECTROSCOPY—Identifying a substance by photography of spectrum lines in the ultraviolet region (wavelength) through a spectroscope. See ULTRAVIOLET RADIATION.

ULTRAVIOLET RADIATION—Abbreviated UV, electromagnetic radiation (light) between 200–400 nanometers, which lies immediately below the visible spectrum. UV light is used to cure special inks and adhesives that contain a photochemical initiator.

UNBLEACHED—A term applied to paper or pulp which has not been treated with bleaching agents.

UNCALENDERED—A term applied to paper that is reeled directly from the drying cylinder without passing through the calenders.

UNCOATED PLAYING CARD STOCK—A heavyweight paper designed as a base stock for conversion into playing cards. See PLAYING CARD STOCK.

UNCOATED POSTAL-CARD STOCK—An uncoated cardboard used primarily for the manufacture of postal cards. See POSTCARD PAPER; UNCOATED PLAYING CARD STOCK.

UNCOATED PRINTING PAPER—See BOOK PAPER (UNCOATED).

UNCOATED WEIGHT—The weight of coating raw stock expressed in pounds per ream or 1000 ft^2 before the coating material has been applied.

UNCOATED MILL BLANKS—A card stock with a news center and a book liner, the latter being applied in the vat of the paper machine. See UNCOATED BLANKS.

UNCOATED FREE SHEET—An uncoated paper used for printing, writing, and related applications, and made almost wholly from chemical woodpulps.

UNCOATED COVER PAPER—A term used to distinguish plain cover from coated cover. It may be embossed or decorated, but not by a coating process.

UNCOATED BOOK PAPER—See BOOK PAPER (UNCOATED).

UNCOATED BLANKS—A term applied to nonpasted or pasted blanks to which a coating of clay has not been applied. The term is no longer in use.

UNDER-RUNS—Production and delivery short of the quantity ordered. Trade custom, and sometimes, specifications allow definite tolerances for under-runs.

UNDERCOATING—The application of a self-leveling coating immediately preceding automatic application of glossy-finish coats. The term undercoating has been replaced by base coating (q.v.), precoating, and prime coating.

UNDERLAY—A piece of paper or board placed under a printing form or type in a printing press to bring it up to proper height for printing.

UNDERLINER—The layer between the top liner and the filler stock in a multilayer board, if present. It is usually intermediate in cost, brightness, and freeness between the top liner (q.v.) and the filler stock. Sometimes back liner (q.v.) stock, a blend of back liner and filler, or a blend of top liner and filler is used.

UNIDIRECTIONAL—A term indicating that there is one direction orientation, physical make-up or property.

UNIFORMITY—The quality of being uniform in some property, such as color, finish, or especially, formation (q.v.) and evenness of fiber distribution.

UNLINED CORRUGATED PAPER—Corrugated paper made from 0.009 corrugating medium and used for lining boxes, covering, padding, etc., and as wrappings for glassware, other fragile articles, or glass-packed goods. See CORRUGATED WRAPPING.

UNORUN—See DRYER SECTION (serpentine dryer).

UNSIZED—Not having been treated with size (q.v.) either during or after manufacture.

UNVULCANIZED ELECTRICAL BOARD—A paperboard made for use in panels upon which electrical instruments may be mounted. It is also used for tabletops and as a substitute for tiling. The board is made of wastepaper stock to which is added up to 30% of resin (usually phenolic). The board is subsequently subjected to heat and heavy pressure to cure the resin and harden the board.

UNWIND—The device that supports a roll on its axis and allows it to rotate and dispense the web to the process. The unwind can incorporate equipment to provide torque or braking. It can also be designed to square the web to the process with a "side-lay" which changes the parallelism of its axis to subsequent rolls. An unwind can also be designed with features to facilitate loading and removal of the core shaft.

UP-ENDER—A mechanical device utilized to change the orientation of a roll of paper from the horizontal (lying on its side) to the vertical (lying on end) position.

UPFLOW TOWER—A retention vessel for bleaching in which the stock enters at the bottom, is pumped upward, and is then fed by a launderer to an outlet through which the stock moves by gravity to a washer.

UPFLOW CLARIFIER—A special type of clarifier which extends the stilling well below the sludge blanket so that the clarified water is filtered through the settled sludge.

UPFLOW-DOWNFLOW TOWER—A combination of an upflow and downflow tower which combines the advantage of a bleaching agent moving upward under pressure with the stock,

and a flexible retention time for the bleaching operation in the downflow tower.

UTM BROKE PULPER—Under the machine pulper, a broke pulper that is located under the paper machine. UTM pulpers are normally capable of forming a slurry continuously at the paper machine's maximum production rate. They are fed by gravity, from one or two broke generating locations, or by conveyor, potentially from a large number of broke generating locations.

UV—See ULTRAVIOLET RADIATION.

UV (ULTRAVIOLET) CURE INKS—See ULTRAVIOLET RADIATION.

UV LIGNIN—See KLASON LIGNIN.

V

V BOARD—Either solid or corrugated fiberboard having certain designated properties as defined by U.S. government specifications (PPP-F-320). This board is characterized by the unusually high percentage of its dry bursting strength and effective lamination of components remaining after immersion in water for 24 hours, as compared with corresponding grades of commercial domestic board. Boards of this type are used primarily for external packaging of commodities which will be subjected to storage or transportation hazards involving exposure to severe atmospheric conditions. The calipers of the existing established grades are: V2S 0.090 inch, V3S 0.090 inch; V4S 0.080 inch; V3C boards which have facings of 0.023 inch and corrugating material of 0.010 inch in caliper; and V11C, V13C and V15C with facings of 0.023, 0.016 and 0.010 respectively in caliper. See also W BOARD.

VACUUM METALIZING—See METALLIC PAPER (3).

VACUUM FILTRATION—Use of suction to deposit solids on the surface of a filter as the liquid flows through.

VACUUM CLEANER BAG PAPER—Porous paper fabricated into bags for vacuum cleaner filters. Generally, the paper manufactured for branded vacuum cleaner manufacturers is made to a higher quality specification than papers made for replacement bags sold in supermarkets.

VACUUM DECKER—See VACUUM DRUM FILTER.

VACUUM DRUM FILTER—A stock thickening device that consists of a large, hollow drum covered with a fine filter media, and mounted in a vat. During operation, the stock is introduced into the vat, and it forms a mat on the fine filter media, while most of the water penetrates the filter media, and exits out the ends of the drum. A vacuum drum filter is distinguished from a gravity decker in that the vacuum drum thickener uses a barometric dropleg to thicken to a higher final consistency, and also often uses a finer filter media. Also called a drum filter or a vacuum decker.

VACUUM DISC FILTER—See DISC FILTER.

VACUUM FELT ROLLS—Specialized rolls used between drying cylinders in a single-tier dryer (q.v.) geometry. They have a perforated shell and operate at low-vacuum levels to prevent sheet blowing as the web passes around the roll.

VACUUM BOXES—See HYDROFOILS.

VALVE (SHIPPING SACK)—The small opening for filling in one corner of a shipping sack whose top and bottom have been closed by the manufacturer.

VALVELESS DECKERS—Cylindrical washers without external barometric legs (droplegs). In the newer designs, the vacuum is generated in a series of small dropleg devices built inside the cylinder, which eliminates the problem of providing a seal valve with a stationary dropleg (the older devices consisting only of a cylinder with seal bands at each end are usually referred to as gravity deckers).

VAPOR PERMEABILITY—That property of paper or paperboard which allows the passage of a vapor. This property must be measured under carefully specified conditions of total pressure, partial pressures of the vapor on the two sides of the sheet, temperature, and relative humidity. Because of the fact that paper has specific affinity for water vapor, vapor permeability should not be confused with air permeability or porosity. See PERMEABILITY; WATER-VAPOR PERMEABILITY.

VAPOR SPHERES—Special vapor collection tanks which are used to accumulate odorous or toxic emissions. They are usually spherical in shape.

VAPORPROOF—Treated to resist penetration by gases or vapors.

VARIABLE CROWN ROLL—To avoid the need for changing the crown (q.v.) on press rolls when the loading is increased or decreased beyond a small range, variable crown rolls allow their crown to be changed by internal means. Although there are several different types in use, most designs use hydraulic pressure to shift deflection of an outer shell to an internal stationary shaft or beam. These rolls may also be covered with an elastomeric composition.

VARIABLES DATA—Measurement information on a characteristic where the result can assume any value over some interval.

VARNISH—A solution of a resin, such as copal, dammar, and shellac, in a solvent, such as turpentine, and boiled linseed oil, containing a dryer, which, after evaporation of the volatile constituents of the vehicle and oxidation of the nonvolatile vehicle, leaves a thin shiny layer of the dissolved bodies. See also LACQUER.

VARNISH LABEL PAPER—A coated or uncoated smooth printing paper made from chemical woodpulps, designed for use in varnished labels. See LABEL PAPER.

VARNISHABILITY—The measure or ability of a sheet of paper to accept varnish. A sheet with a high degree of varnish ability would generally be smooth, of low absorbency, and would have a minimum amount of color change on varnishing.

VARNISHED WALLPAPER—A wallpaper which has been varnished after the design has been printed. It has a washable surface.

VAT PAPERS—(1) Handmade papers. (2) British term for papers made on a cylinder machine.

VAT—(1) A term for handmade papers. (2) The receptacle that holds pulp from which handmade sheets are formed. (3) The oblong tank in which the cylinder of a cylinder machine is mounted and which contains a stock suspension from which the sheet is formed. (4) The tank used for tub sizing of paper.

VAT MACHINE—See CYLINDER MACHINE; STEVENS FORMER.

VAT LINER—An adjective describing a paperboard that has been lined in the process of manufacture on a cylinder machine by having a stock or color in the first or last vats which differs from that in the other vats and which becomes an integral part of the sheet when the plies are squeezed together by the press rolls. If the stock in both the first and last vats is different from that in the other vats the sheet is known as double vat lined.

VAT DYES—Vat dyes applied in the paper industry in the form of pigments, which are characterized by having excellent light fastness and, due to cost, are almost universally restricted to the tinting of high-grade white papers.

VDT—See VIDEO DISPLAY TERMINAL.

VEGETABLE CRATE LINER—A vegetable parchment paper or heavy waxed, resin-treated, or parchmentized kraft or greaseproof paper with sufficient strength to withstand water and the pressure used in packing vegetable crates. The paper may be creped or wrinkled. See also CRATE LINERS.

VEGETABLE PARCHMENT PAPER—A paper, resembling animal parchment, which is made by passing a waterleaf sheet prepared from cotton fiber and/or pure chemical woodpulps through a bath of sulfuric acid, after which the sheet is thoroughly washed and dried. The properties of the finished sheet are dependent upon the furnish, the papermaking procedure used to make the waterleaf sheet, and the variations in the parchmentizing process. It is odorless and tasteless, either greaseproof or grease resistant and has high wet strength which is substantially maintained over a long period. It does not disintegrate in water or salt solutions either hot or cold and has high resistance to disintegration by many other solutions. The sheet may be softened by the use of plasticizers. It may be waxed, coated, embossed, or crinkled. It has the following uses: (1) As a printing parchment, it is used for advertising brochures, greeting cards, letterheads, box liners, stationery, etc., and may be printed by offset, letterpress, gravure, thermography, die stamping, or silk screen; (2) in various weights and degrees of translucency or opaqueness it is used in packaging frozen, moist, greasy, or dry food products, such as: butter, margarine, dry pet food, meat, and poultry; (3) as a carton liner for bake and serve goods; (4) as a pan liner (q.v.); (5) translucent parchment is used for diazotype copymasters; and (6) special types are used as release paper and interleavers for the food, plastic, and rubber industries.

VEHICLE—The liquid portion of a material such as ink, paint, or coating composition, including the binders or adhesives and modifiers.

VELLUM DRAWING PAPER—Drawing paper with a vellum finish.

VELLUM—(1) A strong, high-grade natural or cream-colored paper made to resemble the fine parchment originally made from calf skin. A term applied to a finish rather than a grade. Social and personal stationery are often called vellums. (3) Tracing papers, both natural and those rendered transparent by suitable treatment.

VELLUM FINISH—A finish similar to eggshell, usually produced on a paper made from a harder stock than eggshell book paper. The surface is finer grained than in eggshell finish. It is produced by the use of special felts on the presses. See also CALENDER VELLUM FINISH; PLATER VELLUM FINISH.

VELOUR COVERS—Cover papers coated with cotton, wool, or rayon flock and made in box cover or heavier weights of cover paper. The effect is a suede leather appearance.

VELOUR PAPER—One of several names for flock-dusted paper. An adhesive waterproof varnish or lacquer is first applied to paper. While this is in a tacky state, before it is dried, cotton, rayon, or wool flocks are dusted on the surface. The product is then dried and wound in rolls. Various decorative effects may be obtained by embossing in various patterns. This product finds use in box covering and as a cover stock for folders. Stocks employed are usually of chemical woodpulp. For box covering a stock of 25 pounds (24 x 36 inches – 500) is commonly used, giving a finished weight of around 45 to 50 pounds (24 x 36 inches – 500). For cover use, stocks of from 65 pounds upward may be employed.

VELVET FINISH—A finish suggesting the feel of velvet. It has a dull surface.

VELVET PRINTINGS—A general term applicable to high grade printing papers having a soft surface texture.

VENEER TAPE—See GUMMED VENEER TAPE.

VENEER PAPER—See GUMMED VENEER TAPE.

VENEER—Thin fiberboard, or heavy ledger, highly glazed and used for coverings.

VENT HOLES—Holes punched in multiwall plies to allow the sack to transmit entrapped air when the bag is filled.

VENT STACKS—Stacks used to purge gaseous material to the atmosphere.

VENTA-GROOVE PATTERN—A grooving pattern on the outer cylindrical surface of a web support roll or nip roll that consists of a multiplicity of closely spaced narrow and deep grooves to provide an alternate path of flow of the boundary film.

VENTA-GROOVE—A term referring to a specific family of air film or water film elimination grooving machined into various nip rolls or paper support rolls throughout web handling processes. See VENTA-GROOVE PATTERN.

VENTURI SCRUBBER—A scrubbing system designed so that the scrubbing fluid is introduced into the gas by means of injection into a venturi section (q.v.).

VENTURI SECTION—A gradual constriction placed in a pipe that causes a drop in pressure as fluid flows through it. It consists of essentially a short straight pipe section or throat between two tapered sections. In pulp bleaching, it has been used in the chlorination stage as a device to disperse chlorine gas in water before it is added into the pulp stream.

VERDOL PAPER—See JACQUARD PAPER.

VESSEL—A tubelike system of individual cells in the angiosperms (q.v.) or hardwood species of trees. The individual vessels have peculiarities of anatomy which permit the identification of the hardwood genera.

VIBRATING SCREEN—A non-pressurized screen which uses a gyratory or vibratory motion to draw stock through the screen plate openings, and to clear debris away from the screen plate openings.

VIDEO DISPLAY TERMINAL—Television-like device used to display computer processed information or data.

VIRGIN STOCK—Pulp which has not previously been used in the papermaking process. It is to be distinguished from secondary stock (q.v.).

VIRGIN BLACK LIQUOR—See CONCEN-TRATED BLACK LIQUOR.

VISCOSE—A solution of cellulose xanthate prepared by dissolving the reaction product of carbon disulfide and alkali cellulose in an aqueous solution of sodium hydroxide. When forced into a coagulating bath, it yields regenerated cellulose in the familiar forms of viscose rayon and cellophane. Viscose has been used as a wet-strength agent and in coating or impregnation of papers.

VISCOSITY—(1) The property of a fluid that is shown in resistance to deformation or flow, and in which the stresses are related to the rate of deformation. (2) A measure of the degradation by pulping and bleaching of high-molecular-weight carbohydrates, mainly cellulose, in pulp fibers. The fibers are dissolved in a 0.5% (at times 1%) solution of cupriethylenediamine, and the viscosity of the solution is determined under standard conditions in a capillary viscometer. There is a relationship between viscosity and length of the solute molecules. Ideally, bleaching chemicals should remove or bleach only the lignin and colored compounds of the pulp without degrading the cellulose fibers. But pulping and bleaching chemicals also attack the cellulose molecules, lowering the pulp viscosity. Lowering the viscosity to a particular value for each type of pulp does not lower the strength, but the further the viscosity is lowered below that value, the greater is the strength loss. See also INTRINSIC VISCOSITY.

VISUAL EFFICIENCY—See LUMINOUS REFLECTIVITY.

VOC—An acronym for volatile organic carbon, a catch-all classification referring to all organic air emissions.

VOID FRACTION—The ratio of the volume occupied by voids or air spaces to the gross volume of a sheet of paper. It may also be expressed as one minus the solid fraction (q.v.).

VOIDS—See VOID FRACTION.

VOLT—A unit of electrical potential difference equal to the difference in potential between two points in a conducting wire carrying a current of one ampere when the power dissipated between these points is equal to one watt; it is the potential difference across a resistance of one ohm when one ampere of current is flowing through it.

VOLUME HEAT RELEASE RATE—Ratio of the total heating value input to a recovery boiler furnace to the furnace volume. Often used as design guideline and a measure of the boiler loading.

VOLUMETRIC COMPOSITION—The combination of the fractional volumes of the phases that form a sheet of paper—*viz.,* air, solids, and liquids.

VPI PAPER—A grade of paper treated with vapor-phase inhibitors to provide antirust, antitarnish, and anticorrosion qualities. Such paper is widely used for wrapping metal parts, machinery, cutlery, and similar items which must be protected from corrosion during shipment, storage, etc. See ANTITARNISH PAPER.

VULCANIZED FIBER—Vulcanized fiber is made by combining layers of chemically gelled paper. The paper used is an absorbent grade of cotton or alpha wood cellulose. For most grades of vulcanized fiber, cotton cellulose paper is used. The chemical compound used in gelling the paper, which is normally a zinc chloride solution, is subsequently removed by leaching, and the resulting product, after being dried and finished by calendering, is a dense material of partially regenerated cellulose in which the fibrous structure is retained in varying degrees, depending upon the grade of fiber. It has high strength per unit weight, excellent impact strength, and exceptional resistance to electric arc tracking. It can be machined, formed, embossed, painted, or combined with other materials. It is used as electrical insulation, fuse, tubes, lightning arrestor tubes, abrasive disc base, gaskets, textile bobbin heads, patterns, tags, textile shuttle armor, trunks, wastebaskets, and welders' shields. It is made in the form of sheets, rolls, tubes, and rods. Tubes are made by winding chemically gelled paper on mandrels of the desired inside diameter, leaching out the chemical, drying, calendering, and finishing by grinding and sanding to the desired outside diameter. Rods are made by machining strips cut from the sheets so that the grain runs lengthwise of the rod.

VULCANIZING—The process of treating a fibrous cellulosic composition with a solution of zinc chloride or other chemical to gelatinize the surface and convert it into a colloidal semifibrous mass, removing the gelatinizing agent by leaching and drying the sheet to give a hard, tough, dense product. See VULCANIZED FIBER.

VULCANIZING PAPER—A waterleaf paper made of rag (cotton) or chemical woodpulp and used as a raw material for manufacturing vulcanized fiber. It is usually made in a basis weight range of approximately 30 to 110 pounds (24 x 36 inches – 500), depending on the thickness and particular property desired of the vulcanized fiber product. Important properties are high absorbency, chemical purity, low cupriethylenediamine viscosity controlled to proper limits for the respective fiber grade, uniform formation, and relative freedom from foreign materials.

W

WADDING—See CELLULOSE WADDING.

WAD STOCK—A paperboard spun into a small, tightly wound, compact coil to form the base of a shotgun shell. It is made of reclaimed paper stock or chemical woodpulp, usually 0.01 to 0.012 of an inch in thickness. It has uniform thickness and density, a high water finish, and water resistance. See also SHOT-SHELL TOP BOARD.

WALLBOARD—(1) A type of fiberboard composed of a number of layers of chip, binders, or pulpboard, molded or pasted together and generally sized, either throughout or on the surface. It may also be nonlaminated and homogeneous in nature. Wallboard is generally 3/16 or 1/4 of an inch in thickness. (2) A general term used to indicate a composition material used in the construction of partitions, side walls, and ceilings in interior construction; it is made generally of recovered papers, woodpulp, or wood or other materials. See also GYPSUM BOARD; INSULATING BOARD.

WALLPAPER—A hanging paper or tile stock (q.v.) which has been suitably printed or decorated for wall coverings. The paper is usually given a ground coat of clay or of casein and clay, which may contain pigments to form a part of the final design, and then it is printed with oil inks, casein inks, or paints to complete the design. The paper may also be embossed or plastic coated. See DUPLEX WALLPAPER; OATMEAL PAPER; SANITARY WALLPAPER; VARNISHED WALLPAPER; WASHABLE WALLPAPER.

WALLPAPER (BASE STOCK)—See HANGING PAPER.

WALL (SHIPPING SACK)—One of the sheets of paper, foil, plastic film in a shipping sack.

WARNING INDICIA—A chemical compound or hidden printed design in safety paper, which changes color or becomes visible when ink eradicators are used to alter the writing thereon.

WARP—The curvature of sheets of corrugated board created during the corrugating process or upon exposure to the atmosphere after corrugating.

WARPING—Loss in flatness, particularly in paperboard.

WARP STRAND—See FORMING FABRIC.

WARP WIRE—See FORMING FABRIC.

WASHABLE WALLPAPER—Wallpaper made water and abrasion resistant by the use of coating materials such as casein and soya flour,

325

which are then chemically hardened. Synthetic resins or paint may also be used. The raw stock for such paper is sometimes made as a set-strength paper.

WASHBOARD MARKS—Soft sections in a roll of paper, which appear as corrugations at angles to the axis of the roll. They are caused by uneven caliper across the web. See CHAIN LINES (2).

WASHER—In bleaching, equipment used to wash out chemicals and reaction products from a pulp suspension after a bleaching stage. It is usually a rotary drum or cylinder covered with a mesh screen. As the pulp stock goes over the drum, the spent liquor is withdrawn by vacuum, and the mat is sprayed once or several times with fresh water or recycled filtrate the excess of which is removed by vacuum before the mat drops from the drum into a shredder at 10–20% consistency.

WASHER SHOWERS—Equipment used to spray wash water onto a pulp mat as it goes over the washer drum.

WASHING—(1) At the end of each bleaching stage, the pulp mat is washed to remove as many soluble impurities as possible formed during the bleaching reaction. The soluble impurities impede the efficiency of the next bleaching stage by increasing the consumption of chemicals and aggravate brightness reversion if they are not removed after a final stage. Washing in a bleach plant takes place by dilution, whereby the pulp is alternately diluted and thickened; or by displacement, in which the liquor containing the impurities is displaced by adding a limited amount of water on a thickened pulp mat. (2) A mechanical process of rinsing ink, ash and dirt particles from pulp. It can be performed over a wide range of stock consistencies and operating conditions using different types of equipment. Effective washing depends mainly on the particle size of the contaminant to be removed. It is most suitable for dispersible inks or when a high percentage of inorganic fillers must be removed. See BROWNSTOCK WASHING; DECKER; SIDEHILL SCREEN.

WASTE DISPOSAL—The process of getting rid of waste material.

WASTE HEAT BOILER—See HEAT RECOVERY STEAM GENERATOR.

WASTE MANAGEMENT—The process of deciding what to do with a waste material. Can include reducing the amount of waste produced, eliminating the production of a specific waste material, or reclaiming the waste for beneficial use.

WASTEPAPER—Paper stock that is not recovered for recycling and is otherwise disposed of by landfilling or waste-to-energy.

WASTE SHEET—A term used in bookbinding to designate a piece of recovered paper that is tipped on the outside and over the regular book end sheets on single books and on the top and bottom of a pile of books to protect them from soiling during the progress of the book (or books) through the various binding operations. This sheet is also used for special instructions regarding the binding.

WASTEWATER—Any excess water which is discharged from an industrial process or a municipality.

WASTEWATER TREATMENT—The process of treating wastewater from an industrial process or a municipality.

WATCHMAKERS TISSUE—A tissue paper used by a watchmaker or repairer for holding small parts of a watch, this may be twisted into a sort of bag to keep the parts together. It is made of white cotton rags in a basis weight of 10 pounds (20 x 30 inches – 480) and is merchandised in two sizes, 4 1/4 x 4 1/4 and 3 x 3 inches, 1000 sheets to a bundle. The paper is strong and has a high, white color so that small parts lying upon it are easily discernible.

WATER ABSORBENT—Unsized. See ABSORBENT PAPERS.

WATER-BASE INKS—An ink based on aqueous soluble or dispersable resins or varnish. Air emission regulations on petroleum-base materials have driven the development of water-base inks.

WATER BLEEDING—See BLEEDING.

WATER BOX—A device used for application of calender sizing (q.v.). It consists of an inlet pipe, a flow control valve, and a pond formed by a reinforced rubber lip contacting the machine calender roll and a dam to control pond level; the overflow over the dam is drained back to a recirculation tank. The sizing material is carried on the roll surface into the calender nip where it is applied to the sheet surface. One or more boxes may be fitted to one side of the calender stack, and the same number to the other side if equal pick up is desired.

WATER-COLOR PAPER—A typical drawing paper, tub sized, machine, air or loft dried, and sometimes handmade. The chief characteristic is its hard-sized surface and surface texture suitable for water colors so that they will not run and yet not penetrate too deeply.

WATER-COOLED ROLL—The elastomeric cover on rolls operating under high pressure can develop a considerable amount of heat causing the cover temperature to rise to a dangerous level. Although much can be accomplished with the use of new compositions and by limiting cover thickness, water must be circulated through the roll core to stabilize the cover temperature. The water circulation mechanism must be carefully designed to provide efficient cooling.

WATER-COOLED SPRING ROLL—A spring roll (q.v.) which is designed to allow circulation of cooling water or brine within the roll shell to cool the paper web passing over it.

WATER CREPED—See CREPE PAPER; CREPING TISSUE.

WATER FINISH—A high finish produced on paper or paperboard as it passes through the calender stack by moistening either one or both sides with a fine spray of water, or by troughs or boxes that supply a film of water to one or more calender rolls. The surface is more compact and more glossy than with a dry finish. It is not as uniform as a supercalender finish.

WATER GLASS—See SODIUM SILICATE.

WATER JET SLITTING—A method of slitting a web of paper utilizing high-pressure water. The typical nozzle is about 0.005 inch diameter and, depending upon the paper grade, the water pressure may extend from 3,000 psi to 50,000 psi.

WATERLEAF—An unsized paper. See ABSORBENT PAPERS.

WATER LINED—See WATERMARK.

WATER-LINED PAPER—(1) A book or tablet paper having dandy roll watermarked lines running parallel to the grain of the paper. The distance between these lines varies according to the customer's requirements but usually ranges from one-half to two inches. (2) Paper usually for export to comply with foreign customs law provisions, marked with continuous lines in the machine direction by slight displacement of fibers (as in watermarking) by means other than the use of a dandy roll.

WATERMARK—A true watermark is a localized modification of the formation and opacity of a sheet of paper while it is still quite wet, so that a pattern, design, or work group can be seen in the dried sheet when held up to the light. Such modifications can be accomplished in several ways. The most common method is through the use of a bronze letterpress-type dandy roll (q.v.) riding on top of the sheet at or near the suction-box position of the forming fabric. When an impression of such a roll is made on the top surface of a wet sheet, such an impression will contain less fiber at the point of impress and hence, more translucency. Another method is through the use of a bronze, intaglio-type dandy roll at the same location on the forming fabric. The image resulting from such a dandy roll is characterized by hanging more fiber at the point of impress and hence, more opacity. Such a watermark is often referred to as a "Shadecraft" watermark. A third method is a variation of the letterpress marking method except that a soft rubber, letterpress-type device is used on the first press of the paper machine with the result that a watermark is produced on the underside (i.e., wire side) of the sheet, so that it is visible and readable looking through the sheet from the top side. So-called impressed watermarks are made with metal or rubber letterpress-type marking devices on the top side of the sheet at a point beyond the press section of the paper machine where the water content of the paper has been reduced below the level required for a true watermark. Impressed watermarks are actually a form of embossing and are not true watermarks. Simulated watermarks can also be produced by printing dry paper with transparentizing compounds. See also LAID.

WATERMARKING DANDY—See WATERMARKING DANDY ROLL.

WATERMARKING DANDY ROLL—A dandy roll (q.v.) used to mark the sheet with a design carried on the surface of the roll. The arrangement of the wires on the dandy roll produce a pattern effect on the sheet. When letters, figures, or other devices are attached to the surface of the dandy roll, a watermark is produced.

WATERMARK LAID DANDY ROLL—A laid dandy roll carrying a design to produce a watermark.

WATERMARK WOVE DANDY—A wove dandy roll (q.v.) carrying a design to produce a watermark.

WATER POLLUTION—Discharge of any material which imparts any undesirable characteristic into a lake or stream or a process stream.

WATERPROOF—Heavily sized, coated, or impregnated to resist water penetration.

WATERPROOF BOARD—Board into which water and moisture will not penetrate within a limited time under ordinary conditions. It is produced by including sufficient sizing in the furnish, or by surface treatment.

WATERPROOF PAPER—A water-repellent paper made by combining two sheets of paper by means of asphalt or by impregnating or coating the paper with a suitable waterproofing material.

WATERPROOF SHEATHING—See ASPHALT FELT.

WATERPROOF TAG—See FIBER WATERPROOF PAPER.

WATER QUALITY—An assessment of the state of purity of a water sample. Usually measured in terms of both chemical characteristics and biological parameters.

WATER RESISTANCE—(1) The resistance of the adhesive bond between laminations of paper or board to delamination after prolonged soaking in water under specified conditions. (2) See SIZING. (3) The ability of a shipping container to resist penetration by water in the form of rain.

WATER RESOURCES—An assessment of all available water sources in a given area, usually including lakes, streams, and ground water.

WATER STREAKS—Streaks in paper, appearing as long, light areas which run with the grain. They may be observed by holding the paper to the light.

WATER-VAPOR PERMEABILITY—A permeability specific to water vapor. Because of the unusually high affinity of cellulose for water, water-vapor permeability does not correlate, in general, with the permeability of other vapors and gases. See PERMEABILITY; VAPOR PERMEABILITY.

WATERWALLS—Furnace walls constructed of tubes arranged in flat panels to form a gas tight enclosure and provide a heat exchange surface for transferring heat from combustion within the furnace to water flowing on the inside of the tubes.

WATT—A unit of power equal to the work done at the rate of one joule per second; it is the power produced by a current of one ampere across a potential difference of one volt.

WAVINESS—See WAVY EDGES.

WAVY EDGES—Edges of printing paper that show irregular undulations, similar to warping in boards.

WAX COATING—The operation of applying a coating of paraffin or other wax to a sheet of paper. Wax coating has been largely replaced by plastics and films.

WAXED BOARD—Paperboard that has been waxed either by the hot or the cold process. In the manufacture of hot waxed board, the board is waxed with hot paraffin without any attempt to congeal the paraffin on the surface of the board. In the preparation of cold waxed board, the board is treated with hot paraffin and immediately plunged into ice water to congeal the paraffin on the surface. One-pound butter carton is a typical example of the latter process.

WAXED BUTCHERS—An extra strong paper made of unbleached, semi-bleached, or bleached sulfate or sulfite pulps and heavily waxed on one or both sides. It is highly resistant to meat juices and grease. It is generally made in 40 pound basis weight (24 x 36 inches – 500).

WAXED CONTINUOUS HOUSEHOLD ROLLS—Waxed paper used for wrapping food materials for storage in home refrigerators to reduce moisture loss; also for picnic and lunch food wrapping. The base stock for these rolls is normally a fully bleached chemical woodpulp MG or MF tissue in weights of 14 and 18 pounds (24 x 36 inches – 500). The finished weight range is 18.5 to 28.0 pounds when waxed both sides with a white wax melting at not less than 125°F. The customary roll size is 125 feet by 12 inches width; however, roll lengths vary between 100 and 200 feet.

WAXED GLASSINE—Glassine paper treated with paraffin wax, which renders it more transparent and more resistant to moisture and moisture vapor. See GLASSINE PAPER.

WAXED KRAFT—Kraft paper in various weights which has been impregnated with wax or coated on one or both sides with a wax.

WAXED PAPER—A general term applied to sized or unsized paper which has been impregnated or coated with molten wax in a separate converting operation. When the wax is impregnated into the paper, the product is referred to as "dry-waxed." When the wax remains on the paper surface as a coating, it is then referred to as "wet-waxed." It is generally used for wrapping and packaging purposes. Examples are: bread wrapper, carton liners, cracker box liner, etc. See DRY-WAXED PAPER; WAXED BOARD; WET-WAXED PAPER.

WAXED STENCIL PAPER—See STENCIL; STENCIL TISSUE.

WAXED TISSUE—A tissue of a basis weight of 9 pounds (24 x 36 inches – 500) or heavier which has been waxed by standard waxing methods to resist moisture and odors. It is used for making waxed sheet lunch rolls, waxed lunch envelopes, and waxed butter wraps to protect food products against dust and vermin.

WAXED TISSUE EXCELSIOR—Twelve- to 15-pound waxed tissue which is slit into widths of 1/32 of an inch to make grades known as Easter grass, etc. It is made in various colors; green and purple are the most common.

WAX EMULSION—A stable aqueous emulsion, usually of paraffin or microcrystalline wax, and sometimes containing rosin, prepared by the use of suitable emulsifying agents and mechanical agitation such as provided by a colloid mill. Either acid-stable or alkali-stable products may be obtained depending upon the nature of the emulsifying agent employed. Wax emulsions are used for sizing or waxing paper.

WAXES—Two forms of waxes are in common use: natural waxes such as beeswax and carnauba, and petroleum waxes, such as paraffin. Their most common use in the paper industry is in the manufacture of waxed papers, but waxes are also used in wax-size emulsions and other applications.

WAX IMPREGNATED—A term usually applied to containerboard or corrugated board for a process in which molten paraffin is applied to the hot web so that the paraffin penetrates the sheet. See WAXED BOARD.

WAXING MANILA—A chemical woodpulp paper approximately 20 to 30 pounds (24 x 36 inches – 500) tinted with a yellow dye. After waxing, it is used as a cracker carton liner or for similar purposes. Desirable properties are good formation, finish, and appearance, and uniformity of paraffin absorption.

WAXING PAPER—A paper for impregnation or for coating with wax or paraffin on one or both sides. It may be made of any furnish and the basis weights generally range from 18 to 40 pounds (24 x 36 inches – 500). See DRY-WAXED PAPER; WET-WAXED PAPER.

WAXING TISSUE—(1) A specially sized, moisture-resistant paper used as the base stock for the production of waxed paper. It is manufactured from bleached and unbleached chemical woodpulps in basis weights of 8 to 17 pounds (24 x 36 inches – 500) and is produced in a range of colors. (2) The term is also loosely applied to any tissue which is to be waxed.

WAX PICK TEST— Measures surface strength using a series of calibrated waxes. See SURFACE STRENGTH.

WAX SIZE—A sizing agent containing paraffin or a similar wax as a major component.

WAX-SIZED PAPER—A paper prepared from pulp which has been treated in the beater with an emulsified wax size or in which the sheet has a wax size added to it on the paper machine, either as a tub size or on the calenders. It contains less wax than a waxed paper.

WAX SPOTS—Transparent spots in a paper caused by undispersed particles of wax.

WAYBILL MANILA—A low grade tablet-type paper made largely of mechanical pulp and designed for the printing of railroad and express waybills and the like. It is normally made in basis weights ranging from 25 to 30 pounds (24 x 36 inches – 500).

WAYBILL PAPER—See WAYBILL MANILA.

W BOARD—Solid or corrugated fiberboard which is basically similar to V board (q.v.) except for lower bursting strength and caliper requirements. Board of this type is used primarily for interior packaging of items which will be subjected to unusual storage or transportation hazards involving exposure to severe atmospheric conditions. Thicknesses of the existing established grades are: W5S 0.075 inch, W6S 0.060 inch; W5C and W6C boards have facings of 0.016 and 0.010 inch, respectively; the corrugating material has a caliper of 0.010 inch.

WEAK BLACK LIQUOR—Black liquor (q.v.) at a total concentration of about 15% dry solids.

WEAK WASH—Diluted white liquor (q.v.) resulting from washing the lime mud (q.v.) slurry underflow from a white liquor clarification unit. Used to dissolve smelt (q.v.) from the recovery boiler to form raw green liquor (q.v.). Also referred to as weak white liquor. See also WHITE LIQUOR.

WEB—(1) The sheet of paper coming from the paper machine in its full width or from a roll of paper in any converting operation. (2) The delicate sheet of fibers that exits a carding or air-lay machine.

WEB BREAK—A break in a web of paper.

WEB CALENDERING—The process of finishing paper by passing it through the calender in web form, as distinguished from sheet calendering.

WEB DRAW—The condition of a web with respect to tension as it spans two supporting or driven sections of a machine. The web draw can be expressed as a speed differential or as a percent, which is normally positive. It can be used to describe a discreet portion of a process or the entire process. See DRAW; WEB TENSION.

WEB EMBOSSING—The embossing process that uses a continuous web from an unwinding roll in contrast to plate embossing, which uses sheets. See EMBOSSING.

WEB GLAZING—The glazing or finishing of paper in web form.

WEB NEWS—A term for newsprint in rolls, i.e., a continuous web. This term is now obsolete.

WEB OFFSET PAPER—Coated or uncoated offset paper made for printing in roll form. It requires more strength than sheet-fed offset paper, and if coated, some degree of water resistance, but less than for sheet-fed offset paper. Rolls must be uniformly wound and have no slack areas. To prevent heat blistering, a porous sheet of low moisture content is desirable.

WEB OFFSET PRINTING—The process of printing paper in roll form by the offset printing process.

WEB PAPER—A printing paper in a continuous web or roll.

WEB PRINTING—The process of printing on material that is fed into the press from a continuous roll. Web feeding is often indicated by placing the word "web" before the name of the basic printing process used, as "web offset"; but in the case of gravure the word "rotogravure" is used for this indication. See SHEET-FED PRINTING.

WEB SIZING—See SURFACE SIZED; TUB SIZED.

WEB SPREADING—The process of stretching or widening the traveling sheet of paper in the cross machine direction to counteract the normal tendency of a web under tension to narrow and wrinkle.

WEB TENSION—The tightening or stretching of a web of paper expressed as force per unit of width. Various laboratory machines measure the tensile strength and the stretch (percent elongation) of a web. Some degree of web tension is required to maintain traction and tracking of a web during processing. See WEB DRAW.

WEB TRACTION—A measure of the positive contact that a web has to a conducting member, such as a rotating roll, during its travel through the process. The film of air which travels with a web is a perfect lubricant and has a tendency to destroy the traction of the web to the rotating element at higher speeds, depending on the porosity of the paper and the grooving or surface design of the mating roll and the web tension (q.v.).

WEB WEAVING—The tendency of a web to oscillate laterally when web traction or tension to the conducting rolls change and misalignment factors are affected.

WEDDING BRISTOL—A term applied to a group of high-grade bristols made by pasting together two or more sheets of finished or unfinished paper in different thicknesses or plies. It may be plated to give various finishes; it is used for cards, announcements, etc. The basis weights are: 2-ply, 120 pounds; 3-ply, 180 pounds; 4-ply, 240 pounds (22 1/2 x 28 1/2 inches – 500). See BRISTOLS.

WEDDINGS—A kind of superfine writing paper of medium to heavy substance, slightly modified for better folding. Appearance is the most important factor to be considered. The finish is usually vellum or kid style, i.e., smooth to the touch, but lacking in sheen or glare. It may carry other finishes, as high plate or linen. The basis weights vary from 60 to 160 pounds (21 x 33 inches – 500). It is particularly adapted to steel-engraved wedding announcements, but is frequently used for social correspondence.

WEEPING (WHISKERS)—Thin fingers of coating that bleed through at the blade tip.

WEFT STRAND—See FORMING FABRIC.

WEFT WIRE—See FORMING FABRIC.

WEIGHT—See BASIS WEIGHT.

WEIGHT TOLERANCE—The allowable variation from the weight of paper or board ordered; usually established by trade custom but sometimes by specifications.

WELDING PAPER—A special construction of asbestos paper used by welders to protect parts of castings that are immediately adjacent to the section to be welded. The paper is made of longer fiber than commercial asbestos paper. It should not emit obnoxious fumes at welding temperatures. This term is now obsolete.

WELL-CLOSED FORMATION—An even, regular formation of fibers, producing a uniform appearance in a sheet, as opposed to a wild formation, which gives a mottled, cloudy appearance.

WELL SIZED—See HARD SIZED.

WELTS—Elongated deformations, appearing as a continuous hump or series of alternate humps and depressions parallel to the machine direction.

WET BROKE—Any broke produced before the dryer section of the paper machine. It is sometimes further classified as couch pit broke, or press pit broke. These types of broke may be combined for repulping, or repulped in separate broke pulpers, before being sent to the broke chest. See also BROKE.

WET DRAW—The condition of the web of paper between the sections at the wet end of the paper machine, i.e., whether it is tight or slack.

WET END—That portion of the paper machine comprising the headbox and the sheet forming section. See FOURDRINIER MACHINE. (2) Wet end also designates the entire section of the stock flow from beaters, refiners, or machine chest to the paper machine headbox. Chemical additives are introduced at more than one point of the stock flow.

WET-END ADDITIVE—Any material which is added to the paper furnish after the beating and refining process but before actual sheet formation on the paper machine. Typical wet-end additives include defoamers, retention aids, pitch control agents, slimicides, sizing compounds, wet- and dry-strength resins.

WET END CHEMISTRY—The field of paper science that is primarily concerned with surface and colloid chemistry aspects of papermaking. See WET END.

WET-END FINISH—A finish produced by treatment of the paper at the wet end of the paper machine. Antique, eggshell, vellum, and English finishes are produced by proper control of the felts and presses at the wet end. See also FELT FINISH.

WET FELT—This term usually refers to a press felt (q.v.)..

WET FINISH—See STEAM FINISH; WATER FINISH.

WET-LAID—The process of forming a fibrous sheet using papermaking equipment and water as a fiber carrying medium.

WET LAP—Pulp that has been formed into laps and pressed until it is 35–50% by weight solids, but has not been dried.

WET LAP MACHINE—A machine used to form pulp into thick, rough sheets sufficiently dry to permit handling and folding into bundles (laps) convenient for storage or transportation. It consists essentially of a wire-covered cylinder which rotates in a vat containing the stock and collects the pulp on its surface from which the pulp is removed by a couch roll and felt. The wet web is allowed to wind around the upper press roll until the pulp or board reaches the desired thickness, when it is stripped off the roll in the form of a thick sheet. The wet machine is sometimes used to manufacture certain types of boards which are dried and finished; among these are binders, book, coaster, counter, dobby, electrical pressboard, filter, friction, fuller, genuine pressboard, heeling, innersole, leather, matrix, middlesole, panel, shank, shoe, and truck fiberboard.

WET LAPPED—See LAP.

WET MACHINE BOARD—See BINDERS BOARD; SHOE BOARD.

WET MULLEN—The Mullen bursting strength of paper or paperboard after complete saturation with water.

WETNESS—See FREENESS; HYDRATION.

WET PRESS—See PRESS SECTION.

WET ROLL—Stuck web caused by moisture, with discoloration or ply separation after the roll is dried out.

WET RUB—The resistance of wet paper to scuffing.

WET-SPUN—A method of fiber formation in which the fibers exit the spinnerette into a liquid coagulating bath.

WET STRENGTH—See WET TENSILE STRENGTH.

WET-STRENGTH BROKE—Paper which has been treated chemically to increase its wet strength and which has been discarded in the process of manufacture. Because of its high wet strength, the paper presents special problems in defibering. When the wet strength is produced by urea-formaldehyde or melamine-formaldehyde resins, it is normally defibered by means of mild acid and elevated temperatures. Some high wet-strength papers (e.g., vegetable parchment) cannot be defibered by any feasible means. Similar defibering problems occur in the recovery of wastepapers containing papers of high wet strength.

WET-STRENGTH PAPER—A paper which has extraordinary resistance to rupture or disintegration when saturated with water. This property is produced by chemical treatment of the paper or of the fibers from which it is made. Wet strength is to be distinguished from water repellency or the resistance of a paper to wetting when exposed to water. Wet strength is

most evident and most significant when it occurs in absorbent papers. Normally, a paper loses most of its strength when truly wetted with water. A paper which retains more than 15% of its dry strength when completely wetted with water may properly be called a wet-strength paper.

WET-STRENGTH RESIN—A synthetic material which is incorporated in paper or paperboard to improve its physical strength when wet. Urea and melamine formaldehyde resins and certain related materials are commonly used for this purpose.

WETTABILITY—A term describing the relative affinity of a liquid for a solid surface. Wettability increases with increasing affinity and is measured in terms of the contact angle formed between the liquid and the solid. If the contact angle is zero, complete wettability is said to occur. If the contact angle is greater than 90°, non-wettability exists.

WET TENSILE STRENGTH—The tensile strength of paper after it has been wetted with water under specified conditions.

WET-WAXED PAPER—Paper that has been waxed by passing it through a bath of wax and immediately chilling the wax through the use of cooling rolls or, as is more common practice, by passing the paper through cold water so that the bulk of the wax will not penetrate the sheet but will remain on the surface. The usual weight of the paper before waxing is 18 to 30 pounds (24 x 36 inches – 500). See also DRY-WAXED PAPER; WAXING PAPER.

WET WEIGHT—The weight of pulp as it is lapped or pressed, as distinguished from its airdry or ovendry weight.

WET WRINKLE—A wrinkle or mark in a web of paper produced at the wet end of the paper machine.

WETTING AGENT—A substance, usually a surfactant (q.v.), which improves the wettability of a solid surface by a liquid. Wetting agents are often used to improve absorbency and to improve pigment dispersion.

WF—Water finish.

WHISKERS—See WEEPING.

WHITE—Having that color produced by reflectance of all wavelengths of light in the proportion in which they exist in the complete visible spectrum, or nearly in that proportion; devoid of any distinctive hue. See TINTED WHITE.

WHITE BOX COVER—Descriptive name for plain, embossed, coated, glazed, or decorated white paper used for box covers.

WHITE FIBER SHEATHING—A sheathing paper used for special purposes where a clean white paper is required. It is commonly made of selected white paper stock. See SHEATHING PAPER.

WHITE LIQUOR—An aqueous solution of sodium salts derived from causticizing green liquor (q.v.). The main chemical ingredients are caustic, or sodium hydroxide, and sodium sulfide.

WHITE LIQUOR CLARIFIER—A clarifier used to settle lime mud (q.v.) out of a white liquor/lime mud slurry in order to produce clear white liquor and remove lime mud as an underflow slurry.

WHITE MANILA PAPER—Siderun or defective news in nine-inch counter rolls or in sheets. It is also known as number 2 white manila. It is used in retail stores for auxiliary wrapping purposes.

WHITENERS—See FLUORESCENT DYES.

WHITENESS—The degree of approach of the color to that of the ideal white (q.v.). Since so-called "white" papers have definite hue and colorimetric purity, their color is often specified in some manner; the most common is a requirement for brightness, which is reflectance of a certain wavelength of blue light.

WHITE PAPER—A printer's term for unprinted paper, even though it be colored.

WHITE PITCH—The white deposit that most often appears at the press section and occurs when broke containing polyvinyl acetate coating binder is part of the furnish, and calcium carbonate filler is also present. White pitch is often a problem when alkaline papermaking is being practiced because of the presence of calcium carbonate filler.

WHITEPRINT BASE STOCK—See DIAZOTYPE BASE STOCK.

WHITEPRINT PAPER—See DIAZOTYPE PAPER.

WHITE-ROT FUNGI—Fungi that are able to completely decompose the structural components of wood to carbon dioxide and water; they are apparently unique in their ability to decompose lignin. The white-rot fungi are a large and heterogeneous group of basidiomycetes (higher fungi).

WHITE SHAVINGS—See SHAVINGS.

WHITE TOP LINERBOARD—A duplex linerboard having its top ply composed of bleached fiber with sufficient coverage to produce an all white surface.

WHITE VAT-LINED CHIP—A paperboard used for fabricating setup boxes. It is made from woodpulp and/or recovered paper as a combination board, in thicknesses of 0.024 inch and higher. The board is rigid, and the white surface is suitable for the inside box lining.

WHITE WASTE—See PAPER WASTE.

WHITE WATER—All waters of a paper mill which have been separated from the stock or pulp suspension, either on the paper machine or accessory equipment, such as thickeners, washers, and savealls, and also from pulp grinders. It carries a certain amount of fiber and may contain varying amounts of fillers, dyestuffs, etc.

WHITING—See CALCIUM CARBONATE.

WHOLE STUFF—A term applied to rag pulp as it leaves the beater.

WHOLE TREE CHIPS—Pulp chips produced in the woods using mobile chippers. No debarking is done, but some limbs and tops may be removed prior to chipping the unbarked stems. Whole tree chips contain a high bark content (5–20% by weight), foliage, dirt, and grit. This contamination decreases pulp yield, increases pulping chemical consumption and produces pulp with a higher dirt count. The practice of whole tree chipping is being replaced with satellite woodyards that incorporate debarking ahead of the mobile chipper.

WIDOW—Any objectionably short line or word at the head or tail of a column or page. Also called an orphan.

WILD FORMATION—Having irregular formation (q.v.) or poor distribution of fibers, with a mottled appearance on look-through.

WILLESDEN PAPER—A strong parchmentlike product made by gelatinizing the surface of paper with cuprammonium hydroxide, which like sulfuric acid used in making vegetable parchment, or zinc chloride used in making vulcanized fiber, is a solvent for cellulose. If a thick sheet is desired, several gelatinized sheets are brought together and pressed before washing or drying. It is tough, water- and fire-resistant, and because of the presence of copper salts is resistant to rot, bacteria, fungi, and insect pests. It is also employed for insulating purposes, roof covering, etc. While this paper has been principally manufactured abroad, deriving its name from a London suburb, the term is considered important in international trade.

WINDER—Processes a continuous running web (or webs) into a roll of paper onto a core shaft. The incoming web may be slit into a number of narrow rolls to accommodate the customer or the subsequent converting process. Variations in winder designs are available for specific types of papers, roll widths, roll diameters, and roll structures desired. See CENTER WIND; COMBINING CALENDER; DUPLEX WINDER; REEL; RE-REELER; SURFACE WIND; TISSUE WINDER; TWO-DRUM WINDER.

WINDER TRIM—Trim produced at the paper machine winder.

WINDER WELTS—Grain-direction ridges in the paper, sometimes formed on the surface of the paper roll in the process of winding, which are caused by uneven expansion of the paper due to moisture variation or excessive tension of the web or both. Winder welts often remain in the paper after sheeting. In certain instances, they disappear after conditioning or printing. Winder welts can run into winder wrinkles (q.v.). See ROPE MARKS

WINDER WRINKLES—Long grain-direction crease marks sometimes formed in the surface of the paper in the processes of winding. These marks are due to various causes, such as uneven moisture content in the paper, improper tension in the paper web, or imperfect alignment of the roll shaft. Since these marks are set in the paper, it is seldom possible to eliminate them.

WINDING OR WINDER DEFECTS—See BAGGY ROLL; BURST; CORRUGATIONS (DEFECTS); ROPE MARKS; WINDER WELTS; WINDER WRINKLES; WRINKLES.

WINDOW—A four-sided section of a computer screen in which the user can conduct various operations. Windows can be enlarged, shrunk, or overlaid.

WINDOW-ENVELOPE PAPER—Envelope paper with a portion so treated with a transparentizing material that the address on the letterhead is visible through it, which eliminates typing the address on the envelope. The same effect is produced by cutting out a portion of the envelope and pasting over the opening a piece of translucent material such as glassine.

WINDOWS—Name of the Microsoft trademark computer program that provides a window environment that is icon based, with a mouse interface and pulldown menus.

WINDSHIELD TOWEL—A towel made of windshield wiping paper (q.v.). It is usually folded and dispensed in sheets. It is also called windshield wipe.

WINDSHIELD WIPING PAPER—A creped, absorbent, lint-free paper with good wet-strength, made especially for conversion into towels for cleaning windshields, windows, etc. See WINDSHIELD TOWEL.

WIPE-OFF PAPER—See WIPING-OFF PAPER.

WIPES—A disposable nonwoven cut to size and used for cleaning purposes.

WIPING PAPER—See WIPING-OFF PAPER.

WIPING-OFF PAPER—Usually MG kraft, soft and unsized, used in die-stamping machines to wipe away surplus ink. See DIE-WIPING PAPER.

WIRE—(1) The woven screen made from stainless steel or plastic which forms the surface of a washer drum or cylinder. (2) Forming fabric (q.v.).

WIRE END—The forming section of a paper machine. The term is now obsolete. See FOURDRINIER MACHINE; WET END.

WIRE HOLE—A hole in a paper web caused by a hole in the fourdrinier fabric, which prevents retention of the furnish at that spot. The term is in limited usage.

WIRE LINES—See CHAIN LINES; FABRIC MARK; LAID LINES.

WIRE LOADING—The process of applying a filler to the paper web while on the fourdrinier fabric. The term is in limited usage.

WIRE MARK—See FABRIC MARK.

WIRE SIDE—See FABRIC SIDE.

WIRE SPOT—See FABRIC SPOT.

WIRE WASH—A low volume water shower, just below the take-off roll or doctor board of a washer. It removes fibers trapped in the mesh face. It may be continuous or intermittent, stationary, or oscillating.

WIRE-WOUND ROD COATER—A device for coating which meters and spreads the coating on paper and paperboard by means of a small rotating rod which is spirally wound with wire.

WIRE-WRAP PAPER—(1) See INSULATING TISSUE. (2) A flat, creped, or extensible single-ply or duplex kraft paper, applied either by hand or by machine, for wrapping all types of bare and insulated wire or, more commonly, reels of wire for protection during shipment. It is made in a wide range of basis weights; it may also be waterproofed.

WIT—Abbreviation for wound-in-tension.

WITHERITE—A naturally occurring barium carbonate (q.v.), used in coating for paper.

WIT-WOT—Short for a wound-in-tension, wound-off-tension laboratory winder for measuring roll tensions.

WOOD—The hard fibrous portion of a tree lying between the pith and bark which is the main source of fibers used in the manufacture of papermaking pulps.

WOOD CHIPS—See CHIPS.

WOODCUT—A method of printing in which the non-image areas of a design are cut into the side-grain of a wooden block.

WOOD ENGRAVING—Similar in process to woodcut (q.v.), but using the end grain of the wooden block.

WOODEN PLUGS—See PLUGS.

WOOD FIBER—Elongated, hollow cells comprising the structural units of woody plants. See also FIBER.

WOOD FLOUR—Finely ground wood or fine sawdust used chiefly as a filler in plastics, linoleum, etc., and as absorbent in dynamite. See also FINES.

WOOD-FREE—See GROUNDWOOD FREE.

WOODPULP—Pulp produced by pulping processes in which wood is the raw material. By contrast, pulps produced from grasses, straw, and other annual plants are called nonwood pulps.

WOODPULP BOARD—Paperboard made of woodpulp or a combination of virgin woodpulp and reclaimed paperstock. See PAPERBOARD.

WOODROOM OR WOODYARD—The area of a pulp mill where wood in log, stem, shortwood, and chip form is stored and processed prior to conveying it to the pulping system. Operations that are common to most woodyards or woodrooms are log and chip receiving and storage, debarking, chipping, chip screening and cleaning, and chip quality control.

WOOD ROSIN—See ROSIN.

WOOD VAT-LINED CHIPBOARD—A single vat-lined chipboard made with a liner of mechanical pulp which is used in the manufacture of tubes, small cartons, etc.

WOOD WOOL—A product manufactured by combing wood to a fine fiber; this wool is spread onto paper with a fireproof material and an adhesive, such as asphalt, to obtain an insulating felt.

WOOLEN CARD—A carding machine originally designed for processing short wool fibers and distinguished from other cards by the inclusion of a tape condenser and rub-aprons for the purpose of forming a multitude of rovings at the output end of the card. Upon removal of the tape condenser and rub-aprons, this type of card can be used to make nonwoven webs.

WOOLEN PAPER—An English term for a special paper free from metallic or other foreign matter, used for making supercalender rolls or bowls. See CALENDER-ROLL PAPER.

WORKER—A roll covered with card clothing found on roller-top cards and used to open, disentangle, and blend fibers.

WORKING—The process of operating card clothing point-against-point to open, disentangle, and blend fibers.

WORM—Acronym for write once, read many. An electronic storage device that cannot be erased, but read many times.

WORSTED CARD—A roller top carding machine originally designed to open, disentangle, and blend long wool fibers and deliver one sliver at the output end of the card. The removal of the sliver coiler permits the use of this type of carding machine for making nonwoven webs.

WOUND-IN-TENSION—Sometimes abbreviated WIT, it is the single outer boundary condition that results from the TNTs (q.v.) of winding. It is the tension on the current outer layer of the winding roll.

WOVE DANDY ROLL—A dandy roll (q.v.) that is covered with a wire cloth so as to permit the manufacture of wove (q.v.) paper.

WOVE ENVELOPE—See COMMERCIAL WOVE ENVELOPE.

WOVE PAPER—The usual type of wire mark on a sheet of paper. Wove papers do not exhibit the wiremarks known as laid lines. See LAID.

WP—An abbreviation sometimes used for waterproof.

WRAP—Fabric: the amount of roll surface, in length or degrees, covered by the drying fabric. Paper: the amount of roll surface, in length or degrees, covered by the paper web.

WRAPPING CREPE—A machine or secondary creped paper, usually unbleached kraft, made in a variety of basis weights. It is used primarily for wrapping objects of irregular shapes.

WRAPPING MANILA—A wrapping paper used in meat markets, grocery stores, and the like. The significant properties are moderately high strength and a uniformly high water finish.

WRAPPING PAPER—A general term applied to a class of papers made of a large variety of furnishes on any type of paper machine and used for wrapping purposes. Strength and toughness are predominant qualities.

WRAPPING TISSUE—A term applied to a variety of tissues made from wrapping and packing of merchandise. Basis weights run from 10 to 17 pounds (24 x 36 inches – 480). Qualities vary in accordance with particular uses, but most requirements call for a paper that is strong, well-formed, and clean.

WRINKLE—A creaselike defect in paper produced during manufacturing or converting operations. It is classified as wet or dry depending on the moisture content of the sheet when the wrinkle is formed. See ROPE MARKS; WINDER WELTS; WINDER WRINKLES.

WRITING BRISTOLS—(1) Any bristol that has been sized for writing purposes. (2) An index bristol as opposed to a mill bristol. It is commonly referred to as a printing bristol.

WRITING PAPER—A paper suitable for pen and ink, pencil, typewriter, or printing. It is made in a wide range of qualities from chemical and mechanical wood and rag pulp, or mixtures of rag and chemical pulp or chemical and mechanical pulp. Distinctive finishes and colors produce variations in this class of paper which through long usage have established them as well-known grades of paper. Thus, there is fine and extra-fine writing, azure laid, azure wove, boxed, chemical manila, commercial flat, folded, industrial, laid, machine-dried, manila, railroad, superfine, tablet, etc., each in a form, finish, or color to meet a particular use, but all

fairly typical of this class of paper. It is made in basis weights of 13 to 24 pounds (17 x 22 inches – 500). The most significant class property is good writing and ruling surface. For some uses, good strength and erasability are also necessary.

WRONG SIDE—See RIGHT SIDE OF PAPER.

WVP—Water vapor permeability.

WYSIWYG—Acronym for "what you see is what you get." The property that the image on the computer screen is exactly the image produced on the output device.

X

X BAR CHART—See AVERAGE CHART.

XEROCOPY PAPER—A general term for any grade of paper suitable for copying by the xerographic process. However, commercial xerocopy papers are usually modified bond grades made from chemical woodpulps in basis weights ranging from 16 to 24 pounds (17 x 22 inches – 500), and characterized by a smooth finish, heat stability, non-curing qualities, and good aesthetic properties such as color, brightness, and cleanliness.

XEROGRAPHIC PAPER—A grade of paper, also known as reprographic paper or copy paper, that is suitable for copying by the xerographic or electrographic process. Xerographic papers are bond grades made from chemical woodpulps, mechanical pulps, recycled fiber, cotton, or a combination thereof. See also XEROCOPY PAPER.

XEROGRAPHY—A dry method of reproduction of graphic matter, in which an image is produced in the form of electrostatic charges by reflecting the image onto the surface of a charged photoconductor, which holds its charges in the dark but dissipates them when exposed to light; the image on the photoconductor is developed by bringing it into contact with an ink powder, called a toner; the powder image is then transferred to copy paper and fixed to the sheet by heat fusion.

X PAPER—See PATTERN PAPER.

XYLAN—A hemicellulose with a straight chain backbone composed of xylose monomeric units. The major hemicellulose of hardwoods is an *O*-acetyl-4-*O*-methylglucuronoxylan which has acetyl and 4-*O*-methylglucuronic acid groups substituted periodically on the xylan backbone. The major softwood xylan is a 4-*O*-methylglucuronoarabinoxylan.

XYLANASE—Enzyme (q.v.) capable of hydrolyzing Beta-1,4-xylans. See HEMICELLULASES.

XYLEM—The woody portion of a tree stem or root.

Y

YANKEE DRYER—A large steam heated dryer cylinder from 9 to 22 feet (2743 to 6705 mm) in diameter and up to 300 inches (7620 mm) wide used for drying paper such as tissue, towel, and board. See YANKEE (MG) MACHINE; YANKEE TISSUE MACHINE.

YANKEE MACHINE—A paper machine making lightweight grades (tissue, towel, napkin); it incorporates a Yankee dryer (q.v.). The older style Yankee machine had a single wire suction breast roll former, a single press section, and a Yankee dryer. More recent configurations have a twin wire former, single felted (no press), followed by the Yankee dryer. Some machines incorporate a through-air-dryer section before the Yankee. See also YANKEE (MG) MACHINE.

YANKEE (MG) MACHINE—A paper machine using one large steam-heated dryer cylinder for drying the sheet instead of many smaller ones. The wet sheet is pressed against the surface of the Yankee dryer and may be held in place by a canvas dryer felt as the dryer revolves. It pro-

duces a glazed finish (machine glazed) on the side of the sheet next to the dryer. The machine may have a cylinder or a fourdrinier or twin wire former wet end (q.v.) and may have any number of presses or auxiliary dryers of the usual type; its characteristic feature, however, is the large dryer cylinder (from 9 to 22 feet—2743 to 6705 mm—in diameter).

YANKEE TISSUE MACHINE—A tissue machine that uses a Yankee dryer (q.v.) to evaporate water from the paper. The Yankee, typically, would be equipped with an air cap (q.v.).

YARDAGE—The number of linear yards or area contained in any given roll of paper or board.

YARN—A continuous strand of fiber used for knitting, weaving, braiding, and sometimes other applications. A staple fiber yarn must have twist to achieve adequate strength, whereas a continuous filament yarn may or may not have twist.

YELLOW—The lightest of the three subtractive, transparent process colors.

YELLOW COPYING—See RAILROAD MANILA.

YELLOWING—A gradual change from the original appearance of a pulp or a paper as a result of environment or aging. It is especially pronounced in mechanical pulp or paper but will occur to varying degrees in all types of vegetable fiber. It is sometimes called color reversion.

YELLOW WAYBILL COPYING TISSUE—Yellow copying tissue of sulfite, rope, or mechanical pulp, used for duplicating railroad waybills.

YIELD—A percentage of the original raw material that was converted to product by a process. (1) In pulping, if 100 tons of wood produced 50 tons of pulp from a digester, the pulp yield would be 50%. This is the ovendry basis yield. (2) In papermaking, the amount of paper from pulp or the amount of shipped paper cut stock from manufactured paper. (3) In high-grade deinking plants, it is the weight of 10% moisture content pulp produced from 100 pounds of as received recycled paper. For newsprint deinking, it is the weight of 5–6% moisture pulp from 100 pounds of as-received recycled newsprint. Research laboratories and some mills use the ovendry basis for determining yield. (4) In bleaching it is the ratio, expressed as a percentage, of the weight of pulp fibers remaining after a treatment to the weight of the pulp fibers before treatment.

YIELD VALUE—In plastics, the stress necessary to deform a plastic material.

Z

ZAHN CUP—A dip cup with a hole in the bottom center, used for measuring viscosity of thin inks, like flexographic or gravure.

Z BLEACHING STAGE—See OZONE STAGE (Z).

Z-DIRECTION—A term referring to the direction perpendicular or normal to the plane of the sheet of paper or board, usually within the sheet.

Z-DIRECTION TENSILE STRENGTH—The tensile strength perpendicular to the plane of the sheet. It is used as a measure of bonding strength.

ZEIN—An alcohol-soluble protein present in corn. It was once used in paper sizing, coatings, and adhesive formulations.

ZEOLITE—A hydrated alkali-aluminum silicate (Na_2O Al_2O_5 $(SiO_2)_x$ H_2O) capable of exchanging alkali for calcium and magnesium, and thus used for water softening.

ZERO-SPAN TENSILE STRENGTH—The tensile strength of a sheet of fibrous material, measured with special jaws, at an apparent initial span of zero. It indicates the strength of the material comprising the fibers.

ZETA POTENTIAL—See ELECTROKINETIC CHARGE.

ZIEGLER-NICHOLS TUNING TECHNIQUE—An experimental closed loop procedure which allows for an organized means of arriving at controller (PID) settings. It allows process overshoot above setpoint with 1/4 amplitude decay of the offset until setpoint is reached. Does not provide a smooth control condition because of the overshoot required for adjustment. (Ziegler-Nichols also had an open loop procedure.)

ZINC FINISH—Paper plated between zincs. See ZINC LINER; ZINCS.

ZINC HYDROSULFITE—ZnS_2O_3. A chemical which is formed by the reaction of sulfur dioxide with an aqueous suspension of powered zinc. It was used in bleaching mechanical pulp. This chemical is no longer used because of the environmental problem it causes.

ZINC LINER—A metal plate applied to the top and bottom of each sheet of paper in making a platerbook (q.v.). A typical combination would include a top zinc liner, a sheet of textured cloth or paper followed by the paper to be plated, followed by a bottom zinc plate.

ZINC SULFIDE—(ZnS) A pigment used as such or as a component of lithopone (q.v.) as a loading material or a coating pigment.

ZINC WHITE—Zinc oxide (ZnO), a product made in high-temperature furnaces, is used as a filler to impart opacity and color. ZnO is also used in certain duplicating papers due to its ability to pick up, retain, and release an electrostatic charge.

ZINCS—Polished sheets of zinc used in plating. See ZINC LINER.

Z-TRANSFORMS—A mathematical tool for solution of discrete-time systems analysis similar to what Laplace transforms (q.v.) accomplish for continuous process dynamic analysis.

TABLE 1

METRIC UNITS OF MEASURE

PREFIX	SYMBOL	FACTOR	UNIT	SYMBOL
mega	M	1 000 000	centimeter	cm
kilo	k	1 000	meter	m
hecto	h	100	kilometer	km
deca	da	10	square centimeter	cm^2
deci	d	0.1	square meter	m^2
centi	c	0.01	hectare	ha
milli	m	0.001	milligram	mg
micro	μ	0.000 001	gram	g
			kilogram	kg
			metric ton or tonne	t
			gram per square meter	g/m^2
			liter	L

TABLE 2

CONVENIENT RELATIONSHIPS
BETWEEN METRIC AND TRADITIONAL UNITS
FOR SOME FOREST PRODUCT MEASUREMENTS

To Convert Column 1 into Column 2 Multiply by	Column 1	Column 2	To Convert Column 2 into Column 1 Multiply by
LENGTH			
0.039	micrometer, μm	point (0.001 in.)	25.4
0.039	millimeter, mm	inch, in.	25.4
3.37	meter, m.	foot, ft.	0.305
1.09	meter, m.	yard, yd.	0.914
0.051	meter, m.	chain (66 ft.), ch.	20.1
0.621	kilometer, km	mile, mi.	1.61
AREA			
10.8	square meter, m^2	square feet, ft^2	0.093
1.20	square meter, m^2	square yard, yd^2	0.836
0.386	square kilometer, km^2	square mile, mi^2	2.59
2.47	hectare, ha (10,000 m^2)	acre, A	0.405
VOLUME			
0.264	liter, L (0.001 m^3)	U.S. gallon, gal.	3.79
35.3	cubic meter, m^3	cubic foot, ft^3	0.028
1.31	cubic meter, m^3	cubic yard, yd^3	0.765
0.276	cubic meter, m^3	cord (128 stacked ft^3)	3.63
0.177	cubic meter, m^3	unit (200 ft^3 loose packed)	5.66
0.353	cubic meter, m^3	cunit (100 ft^3 solid wood)	2.83
0.424	cubic meter, m^3	thous. board feet, MBF (nominal)	2.36
1.13	cubic meter, m^3	plywood (1,000 sq. ft. ⅜″)	0.883
MASS			
2.205	kilogram, kg	pound, lb	0.454
1.102	metric ton, t	ton (2,000 lb)	0.907
FORCE			
0.225	newton, N	pound force, lb f	4.45

342

TABLE 3

CONVENIENT RELATIONSHIPS BETWEEN METRIC AND TRADITIONAL UNITS FOR SOME FOREST PRODUCT MEASUREMENTS

To Convert Column 1 into Column 2 Multiply by	Column 1	Column 2	To Convert Column 2 into Column 1 Multiply by
	PRESSURE		
7.52	kilopascal, kPa	millimeter mercury, mm Hg	0.133
0.295	kilopascal, kPa	inch mercury, in. Hg	3.38
4.02	kilopascal, kPa	inch water, in. H_2O	0.249
0.145	kilopascal, kPa	pound per square inch, psi	6.90
0.010	kilopascal, kPa	bar (or 1 atmosphere)	100
	RATIOS		
0.205	gram per square meter, g/m^2	pound per 1,000 square feet	4.88
0.437	gram per cubic meter, g/m^3	grain per cubic foot, gr/ft^3	2.29
0.893	kilogram per hectare, kg/ha	pound per acre, lb/A	1.12
0.446	metric ton per hectare, t/ha	ton (2,000 lb) per acre	2.24
4.36	square meter per hectare, m^2/ha	square feet per acre, ft^2/A	0.230
	ENERGY		
0.943	kilojoule, kJ	British thermal unit, Btu	1.06
0.28	megajoule, MJ	kilowatt hour, kWh	3.60
0.009	megajoule, MJ	therm (100,000 Btu)	105.5
0.430	kilojoule per kilogram, kJ/kg	Btu per pound, Btu/lb	2.33
0.027	kilojoule per cubic meter, kJ/m^3	Btu per cubic foot, Btu/ft^3	37.3
	MISCELLANEOUS		
0.093	lux, lx	foot candle, fc	10.76
0.738	newton meter, N·m	foot pound, ft. lb.	1.36
	TEMPERATURE		
1.8 (°C) + 32	Degree Celsius C°	Degree Fahrenheit, F°	0.56 (F − 32)

TABLE 4

GRAMMAGE OF PAPER

Conversion: From basis weight in pounds to grams per square meter, multiply the basis weight by 1406.13 and divide by the square inches in the base sheet.
(e.g., $17'' \times 22''$ — 20 lb. per ream. $20 \times 1406.13 = 28122.60 \div 374 = 75$ g/m²)

Basis Weight Per 500 Sheets						Grammage, Grams Per Square Meter (g/m²)
$17''$ $\times$ $22''$	$20''$ $\times$ $26''$	$22\frac{1}{2}''$ $\times$ $28\frac{1}{2}''$	$25\frac{1}{2}''$ $\times$ $30\frac{1}{2}''$	$24''$ $\times$ $36''$	$25''$ $\times$ $38''$	
Writing Paper including Bond, Form Bond & Ledger	Cover	Printing Bristols/ Postcard	Index Bristols	Newsprint/ Industrial Papers/Tag and Tab Card/ Sanitary Tissue	Printing Papers including Book, Offset and Clay Coated Papers	All Types of Paper
x				18	x	29
8				18.4	20	30
x				19	x	31
x				20	x	33
9				21	x	34
x				x	24	36
10				x	x	38
11				x	x	41
12				x	x	45
x				28	x	46
13				30	33	49(a)
x				x	34	50
x				32	35	52
14				x	36	53
15				x	38	56
x				x	40	59
16				x	x	60
x				x	42	62
x				40	x	65
x				x	45	67
18				x	x	68
x				x	50	74
20				x	x	75
x				50	55	81

a) In the newsprint industry 30 lb. paper ($24'' \times 26''$ — 500) is quoted at 48.8 g/m²

344

Table 4 (Continued)

GRAMMAGE OF PAPER

Basis Weight Per 500 Sheets						Grammage, Grams Per Square Meter (g/m²)
17"×22"	20"×26"	22½"×28½"	25½"×30½"	24"×36"	25"×38"	
Writing Paper including Bond, Form Bond & Ledger	Cover	Printing Bristols/Postcard	Index Bristols	Newsprint/Industrial Papers/Tag and Tab Card/Sanitary Tissue	Printing Papers including Book, Offset and Clay Coated Papers	All Types of Paper
x				x	60	89
24				x	x	90
x				60	x	98
x				x	70	104
28	x			x	x	105
x	40			x	x	108
x	x			x	75	111
x	x			70	x	114
x	x			x	80	118
32	x		x	x	x	120
x	x		72	80	x	130
x	x		x	x	90	133
36	50		x	x	x	135
x	x		x	90	x	146
x	x	67	x	x	x	147
x	x	x	x	x	100	148
40	x	x	x	x	x	150
x	60	x	x	x	x	162
x	x	x	90	100	x	163
44	x	x	x	x	x	165
x	x	80	x	x	x	175
x	65	x	x	x	x	176
x	x	x	x	x	120	178
x	70	x	x	x	x	189
x	72	x	x	x	x	195
x	x	90	x	x	x	197
x	x	110	x			199
x	x	x	125			203
80	x	x	x			216
x	100	x	x			219
x	x	125	x			226
x	x	x	150			244
x	x	140	x			253
x	120	x	x			263
100	x	x	x			270
x	125	x	x			274

Table 4 (Continued)

GRAMMAGE OF PAPER

Basis Weight Per 500 Sheets						Grammage, Grams Per Square Meter (g/m²)
17″ × 22″	20″ × 26″	22½″ × 28½″	25½″ × 30½″	24″ × 36″	25″ × 38″	All Types of Paper
Writing Paper including Bond, Form Bond & Ledger	Cover	Printing Bristols/ Postcard	Index Bristols	Newsprint/ Industrial Papers/Tag and Tab Card/ Sanitary Tissue	Printing Papers including Book, Offset and Clay Coated Papers	All Types of Paper
	x	x	x	175		285
	x	140	170	x		307
	x	x	x	200		325
	x	150	x	x		329
	x	160	x	x		351
	130	x	x	x		352
	x	175	x	x		384
	x	180	x	x		395
	x	x	220	x		398
	x	x	x	250		407
	x	200	x	x		439

NOTE: Conversion of uncoated carbonizing, 20″ × 30″, is as follows:

Basis Weight/500 Sheets	g/m²
5.5	13
6	14
6.5	15
7	16
7.5	18
8	19
8.65	20
9	21
10	23

346

TABLE 5

GRAMMAGE OF PAPERBOARD

Lbs. Per 1000 Sq. Ft.	Grams Per Sq. Meter g/m²	Lbs. Per 1000 Sq. Ft.	Grams Per Sq. Meter g/m²	Lbs. Per 1000 Sq. Ft.	Grams Per Sq. Meter g/m²	Lbs. Per 1000 Sq. Ft.	Grams Per Sq. Meter g/m²
26	127	72	352	108	527	156	762
33	161	76	371	111	542	160	781
38	186	77	376	112	547	164	801
42	205	80	391	116	566	180	879
47	229	82	400	120	586	206	1006
51	249	85	415	124	605	240	1172
53	259	88	430	128	625	277	1352
56	273	90	439	131	640	288	1406
58	283	92	449	132	644	313	1528
60	293	96	469	140	684	327	1597
63	308	100	488	144	703	343	1675
64	312	103	503	148	723	379	1850
65	317	104	508	152	742	423	2065
69	337					514	2510